Digital Transformation

Birgit Vogel-Heuser · Manuel Wimmer
Editors

Digital Transformation

Core Technologies and Emerging Topics
from a Computer Science Perspective

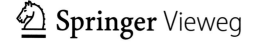

Editors
Birgit Vogel-Heuser
Technische Universität München
Garching b. München, Bayern, Germany

Manuel Wimmer
Johannes Kepler University Linz
Linz, Oberösterreich, Austria

ISBN 978-3-662-65006-6   ISBN 978-3-662-65004-2   (eBook)
https://doi.org/10.1007/978-3-662-65004-2

Responsible Editor: Alexander Gruen
This Springer Vieweg imprint is published by the registered company Springer-Verlag GmbH, DE, part of Springer Nature.
The registered company address is: Heidelberger Platz 3, 14197 Berlin, Germany

Preface to "Digital Transformation: Core Technologies and Emerging Topics from a Computer Science Perspective"

Scope of the Book

The terms "digitalization" and "digital transformations" are nowadays used on a daily basis in media, enterprise, research, and societal contexts. In particular, the competitiveness of the industrial sector is considered to depend on the level of digitalization and potential digital transformations of the products, enterprises, business models, etc. This is mostly due to the ever-growing importance of flexibility in industry, e.g., shorter innovation cycles, rapidly changing customer needs, changes in legislation, and more emphasis on sustainability such as resource efficiency, to mention just a few reasons, as is also highlighted by initiatives such as Industry 4.0/5.0. However, this requested flexibility puts additional challenges on the design, realization, operation, maintenance, and reuse of industrial systems. The hope is that digitalization allows to deal with these challenges, as it is currently stressed by the emerging topic of Cyber-Physical Production Systems (CPPS). CPPS highlight that both physical and virtual entities and processes are nowadays required to realize modern production systems which are able to deal with the current challenges. At the same time, by having this new type of systems, we are currently facing a dramatically increasing complexity in engineering, operation, and management of such CPPS as they are complex, heterogeneous, and networked socio-technical systems. The latter aspect is of particular importance to drive the digital transformation processes within an enterprise considering not only the technological aspects but also organizational ones such as business models, staff, or general societal needs and challenges.

As an answer, several digital technologies are emerging and gaining attention to support different phases of the system life-cycle such as design-space exploration, runtime adaptation, and predictive maintenance to mention just a few examples. For instance, the term Digital Twin refers to the capability to partially copy an actual system into a virtual counterpart, that reflects the important properties of the system for both engineering as well as operation purposes. Thus, digitalization is not only a key enabler for innovation, but is nowadays deeply integrated in the engineering and operational processes, and thus, into the aforementioned socio-technical systems.

This book is focussing on the digitalization topics in Industry 4.0/5.0 from a computer science perspective. In particular, the book introduces a rich set of different concepts, techniques, and methods from the computer science discipline and provides insights on their application, especially for the domain of CPPS, and thus, complements other viewpoints which are equally important in this area such as automation, mechatronics, and business engineering.

Having this focus, the book is in particularly intended as an entry point to digitalization for disciplines outside computer science by giving an orientation and providing the basics of the different sub-domains of computer science relevant for Industry 4.0/5.0 as well as giving pointers to further literature to find more information on the presented topics.

Content of the Book

The book "Digital Transformation: Core Technologies and Emerging Topics from a Computer Science Perspective" is structured into five parts. While Part 1 is about why and how industrial systems are represented as digital artefacts, Part 2 is focussing on emerging digital infrastructures to run industrial systems by showing their potential and applications, as well as how they can be integrated and secured in larger settings. Parts 3 and 4 are focusing on the data-driven paradigm, especially on data management and analytics which opens the door for a multitude of innovation potentials for industrial systems. Finally, Part 5 is giving an outlook on particular digital transformations aspects such as improved human-machine interactions possibilities as well as driving and managing organizational issues such as enterprise model transformations. By this, the book gives an outlook on the emerging Industry 5.0 paradigm which has been presented recently by the European Commission.

We would like to thank all the authors who contributed chapters to this book and hope you will enjoy reading about the core technologies and emerging topics for digital transformation from a computer science perspective.

January, 2022   Birgit Vogel-Heuser
                                                          Manuel Wimmer

Contents

Digital Transformation towards Industry 5.0

Digital Representation

Engineering Digital Twins and Digital Shadows as Key Enablers for Industry 4.0

Stefan Braun⃝, Manuela Dalibor⃝, Nico Jansen⃝, Matthias Jarke⃝, István Koren⃝, Christoph Quix⃝, Bernhard Rumpe⃝, Manuel Wimmer⃝ and Andreas Wortmann⃝

Abstract

Industry 4.0 opens up new potentials for the automation and improvement of production processes, but the associated digitization also increases the complexity of this development. Monitoring and maintenance activities in production processes still require high manual effort and are only partially automated due to immature data aggregation and analysis, resulting in expensive downtimes, inefficient use of machines, and too much production of waste. To maintain control over the growing complexity and to provide insight into the production, concepts such as Digital Twins, Digital Shadows, and model-based systems engineering for Industry 4.0 emerge. Digital Shadows consist of data

S. Braun · M. Jarke
Information Systems and Databases, RWTH Aachen University, Aachen, Germany
e-mail: braun@dbis.rwth-aachen.de

M. Jarke
e-mail: jarke@dbis.rwth-aachen.de

M. Dalibor · N. Jansen · B. Rumpe
Software Engineering, RWTH Aachen University, Aachen, Germany
e-mail: dalibor@se-rwth.de

N. Jansen
e-mail: jansen@se-rwth.de

B. Rumpe
e-mail: rumpe@se-rwth.de

I. Koren
Process and Data Science, RWTH Aachen University, Aachen, Germany
e-mail: koren@pads.rwth-aachen.de

B. Vogel-Heuser and M. Wimmer (eds.), *Digital Transformation*,
https://doi.org/10.1007/978-3-662-65004-2_1

traces of an observed Cyber-Physical Production System. Digital Twins operate on Digital Shadows to enable novel analysis, monitoring, and optimization. We present a general overview of the concepts of Digital Twins, Digital Shadows, their usage and realization in Data Lakes, their development based on engineering models, and corresponding engineering challenges. This provides a foundation for implementing Digital Twins, which constitute a main driver for future innovations in Industry 4.0 digitization.

Keywords

Digital Twin • Digital Shadow • Industry 4.0 • Model-based Systems Engineering • Data Lake

1   Introduction

The fourth industrial revolution, Industry 4.0, is a fundamental driver for agile manufacturing through integration and communication of production systems. It aims at integrating Cyber-Physical Production System (CPPS) to optimize the complete value chain [74]. Initially announced by the German Federal Ministry for Education and Research in the year 2011 [6], Industry 4.0 has become an international phenomenon as the next big step towards future development and manufacturing [7, 32, 48, 52]. To this end, it leverages state-of-the-art research results from a variety of fields, including the Internet of Things (IoT), big data, and machine learning. Combining these approaches, Digital Twins are envisioned as digital duplicates of CPPS that represent, control, and monitor their physical counterpart to make better use of resources. In this vision, Digital Twins need to rely on detailed knowledge of the system, including its requirements and operation data. Since an exact digital replication of all parameters down to the atomic level is not feasible, the concept of Digital Shadows was introduced, which promotes purpose-driven compilations of production data. Moreover, while simulations approximate the behavior and effects, the calculated results often diverge from reality due to external influences such as wear, tear, pollution, or environmental impacts. Remedial actions require extensive manual effort by experienced operators performing measures on real-world counterparts. In the following, we introduce the main

C. Quix
Information Systems and Data Science, Hochschule Niederrhein, Krefeld, Germany
e-mail: christoph.quix@hs-niederrhein.de

M. Wimmer
Business Informatics – SW Engineering, JKU Linz, Linz, Austria
e-mail: manuel.wimmer@jku.at

A. Wortmann (✉)
Institute for Control Engineering of Machine Tools and Manufacturing Units, University of Stuttgart, Stuttgart, Germany
e-mail: andreas.wortmann@isw.uni-stuttgart.de

themes of this chapter, including the aforementioned engineering models, Digital Shadows and Digital Twins.

1.1   Engineering Models in Industry 4.0

Modern product development processes employ methods of model-based systems engineering (MBSE) [28, 61] and model-driven development (MDD), in which models become the primary development artifacts [15]. MBSE raises abstraction in traditional systems engineering approach by harnessing structured models over unstructured documents. MDD extends the classic model-based approach even further by establishing models as primary drivers within the development process. Adhering to explicit modeling languages, with well-defined semantics (meaning) [29], these promote understanding and transparency in development, fostering intra- and interdisciplinary communication [41], as well as automated analysis and synthesis [31] of (parts of) the system [57] under development.

Managing complexity in the interdisciplinary engineering process requires domain-specific views on the overall system, filtering essential information. Thus, by leveraging view-based modeling [18, 55], experts of participating domains are provided with the information relevant to their specific concerns in suitable modeling languages (the views), which is anchored in the overall system's context. Systems Modeling Language (SysML) [23] is a prominent collection of modeling languages, to describe the relationships between the concerns of a system and thus provides the glue between participating engineering domains and their domain-specific models.

Engineering models [8] are typically used constructively, i.e., to prescribe a system under development, and contribute directly to the CPPS development process. As CPPS development is a highly interdisciplinary effort, different models exist across different domains. These are also relevant in engineering Digital Twins since they contain essential information about the CPPS. Thus, engineering models do not only contribute to system development, but also to the development of its twin by integrating important information as well as runtime simulations.

1.2   Digital Shadows

Modern CPPSs are typically equipped with sensors that capture tremendous amounts of raw data while these CPPSs are running. These large amounts of data can no longer be transmitted in real-time or sensibly processed to gain insights into the system's state. Thus, a software system running in this context must provide mechanisms to reduce the sheer data volume and its level of detail, while also coping with obsolete and incomplete data. Data must be provided in a reduced and purpose-oriented fashion to achieve better performance and more context adaptation. To address this, we conceived a notion of Digital Shadows that provide compact views on dynamic processes, usually combining condensed measurement data with highly efficient simplified mathematical models [42]. Therefore, we define digital twins as follows [3, 13]:

Definition 1 *(Digital Shadow) A Digital Shadow is a set of temporal data traces and/or their aggregation and abstraction collected concerning a system for a specific purpose with respect to the original system.*

Thus, the Digital Shadow is a set of data observed from the CPPS. This data is usually captured using sensors of various forms but also data of the state of computation, actions executed by control devices, or even input from human operators. A Digital Shadow contains purpose-oriented data for a specific point in time, provided in a transformed (e.g., reduced or augmented) form. Additionally, this data can be enriched with quality information, such as origin, fidelity, accuracy, and more. Consequently, this implies the existence of multiple Digital Shadow at different times and for a variety of objectives that may reference another. The complete history of Digital Shadows is available to a Digital Shadow, enabling temporal analyses such as predictive maintenance based on variations in the collected data.

1.3   Digital Twins

While there is some consensus that a Digital Twin is a sort of digital duplicate of a physical entity [67], there is still no generally accepted definition that specifies what this means and entails. There are numerous differing interpretations of what a Digital Twin actually is and should be capable of. Many realizations of this concept heavily depend on their specific domain or application purpose. Digital Twin applications range from simple data acquisitions across virtual models of the physical system to an integrated twin with optimization capabilities. Thus, they can serve observation and monitoring purposes purely or directly support management and controlling to support the development process of CPPSs actively. Engineering in Industry 4.0 is highly interdisciplinary, resulting in modern trends in MBSE striving for integrated models to bridge the gap between different domains. Therefore, we use a definition of the Digital Twin based on models that was developed within the German Cluster of Excellence "Internet of Production",[1] which comprises 200 researchers in 25 departments from different domains, including mechanical engineering, electrical engineering, automation, factory construction, software engineering, systems engineering:

Definition 2 *(Digital Twin) A Digital Twin of a system consists of a set of models of the system, a set of Digital Shadows, and a set of services that allow using the data and models purposefully with respect to the original system.*

---

[1] Internet of Production: https://iop.rwth-aachen.de.

Hence, a Digital Twin can be engineered based on heterogeneous sets of engineering models of different domains. It relies on the Digital Shadows as highly optimized representations of cohesive, purpose-driven, data of the CPPS to conduct analyses for monitoring, decision making, prediction, and performing operations with the system. Additionally, the Digital Twin of a system provides services for handling user input, controlling its real-world counterpart, and interacting with other systems (e.g., product life-cycle management systems or other Digital Twins). We assume that each CPPS will, in the future, have its own Digital Twin and both closely interact for monitoring but potentially also for controlling purposes.

To conclude, Fig. 1 illustrates the triangular relationship between the production site, Digital Shadows, and the Digital Twin. Starting on the left-hand side of the figure, application-specific viewpoints illuminate certain aspects of the production area to collect specific data from the machines and processes. These are then collected figuratively by the Digital Shadows and flow into the Data Lake. From there, the data diffuses into the Digital Twin. Engineering models contribute to the digital replicate, releasing both analysis results back to the Digital Shadows, as well as configuration and control parameters into the production site.

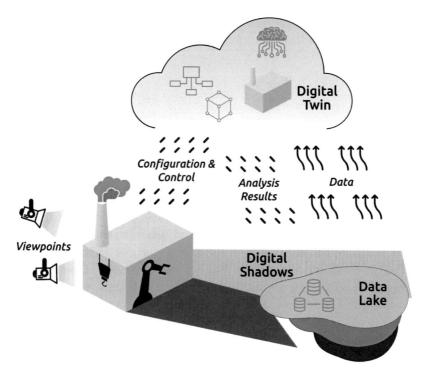

Fig. 1 Continuous Data Cycle in Industry 4.0

1.4   Outline

The remainder is structured as follows: Sect. 2 discusses challenges in engineering Digital Twins with respect to their construction, operation, and services. Section 3 introduces a conceptual structure of Digital Twins and Digital Shadows based on an underlying Data Lake infrastructure, while Sect. 4 introduces specific engineering approaches based on state-of-the-art technologies. Sect. 5 elaborates on data processing techniques (e.g., artificial intelligence) for Digital Shadows. Sect. 6 concludes with an outlook on the future of Digital Twins and Digital Shadows and their potential in Industry 4.0.

2   Challenges in Engineering a Digital Twin and Its Digital Shadows

This section discusses challenges in the model-driven engineering and operation of Digital Twins.

2.1   Challenges in Engineering Digital Twins

Integrated engineering of a digital twin and its represented system Engineering Digital Twins either coincides with the engineering of the represented system, i.e., both systems are conceived and developed together ("greenfield" engineering), or takes place after the represented system has been deployed ("brownfield" engineering) [26]. In both cases, the purpose of the Digital Twin and its information management requirements must be made explicit. However, during the greenfield engineering of Digital Twins, development of the represented system can consider these requirements, whereas in the brownfield case, the required information might not be available, demand augmentation of the represented system, or synthesis from other sources. Moreover, most Digital Twin operations demand information with specific quality requirements, such as frequency, precision, reliability and augmented with additional contextual information (e.g., time and source of origin). Where greenfield approaches can consider this during development of the represented system, brownfield approaches often cannot. This can lead to less precise, out-of-date, and unreliable Digital Twins. Consequently, brownfield approaches can severely limit the capabilities of potential Digital Twins and, to achieve effective Digital Twins, they should be considered an integral part within the development of the represented system.

The systematic identification of Digital Twin purposes in the context of available, producible, and synthesizable information is a complicated task. Consequently, its

systematic integration into the development of the represented system is a major
challenge in engineering useful Digital Twins.

Data quality in context of sensor data and data streams has been addressed in various applications [20, 38]. In the context of Digital Twins, this challenge has been recognized, but not
yet explicitly addressed [72]. The parallel development of a Digital Twin and its physical
twin is discussed in [51], where different levels of a Digital Twin are presented.

Systematic specification and implementation of digital shadows Another quintessential
challenge in engineering Digital Twins is enabling these to leverage the data retrieved from
the represented system and making sense out of it. In order to retrieve data, both the aspect
of data streams originating from different sources as well as heterogeneous formats of data
have to be addressed. in (domain-specific) real-time and ultimately hamper Digital Twins
to fulfill their purposes: On the one hand, the three Vs of big data—Volume, Velocity, and
Variety—pose a challenge for handling the data. At the same time, all possibly relevant data
shall be captured as one wants to keep all data for later possible usage. To mitigate this, data,
information, and models of the represented system and related systems can be aggregated and
abstracted prior to processing, resulting in Digital Shadows [36]. To be more exact, it might
be even not possible to collect all data initially, as the volume is too large, the systems are not
capable of analyzing the data at the required speed, or there is no sensor to capture specific
information. As illustrated in Fig. 1, Digital Shadows are produced by different viewpoints
on the represented object, i.e., only data relevant for a specific purpose or application is
considered. Still, it is not possible to foresee all potential use cases in Digital Twins when
designing the CPPS or the corresponding data management systems. Therefore, we propose
to use a Data Lake [60] as a repository for raw data that supports functions to prepare data
as required by the Digital Twins, integrates models and other sources of information, and
provides the right amount, granularity, and precision of data at the right time.

For the definition of Digital Shadows, suitable data modeling and integration techniques are needed. These must enable the implementation of Digital Shadows consisting of required data integrated from multiple sources, including data transformation
functions to clean, harmonize, and restructure the data into the desired format.

Some case studies that apply the concept of a Digital Shadow have been described in [42].
Data Lakes have been proposed also in [70] as a solution to the data challenges that are
present in the context of Digital Twins, e.g., heterogeneity, volume.

Integration of domain expert solutions Engineering Digital Twins is an interdisciplinary
effort that, depending on their purpose, can demand the collaboration of experts from automa-

tion engineering, human-machine interaction, software engineering, and more. These experts employ a variety of different solution paradigms, methods, and technologies that need to be properly integrated to produce and deploy Digital Twins. To this end, often solutions following geometric and functional paradigms, continuous and discrete perspectives, heavily front-loaded and agile methods, which employ a large variety of different implementation techniques (CAD, knowledge representation, physical modeling, state-based modeling, programming, etc.) must be integrated. Due to the required technical complexity and level of detail of the different experts' solution implementations, their integration and joint use in Digital Twins introduces another level of complexity.

MDD is a solution paradigm that aims to facilitate the integration of these different domains by overcompensating the growth in complexity of their solutions. To this end, MDD lifts abstract models (in contrast to technically detailed implementations) to primary development artifacts that can leverage terminology and concepts of the participating experts' domains. Through automated analyses and syntheses that reify domain expertise, these models can greatly facilitate the integration of solutions and their joint use in Digital Twins. Consequently, the efficient inter-disciplinary engineering of Digital Twins demands suitable modeling techniques and tools for the different domains and means to their integration. While appropriate domain-specific modeling techniques and tools have been brought forth, such as Simulink [11] or Modelica [16] for physical modeling, various CAD variants [21, 22, 43] for geometric modeling, and different software modeling techniques [14, 43, 54], means for their integration are rare and either do not consider how the different domain-specific solutions shall be integrated, nor consider the models' semantics (meaning) [15], but consider their syntactic structure [9] only.

> The systematic, syntactic, and semantic integration of the Digital Twin parts contributed by experts from participating domains demands suitable and precise modeling techniques for these experts that can be integrated easily and semantically meaningfully.

Research in software language engineering [31, 39] investigates the conception, engineering, and evolution of suitable modeling techniques; on the basis of SysML [14, 75], their integration can be achieved.

Composable digital twins Systems of systems are ubiquitous in manufacturing. Through well-defined interfaces, standards, or handcrafted integration, these collaborative system groups achieve goals unachievable alone. While there are various standards on the integration of systems through joint interfaces or shared data structures in manufacturing [74], research in Digital Twins rarely considers their collaboration, integration, or composition. With Digital Twins often being data-intensive applications that conduct complex analyses relying on (domain-specific) real-time data acquisition, operating many Digital Twins on the same data sources (e.g., Data Lakes) can lead to bottlenecks rendering the data acquisition in real-time impossible. Collaborative Digital Twin groups or composed Digital Twins that share data, reduce redundant data acquisition, could mitigate this. For achieving this, reuseability and interoperability of Digital Shadows is desirable. (8) Moreover, research on composable Digital Twins can facilitate their engineering through reusing off-the-shelf Digital Twins provided by third-party vendors.

> The systematic, syntactic, and semantic integration of the Digital Twin parts contributed by experts from participating domains demands suitable and precise modeling techniques for these experts that can be integrated easily and semantically meaningfully.

Leveraging the time-honored concepts of component-based software engineering [3], such as encapsulation of functionality behind stable interfaces or substituability of implementations, and applying these through modern architecture description languages [50] can greatly facilitate engineering Digital Twins through composition.

Measurable digital twin and digital shadow fidelity Digital Twins intrinsically serve to represent the system they observe. As such, the usefulness of a Digital Twin depends on the fidelity, i.e., the precision of this representation, with respect to its purpose. For instance, if a Digital Twin controlling an injection molding machine [3] is subject to a deviation of several millimeters, the resulting products might be rendered unusable, whereas for a Digital Twin monitoring an automated vehicle [40], this relatively small discrepancy might be tolerable. Consequently, the quality of representation and its possible degradation leading to divergence between the represented system and the Digital Twin must be considered when engineering Digital Twins. As the possible quality of representation directly depends on the quality of data in the Digital Shadow, the data quality requirements of Digital Twins need to be made explicit, such they can be considered during design, implementation, and runtime of the Digital Shadows. Moreover, Digital Twins must anticipate degradation of that quality and provide metrics to gauge that quality during runtime. Similarly, as Digital Twins are meant to strategically contribute added-value to manufacturing, these expectations should be made explicit and controlled at runtime, e.g., through simulation-based benchmarks applied to the Digital Twin. Data quality requirements, degradation metrics, and benchmarks need to be formulated relative to the heterogeneous models the Digital Twins consist of and, thus, demand suitable modeling techniques as well.

A Digital Twin and its Digital Shadows are subject to representation quality that depends on the quality of available data and aim to produce added-value through optimization. Measuring all of these must be anticipated at design-time measured to prevent diverging of the represented system, its Digital Twin, and its strategic goals.

As mentioned before, data quality issues have not been explicitly addressed for Digital Twins, but more general frameworks have been developed for representing data quality requirements for sensor data and data streams. For example, [20] proposes an ontology-based approach to model data quality requirements, dimensions, metrics, and measurements.

2.2   Challenges in Operating Digital Twins

Mind the gap between design-time and runtime Digital Twins can be used at system design-time as well as during their operation and even after their decommissioning. At design-time, they can support exploration of the design space through the rapid creation of system variants for dimensioning, testing, and simulation. To this end, the design-time Digital Twins need to operate on data and models from a system yet to be created. Both, data and models, can be provided from observations of similar or predecessor systems or synthesized from simulation of similar systems. When migrating from design-time to runtime, not only the sources of data that the Digital Twin operates upon must change, but changes in quality, precision, frequency, etc. of available real data might render observations made from design-time data invalid. For instance, a Digital Twin at runtime, controlling a real-time process, might not be able to reproduce behavior previously anticipated in design-time. Moreover, migrating from design-time to runtime might change the form of models the Digital Twin relies upon: while design-time models may be of arbitrary complexity due to the availability of sufficient computing capacity, deploying the same Digital Twin on a CPPS of less capacity might demand reductions in the used models' orders to ensure operation of the Digital Twin.

Migrating a Digital Twin from design-time to runtime demands changes in the sources of data, the form of models, and the interpretation of both. This may lead to additional engineering challenges for a Digital Twin. A careful separation of data acquisition and the conclusions drawn from design-time data can prevent this gap.

For numerical models, research in model order reduction [71] can contribute to the (partially) automated and deployment-specific reduction of design-time models into run-time abstractions. For symbolic (knowledge bases, systems engineering models) and subsymbolic (artificial neural networks) models, this is subject of ongoing research [62].

Extensible digital twins Digital Twins are expected to chaperone their represented systems during their lifecycles. Therefore, they need to evolve with changes to the represented systems and in their environment. When new systems, subsystems, Digital Twins, sensors, or actuators become available, their data sources and models might be useful for optimizing the behavior of existing Digital Twins. Making these new data sources and models available in a Digital Shadow, without stopping the corresponding Digital Twin, and without major software engineering efforts either demands means to automatically discover and incorporate these or configuration mechanisms to adjust existing Digital Twins during their runtime. The former not only requires inventing mechanisms to discover novel data sources and models within the Digital Shadow, but also to integrate this new information meaningfully and safely without causing operational issues. The latter requires means to describe the availability of new data sources and models and their integration in Digital Shadows and Digital Twins tailored to non-programmers.

> The evolution of represented CPPS requires a Digital Twin to incorporate new data sources and models during its runtime. Foreseeing this evolution during design is crucial to ensure the future extension of a Digital Twin.

For the syntactic and semantic representation of data sources, models, their interfaces and relations leveraging data structure modeling techniques, such as Unified Modeling Language (UML) class diagrams [64], and knowledge representation techniques, such as ontologies [45, 59], can facilitate engineering extensible Digital Shadows and Digital Twins. For both paradigms, (semi-)automated matching techniques that can facilitate discovering and integrating new data sources and models [1, 45] as well as service discovery mechanisms [37, 46] are available. For the non-invasive configuration of Digital Twins during runtime, low code modeling techniques [68] and results from research on models at runtime [2] might facilitate the integration of new data sources and models.

Adaptive digital twins Digital Twins are used to predict the behavior or evolution of the represented system, e.g., for predictive maintenance, future resource consumption, and more. To this end, they simulate possible behaviors based on currently available data and models. Depending on the frequency of changes of data and models as well as on the duration of computing predictions, these predictions might become outdated while being computed. Hence, computational resources are wasted and the Digital Twin's chances to react on properly predicted results are missed. If prediction is successful, but prediction and reality diverge, either the Digital Twins underlying data, models, knowledge bases, etc.

must be changed or the Digital Twins must use the predictions results to change reality. The Digital Twin's reaction to this divergence are highly application-specific and demand suitable modeling techniques to describe these reactions by domain experts.

> A predictive Digital Twin must support domain-specific actions for reacting on the divergence of expectations and reality. These reactions can include applying domain-specific expertise or adjusting its models of reality.

Modeling can facilitate encapsulating domain-specific expertise in a machine-processable form that the Digital Twins can leverage. This can be continuous Simulink [63] models, discrete state-based [64] descriptions, or rule-based specifications [3]. To enable domain experts in efficiently reifying their expertise, all of these must be tailored to their domains, e.g., by applying techniques from software language engineering [31].

3 Engineering a Digital Twin and Its Digital Shadows

For engineering Digital Twins and Digital Shadows, we first have a closer look at their relationship and their components, which is depicted in Fig. 2. With respect to the definitions introduced in Sect. 1, we identify a fundamental set of conceptual constituents that most Digital Twin applications should comprise, even though concrete realizations might differ.

Modern CPPSs are usually equipped with sensors that monitor the system behavior. In addition, there are also cameras and external sensors in production plants, for example to monitor the activities of employees or to be able to react in time to smoke detection. These data are usually managed in different databases but can be combined with the help of a Data Lake.

The Digital Twin accesses the Data Lake and extracts Digital Shadows from it, combining exactly the information it requires for a specific purpose. A dedicated component within the Digital Twin, the *Shadow System*, performs this task. The Shadow System encapsulates the *Shadow Caster* and the *Shadow Controller*. The Shadow Caster processes data of the Data Lake and calculates Digital Shadows by combining insights and information of heterogeneous data sources within the Data Lake. It creates task-specific Digital Shadows, that can support the Digital Twin in performing its tasks. For example, a Digital Shadow can be the temporal evolution of a parameter within the CPPS to detect wear of CPPS components. Combined with material and maintenance information about this component, the Digital Shadow can enable predictive maintenance tasks. The Shadow Controller decides, when to create and destroy Digital Shadows. It controls the storing process of digital shadow insights and analysis results within the Data Lake and decides which data to be stored or ignored. The Shadow Controller also manages the visibility and access rights of Digital Shadows.

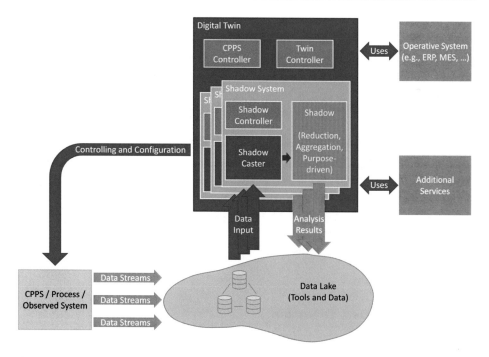

Fig. 2 Schema of the Relationship of Digital Twins and Digital Shadows

Other operative systems, such as ERP or MES, can access information provided by the Digital Twin, or leverage the Digital Twin to understand or influence the behavior of the CPPS. The same applies to other, domain-specific services. For applying reconfigurations to the system and controlling its activities the Digital Twin has a control flow to the CPPS. When the CPPS behavior needs to be adjusted and how its communication interface can be accessed is managed by the *CPPS Controller*, which is part of the Digital Twin. The *Twin Controller* encapsulates the behavior of the Digital Twin and decides which actions the Digital Twin should take, which tasks it should perform, and which Digital Shadow it requires for these tasks.

4   From Engineering Models to a Digital Twin

Engineering models are models that promote the design and construction of systems. Typical engineering models are technical or CAD drawings, circuit diagrams, or UML models. They support design and engineering activities by capturing a system's properties and giving insight about the system even before it is constructed, i.e., they prescribe properties of the things to be. Therefore, their application domain is usually in constructive domains where their prescriptive capabilities can support design decisions and design space exploration of different variants of a system before it is constructed. Engineering models are created by domain experts, e.g., mechanical engineers, electrical engineers, software developers, administration, UI designers, etc. to master the complexity of CPPS systems. As such, they encapsulate domain knowledge that can support the realization of Digital Twins. In addition, engineering models can also support the Digital Twin at runtime, as they provide the Digital Twin the opportunity to understand and reason about the internal structure and behavior of the represented systems to detect anomalies, divergence of expected (e.g., simulated) behavior and real behavior, or other perils to operations.

Typically, engineering models are developed during the design phase of a CPPS and are of limited use after deploying the system, as they are hardly updated and, therefore, cannot represent the CPPS precisely. Their lifetime can be extended by deriving parts of the Digital Twin's functionality or providing a knowledge base for the Digital Twin. For this purpose, the essential information must be extracted from the engineering models and prepared in such a way that it can be further processed by other software systems and utilized by the Digital Twin.

4.1   Semantic Data Extraction

Each phase of a system's lifecycle contains specific activities, creates characteristic data, and relies on different services of the Digital Twin. When integrating model information, we distinguish between horizontal and vertical integration. Horizontal integration focuses on combining information from different lifecycle phases of the CPPS. For example, a CAD model of the design phase can help to evaluate whether the assembled product conforms to the specification during the implementation phase. Vertical integration combines information of models of different hierarchical levels. For example, a CPPS' SysML model consists of models representing its individual components. Vertical integration also refers to a stepwise refinement, where higher-resolution models refine a coarse system description.

Figure 3 shows the lifecycle phases of a CPPS and its Digital Twin. During *Design* the system's concept, its concrete functions and appearance are developed, based on customer requirements and market information. Data aggregated during this phase consists of customer requirements specification, often in unformalized mediums, such as plain text. However, it can also involve requirements models, e.g., specified in DOORS, where requirements are recorded and managed. Typical models that support the design phase are CAD models, describing the appearance and dimensions of the CPPS, SysML models that translate customer requirements into system functionality descriptions, and also simulations, e.g., in Matlab that evaluate the system's behavior. The Digital Twin supports the design phase by evaluating different variants of the CPPS and providing the optimal one concerning customer requirements and estimated working conditions. The *Implementation* phase combines all activities for creating the CPPS, including software development and hardware construction. During this phase, the different components of the CPPS are created, combined, and

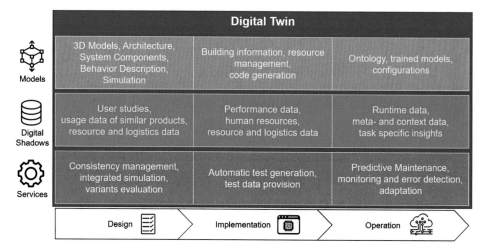

Fig. 3 The DT contains data, models, and services for every lifecycle phase of the underlying CPPS

tested to detect errors. Data in this phase include human performance, use of resources, material properties and test results. During the implementation, software architects create models specifying the internal structure of the software components and interfaces to other software systems. Also, models for Building Information Modeling (BIM) and Engineering Change Requests (ECR) are employed to support the realization of hardware components. The Digital Twin could support this phase by ensuring consistency between models of the design phase and the implementation, or by generating test cases to evaluate the implementation. During the *Operation* phase, the system is assessed and evaluated to ensure it functions as intended and does not become obsolete. In this phase, large amounts of data are created, since modern CPPS are equipped with sensors that capture changes within the system itself and its operating context likewise. The Digital Twin can monitor the system and check whether the expected parameters from the design phase conform to the real operating data. It accesses the Digital Shadows that provide information about estimated and real values and evaluate these to perform e.g., predictive maintenance.

4.2   Technologies for Connecting Digital Twins and Engineering Models

For the further use of engineering models beyond their intended life phase, special software technologies are required. During the realization of a CPPS different domains collaborate and develop individual model artifacts that in combination describe the CPPS. Combining information from these different models requires model integration mechanisms [15]. Through model integration each discipline involved in the CPPS and product development process can develop using its own terminology and tools while the combination of these models describes the entire CPPS . Language integration [27, 44] focuses on integration of models on the language level, thus, also affects the tooling provided for describing CPPS. Language aggregation employs several artifacts of different languages to describe aspects of the target domain. The processed artifacts remain separated but describe the same CPPS. Language embedding combines models into one common artefact, while keeping the languages and tooling for these languages separated. Language inheritance also combines different languages into one artifact but in addition customizes elements of these languages [30].

Another approach for combining knowledge is tool integration [12]. Tool integration realizes the exchange of model data between specific tools. Thus, it reduces gaps during the development process and reduces inconsistencies.

Model-Driven Software Development Generating software is crucial to successfully integrate modeling in development processes [65]. Code for production systems or for test drivers can be efficiently generated and thus, improve the consistency between model and implementation and save resources. By deriving software code from engineering models, the development effort for Digital Twins can be reduced even further. Engineering models incorporate domain knowledge from which components of the Digital Twin can be derived. One challenge is the extraction of the relevant information from these models. In model-based software engineering symbols are an established way of extracting the relevant information from a model. Accordingly, it might be interesting to closely analyze engineering models according to this aspect and identify the significant symbols, that can be useful in other life phases, too.

Tagging One possibility to add required data to already existing engineering models is tagging. A tag model logically adds information to the tagged model while technically keeping the artifacts separated. Thus, the engineering model can stay clean of this additional information, and domain experts are not confused by additional model elements that do not directly correspond to the domain. A tag model could e.g., connect model elements from different life phases, thus ensure horizontal integration. It could also add information about the structural decomposition of a model to support vertical integration. It could also add data retrieval information to models to connect the engineering model with data about the physical system, its behavior or its development process.

Models at Runtime In contrast to those development methods for Digital Twins, models at runtime focuses the utilization of models during the CPPS runtime. Since engineering models encapsulate domain knowledge and information about the underlying CPPSs they can build a knowledge base for Digital Twins to rely on while they are running [4]. Thus, they support the Digital Twin to cope with new challenges and to adapt the twin to even to unanticipated changes. Runtime models are reflective, meaning that they are causally connected with the underlying CPPS. Thus, every change in the runtime model leads to a change in the reflected system and vice versa [47]. A structural model, that describes the components of the CPPS, reflects the CPPS's constituents and can be update if new physical assets are added, or components are exchanged [24]. Physical models describe the dynamics and current state of physical phenomena [69]. The Digital Twin can rely on physical models to assure selected properties of the running CPPS. Behavioral denotes a runtime model capturing the dynamics of the systems, i.e., what the system can or will do based on its current state. Behavioral models can support run-time verification and sensitivity analysis, to determine how CPPS parameters are affected based on reconfigurations. For example, behavioral models may support the Digital Twin in predicting the CPPS's behavior or evaluating several possible reconfigurations for identifying the best solution for a present problem [17].

View-based Modeling Often, while applying model-driven development leads to redundancies and inconsistencies if models share a semantic overlap and describe the same CPPS. View-based modeling is an approach to tackle fragmentation of information across instances of different metamodels and can support the Digital Twin in combining knowledge from heterogeneous models. A view type defines the set of meta-classes whose instances can be displayed by a view [5]. Different view types can support the visualization of relevant model elements. Combined with a query language, they build a tool for describing and creating Digital Shadows and thus can also support the Digital Twin in providing insight about the current CPPS state.

Executable Metamodeling A metamodel is a model of a model and specifies the concepts and model elements that are useful for solving a specific class of problems. By means of Kermeta workbench [49], Ecore [66] meta models can also be executed. Kermeta provides an action language to implement the execution semantics of Ecore metamodels. It supports adding new methods to existing Ecore meta-classes, that define the execution semantics of the corresponding metamodel in the form of an interpreter [10].

5   Digital Shadows and Data Processing

For engineering Digital Shadows, the task of data processing is essential, as the Digital Shadow is closely linked to the data providing infrastructure. Regarding the standard data processing steps of ETL (*Extract-Transform-Load*) [34], the Digital Shadow has strong ties to the *transform* task: in the traditional sense, data is adjusted in this step, in order to be stored and used later. Digital Shadows exceed traditional data adjustment and aim to embed further intelligence in this step, making the *transform* task more sophisticated and therefore introducing a new smart data layer. For this to be possible, incoming data streams have to be steered—which is taken care of by the Data Lake—and relevant data must be identified, specified, and processed—which is the task of the Digital Shadow. Since Digital Shadows serve specific purposes, their data granularity can differ significantly; for some

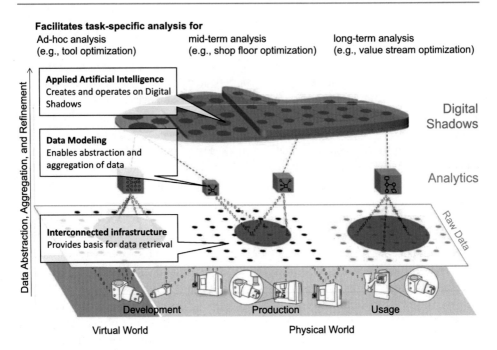

Fig. 4 Purpose-Driven Digital Shadows

purposes, a very detailed view on data is necessary, whereas for other purposes, aggregated or intelligently processed data is of interest. The structure of different granularity of Digital Shadows is depicted in Fig. 4, as envisioned in the Internet of Production [33, 36, 58].

From the bottom to the top, we observe the data abstraction, aggregation, and refinement process: In the bottom, data is in its raw format, corresponding to data gathering in the interconnected infrastructure. With each subsequent step up, the data can increasingly be refined, e.g., by employing data modeling, such that in the top we obtain task-specific Digital Shadows where artificial intelligence is applied. Depending on the task, we may need more refined or more raw data. Use cases regarding long-term analysis, e.g., value stream optimization, require more refined data, whereas, for ad-hoc analysis, e.g., tool optimization, more granular data is required. Mid-term analysis, e.g., shop floor optimization, might demand a mixture of both raw and already partly refined data.

The different levels of granularity and the different levels of data abstraction, aggregation, and refinement of Digital Shadows correspond well with the concept of Data Lakes and their levels of different data layers.

5.1   Data Lakes as a Serving Infrastructure

A Data Lake, as described in [60] is a repository where data from various sources can be stored in their original structures. In this regard, the concept of a Data Lake significantly differs to the data-handling approach of traditional data storage systems, with the most prominent concept being the data warehouse: The Data Lake introduces an ELT (*Extract-Load-Transform*) approach instead of the traditional ETL. Raw data is stored in its raw format without prior integration or aggregation to avoid restricting data usage and analysis to a predefined integrated schema. This especially means that we do not—as is common in data warehouses—have a schema-on-write paradigm, but delay the importance of schemas to the time of reading the data, i.e., we follow a schema-on-read approach. This also contributes to the challenge discussed in Sect. 2, that Digital Shadows should be flexible and should be able to incorporate new data sources at least semi-automatically, without requiring a software engineering process to define ETL workflows and integrated schemas.

In general, Data Lakes provide tools to extract data and metadata from heterogeneous sources, offer a data transformation engine that can transform or clean data and integrate it with other data, and provide interfaces to explore and to query data and metadata it contains. According to [60], a Data Lake is distinguishable into four components serving those purposes: *Ingestion Layer, Storage Layer, Transformation Layer,* and *Interaction Layer.* The composition of a typical Data Lake is shown in Fig. 5 and described in the following.

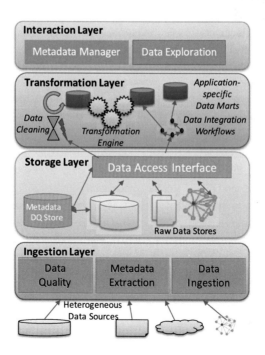

Fig. 5 Data Lake Architecture

Ingestion Layer: To make data available in Digital Shadows or Digital Twins, it is important that the data collection is as easy as possible. The ingestion of new data sources into the Data Lake should be as simple as a copy operation in a file system, as data from the sources is copied to the storage layer in its original format. To ensure sufficient data quality, the ingestion should extract metadata from the sources automatically and enforce data quality rules to prevent that data with insufficient quality pollutes the Data Lake.

Storage Layer: This layer is comprised of common big data tools, such as Apache Hadoop or Apache Spark. They provide storage and query mechanisms for heterogeneous data. Metadata management is usually limited to structural information, but should be enriched with semantic and data quality information to make the data more usable.

Transformation Layer: Raw data needs to be transformed and cleaned if it should be usable for an application. This functionality is provided by the Transformation Layer, which might also be implemented with big data tools as Apache Spark. This layer can be also considered as the Shadow Caster, as it creates Digital Shadows (or *Application-Specific Data Marts*) by aggregating, transforming, reducing, etc. the raw data.

Interaction Layer: Finally, the Interaction Layer provides means to interact with the data in the Data Lake or Digital Shadows and to make it available for a certain purpose in a Digital Twin.

An example reference model of a Data Lake System is provided by Constance, introduced in [25]. There are various benefits of Data Lakes compared to more traditional approaches. These include easier integration of heterogeneous data (e.g., relational data, JSON, XML, graphs, data streams, documents), easy exploration of heterogeneous data on data level instead of application level, and avoidance of data transformation during data ingestion. The latter has the effect of making it easier to handle data with a purpose that is unclear at design time.

On the other hand, Data Lakes pose the threat of deteriorating into data swamps. A data swamp is a Data Lake that is overwhelmed by unusable data to the degree of becoming unusable itself, because data of interest cannot be found or data cannot be interpreted anymore. A countermeasure for data swamps is to ensure metadata ingestion and data quality management when loading the data. An additional approach to ensure data comprehensibility is to adopt ontologies and other rich data models in Data Lakes. The relationship of Data Lakes and traditional data management systems is further addressed in [35].

For the use case of Industry 4.0 and the employment of Digital Shadows, the principle of Data Lakes to store data in its raw format, in particular, gives us the following advantages: The immense heterogeneity regarding the involved sensors and machinery, which often characterizes industrial production settings, can be handled well since it perfectly fits the paradigm of Data Lakes. The characteristic of Data Lakes to keep all data enables us to create multiple purpose-specific Digital Shadows based on data located in the Data Lake. In general, Digital Shadows fit well onto the idea that we have an extensive repository of data and may select different traces of data, depending on the application of the specific Digital Shadow.

Addressing the challenges introduced in Sect. 2, we conclude: the three Vs of big data, are addressed as follows: Volume is addressed by the distributed nature of Data Lakes. Variety can be managed by the heterogeneous nature of the storage layer in Data Lakes. Velocity is most challenging aspect for Data Lakes as they are more 'static' repositories for long-term collection of data rather than real-time processing. However, it is also important to store streaming data in the Data Lake, which can be used, for example, for the training of machine learning algorithms. The challenges to capture all possibly relevant data and still keep all other data for later possible usage are implicitly handled by Data Lakes because all data is stored in its raw format.

With introducing a Data Lake as serving infrastructure for Digital Shadows, we face another non-technical issue: in a conglomerate of cross-company contributors, each stake-holder needs to collaborate ideally using a unified interface to access and share their data and Digital Shadows. The Data Lake provides us with such a central repository instead of many separate data silos. However, the nature of the collaboration may impose that not all data may be kept in the Data Lake. Reasons may be data access rights or policies of individual organizations. We address this issue by proposing a hybrid Data Lake approach: We introduce one central Data Lake, that everybody has access to. At the same time, we keep some data landscapes (e.g., Data Lakes) at participants' sites where confidential data that cannot be shared with everybody resides. Still, we keep the central Data Lake and its interface as the primary instance in questions of data. This entails that the data which is not in the central Data Lake has to be retrieved from the private data repository—if allowed for the specific participant. With this, we face a new challenge—the challenge of integrating private Data Lakes. One Solution for this is to employ data virtualization. Data virtualization may take place on different levels, but for the architecture of a hybrid approach and to hide the virtualization from the users, the best choice is to employ it on the lowest level possible [73]. In this case, data governance becomes an interesting and at least partially open problem. An alliance-like approach to data governance is proposed by the Industrial Data Space initiative [56]. Concluding, we see Data Lakes as an essential component to enable Digital Shadows, already solving a lot of presented challenges.

5.2   From Data Processing to Digital Shadows

In order to engineer Digital Shadows, both application domain and computer science knowl-edge become a necessity resulting in cross-domain collaboration. The added value of cross-domain collaborations to engineer Digital Shadows is best illustrated by an example [36]: In [53] the combination of domain knowledge and artificial intelligence expertise provided by an interdisciplinary team allowed for the evaluation time of an engineering process simula-tion to be reduced by six orders of magnitude: while traditionally finite element simulation is used to evaluate manually designed schedules for heavy plate rolling, [53] were able to adequately approximate this simulation by six reduced analytical models operating on data provided by Digital Shadows, therefore reducing the computation time from 30–240 min to less than 50 milliseconds. This example of significantly influencing a simulation's computability demonstrates the advantages for diagnosis and predictions in domain-specific real-time if they are based on Digital Shadows.

The challenges formulated in Sect. 2 are tackled by Digital Shadows as follows: The reduction of the three Vs of big data is partially handled by the Data Lake system. Unsolved by the Data Lake is the reduction and refinement from massive data volumes to manageable ones. The Digital Shadow addresses this issue for e.g., computability reasons—as opposed to using the whole data in the Data Lake. The challenge of interoperability and traceability of data traces in the data layers is addressed by [19], introducing the concept of FACTDAG. Closely related to interoperability and traceability is the desired reusability of Digital Shadows. Reusability is particularly powerful when the groundwork of agreed-upon ontologies has been established beforehand. Using Digital Shadows, not only reusability of the data is to be considered, but also the reusability of the component extracting this data, the Shadow Caster. As Digital Shadows are digital artifacts describing real objects, they serve a specific purpose regarding this real object. Beyond this, they can be used as input for other systems or for analysis exceeding the extent of the primary use of the Digital Shadow. This is typically achieved by conforming to best practices of development, e.g., modularization and reuse. The difficulty to know which data is relevant and the data selection process that shall reduce the used data for computability, are addressed by on the one hand collaborating with application domain experts—for Industry 4.0 mostly mechanical engineers—and on the other hand with the approach to systematically use artificial intelligence methods to first find structure and insights in the data and second to automatize the process of building Digital Shadows, yielding a combination of the two. The commitment of whether data is relevant usually has to be made during the data gathering and design phase. However, based on the Data Lake infrastructure, we obtain the possibility to gather all data and only with the engineering of the Digital Shadow commit to a specific data model as input. This leads us to the importance of Artificial Intelligence in Digital Shadows.

5.3　Artificial Intelligence in Digital Shadows

In order to understand how data that is about to be incorporated into the Digital Shadow has to be prepared and handled, we first must have a closer look at the composition of the *Digital Shadow System* (cf. Fig. 2). In general, the Digital Shadow System consists of two main components—a passive part, the *Shadow Data*, and an active part, the *Shadow Caster* that may be executed and manipulates the passive part. The first component Shadow Data is typically located in a Data Lake. The second component, Shadow Caster, can embrace advanced techniques, especially from the domain of artificial intelligence, transcending traditional transformations and thus incorporating intelligence. From a data-flow perspective, intelligence is applied earlier than before. This especially entails that Digital Shadows themselves are already incorporating domain knowledge and intelligence, establishing a new smart data layer.

Regarding the employment of AI in Digital Shadows, we distinguish between two approaches: Digital Shadows repeatedly using AI methods to generate the Shadows' data on the one hand and Digital Shadows using AI initially and only once to determine the structure of the Digital Shadow. The first approach employs continuous learning and employment of AI, whereas the second employs one-time learning of the structure. An example of the first case would be to have some parameterized machine learning component in the Digital Shadow, e.g., a neural network, where the parameters, i.e., in the case of the neural network the weights of the edges, are continuously recalculated. An example of the second approach would be to determine the inputs for the Digital Shadow, e.g., the input layer of a neural network. Naturally, the two approaches are not exclusive and can be combined. The integration of intelligence into the Digital Shadow raises the question at which point a Digital Shadow ends, and the application using the Digital Shadow begins: depending on the complexity of the Digital Shadow, intelligence that previously was contained in the application layer, may now be shifted into the Digital Shadow instead. This can have the advantage that this artifact can be used by several applications. In this case, however, they have to share the same ontology to be able to use the same Digital Shadow, as they would otherwise use the same data with different interpretations. Finding the right degree of transferring logic into the Digital Shadow, and therefore deciding how much logic should remain in the application layer and how much should instead be located in the Digital Shadow, poses an interesting research question of its own.

6   The Future of Digital Twins and Digital Shadows

This chapter has covered the concepts of Digital Shadows and Digital Twins, with a particular focus on engineering methods. Digital Shadows consist of data traces of an observed CPPS. A Digital Twin operates on them to enable novel analyses, monitoring, and optimization. They both offer extensive support for production companies throughout the development, production, and usage life cycles, while harnessing the interconnectivity of manufacturing devices with processes and domain knowledge. This lifts the Internet of Things to an Internet of Production, particularly considering the aspects of Industry 4.0. Integrated applications on top then allow to specifically plan, simulate, inspect, and control processes. This is an essential requirement to manage the ever-increasing complexity of modern production processes.

Depending on the levels of sophistication and integration, these concepts elevate the operability of CPPS. Low-level realizations, e.g. for real-time analysis, allow deployment directly on the CPPS. In contrast, a variety of solutions rely on cloud-based systems, as these can dynamically provide necessary storage and computation capacities, often required for complex operations. Following the platform as a service (PaaS) paradigm, virtual platforms can be provided, which offer dedicated services for engineering individual Digital Twin applications. In turn, this fosters general interchangeability and extension of these frameworks. After all, complete and ready-made systems may be provided, which companies can configure and directly put into operation, similar to the software as a service (SaaS) approach. The conceptual separation of the Digital Shadow particularly offers enormous application potential for sustainability in Industry 4.0. The archival storage of raw data in Data Lakes ensures its long-term usability. This is an important factor in commissioning production systems; the infrastructure must be able to sustain for years or even decades. For instance, machine learning algorithms may operate on years of acquired data to enable more precise decision support. As a challenge, interoperability between various providers needs to be ensured; the potential withdrawal of an operator may not cause disruptions. Similarly, switching providers should not infer severe migration costs.

Developing corresponding frameworks offers exceptional possibilities for both research and industry in the near future. The data collections can be used to perform sustainable analyses on raw data collected over many years, to provide insights into CPPSs and automatically optimize them by integrating domain expertise. It will also enable new data-driven business model opportunities. It is particularly in cross-company collaboration that the automatic exchange of data opens up entirely new perspectives. First and foremost, challenges regarding data sovereignty must be solved [33]. Ultimately, however, a data-driven worldwide integration of CPPS leads to the emergence of synergy effects. While the advantages for machine manufacturers through distributed data collection are most obvious here, it also offers new possibilities for the production industry in general, e.g. to mutually benefit from the experience of other companies, for instance in terms of concrete machine parameters. Similar effects can also be anticipated through the exchange of engineering models. Therefore, Digital Shadows and Digital Twins will continue to be the key enablers of Industry 4.0, offering exciting new avenues regarding sustainability and data sovereignty.

Acknowledgements Funded by the Deutsche Forschungsgemeinschaft (DFG, German Research Foundation) under Germany's Excellence Strategy—EXC-2023 Internet of Production—390621612

References

1. Mojeeb Al-Rhman Al-Khiaty and Moataz Ahmed. UML class diagrams: Similarity aspects and matching. *Lecture Notes on Software Engineering*, 4(1):41, 2016.
2. Gordon Blair, Nelly Bencomo, and Robert B France. Models@run.time. *Computer*, 42(10):22–27, 2009.
3. Pascal Bibow, Manuela Dalibor, Christian Hopmann, Ben Mainz, Bernhard Rumpe, David Schmalzing, Mauritius Schmitz, and Andreas Wortmann. Model-Driven Development of a Digital Twin for Injection Molding. In Schahram Dustdar, Eric Yu, Camille Salinesi, Dominique Rieu,

and Vik Pant, editors, *International Conference on Advanced Information Systems Engineering (CAiSE'20)*, volume 12127 of *Lecture Notes in Computer Science*, pages 85–100. Springer International Publishing, June 2020.

4. Nelly Bencomo, Sebastian Götz, and Hui Song. Models@run.time: a guided tour of the state of the art and research challenges. *Software & Systems Modeling*, 18(5):3049–3082, 2019.

5. Erik Burger, Jörg Henss, Martin Küster, Steffen Kruse, and Lucia Happe. View-based model-driven software development with modeljoin. *Software & Systems Modeling*, 15(2):473–496, 2016.

6. Bundesministerium für Bildung und Forschung. Zukunftsprojekt Industrie 4.0. https://www.bmbf.de/de/zukunftsprojekt-industrie-4-0-848.html. Accessed: 2020-05-29.

7. High Value Manufacturing Catapult. https://hvm.catapult.org.uk/. Accessed: 2018-06-05.

8. Benoit Combemale, Robert France, Jean-Marc Jézéquel, Bernhard Rumpe, James Steel, and Didier Vojtisek. *Engineering Modeling Languages: Turning Domain Knowledge into Tools.* Chapman & Hall/CRC Innovations in Software Engineering and Software Development Series, November 2016.

9. María Victoria Cengarle, Hans Grönniger, and Bernhard Rumpe. Variability within Modeling Language Definitions. In *Conference on Model Driven Engineering Languages and Systems (MODELS'09)*, LNCS 5795, pages 670–684. Springer, 2009.

10. Benoit Combemale, Cécile Hardebolle, Christophe Jacquet, Frédéric Boulanger, and Benoit Baudry. Bridging the chasm between executable metamodeling and models of computation. 09 2012.

11. James B Dabney and Thomas L Harman. *Mastering simulink.* Pearson, 2004.

12. Manuela Dalibor, Nico Jansen, Judith Michael, Bernhard Rumpe, and Andreas Wortmann. Towards Sustainable Systems Engineering-Integrating Tools via Component and Connector Architectures. In Georg Jacobs and Jonas Marheineke, editors, *Antriebstechnisches Kolloquium 2019: Tagungsband zur Konferenz*, pages 121–133. Books on Demand, February 2019.

13. Manuela Dalibor, Judith Michael, Bernhard Rumpe, Simon Varga, and Andreas Wortmann. Towards a Model-Driven Architecture for Interactive Digital Twin Cockpits. In Gillian Dobbie, Ulrich Frank, Gerti Kappel, Stephen W. Liddle, and Heinrich C. Mayr, editors, *Conceptual Modeling*, pages 377–387. Springer International Publishing, October 2020.

14. Sanford Friedenthal, Alan Moore, and Rick Steiner. *A practical guide to SysML: the systems modeling language.* Morgan Kaufmann, 2014.

15. Robert France and Bernhard Rumpe. Model-driven Development of Complex Software: A Research Roadmap. *Future of Software Engineering (FOSE '07)*, pages 37–54, May 2007.

16. Peter Fritzson. *Principles of object-oriented modeling and simulation with Modelica 2.1.* John Wiley & Sons, 2010.

17. A. Filieri, G. Tamburrelli, and C. Ghezzi. Supporting self-adaptation via quantitative verification and sensitivity analysis at run time. *IEEE Transactions on Software Engineering*, 42(1):75–99, 2016.

18. Hans Grönniger, Jochen Hartmann, Holger Krahn, Stefan Kriebel, and Bernhard Rumpe. View-based Modeling of Function Nets. In *Object-oriented Modelling of Embedded Real-Time Systems Workshop (OMER4'07)*, 2007.

19. L. Gleim, J. Pennekamp, M. Liebenberg, M. Buchsbaum, P. Niemietz, S. Knape, A. Epple, S. Storms, D. Trauth, T. Bergs, C. Brecher, S. Decker, G. Lakemeyer, and K. Wehrle. Factdag: Formalizing data interoperability in an internet of production. *IEEE Internet of Things Journal*, 7(4):3243–3253, 2020.

20. Sandra Geisler, Christoph Quix, Sven Weber, and Matthias Jarke. Ontology-based data quality management for data streams. *ACM Journal of Data and Information Quality*, 7(4):18:1–18:34, 2016.

21. Georges GE Gielen and Rob A Rutenbar. Computer-aided design of analog and mixed-signal integrated circuits. *Proceedings of the IEEE*, 88(12):1825–1854, 2000.
22. Mikell Groover and EWJR Zimmers. *CAD/CAM: computer-aided design and manufacturing*. Pearson Education, 1983.
23. Matthew Hause et al. The SysML Modelling Language. In *Fifteenth European Systems Engineering Conference*, volume 9, pages 1–12, 2006.
24. Christian Heinzemann, Steffen Becker, and Andreas Volk. Transactional execution of hierarchical reconfigurations in cyber-physical systems. *Software and Systems Modeling*, 18:157–189, 02 2017.
25. Rihan Hai, Sandra Geisler, and Christoph Quix. Constance: An Intelligent Data Lake System. In Fatma Özcan, Georgia Koutrika, and Sam Madden, editors, *Proceedings of the 2016 International Conference on Management of Data - SIGMOD '16*, pages 2097–2100, New York, New York, USA, 2016. ACM Press.
26. Richard Hopkins and Kevin Jenkins. *Eating the IT elephant: Moving from greenfield development to brownfield*. Addison-Wesley Professional, 2008.
27. Arne Haber, Markus Look, Pedram Mir Seyed Nazari, Antonio Navarro Perez, Bernhard Rumpe, Steven Völkel, and Andreas Wortmann. Integration of Heterogeneous Modeling Languages via Extensible and Composable Language Components. In *Model-Driven Engineering and Software Development Conference (MODELSWARD'15)*, pages 19–31. SciTePress, 2015.
28. Katrin Hölldobler, Judith Michael, Jan Oliver Ringert, Bernhard Rumpe, and Andreas Wortmann. Innovations in Model-based Software And Systems Engineering. *The Journal of Object Technology*, 18(1):1–60, July 2019.
29. David Harel and Bernhard Rumpe. Meaningful Modeling: What's the Semantics of „Semantics"? *IEEE Computer*, 37(10):64–72, 2004.
30. Katrin Hölldobler and Bernhard Rumpe. *MontiCore 5 Language Workbench Edition 2017*. Aachener Informatik-Berichte, Software Engineering, Band 32. Shaker Verlag, December 2017.
31. Katrin Hölldobler, Bernhard Rumpe, and Andreas Wortmann. Software Language Engineering in the Large: Towards Composing and Deriving Languages. *Computer Languages, Systems & Structures*, 54:386–405, 2018.
32. The Industrial Value Chain Initiative. https://iv-i.org/wp/en/about-us/whatsivi/. Accessed: 2018-06-04.
33. Matthias Jarke. Data Sovereignty and the Internet of Production. In *Advanced Information Systems Engineering*, Lecture Notes in Computer Science, Cham, Switzerland, 2020. Springer International Publishing.
34. Matthias Jarke, Maurizio Lenzerini, Yannis Vassiliou, and Panos Vassiliadis. *Fundamentals of Data Warehouses*. Springer Berlin Heidelberg, Berlin, Heidelberg, 2003.
35. Matthias Jarke and Christoph Quix. *On Warehouses, Lakes, and Spaces: The Changing Role of Conceptual Modeling for Data Integration*, pages 231–245. Springer International Publishing, Cham, 2017.
36. Matthias Jarke, Günther Schuh, Christian Brecher, Matthias Brockmann, and Jan-Philipp Prote. Digital shadows in the internet of production. *ERCIM News*, 2018(115), 2018.
37. Hejun Jiao, Jing Zhang, Jun Huai Li, and Jinfa Shi. Research on cloud manufacturing service discovery based on latent semantic preference about owl-s. *International Journal of Computer Integrated Manufacturing*, 30(4-5):433–441, 2017.
38. Anja Klein and Wolfgang Lehner. Representing data quality in sensor data streaming environments. *J. Data and Information Quality*, 1(2):10:1–10:28, 2009.
39. Anneke Kleppe. *Software Language Engineering: Creating Domain-Specific Languages using Metamodels*. Pearson Education, 2008.

40. Sathish AP Kumar, R Madhumathi, Pethuru Raj Chelliah, Lei Tao, and Shangguang Wang. A novel digital twin-centric approach for driver intention prediction and traffic congestion avoidance. *Journal of Reliable Intelligent Environments*, 4(4):199–209, 2018.

41. Johannes Kößler, Kristin Paetzold, et al. Support of the System Integration with Automatically Generated Behaviour Models. In *DS 80-11 Proceedings of the 20th International Conference on Engineering Design (ICED 15) Vol 11: Human Behaviour in Design, Design Education; Milan, Italy, 27-30.07. 15*, pages 021–030, 2015.

42. Martin Liebenberg and Matthias Jarke. Information Systems Engineering with Digital Shadows: Concept and Case Studies. In *Advanced Information Systems Engineering*, Lecture Notes in Computer Science, Cham, Switzerland, 2020. Springer International Publishing.

43. WD Li, Wen Feng Lu, Jerry YH Fuh, and YS Wong. Collaborative computer-aided design-research and development status. *Computer-aided design*, 37(9):931–940, 2005.

44. Markus Look, Antonio Navarro Pérez, Jan Oliver Ringert, Bernhard Rumpe, and Andreas Wortmann. Black-box Integration of Heterogeneous Modeling Languages for Cyber-Physical Systems. In *Globalization of Modeling Languages Workshop (GEMOC'13)*, volume 1102 of *CEUR Workshop Proceedings*, 2013.

45. Severin Lemaignan, Ali Siadat, J-Y Dantan, and Anatoli Semenenko. MASON: A proposal for an ontology of manufacturing domain. In *IEEE Workshop on Distributed Intelligent Systems: Collective Intelligence and Its Applications (DIS'06)*, pages 195–200. IEEE, 2006.

46. Matthias Loskyll, Jochen Schlick, Stefan Hodek, Lisa Ollinger, Tobias Gerber, and Bogdan Pîrvu. Semantic service discovery and orchestration for manufacturing processes. In *ETFA2011*, pages 1–8. IEEE, 2011.

47. Pattie Maes. Concepts and experiments in computational reflection. In *Conference Proceedings on Object-Oriented Programming Systems, Languages and Applications*, OOPSLA '87, page 147-155, New York, NY, USA, 1987. Association for Computing Machinery.

48. merics - Mercator Institute for China Studies. Made in China 2025. https://www.merics.org/sites/default/files/2017-09/MPOC_No.2_MadeinChina2025.pdf. Accessed: 2018-06-06.

49. Pierre-Alain Muller, Franck Fleurey, and Jean-Marc Jézéquel. Weaving executability into object-oriented meta-languages. In *International Conference on Model Driven Engineering Languages and Systems*, pages 264–278. Springer, 2005.

50. Ivano Malavolta, Patricia Lago, Henry Muccini, Patrizio Pelliccione, and Antony Tang. What Industry Needs from Architectural Languages: A Survey. *IEEE Transactions on Software Engineering*, 39(6):869–891, 2013.

51. Azad M. Madni, Carla C. Madni, and Scott D. Lucero. Leveraging Digital Twin Technology in Model-Based Systems Engineering. *Systems*, 7(1):7, 2019.

52. Michael F. Molnar. The U.S. Advanced Manufacturing Initiative. https://www.nist.gov/system/files/documents/2017/04/28/Molnar_091211.pdf, 2017. Accessed: 2018-06-06.

53. R. Meyes, H. Tercan, T. Thiele, A. Krämer, J. Heinisch, M. Liebenberg, G. Hirt, Ch. Hopmann, G. Lakemeyer, T. Meisen, and S. Jeschke. Interdisciplinary data driven production process analysis for the internet of production. *Procedia Manufacturing*, 26:1065 – 1076, 2018. 46th SME North American Manufacturing Research Conference, NAMRC 46, Texas, USA.

54. Arne Nordmann, Nico Hochgeschwender, and Sebastian Wrede. A survey on domain-specific languages in robotics. In *International Conference on Simulation, Modeling, and Programming for Autonomous Robots*, pages 195–206. Springer, 2014.

55. Petru Nicolaescu, Mario Rosenstengel, Michael Derntl, Ralf Klamma, and Matthias Jarke. View-Based Near Real-Time Collaborative Modeling for Information Systems Engineering. In *International Conference on Advanced Information Systems Engineering*, pages 3–17. Springer, 2016.

56. Boris Otto and Matthias Jarke. Designing a multi-sided data platform: findings from the international data spaces case. *Electron. Mark.*, 29(4):561–580, 2019.

57. Eldad Palachi, Chaim Cohen, and Sakairi Takashi. Simulation of cyber physical models using SysML and numerical solvers. In *2013 IEEE International Systems Conference (SysCon)*, pages 671–675. IEEE, 2013.
58. Jan Pennekamp, René Glebke, Martin Henze, Tobias Meisen, Christoph Quix, Rihan Hai, Lars Gleim, Philipp Niemietz, Maximilian Markus Rudack, Simon Knape, Alexander Epple, Daniel Trauth, Uwe Vroomen, Thomas Bergs, Christian Brecher, Andreas Bührig-Polaczek, Matthias Jarke, and Klaus Wehrle. Towards an Infrastructure Enabling the Internet of Production. In *2019 IEEE International Conference on Industrial Cyber Physical Systems (ICPS 2019) : Howards Plaza Hotel Taipei, Taiwan, 06-09 May, 2019 / publisher: IEEE*, pages 31–37, Piscataway, USA, May 2019. 2nd IEEE International Conference on Industrial Cyber-Physical Systems, Taipei (Taiwan), 6 May 2019 - 9 May 2019, IEEE.
59. André Pomp, Johannes Lipp, and Tobias Meisen. You are missing a concept! enhancing ontology-based data access with evolving ontologies. In *13th IEEE International Conference on Semantic Computing, ICSC 2019, Newport Beach, CA, USA, January 30 - February 1, 2019*, pages 98–105. IEEE, 2019.
60. Christoph Quix and Rihan Hai. Data Lake. In Sherif Sakr and Albert Zomaya, editors, *Encyclopedia of Big Data Technologies*, pages 1–8. Springer International Publishing, Cham, 2018.
61. Ana Luísa Ramos, José Vasconcelos Ferreira, and Jaume Barceló. Model-Based Systems Engineering: An Emerging Approach for Modern Systems. *IEEE Transactions on Systems, Man, and Cybernetics, Part C (Applications and Reviews)*, 42(1):101–111, 2011.
62. K Ramesh, R Girish Ganesan, and K Mahalakshmi. Approximation and optimization of discrete systems using order reduction technique. *Energy Procedia*, 117:761–768, 2017.
63. Bernhard Rumpe, Christoph Schulze, Michael von Wenckstern, Jan Oliver Ringert, and Peter Manhart. Behavioral Compatibility of Simulink Models for Product Line Maintenance and Evolution. In *Software Product Line Conference (SPLC'15)*, pages 141–150. ACM, 2015.
64. Bernhard Rumpe. *Modeling with UML: Language, Concepts, Methods*. Springer International, July 2016.
65. Bernhard Rumpe. *Agile Modeling with UML: Code Generation, Testing, Refactoring*. Springer International, May 2017.
66. Dave Steinberg, Frank Budinsky, Ed Merks, and Marcelo Paternostro. *EMF: eclipse modeling framework*. Pearson Education, 2008.
67. Duansen Shangguan, Liping Chen, and Jianwan Ding. A Hierarchical Digital Twin Model Framework for Dynamic Cyber-Physical System Design. In Unknown, editor, *Proceedings of the 5th International Conference on Mechatronics and Robotics Engineering - ICMRE'19*, pages 123–129, New York, New York, USA, 2019. ACM Press.
68. Raquel Sanchis, Óscar García-Perales, Francisco Fraile, and Raul Poler. Low-Code as Enabler of Digital Transformation in Manufacturing Industry. *Applied Sciences*, 10(1):12, 2020.
69. Adalberto Sampaio Junior, Fabio Costa, and Peter Clarke. A model-driven approach to develop and manage cyber-physical systems. volume 1079, 09 2013.
70. Sumit Singh, Essam Shehab, Nigel Higgins, Kevin Fowler, Tetsuo Tomiyama, and Chris Fowler. Challenges of digital twin in high value manufacturing. In *SAE Technical Paper*. SAE International, 10 2018.
71. Wilhelmus HA Schilders, Henk A Van der Vorst, and Joost Rommes. *Model order reduction: theory, research aspects and applications*, volume 13. Springer, 2008.
72. Thomas Uhlemann, Christian Lehmann, and Rolf Steinhilper. The digital twin: Realizing the cyber-physical production system for industry 4.0. *Procedia CIRP*, 61:335–340, 12 2017.
73. Rick F. van der Lans. The Fusion of Distributed Data Lakes Developing Modern Data Lakes A Technical Whitepaper. https://itlligenze.conthub.io/uploads/5/241/files/6430_Whitepaper %20TIBCO%20BigData%20V2%20(1)_0.pdf, 2019. Accessed: 2020-05-13.

74. Andreas Wortmann, Olivier Barais, Benoit Combemale, and Manuel Wimmer. Modeling Languages in Industry 4.0: an Extended Systematic Mapping Study. *Software and Systems Modeling*, 19(1):67–94, January 2020.
75. Tim Weilkiens. *Systems engineering with SysML/UML: modeling, analysis, design*. Elsevier, 2011.

Designing Strongly-decoupled Industry 4.0 Applications Across the Stack: A Use Case

Christoph Mayr-Dorn⬤, Alois Zoitl⬤, Georg Weichhart⬤,
Michael Mayrhofer and Alexander Egyed

Abstract

Loose coupling of system components on all levels of automated production systems enables vital systems-of-systems properties such as simplified composition, variability, testing, reuse, maintenance, and adaptation. All these are crucial aspects needed to realize

The research reported in this chapter has been funded in part by LIT-ARTI-2019-019, and the LIT Secure and Correct Systems Lab, and the FFG, Contract No. 881844: Pro^{2}Future is funded within the Austrian COMET Program Competence Centers for Excellent Technologies under the auspices of the Austrian Federal Ministry for Climate Action, Environment, Energy, Mobility, Innovation and Technology, the Austrian Federal Ministry for Digital and Economic Affairs and of the Provinces of Upper Austria and Styria. COMET is managed by the Austrian Research Promotion Agency FFG.

C. Mayr-Dorn (✉) · A. Egyed
Institute for Software Systems Engineering, Johannes Kepler University, Linz, Austria
e-mail: christoph.mayr-dorn@jku.at

A. Zoitl
LIT CPS Lab, Johannes Kepler University, Linz, Austria
e-mail: alois.zoitl@jku.at

G. Weichhart
Flexible Production Systems, PROFACTOR GmbH, Steyr-Gleink, Austria
e-mail: georg.weichhart@profactor.at

M. Mayrhofer
Cognitive Robotics and Shopfloors, Pro2Future GmbH, Linz, Austria
e-mail: michael.mayrhofer@pro2future.at

B. Vogel-Heuser and M. Wimmer (eds.), *Digital Transformation*,
https://doi.org/10.1007/978-3-662-65004-2_2

highly flexible and adaptable production systems. Based on traditional software architecture concepts, we describe in this chapter a use case of how message-based communication and appropriate architectural styles can help to realize these properties. A building block is the capabilities that describe what production participants (machines, robots, humans, logistics) are able to do. Capabilities are applied at all levels in our use case: describing the production process, describing machines, transport logistics, down to capabilities of the various functional units within a machine or robot. Based on this use case, this chapter aims to show how such a system can be designed to achieve loosely coupling and what example technologies and methodologies can be applied on the different levels.

Keywords

Software Architecture • Actor Model • Message-centric • OPC UA • Capabilities • Machine-to-machine Communication • Production Process • Composition • Orchestration • Loose Coupling

1   Introduction

Cyber-Physical Systems (CPS) tightly interweave software and physical components for integrating computation, networking, and physical processes in a feedback loop. In this feedback loop, software influences physical processes and vice versa. CPS in the manufacturing context are referred to as Cyber-Physical Production Systems (CPPS). A production cell involving machines, robots, humans, and transport systems such as pick and place units are one example of a CPPS; drones, smart buildings, or medical devices are not. One goal of CPPS is increasing the flexibility and adaptability of industrial production systems to address the need to reconfigure a physical plant layout with little effort and to produce a higher variety of products on the same layout, in smaller lot sizes, at the same costs. The requirements of customized mass production imply that control and integration software needs to be adaptable after deployment in a shop floor (factory), possibly even without interrupting production, at all levels: down at the level of Programmable Logic Controllers (PLCs) all the way up to Manufacturing Execution Systems (MES).

Software, and specifically, software architecture plays a central role in achieving this goal in multiple ways. Fast and cheap reconfiguration can only happen through software specifically designed to enable adaptability and flexibility. The resulting software comes with non-negligible complexity and—with incomplete upfront requirements on what kind of adaptation and flexibility might be desirable at a future stage—needs to be prepared for extensibility.

Ensuring decoupling of system components on all levels of production systems enables vital *systems of systems properties* such as simplified composition, variability, testability, reusability, maintainability, and adaptability. All these are crucial aspects towards realizing highly flexible and adaptable production systems and are highly dependent on the software's architecture [21].

Inspired by traditional *software architecture* concepts (i.e., components, connectors, interfaces, ports, and wires), we describe in this chapter how message-based communication in general, and an architectural style in particular, enable system designs that can help to realize these properties.

Capabilities are the key element that describes what production participants (machines, robots, humans, logistics) are able to do at every level (from MES down to PLC level) without specifying details on how to do it—thus achieving abstraction and information hiding. Capabilities comprise structural interface description as well as a behavioral description (e.g., finite state machines grounded in OPC UA[1] programs).

Capabilities find application at all levels: describing the overall production process, describing machine proxies that enable a scheduler to reason on available shop floor resources (and changes thereof), enabling an MES to interact with production resources, describe material transport (e.g., capabilities of a turntable, conveyors, pick and place units, etc.), down to capabilities of the various components within a machine or robot (which again may be hierarchically composed).

This chapter demonstrates based on a use case of how such systems can be designed and what example technologies and methodologies can be applied on the different levels. Specifically, our use case "Factory in a Box" shows how the C2 architecture style enables decoupling and reuse of components at the shop floor level and the intra-machine level and how IEC61499 and actor frameworks realize such decoupled and reusable architectures in the actual software. We then describe how our use case has individual shop floor participants expose their capabilities in OPC UA and how these capabilities are then reused for production process specifications, process execution, and discovery of shop floor participants.

The remainder of this chapter is structured as follows. We first introduce a running example based on a lab-scale production environment in Sect. 2. Subsequently, in Sect. 3 we introduce the notion of architecture-centric design [41]. After this, we discuss the building blocks for adaptive and flexible productions systems in Sect. 4. Lastly, in Sect. 5 we outline how these concepts can be applied for designing the application logic of soft real-time execution control devices 5.1, of non-time critical machine control 5.2, at the cell level 5.3, at the production process level 5.4, and ultimately at the systems-of-systems level 5.5.

2   Running Example: Factory in a Box

Our lab-scale production cell, as part of a university-funded demonstrator project Factory in a Box (FIAB), aims at illustrating the concepts that enable flexible production and the need for software to achieve this. Our particular demonstrator has the capability to customize the drawings on a piece of paper at multiple plotting stations. The production cell consists of the

---

[1] OPC UA is short for Open Platform Communications Unified Architecture.

Fig. 1 Lab-scale production environment: Factory in a Box

following machine types: input stations that provide pallets with paper, plotters that load the pallets and draw images, turntables that transport the pallets between plotters, and finally, output stations where the finished product (i.e., paper) is placed.

Communication between the machines is purely based on OPC UA (the predominant shop floor communication standard). Plotters and output stations are designed and programmed with the IEC 61499[2] industry standard using the open source IDE Eclipse 4diac [30] and the respective FORTE runtime environment that builds on the Open62541 OPC UA [16] server. The turntables are implemented in Java, using the Eclipse Milo OPC UA framework. Hence, despite the toy character of the setup, the used software and communication infrastructure is industrial grade. We use Lego Mindstorm EV3-based PLCs as a basis as this enables rapid and cheap machine prototyping without the need to test individual subcomponents such as motors, actuators, and sensors. Also, it ensures their seamless integration. The complete physical lab environment can be seen in Fig. 1.

In this chapter, we use this lab-scale production cell as a running example to outline the introduced concepts for decoupling at the soft real-time execution level (i.e., plotter), actor-based non-time-critical execution (i.e., turntable), process-based execution at machine or micro-cell level (i.e., alternative turntable control), scheduler based execution at MES level (i.e., print order scheduling), and the system of system composition (i.e., actors at MES level).

Flexible production aspects in this lab-scale production cell include, but are not limited to, (i) flexibly positioning plotters, input/output stations, and turntables next to other turntables,

---

[2] http://www.iec61499.de/.

(ii) dynamically allocating production jobs to plotters, (iii) dynamic routing of production orders between plotters, and (iv) dynamic discovery of shop floor participants and their capabilities.

Aside from showcasing such flexibility, FIAB supports the demonstration of how to design the various system elements to enable adaptation, how to efficiently test the adaptation mechanisms, how to achieve component reuse, and how to abstract from implementation details.

3 Architecture-centric Design

The primary difference between architecture-centric design and model-driven software development [41] is that the latter focuses heavily on modeling details sufficiently fine granular for code generation or further model generation, while the former focuses on supporting a design that achieves the overall desirable system properties without an explicit claim for executability. Hence, in the following subsections, we focus on the typical architecture modeling elements, and how their utilization in an architecture model can be constrained (i.e., limited) to achieve a system design that is reusable and flexible, and more.

3.1 Architecture Modeling Elements

Conceptually, the following metamodel elements [23] enable the definition of architectures. Various software architecture description languages such as xADL [22] or software design languages (such as UML/SysML) provide these meta model elements but apply a different vocabulary and/or visualizations. The purpose of this subsection is neither to mandate nor to promote one modeling language over another but to convey the necessary concepts in a manner that the readers can map these to the language of their choice.

- *Components* are the loci of computation, that implement the essential capabilities of a system, for example, in Fig. 2 the *ConveyingComponent* (responsible for moving pallets onto the turntable as well as offloading), *Turning Component* (responsible for positioning the conveyor in the right orientation) and *TurntableCoordinator* (responsible for telling the turning component when to turn and in which orientation, and the conveyor component when to engage and in which direction).
- *Connectors.* In contrast to e.g., UML/SysML component diagrams that define only a component element, Taylor et al. [22] motivate the distinction between components and connectors. Connectors are responsible for communication between and coordination of components and often surpass the components in size and complexity. The use of appropriate connectors has a profound impact on an architecture's qualities such as adaptability or robustness. For instance, in Fig. 2 the shaded boxes showing simple *MessageConnec-*

*tor*s for directed message passing, and a *MessageBus* supporting publish/subscribe capabilities. A more complex connector for coordination of transport is discussed in Sect. 5.2. A question that one can ask to determine whether an element is a component or a connector: within the observed system scope, does this element provide business value (i.e., is it necessary for the system's purpose), or does it provide a non-value adding (but still important, quality-inducing) functionality? At the coarse-grained shop floor level in FIAB, for example, plotters are the components and turntables the connectors; within the turntable scope the turning and conveying components are value-adding, and any messages passing infrastructure are the important, loose-coupling achieving (but non-value adding) connectors.

- *Interfaces* describe (similar to UML/SysML) the services of a component or connector. They describe the contract between the consumer (i.e., component or connector) of the interface and the provider of the interface (i.e., components or connectors). In Fig. 2 interfaces are visualized as annotations on components. In FIAB, interfaces describe the messages that are passed between components and connectors.
- *Ports* model the realization of an interface, respectively the usage of an interface (also denoted required interface), by a component or connector. In method-invocation-centric architecture (e.g., UML component diagram), these ports are also known as provided, respectively required interface. In data-flow centric architectures (e.g., SysML Internal Block Diagram (IBD)) provided and required interfaces are also described as in or out proxy ports. In ports model the flow of information, events, or requests into a component or connector, while the out ports have been modeled in reverse. Ports are depicted as boxes with ingoing/outgoing arrows in Fig. 2.
- *Wires* (or links, or connectors in UML/SysML) describe the configuration of how components and connectors interact by connecting two or more ports. This is also often described as a dependency. A wire connects exactly two ports. A wire is visualized as a full line between ports in Fig. 2.
- *Port Mappings* describe for hierarchical defined components (or connectors), how the outermost ports map to the port of the internal components (and connectors) that actually implement an interface (visualized as dashed lines, discussed in Sect. 5.2).

3.2   The *C2myx* Architectural Style for Strong Decoupling

Differences in software architectural styles result in various degrees of adaptability [24]. Combining the very related Components and Connector (C2) [31] and myx architectural style [9], this section describes the *C2myx* style that hence also produces, in particular, strongly decoupled, highly adaptable architectures.

When discussing an architectural style, it is important to note two crucial aspects: Firstly, an architectural style is a set of constraints. Thus the freedom to deviate from a particular style is always given and only bounded by the elements the modeling language provides.

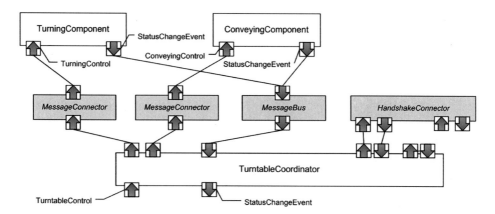

Fig. 2 Excerpt of the internal FIAB turntable components and connectors architecture. (xADL notation)

Secondly, an architectural style is not bound to a particular modeling language as long as its design constraints can be encoded in the language. See below for a brief discussion on how to utilize SysML for modeling a *C2myx* compliant architecture.

We distinguish between two general building blocks: components and connectors (as defined above). Specifically, connectors are limited to message-based communication and coordination, thus no long-blocking requests or shared memory access is foreseen. Components (and asynchronous event/message connectors) typically have their own (virtual) thread of control. Components and connectors are arranged in alternating layers (see e.g., Fig. 2), with ports on the top and/or bottom. A port is then said to be in the top domain or in the bottom domain of a component, respectively connector. Ports may be required or provided regardless of whether they are part of the top and bottom domain.

The concept of top and bottom domain (summarized in Table 1) is essential in obtaining a robust architecture. A component is independent of any other components that are "below", i.e., components that are reachable via ports in its bottom domain. It must not fail if these ports are not wired. Any required ports in the bottom domain thus need to be event publisher endpoints (or potentially synchronous calls to an event connector which, in turn, asynchronously forwards the call as an event to components below it). The component waits on a provided port in the bottom domain for requests. In other words, a component must be completely indifferent whether any call, message, or event leaving via a bottom domain port is received or processed by any lower components.

This implies, that a component may only request services from other components via ports that are part of its top domain, thus accessing services that are located "above" it. If these are not available, it then should degrade gracefully. Required ports in the top domain represent sending of asynchronous requests (i.e., messages) or synchronous invocations with immediate responses, while provided ports represent endpoints for receiving events or response messages (to asynchronous requests).

Table 1 Top and Bottom ports vs Required/Out and Provided/In ports from the perspective of a component

	Provided / In	Required / Out
Top	Incoming messages via this port are required by the component. Missing or ill-timed messages should not cause failure but graceful degradation.	Component sends messages (via connector) to component above. If no such recipient is available or the message is not received, component should degrade service but not fail.
Bottom	Component waits for (request) messages, to start, adapt, continue, or stop its behavior. Represents the main purpose of the component	Component sends message regardless of whether these are processed anywhere below with no influence on its behavior (e.g., publishing of its state or responses to requests)

Note that the resulting layers have to be understood in reverse order to traditional layered architectures. In *C2myx*, the components in the top layer are the most detailed/reusable ones, and the bottom layer contains the most application-specific components.

Connecting connectors and components via wires between ports must obey the following conditions:

- A port in a component's top domain may be wired to a single port in a connector's bottom domain.
- A port in a component's bottom domain may be wired to a single port in a connector's top domain.
- A connector's top and bottom ports may be wired to multiple component ports (and/or connector ports); see, for example, the multiple incoming wires of the *MessageBus*'s top domain port in Fig. 2.
- Wires between directly linked connectors must be between one top port and one bottom port; see, for example, the discussion of the wires between *MessageConnector* and *OPC UA Server* in Sect. 5.2).

3.3   Benefits of Using *C2myx*

As *C2myx* is a pragmatic merge of the already similar C2 and myx styles, it inherits their properties:

- Reusability: components and connectors clearly describe their required and provided interfaces as the only mechanisms of communication (recall: no implicit shared memory or side-channel invocations). A component or connector is thus agnostic to its execution context.

- Flexibility: the strict independence of components and layers below (i.e., substrate independence) and the resulting layering enable replacing or removing at every level the lower part of the architecture without affecting the functionality in the remaining components and connectors above the cut.
- Dynamism: connectors are dedicated loci for enabling run-time replacement and rewiring of components: e.g., via buffering of messages. As components remain agnostic of connector implementation, an engineer can choose very detailed at design-time via the application of different connectors exactly when and where the resulting architecture may be later adapted during run-time.
- Concurrency: asynchronous communication via messages enables components to operate independently, thus promoting concurrent execution across threads, processes, or cores.
- Distributability: given asynchronous messaging as the sole means of interaction combined with the use of connectors facilitates distributing components (and connectors) across different (networked) machines. Note, however, that the interface design (i.e., the granularity of messages) and the timeliness of expected responses from ports in the top domain have an impact on the overall system's performance attributes. In this respect, *C2myx* is not a silver bullet.

Note that an engineer may decide to deviate from these constraints when other qualities become more important. For example, a data-processing component may need to be optimized primarily for throughput, and may then be internally structured along the pipe-and-filter architecture style.

3.4   Architecture Style Vs Modeling Language

In the previous subsections and multiple figures, we utilized xADL as the architecture language to describe various excerpts of the running example. We used xADL as it supports the design along the *C2myx* style with explicit meta modeling elements (e.g., top and bottom ports). However, xADL is not the only possible language, with a plethora of languages available [18]. A viable alternative is a UML component diagram. While it doesn't explicitly support the notion of connectors (in the sense of "boxes" rather than arrows), or a distinction of top or bottom ports, such semantics could be added as profiles or as custom properties. For example, a port could then have simply a custom property "domain" with two possible values: "top" or "bottom". In a similar manner, also textual languages are amenable to designing according to the *C2myx* style. How much effort it is to check whether the resulting models indeed follow the *C2myx* style constraints, is a different aspect. Here xADL provides more support. Such conformance checking [40] is a whole different topic that goes beyond the scope of this chapter.

3.5   Brief Related Work on Architectures in CPPS

Ahmad and Babar [1] show that the last decades have seen the adoption of software development paradigms in robotics. As robots are a specialization of CPS, we expect a similar development for the CPS and CPPS community. For example, see [36], [2] for a "3c" architecture (i.e., communication, computation, control); [4] for a "5c" architecture (i.e., connection, conversion, cyber, cognition, and configuration levels), and [15] for an "8c" architecture ("5c" plus customer, content, and coalition facets). Further, a proposed anthropocentric CPS integrates physical, computational, and human components [26]. The focus of these architectures are systems at a component level and aim to work within the context of the "automation pyramid". This concept describes a hierarchical system of systems, where parent systems aggregate data from child systems, and strict top-down control dominates. We show in our use case that such a rigid pyramid is not necessary to achieve loose coupling but also that a network-centric layout as proposed by the following works provides too little structure. Pisching et al. [27] propose to use service-oriented architectures for CPPS and define a layout for CPS to behave as services. Thramboulidis et al. [32] investigate the usage of CPS as microservices. Others develop architectures, usually based on patterns studied well already in software architectures [12, 14, 29]. Their goal is to improve the compatibility between components.

4   Building Blocks

Next, we need to select a set of concrete models and technologies that are suitable to express the architectural design decisions (i.e., including decision that follow the architectural style) in a CPPS context. We focus primarily on OPC UA as the communication standard across network boundaries and messages in the respectively used programming language (here IEC 61499 and Java) as the inter-component/connector communication mechanism. Other technologies such as MQTT (a message-based communication standard) or ROS (a robot operating system) are equally potent candidates for implementing a system according to the *C2myx* style. In the following subsections, we outline the building blocks that allow the execution and integration of participants at the shop floor level and how these enable the designing according to the *C2myx* style.

4.1   Capabilities

The concepts, that our approach is based on, are defined in a core meta-model developed in multiple applied research projects within the research center $Pro^2 Future$. The two central, and tightly linked concepts in this meta-model are *Capabilities* and *Participants* [20].[3] Capabilities are abstract from machine-specific functions, software method calls, or human tasks and activities, and describe the ability to bring about a particular output. As such *Capabilities* most closely resemble interfaces in traditional architecture description languages such as xADL [22] or interfaces in UML/SysML. In CPPS, capabilities range from representing simple activities such as drilling a hole to more complex activities such as attaching a car power train to the chassis. However, capabilities are not necessarily limited to representing physical activities. A capability may also represent purely software-centric activities (i.e. services) such as planning an optimal route between two machines, sending an alert to a foreman, or updating production statistics in an enterprise resource planning system. Similar to interfaces, a capability describes what to do (with what input and expected output), but not how to achieve the represented activity, thus hiding internal realization details. Input typically includes parameters that configure the desired result—from simple values like the diameter of a hole to complex geometries describing the path of a cutting pattern—and output parameters describing the result. Being targeted specifically to CPPS, our model also includes concepts for parts and resources (such as tools) as part of a capability provisioning or requiring. These, however, are not relevant for the scope of this chapter. *Participants* are then systems (humans, robots, machines) that provide capabilities—somewhat similar to objects/components implementing an interface, and which require (i.e., depend on) other capabilities provided by other participants to do so. For example, a plotter machine (participant) provides plotting (capability) and requires a pallet feeder (another participant, with loading/unloading capabilities) to work properly.

Most important to note is, that capabilities describe who needs the services/capabilities of someone else, but not necessarily in which direction the information and control flow between the providing and requiring participant is realized. Requiring and providing a capability can thus be compared to the top and bottom domain of a component. For example, a pick and place unit (participant) requires the current loading state (capability) from an input stack (participant). The input stack publishes its state updates as events without needing to know who will receive them or if anyone receives them. Note that when modeling this with a UML/SysML component diagram or with a SysML IBD proxy port, such an event source would be modeled as a required interface of the input stack as the stack "requires" another endpoint/interface to send the events to.

4.2   Grounding Capabilities in OPC UA

We intend to use our metamodel in industrial environments which rely on OPC UA interaction protocols. Hence, atomic capabilities include OPC UA method calls, OPC UA readable and writable parameters, OPC UA events, and OPC UA programs. Depending on such a bind-

---

[3] The core meta-model uses the term *Actor* which we replace here with *Participant* to avoid confusing the term with the actor implementation approach outlined in Sect. 5.

ing, other participants make use of a capability by sending an event using a publish-subscribe protocol, invoking a method, or writing to a parameter. Mapping from atomic capabilities to OPC UA nodes is subsequently achieved via browsename[4] matching. Implementing a capability is only one part of the solution: making a shop floor participant discoverable is the other one. A participant needs to describe what capabilities they provide and which ones they require.

Participants may provide or require the same capability *multiple times*. A turntable (participant) as part of a flexible transport system may receive and distribute pallets from four sides, with machines, robots, or other transport systems placed at each side. The turntable needs to synchronize un/loading actions with each neighboring participant separately to ensure a successful handover of a pallet. Such synchronization is typically defined as a capability to ensure that participants from different vendors can seamlessly interact using an interoperable interface. The capability then may consist of a state machine describing the synchronization protocol between the two neighboring participants, the methods to trigger transitions in the protocol state machine, and events to monitor the current protocol state. A turntable participant would then provide a separate synchronization capability for each side. A single neighboring participant then only needs to monitor/interact with its assigned (i.e., wired) capability without any further knowledge needed how many other capabilities (i.e., pallet un/loading sides) the turntable participant exposes or what state their synchronization protocols are in. Internally the turntable participant may want to distinguish among its synchronization capabilities by roles with labels such as "NORTH", "EAST", "SOUTH", and "WEST".

In order for participants to become connected to other participants (or to be integrated in a production process/recipe), they need to expose what capabilities they provide and which they require. To this end, an OPC UA server may contain at any hierarchy level a *Capabilities* folder node. A *Capabilities* folder lists all *Capability* references that this participant provides or requires. Such a reference needs to provide at least the following properties:

- A host-wide unique identifier of this capability instance.
- A URI referencing the capability definitions (similar to how XML identifies schema documents).
- A flag for specifying whether the capability is provided or required.
- An optional actor-centric specific role identifier that may be non-unique across the server, useful when more than one capability instance of the same type is required/provided (see above).

Exposed required and provided capabilities are sufficient to model a distributed architecture of the shop floor or its participants at any desired level of detail. Note that we do not promote the explicit tagging of a capability at the OPC UA level as being in the bottom or top domain as, first, capabilities per-se do not enforce a particular architectural style, and,

---

[4] An OPC UA node has a unique id, but potentially a non-unique browsename.

second, engineers might decide to deviate from a particular architectural style in dedicated, carefully-chosen cases.

4.3 Production Process Modeling

Increasing flexibility of production is already a well-established concept in specific industries. For example, chemical industry knows the concept of process templates, so-called "process recipes" as described in ISA88. Process recipes specify, what needs to be done to realize a specific product. In a manufacturing context, a task as small as "rotate the conveyor belt by 90 degrees" could be a step in such a recipe, and as large as "fold a box". The recipe only specifies the production order, but neither who is executing the task, nor when this is scheduled.

The additional information (who, when, where) is generated during the allocation of the recipe, based on the current state of the shop floor. The boundary conditions for the allocation are the availability of machines, materials, and operators, as well as concurrent production processes. Possible criteria for optimization could be to reduce spatial distances between production steps, or optimal load balancing and maximum throughput of the shop floor.

A process is put together from capability descriptions (as introduced above). The process itself consists of the following elements:

- *ProcessSteps* are the fundamental elements to model process templates, processes allocated to a shop floor layout, and, partially allocated processes. To support multiple levels of granularity, and reuse of processes, *processes* again are ProcessSteps containing a list of ProcessSteps.
- *LogicSteps* like decisions, parallel branches, or loops, structure the flow of processes.
- *CapabilityInvocations* are specialized ProcessSteps, that link the invocation of a capability with a distinct role.
- *EventSources* and *EventSinks*, in turn, are capability invocations based on asynchronous, message-based communication. The most basic way to use message-based communication is an event sink which, when activated, waits for a specific event.
- *Parts* have been foreseen to model products, raw materials, tools, and resources used in manufacturing processes. An important property of parts is their passivity, they require an actor to be part of a process. As an example, a gripper requires a handling robot to be useful in a process. As soon as parts are able to provide capabilities independently—more or less, as they are actuated by their own control software—they become actors.

A process engine executing a process obeys the *C2myx* style as any included shop floor participants remain unaware of each other, react only to requests (i.e., invocations) from the process engine, which in turn receives responses and events. The process engine may, in turn, react to requests (e.g., the triggering of a new process instance) or other external signals to stop, pause, or abort a process from its bottom domain, respectively send events "downwards". To keep in line with the capability-centric design, a process engine and process may be deployed as a dedicated component, exposing capabilities on its bottom domain (i.e., describing the capability of the process), and on its top domain (i.e., describing the capabilities required for the process to be executed).

5   Designing Behavior with Capabilities

In this section, we start with the lowest level of system design: at the soft real-time execution within control devices and then work our way up to system-of-systems that span shop floors and beyond. At each level, we discuss how the *C2myx* style could be applied and which technologies can be used. We are neither suggesting to use these four levels as the only layers in a *C2myx* architecture style nor that the complete system of systems from the lowest level to highest level should always follow the *C2myx* architecture style. Rather, we demonstrate that the same principles are applicable at each of these levels. Figure 3 provides an overview of how a software architecture design following the *C2myx* style may be mapped to implementation at the various levels.

5.1   Soft Real-Time Execution within Control Devices

The demand for more and better software in Industry 4.0 poses especially high demands to the soft real-time control layer of automation systems. At this layer, we have typically non-software engineers working with Domain-specific Languages (DSLs) which are not apt to the demands anymore. This results in a high effort, errors, and cost [34]. New DSLs, architectures, and approaches are needed [3]. One promising language for our context is the domain-specific modeling language defined in the IEC 61499 standard [13]. IEC 61499 defines strongly decoupled components, an event-based execution model which directly reflects the actor model as well as basic support for dynamic reconfiguration. The latter provides the infrastructure to change control systems quickly to the current needs of the production.

As a first step to integrating the soft real-time control layer into a *C2myx* style system we need to represent capabilities in IEC 61499. IEC 61499 offers a very interesting concept for defining component interaction with the adapter interface elements. An adapter interface enables to model the bidirectional interaction between two components. In the terms of IEC 61499 it combines a set of events and data sent between the two components. Therefore, we often speak of adapters as cables and the two ends of the cable are called plugs and sockets. Sockets define provided services, plugs the required services. Components providing or requiring an interface can be developed and later used (i.e. connected) connected without

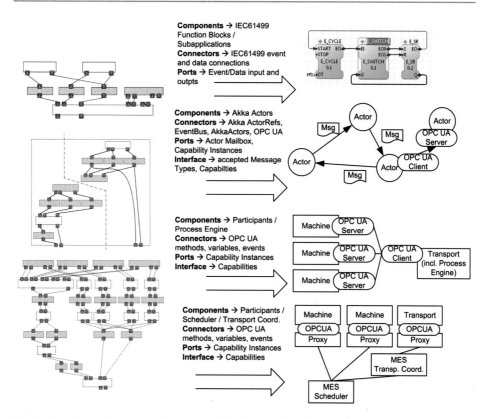

Fig. 3 Overview of the potential mapping of the design to implementation at each level

having to know each other. This is in the intention of the *C2myx* architecture and will be the basis for our proposed approach.

A shortcoming of today's PLCs is the lack of communication support. This is typically done on a variable basis, i.e., the developer has to define which variables to offer and which to expect. Any semantic information is lost. For an approach where each participant is offering and requiring services through message exchange, this is by far not sufficient. It is required that interfaces modeled in form of adapter interfaces can be used to automatically configure the required communication. An approach of mapping adapter interfaces to communication systems is presented by Dorofeev et al. [10]. A first approach for OPC UA has been presented and makes the implementation of capabilities on IEC 61499 very easy. Especially the event-driven nature of IEC 61499 enables to directly react on capability requests and sending the appropriate response.

While this first approach is mainly targeted towards how to provide capabilities in the overall context it is also important to consider how to better implement the soft real-time control software as a whole. The strong encapsulation of IEC 61499 Function Block (FB) is a starting point to support this. Utilizing the adapter interfaces together with the aggregation means sub-applications of IEC 61499 applications can be very well structured. Utilizing these means [42] presented a hierarchical design pattern, where the application hierarchy is following the mechatronical structure of the machine. This enables to build soft real-time control application structures in the *C2myx* style which represent the same level of decoupling as shown in Fig. 2. The main difference is that because of the limitations of IEC 61499, a visualization of an IEC 61499 application would have to be rotated by 90° counterclockwise to accurately reflect top and bottom domains (i.e., a right side plug becomes a top port and a left side socket becomes a bottom port). The "cables" then represent individual point-to-point asynchronous event connectors provided by an IEC 61499 runtime environment such as 4diac FORTE [30]. More sophisticated connectors are then built from FBs in sub-applications. An example of an application designed in that way is shown in Fig. 4.

However, recent work [39] identified drawbacks in the pure hierarchical design pattern. While the hierarchical design pattern leads to well-structured decoupled components, which can be independently easily replaced (especially on the lower levels), changes in the machine structure require a high adaptation effort. In order to overcome this limitation a new design pattern called "distributed hierarchical design pattern" was proposed and analyzed. The first results are very promising as they show much less effort for adapting control solutions to

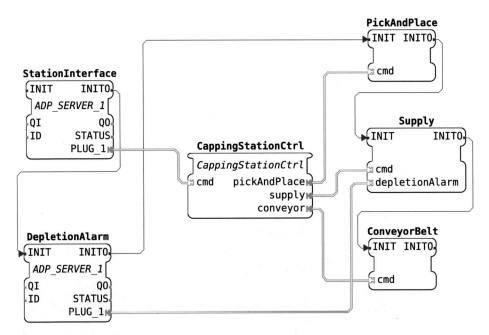

Fig. 4 Example of the top level control of an pick and place unit can be represented in a modular IEC 61499 application following the *C2myx* style. Green double lines represent connectors, red single lines depict FB initialization signals. (IEC 6199 4Diac visualization)

new requirements and production scenarios. However, the approach needs to be applied to more machine structures to identify its potential of generalized applicability, especially for hard real-time applications.

5.2 Actor-based Non-time Critical Execution—PLC Level

Within a machine or robot, the *C2myx* architecture style for controlling non soft real-time critical elements such as a turntable or conveyor may be achieved using actor implementation frameworks such as Akka[5] or CAF [5].[6]

Actor frameworks provide a runtime environment for the actor model, first introduced by Hewitt et al. [11] in 1973. Note that this actor model is different from our (shop floor participant) actor concept introduced above in that our concept describes the allocation of required and provided capabilities to executing entities, rather than prescribing a particular interaction or implementation mechanism.

In the scope of this Chapter, the following properties of the actor model are of relevance:

- Actors are the units of computation and coordination.
- Actors operate independently (i.e., concurrently) of other actors (compare to *C2myx* Components having their own (virtual) thread of control).
- Actors receive messages and send messages, no other mechanism of interaction such as shared memory or method invocation is available. The set of messages that are received and understood make up their in-port, the message types represent the interface. The set of messages sent by actors makes up the out-ports. This achieves a clear information flow from one actor to the next.
- Actors achieve high parallelism as upon sending a message they typically will not wait (i.e., blocking) for a reply but instead immediately continue with processing other queued messages.
- Actors typically have their own state, make decisions based on this state, update their state based on received messages, and create (child) actors (with which they can interact again only via messages).

On top of this, actor implementation frameworks such as Akka or CAF (and many more)[7] often provide additional services and guarantees (e.g., an actor will process at most one message at a time). For our use case, we opted for Akka due to the developers' familiarity with the Java programming language.

Taking the internal architecture of the *HandshakeConnector* in Fig. 5 as an example. The components *ClientHandshakeActor* and *ServerHandshakeActor* are implemented as Akka

---

[5] https://akka.io/.

[6] https://actor-framework.org/.

[7] https://en.wikipedia.org/wiki/Actor_model#Actor_libraries_and_frameworks.

actors, the *MessageConnector* connectors are realized as the Akka ActorRefs, thus the Akka framework takes care of message delivery. The *MessageBus* connector is built on top of the Akka EventBus, which is customizable to control message filtering and delivery and manages subscriptions. The dashed lines in Fig. 5 describe which ports are externally visible, i.e., at runtime, any component or connector below does not connect to these mapping ports but directly to the internal connectors. However, any outside connecting element remains unaware of the HandshakeConnector's internal structure. It is the task of the wiring logic (or dependency injection mechanism) at system startup to link the correct ports.

We are not the first to propose the application of the actor model in cyber-physical production systems, e.g., see [8] or [28]. Yet, applying the actor model itself will not automatically yield strongly decoupled systems (or systems-of-systems).

The attentive reader will notice that the properties of the actor model enable to build architectures along the *C2myx* style, but these properties are not constraining an engineer to

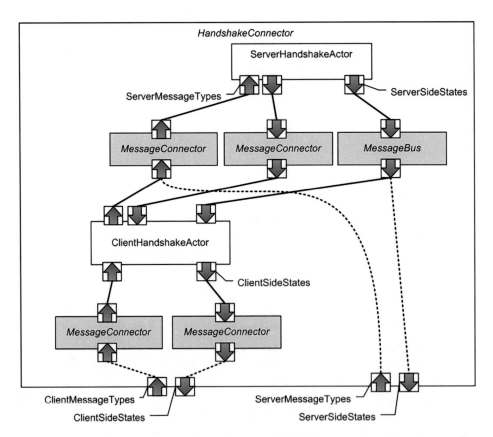

Fig. 5 Internal architecture of the HandshakeConnector; dashed lines are port mappings that associate a high-level component's or connector's port with a port on its internal components and/or connectors; full lines depict the message flow. (xADL notation)

obtain different architectural styles such as peer-to-peer or pipes-and-filter. In other words, the actor model, respectively implementations thereof, is not an *C2myx* architecture implementation framework that guarantees the *C2myx* qualities but provides some important basic building blocks for obtaining a system with these qualities.

The main aspects that provide the freedom for an engineer/architect to deviate from *C2myx* are:

- Actor implementation frameworks typically enable only to explicitly describe which message types an actor may receive (i.e., understand) but not what messages an actor will send. Hence a component or connector implemented via an actor framework will require an additional, external (out of bands) mechanism to describe an actor's required interfaces (i.e., in-ports).
- An actor has a single message inbox. Thus, again, an out of band mechanism is required to distinguish between different ports, especially between ports that are in the bottom domain, and ports that are in the top domain. Using dedicated actors per port or per top/bottom domain has a similar issue: they require an additional mechanism (e.g., different message types) to distinguish between component/connector internal and external messages.

Within a machine (or robot) or production cell, individual actors provide capabilities while other actors consume these capabilities. These actors make up the components. Whenever message passing becomes insufficient for coordinating components, actors will also serve as the implementation mechanisms for connectors.

Designing a machine using actors, wired up in the *C2myx* style, provides all the benefits outlined in Sect. 3.3. We exemplify how these benefits manifested in the FIAB scenario.

Reusability: the ConveyingComponent, as an example, is only concerned with controlling the conveyor, receives messages to reset, load, or unload, and provides its state as messages in return. It remains unaware of its usage context. As such we reuse the component on the turntable system (see Fig. 2) and in the plotter system.

Flexibility: the layering resulting from substrate independence enabled to flexibly run the various components of the turntable system in multiple settings: testing at various levels of granularity. The layers enabled to test the ConveyorComponent and TurningComponent each in isolation (with mocked hardware and with real hardware), the turntable system in isolation (with mocked conveying and turning functionality, with real components and mocked hardware, and on real hardware), and then again at the MES level with turntables using mocked hardware. This reduced testing complexity and time enormously, as hardware needs to be set up only for a very specific, limited number of tests. It also made it possible to test partial hardware integration, e.g., turning without conveying, without requiring changes to the software.

Dynamism: moving the synchronization abilities into a dedicated (distributed) connector enables rewiring of the connector internally (e.g., replacing the plotter on one turntable side with an output station) without having to integrate such logic into the TurntableCo-ordinatorComponent itself. For the connectors below turning and conveying components, we opted from Akka built-in mechanisms, i.e., plain messages amongst actors and an event bus. This clearly reduces the ability to replace components easily at run-time. Due to the hardware-related functionality, we do not expect this to be a requirement at the moment. In future scenarios where the conveying functionality should be temporarily replaced by a robot gripping the pallet (thus no conveying should occur), a different connector may simply be integrated before the system is deployed, which enables the switching between convey-ing and robot gripping. In this case, the actual components, nevertheless, remain unchanged.

Concurrency: in the turntable system, each component is implemented as a separate actor. This enables to have the turning component run totally in parallel to the conveying com-ponent. The turning component thus may check, for example, sensor and actuator states at fine-granular intervals independently from the conveying components monitoring its respec-tive hardware; all the while the TurntableCoordinatorComponent will listen to handshake events at the same time. Given the Akka framework guarantees that each actor processes only one message at a time, we can also avoid thread deadlocking or race conditions within an actor which greatly simplifies their design.

Distributability: selecting a different HandshakeConnector implementation enables switch-ing the whole FIAB shop floor from a single process system to a distributed system where each turntable system, plotter system, and input/output station are placed on a different host. To this end, we duplicated the message connectors and message bus between the ClientHandshakeActor and the ServerHandshakeActor (see Fig. 5) and added a sophisti-cated, distributed connector that communicates via OPC UA (see Fig. 6). This OPC UA based connector then also provides the ability to dynamically rewire the endpoints: e.g., the ServerHandshakeActorProxy, i.e., an OPC UA client connects to a different OPC UA server to enable machines to be physically moved around without having to shutdown either side. The *OPC UA HandshakeConnector* in Fig. 6 exemplifies the direct linking of multiple connectors (top to bottom). However, these connectors are of various complexity. The *OPC UA* connector is typically a large software subsystem while the *MessageConnector* or *Mes-sageBus* on top of it are simpler software elements provided by the actor framework runtime environment. In order to move from a local intra-actor framework connector as depicted in Fig. 5, any connector between the two internal components (i.e., *ClientHandshakeActor* and *ServerHandshakeActor*) are duplicated, and subsequently OPC UA client and server are placed in their midst. Hence the two components remain completely unaware in which configuration they are used: local or distributed. Using the same approach, one could also derive a MQTT based handshake connector. As long as the *C2myx* constraints are not vio-lated (or any deviations encapsulated in a connector or component that externally follows the

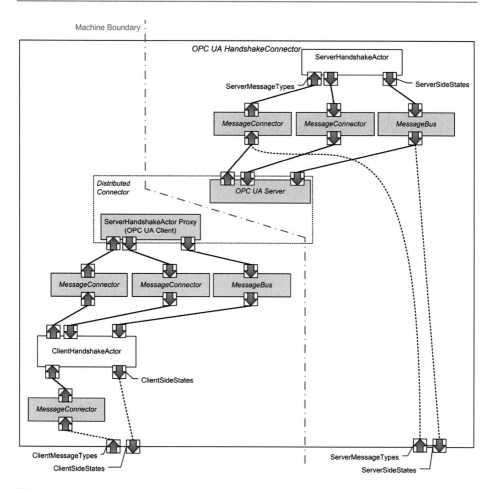

Fig. 6 Distributed HandshakeConnector using OPC UA. (xADL notation)

constraints) the resulting new handshake connector yields the same architectural properties. We provide additional details on how actors are bound to OPC UA in the following section.

As explained above, the actor model does not enforce the *C2myx* style. Similar, we were free (but chose not for the above-listed reasons) to deviate from the style's constraints as we directly build on the Akka framework's API. FIAB thus follows the *C2myx* style only through careful selection of connectors, the appropriate specification of interfaces, and correct wiring. This is a viable approach for small systems. Dedicated, architectural support is needed for more complex setups: either via architecture implementation frameworks that guarantee a style or through run-time monitoring that detects violations upon deployment. One specific approach we are currently exploring is the use of code generators to obtain actor skeletons and messages tagged with bottom and top ports to ensure that actor logic dispatches the right events via the right ports, and inversely, to ensure that actors receive only valid messages from the correct (virtual) ports.

5.3 Process-Based Execution

Process-based execution at the cell level or even within a machine (or robot) enables the quick reconfiguration of behavior without having to deploy new software. Using a process editor, an engineer selects the capabilities and the respective capability instances that will be provided in a cell (or machine) and integrates them into a process. For pre-existing or pre-planed cells, the engineer can already select the actors that these capabilities should be consumed from. Any capabilities that are to be flexibly linked on site remain un-allocated and are exposed by the process engine are required capabilities. An actor internally implemented as a process engine is indistinguishable from an actor that is hard-wired (as in the previous subsection) based on the provided and required capability instance alone.

While several process engines in the CPPS domain exist, for example, [25], in the scope of projects at the $Pro^2Future$ competence center, we focused on implementing a lightweight process engine that is also able to run on a PLC [20]. First, the high-level process description is transformed into a state-transition system. This enables using less complex interpreters that can be integrated easier with the programmable logic controllers widely used in industry. After this transformation, the engine and process description can be used as a standalone application orchestrating the process by calling above mentioned OPC UA mechanisms.

We also support adapters to be used directly with the control of a machine, thus enabling to run the engine in the same system process as a machine's control software. This results in a controller whose product-specific code can be reprogrammed online. At the time of writing, three adapters have been created. The first wraps java functions, the second adapts to the Akka Actor Framework, and the third integrates with a company partner's custom control software. There is no need for an adapter for OPC UA calls, as these are already supported by the core functionality.

In addition to explicitly provided capabilities in an engine's bottom domain, the engine also publishes the currently executed process, together with its active steps, in the OPC UA nodeset and thus makes the process available for monitoring.

Binding Actors to OPC UA

The only means of communication for an actor are messages. Hence, an OPC UA server's variables, methods, and events need to be mapped to messages:

- **Methods** are converted into messages by taking the method name as a message type, and each parameter as a message property. The method handling component of the OPC UA server needs to be instrumented with a reference to the actor responsible for handling this message and forwards a message upon every method invocation. Actor frameworks typically allow the sender of a message to block for a reply message from the actor (if one is needed).

It is best practize to implement an actor to be able to process each incoming message promptly, therefore having all long-running (and potentially blocking) logic placed in child actors. Hence, an actor should be able to immediately reply to a message stemming from a method invocation, subsequently enabling the method invocation handler to provide a prompt response.

- **Variables** require the inverse mapping. An actor having access to the OPC UA server (e.g., managing it, or having obtained a reference to an OPC UA node subset as part of its spawning) receives messages that represent new variable values. Such messages thus need to identify the variable node and its value. Alternatively, to avoid a high message load for variables that are expected to change a lot, a reference to the variable node can be passed (not as a message but upon actor spawning) to the actor that is responsible for determining the variables value. In the same manner, an actor can read the value from an externally writable variable node.

- **Events** are conceptually the simplest to map, as one can either pass on the message as it is (thus exposing the OPC UA data model to the actor implementation) or map it rather straightforward event property by property. Any automatic or supported mapping (for events or methods) would be OPC UA framework specific. To the best of our knowledge, no such mappers exist at the time of writing this chapter, but we found them easy to write during the process of binding Eclipse Milo OPC UA server to our Akka actor implementation.

Compliance with the *C2myx* style The process engine, in combination with a deployed process, obeys the *C2myx* style. All the capabilities a process requires (as provided by an actor) are represented as components (i.e., the shop floor-level actor) at the architectural level. Hence, the interaction means to send messages to these actors (and the messages obtained in return) define the ports on the process engine's top domain. The engine's bottom domain consists then of the messages that the engine accepts for the currently running process (as well as those that allow deploying new processes) and the messages that signal changes to the engine's and process' state. A system implementing one or more components (or connectors) with a process engine is thus following the rules of *C2myx*, and thus will yield the same benefits as outlined in the previous section.

Cell-level process execution in FIAB We used our editor to create a process model that defines the turntable's behavior (i.e., loading parts from one machine, unloading to another machine). To this end, the turntable controlling process makes use of capabilities provided by the turntable's individual submodules. The capabilities describe the following aspects: first, rotating the turntable superstructure; second, conveying parts to/from the superstructure from/to neighboring machines (i.e., input/output stations, drawing stations, and other turntables); and third, a synchronization protocol to coordinate the handover of parts with an adjacent machine. Our process engine integrated in the turntable exposes a capability for loading process models. These capabilities are defined using our core model and as such are independent of a particular turntable implementation. Subsequently, we can model different behaviors such as, but not limited to, process option 1: turn to the source station, complete the synchronization protocol, load, turn to the destination station, complete the synchronization protocol, unload; or process option 2: start the synchronization protocol with source and destination, turn to the source station, wait for the source station synchronization to finish, load, turn to the destination station, wait for the destination station synchronization to finish, unload.

5.4   Scheduler Based Execution and Transport

In contrast to machine or cell-level control where a process engine has tight control over the participating subsystems (i.e., actors) and where behavior is comparatively rigid (e.g., a turntable only transports one pallet at a time, a plotter only plots one image at a time), this assumptions hold no longer true when multiple orders are concurrently under production on the shop floor.

Figure 7 depicts the FIAB shop floor view. Shop floor participants such as Turntable(s), InputStation, OutputStation, and Plotters are found at the top (in accordance with the *C2myx* style) and connected via OPC UA to respective proxies (middle) which are part of the MES. Within the MES we find the Transport System Coordinator that determines which turntable(s) and orientations are needed to move a pallet from one machine to another, and the Order/Process Scheduler (bottom) that determines which order get assigned to which machine at what time. All components in Fig. 7 are implemented as Akka actors.

At the shop floor/MES level, a process modeled in our language (see Sect. 4.3) describes all necessary production steps and their order for a particular product using only capabilities without allocating the individual steps to concrete machines (providing the respective capability). Consequently, neither are the required transport means modeled as part of the process.

A transport routing system similarly benefits from modeling transport systems via capabilities. As individual transport units such as conveyors, turntable, autonomous guided vehicles (AGVs), pick and place units (PPUs), etc. expose their ability to move items around via a standardized capability, a transport scheduler does not need to care about implementation details during route planning or when transport units are replaced and extended. Furthermore, if transport capabilities are wired up similar to the synchronization handshake, the wires represent physical transport handover interfaces (otherwise there would be no synchronization handshake required). Given that these capabilities are integrated into the respective transport and production systems' architecture, one can determine from the overall shop

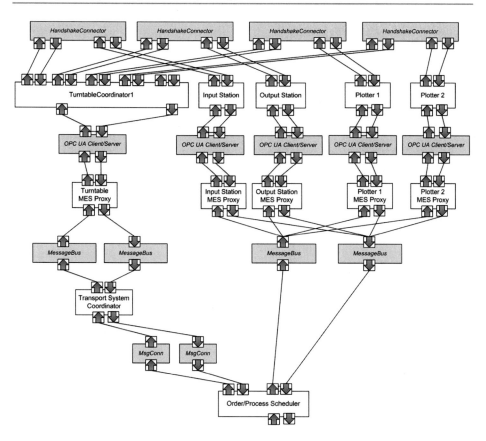

Fig. 7 Minimal FIAB shop floor architecture layout comprising one turntable, two plotters, input station, output station, transport system, and order scheduler (additional components/connectors for Web based monitoring and control not shown). (xADL notation)

floor architecture which transport pathways exist. Hence, the architecture creates an overlay network that can be used for routing items on the shop floor. Such a shop-floor level architecture can be established top-down during planning (e.g., modeled with AutomationML) or gathered during run-time from wiring information exposed by the various shop floor participants [19].

Different strategies for a production scheduler are possible. Some examples across the spectrum from centralized to distributed are the following:

- Centralized scheduler that is aware of production and transport durations which subsequently optimizes the order sequence for maximum throughput, order priority, or order robustness (minimal effect of out-of-service machines). Such a scheduler only needs to know which machine provides which capability, and how input parameters to the capability determine duration.

- Centralized just-in-time scheduler that is merely reacting to machines signalling the end of a production job, respectively their availability for a new job. The scheduler then requests each order that has been completed at one machine to be routed to the next machine that provides (one of) the next capability as defined by the production process. Such a scheduler is maximally flexible for just-in-time arriving orders and shop floor reconfigurations (e.g., failing machines, added machines, removed machines) at the expense of utilizing the available machines at full capacity.
- Decentralized per-order scheduler that exists for its assigned order only, thus many such schedulers request capabilities from machines and transport means. These schedulers behave as autonomous actors bidding for production time or observe load on machines (based on issued requests from other schedulers) to select which machine to reserve for the next production step. A per-order scheduler flexibly deploys different scheduling strategies and reactions to machine availability and demand without having to integrate such behavior in a centralized scheduler.

From an architecture point of view, a centralized production scheduler is not much different from the process engine at the machine level. It receives new production orders from its bottom domain, observes events from available machines and their state through its top domain, and sends production jobs requests to machines as well as transport requests to the transport subsystem (both via a top domain port). Similar decentralized per-order schedulers (actors) also follow the *C2myx* architectural style as long as their coordination protocols are encapsulated in (distributed) connectors that are correctly wired.

Order scheduling in FIAB Two reasons determined our choice to deploy a centralized just-in-time order scheduler in FIAB (see Fig. 7 bottom). First, the goal for FIAB was to demonstrate flexible production systems, hence being able to react to changing available capabilities and changing shop floor layout. Second, due to spatial restrictions, FIAB has no buffers, i.e., pallets with intermediary images cannot be temporarily parked anywhere but block a machine, respectively a turntable, until a machine with one of the next required capabilities becomes available and is transported there.

When the scheduler receives a new order, it checks all available steps (e.g., a single one in a sequential process, multiple ones in a process with parallel branches) whether at least one of these require a capability that is currently provided on the shop floor (otherwise the order is rejected). If no machine is currently idle, the scheduler will pause the order and unpause it once a suitable machine signals ready for production (we check orders in FIFO order to avoid starvation). The scheduler will step through the process as long as provided machine capabilities match the required process capabilities. If the process arrives at a step that requires a capability not provided on the shop floor anymore (or that has not been added yet), then the process is aborted and the order removed from the shop floor (recall that we have no buffers available where an order could wait for a required capability to be provided by a new machine).

Using capabilities, the scheduler remains unaware of whether the received signals and dispatched requests are handled by mock machine and transport actors, whether there are mock hardware machines that communicate via OPC UA, or whether it is running on the real, physical shop floor. Hence, testing different scheduling algorithms and different load profiles becomes very easy and limits the need to deploy actual hardware.

FIAB Shop floor Architecture In the IEC 62264[8] control hierarchy (aka. automation pyramid), level 3 components (MES—manufacturing execution systems) are responsible to provide a plan for the execution of orders (which are dealt with within the Enterprise Resource Planning system (ERP)). The overall responsibilities of that layer involve scheduling, which is the determination of which steps of a production order should be executed at which machine, at what time. This includes logistic steps. To do so, the MES needs a dynamic view of the overall shop floor and its machinery [38]. Dynamic here refers to (a) that the current state needs to be known (which machines are operational) and (b) durations for each step of a production order.

In FIAB, also all elements involved in transport coordination and order scheduling conform to the *C2myx* style. Due to space constraints, Fig. 7 depicts a shop floor configuration with merely a single turntable, two plotters, input station, and output station. These components expose only their external ports (i.e., to receive commands and send status updates from the bottom domain, and conduct the pallet handshake via ports in their top domain). All internal components and connectors are hidden for the sake of clarity. Each shop floor participant has a corresponding MES proxy component residing with the MES that is responsible for the discovery of the participants via OPC UA, forwarding status events downwards, and translating requests upwards: specifically they translate process step information (described in our metamodel) to actual OPC UA calls. Within the MES, the *Transport System Coordinator* and *Order/Process Scheduler* components are actors that dynamically obtain the set of available shop floor participants (via status events from MES proxies) and thus use a message bus also for upward requests to address the individual participants.

5.5 Interoperability and Composition of Systems-of-Systems

The actor concepts and the *C2myx* architecture complement each other in order to reach strong decoupling. We, therefore, discuss the relationship of *C2myx* to the research domain of Enterprise Interoperability (EI)[6]. EI discusses the coupling of systems not only from a technical but also from an organizational point of view. In Cyber-Physical Systems (CPS) like production systems, it is important to consider both aspects when discussing systems' boundaries and interfaces.

---

[8] https://www.iso.org/standard/57308.html.

The concept of interoperability is related to integration [33]. Interoperability (in the general sense) defines a range from ad-hoc federation over loose coupling to tight integration of parts [7]. In a specific sense, interoperability refers to the loose coupling of system parts. Such a loose coupling gives the systems designers more freedom in connecting different parts. One part can more easily be exchanged with a different part. In this view, interoperability is related to systems-of-systems [35]. Here the parts that form a system are systems themselves [17]. Following the definition of Maier [17] a system-of-systems has to show two distinct properties:

- Operational independence of parts: This requires that parts remain operational when removed from a system and put into another system. This underlines the loose coupling.
- Managerial independence of parts: Parts are independently created and then assembled in a system-of-systems.

Both aspects are desirable for building distributed systems where different parts are controlled by different people. It also helps to build more complex systems, as control for the inner complexity is distributed among different systems.

The disadvantage of systems-of-systems is that the different parts can not be put under centralized control. There is also some overhead when communicating through a standard interface with other systems.

Enterprise Interoperability discusses different levels of concern with respect to interoperability [7]. In the following, we use these levels to show the contributions of *C2myx*

Technical Interoperability: Actor systems provide an architectural means for building systems-of-systems [37]. The interface to other parts of the systems are the send and receive methods used for communication. Standards and common libraries enforce interoperability on a technical connectivity level.

Semantic Interoperability: Interoperability on a semantic level, refers to the vocabulary used by different actors. The messages are the content that is exchanged. Actor-based systems have a single ontology that is well known to all. Depending on the level to which the ontology is used only external for communication or internal as a conceptual model for reasoning, it might be a tight or loose coupling. If the internal model of an actor, used for reasoning and decision making, is mapped to the common ontology this is a loose coupling. If the actor's inner processes depend on that model then there is tight coupling.

Organisational Interoperability: With respect to the scenario discussed here, it is important to include the physical aspects of interoperability in CP(P)S. Practize has shown that in production systems, it can be easier to exchange a piece of hardware than to exchange a piece of software. For a similar component, interfaces have to be rewritten, and the detailed behavior in the context of the overall SW has to be tested. *C2myx* support for strong decoupling enables to clearly separate different aspects, by making the connectivity a dedicated connector.

In the case of FIAB, our lab-scale production cell constitutes just one system in a larger system-of-systems that combines, perhaps, multiple production cells involving manual labor for folding paper into boxes, for example, that become then input to a different production application that then fills these boxes with items. At this stage, centralized design is no longer feasible, but by designing individual systems in a manner compatible with the *C2myx* style, then partial, local adaptability, testability, etc. remains easy to achieve.

6   Discussion and Conclusion

Increasing the flexibility and adaptability of industrial production systems requires to design loosely coupled systems at every level. As we have shown in this chapter, achieving such loose coupling through an architecture-centric design approach, at all levels, supports the awareness of how components interact and what connectors are best suited. Even the same architectural style–*C2myx*–is applicable at every level, although implemented by vastly different technologies.

Explicitly choosing an architectural style is more important than choosing the right modeling language, be it UML, SysML, xADL, or others, as all these languages, on purpose, enable a large degree of design freedom. The benefit and qualities of a system emerge from the design constraints that the engineer applies: from simple ones such as using well-defined capabilities (interfaces), to asynchronous, event-driven connectors for inter-component communications, to very specific ones such as the distinction of ports according to top and bottom domain in combination with wiring top to bottom only.

This brings us to the outlook on future CPPS design. As flexibility, adaptability, testability does not happen by chance but careful decisions, we need to focus on supporting the engineer to make the right decisions, to make the engineer aware when decisions are conflicting or jeopardizing the aimed for design qualities. Such support ranges from architecture implementation frameworks that are best suited to target languages and environments such as IEC 61499 that will remain relevant for a considerable time. Such a framework could check in the Eclipse 4diac IDE whether "cables" between sub-applications violate the *C2myx* (or any other chosen) style. Such support is much harder to obtain at higher levels where programming languages and supporting libraries are subject to frequent extensions. As we reach the system-of-systems level, design-time guidance has only limited effect as there is rarely a central point of control that could enforce the constraint apriori to deployment. In this sense, runtime monitoring can help to identify components that are tightly coupled (or which violate a particular style).

One orthogonal measure to prescribing an architectural style is putting architectural information into the systems for discovery at design-time (to obtain better guidance on how to compose, e.g., the various subsystems of a mechatronical system) and at runtime for sophisticated monitoring when unknown actors dynamically interact on the shop floor.

References

1. Ahmad, A., Babar, M.A.: Software architectures for robotic systems: A systematic mapping study. Journal of Systems and Software **122**, 16 – 39 (2016)
2. Ahmadi, A., Sodhro, A.H., Cherifi, C., Cheutet, V., Ouzrout, Y.: Evolution of 3C Cyber-Physical Systems Architecture for Industry 4.0. In: 8th Workshop on Service Orientation in Holonic and Multi-Agent Manufacturing. Bergamo, Italy (Jun 2018), https://hal.archives-ouvertes.fr/hal-01788468
3. Atmojo, U.D., Vyatkin, V.: A review on programming approaches for dynamic industrial cyber physical systems. In: 2018 IEEE 16th International Conference on Industrial Informatics (INDIN). pp. 713–718 (2018)
4. Bagheri, B., Yang, S., Kao, H.A., Lee, J.: Cyber-physical systems architecture for self-aware machines in industry 4.0 environment. IFAC-PapersOnLine **48**(3), 1622 – 1627 (2015). https://doi.org/10.1016/j.ifacol.2015.06.318, http://www.sciencedirect.com/science/article/pii/S2405896315005571, 15th IFAC Symposium onInformation Control Problems inManufacturing
5. Charousset, D., Hiesgen, R., Schmidt, T.C.: Caf-the c++ actor framework for scalable and resource-efficient applications. In: Proceedings of the 4th International Workshop on Programming based on Actors Agents & Decentralized Control. pp. 15–28 (2014)
6. Chen, D., Doumeingts, G., Vernadat, F.: Architectures for enterprise integration and interoperability: Past, present and future. Computers in industry **59**(7), 647–659 (2008)
7. Chen, D., Youssef, J.R., Zacharewicz, G.: Towards an enterprise operating system - requirements for standardisation. In: Zelm, M. (ed.) Proc. 6th Workshop on Enterprise Interoperability (IWEI) (2015)
8. Cruz, S.L.A., Vogel-Heuser, B.: Comparison of agent oriented software methodologies to apply in cyber physical production systems. In: 2017 IEEE 15th International Conference on Industrial Informatics (INDIN). pp. 65–71 (July 2017)
9. Dashofy, E.M.: The Myx Architectural Style. Tech. rep., University of California, Irvine, Institute for Software Research (2006)
10. Dorofeev, K., Profanter, S., Cabral, J., Ferreira, P., Zoitl, A.: Agile operational behavior for the control-level devices in plug&produce production environments. In: 24th IEEE International Conference on Emerging Technologies and Factory Automation. pp. 49–56. IEEE (09 2019)
11. Hewitt, C., Bishop, P., Steiger, R.: A universal modular actor formalism for artificial intelligence. ijcai3. In: Proceedings of the 3rd International Joint Conference on Artificial Intelligence. pp. 235–245 (1973)
12. Hussnain, A., Ferrer, B.R., Lastra, J.L.M.: Towards the deployment of cloud robotics at factory shop floors: A prototype for smart material handling. In: 2018 IEEE Industrial Cyber-Physical Systems (ICPS). pp. 44–50 (2018)
13. IEC TC65/WG6: IEC 61499: Function blocks – Parts 1, 2, & 4. International Electrotechnical Commission (IEC), Geneva, 2 edn. (2012)
14. Ismail, A., Kastner, W.: A middleware architecture for vertical integration. In: 2016 1st International Workshop on Cyber-Physical Production Systems (CPPS). pp. 1–4 (2016)

15. Jiang, J.R.: An improved cyber-physical systems architecture for industry 4.0 smart factories. Advances in Mechanical Engineering **10**(6), 1687814018784192 (2018). 10.1177/1687814018784192, https://doi.org/10.1177/1687814018784192

16. Mahnke, W., Leitner, S.H., Damm, M.: OPC unified architecture. Springer Science & Business Media (2009)

17. Maier, M.W.: Architecting principles for systems-of-systems. Systems Engineering **1**(4), 267–284 (1998)

18. Malavolta, I., Lago, P., Muccini, H., Pelliccione, P., Tang, A.: What Industry Needs from Architectural Languages: A Survey. IEEE Transactions on Software Engineering **39**(6), 869–891 (2013). 10.1109/TSE.2012.74

19. Mayrhofer, M., Mayr-Dorn, C., Bishara, M., Weichhart, G., Egyed, A., Konnerth, M.: Capability-based process modeling and control. In: Proceedings of IEEE International Conference on Industrial Technology (ICIT). IEEE (2021)

20. Mayrhofer, M., Mayr-Dorn, C., Guiza, O., Weichhart, G., Egyed, A.: Capability-based process modeling and control. In: 25th IEEE International Conference on Emerging Technologies and Factory Automation, ETFA 2020, Vienna, Austria, September 8-11, 2020. pp. 45–52. IEEE (2020). 10.1109/ETFA46521.2020.9212013

21. Mayrhofer, M., Mayr-Dorn, C., Zoitl, A., Guiza, O., Weichhart, G., Egyed, A.: Assessing adaptability of software architectures for cyber physical production systems. In: Bures, T., Duchien, L., Inverardi, P. (eds.) Software Architecture - 13th European Conference, ECSA 2019, Paris, France, September 9-13, 2019, Proceedings. Lecture Notes in Computer Science, vol. 11681, pp. 143–158. Springer (2019)

22. Medvidovic, N., Rosenblum, D.S., Taylor, R.N.: A language and environment for architecture-based software development and evolution. In: Proceedings of the 21st International Conference on Software Engineering. pp. 44–53. ICSE '99, ACM, New York, NY, USA (1999)

23. Medvidovic, N., Taylor, R.: A Classification and Comparison Framework for Software Architecture Description Languages. IEEE Transactions on Software Engineering (2000)

24. Oreizy, P., Medvidovic, N., Taylor, R.N.: Runtime software adaptation: framework, approaches, and styles. In: Companion of the 30th international conference on Software engineering. pp. 899–910. ACM (2008)

25. Pauker, F., Mangler, J., Rinderle-Ma, S., Pollak, C.: centurio.work - modular secure manufacturing orchestration. In: 16th International Conference on Business Process Management 2018. pp. 164–171 (2018)

26. Pirvu, B.C., Zamfirescu, C.B., Gorecky, D.: Engineering insights from an anthropocentric cyber-physical system: A case study for an assembly station. Mechatronics **34**, 147–159 (2016). https://doi.org/10.1016/j.mechatronics.2015.08.010, http://www.sciencedirect.com/science/article/pii/S095741581500152X, system-Integrated Intelligence: New Challenges for Product and Production Engineering

27. Pisching, M.A., Junqueira, F., d. S. Filho, D.J., Miyagi, P.E.: An architecture based on iot and cps to organize and locate services. In: 2016 IEEE 21st International Conference on Emerging Technologies and Factory Automation (ETFA). pp. 1–4 (2016)

28. Shen, W., Wang, L., Hao, Q.: Agent-based distributed manufacturing process planning and scheduling: a state-of-the-art survey. IEEE Transactions on Systems, Man, and Cybernetics, Part C (Applications and Reviews) **36**(4), 563–577 (July 2006)

29. Spinelli, S., Cataldo, A., Pallucca, G., Brusaferri, A.: A distributed control architecture for a reconfigurable manufacturing plant. In: 2018 IEEE Industrial Cyber-Physical Systems (ICPS). pp. 673–678 (2018)

30. Strasser, T., Rooker, M., Ebenhofer, G., Zoitl, A., Sunder, C., Valentini, A., Martel, A.: Framework for distributed industrial automation and control (4diac). In: 2008 6th IEEE International Conference on Industrial Informatics. pp. 283–288 (2008)
31. Taylor, R.N., Medvidovic, N., Anderson, K.M., Jr., E.J.W., Robbins, J.E., Nies, K.A., Oreizy, P., Dubrow, D.L.: A component- and message-based architectural style for GUI software. IEEE Trans. Software Eng. **22**(6), 390–406 (1996)
32. Thramboulidis, K., Vachtsevanou, D.C., Solanos, A.: Cyber-physical microservices: An iot-based framework for manufacturing systems. In: 2018 IEEE Industrial Cyber-Physical Systems (ICPS). pp. 232–239 (2018)
33. Vernadat, F.B.: Interoperable enterprise systems: Principles, concepts, and methods. Annual reviews in Control **31**(1), 137–145 (2007)
34. Vogel-Heuser, B., Sardá-Espinosa, A.: Current status of software development in industrial practice: Key results of a large-scale questionnaire. In: 2017 IEEE 15th International Conference on Industrial Informatics (INDIN). pp. 595–600 (2017)
35. Wachholder, D., Stary, C.: Enabling emergent behavior in systems-of-systems through bigraph-based modeling. In: 10th System of Systems Engineering Conference (SoSE) (5 2015)
36. Wan, K., Hughes, D., Man, K.L., Krilavicius, T., Zou, S.: Investigation on composition mechanisms for cyber physical systems. International Journal of Design, Analysis and Tools for Circuits and Systems (2011)
37. Weichhart, G., Guédria, W., Naudet, Y.: Supporting interoperability in complex adaptive enterprise systems: A domain specific language approach. Data and Knowledge Engineering **105**, 90–106 (9 2016)
38. Weichhart, G., Hämmerle, A.: Lagrangian relaxation realised in the ngmpps multi actor architecture. In: Berndt, J.O., Petta, P., Unland, R. (eds.) 15th German Conference on Multiagent System Technologies, MATES 2017. Lecture Notes in Artificial Intelligence, Springer (2017)
39. Wiesmayr, B., Sonnleithner, L., Zoitl, A.: Structuring distributed control applications for adaptability. In: ICPS 2020, Tampere. In Press. (2020)
40. Zheng, Y., Cu, C., Taylor, R.N.: Maintaining architecture-implementation conformance to support architecture centrality: From single system to product line development. ACM Trans. Softw. Eng. Methodol. **27**(2), 8:1–8:52 (2018). 10.1145/3229048
41. Zheng, Y., Taylor, R.N.: A classification and rationalization of model-based software development. Softw. Syst. Model. **12**(4), 669–678 (2013). 10.1007/s10270-013-0355-3
42. Zoitl, A., Prähofer, H.: Guidelines and patterns for building hierarchical automation solutions in the iec 61499 modeling language. IEEE Transactions on Industrial Informatics **9**(4), 2387–2396 (2013)

Variability in Products and Production

Alexander Egyed⊙, Paul Grünbacher⊙, Lukas Linsbauer⊙,
Herbert Prähofer⊙ and Ina Schaefer

Abstract

Products and production are inherently variable. That is, the products themselves often need to be variable—as in a car plant producing many similar, albeit not identical cars. Such flexibility allows a product to be more easily customizable. We speak of variable products. At the same time, production systems typically need to be flexible in supporting the production of different products. Such flexibility allows for a broader use of production systems, supports lower production volumes while remaining economical, or optimizes

A. Egyed · P. Grünbacher (✉)
Institute of Software Systems Engineering, Johannes Kepler University Linz, Linz, Austria
e-mail: paul.gruenbacher@jku.at

A. Egyed
e-mail: alexander.egyed@jku.at

L. Linsbauer · I. Schaefer
Institute of Software Engineering and Automotive Informatics, Technische Universität
Braunschweig, Braunschweig, Germany
e-mail: l.linsbauer@tu-braunschweig.de

I. Schaefer
e-mail: i.schaefer@tu-braunschweig.de; ina.schaefer@kit.edu

H. Prähofer
Institute for System Software, Johannes Kepler University Linz, Linz, Austria
e-mail: herbert.praehofer@jku.at

I. Schaefer
Institut für Informationssicherheit und Verlässlichkeit (KASTEL), Karlsruher Institut für
Technologie, Karlsruhe, Germany

© The Author(s), under exclusive license to Springer-Verlag
GmbH, DE, part of Springer Nature 2023
B. Vogel-Heuser and M. Wimmer (eds.), *Digital Transformation*,
https://doi.org/10.1007/978-3-662-65004-2_3

production resources to avoid delays. We speak of variable production. This chapter explores variability in products and variability during production where product variability needs to be understood together with its implications on production. Special considerations are products that are consequently used during production and the issue of hardware/software variability, which is mostly handled separately today. We provide examples from an injection molding machine and also discuss open research challenges.

Keywords

Variability • Products and production systems • Variability engineering • Product lines

1   Introduction

The term Industry 4.0 is widely used to describe industrial automation systems combining traditional platforms and practices with the latest smart technology in manufacturing and production. Variability is fundamental in Industry 4.0. Many problems such as insufficient product quality, machine downtimes, or schedule and cost overruns are caused by problems of engineering and managing variant-rich systems. Understanding what is varying, why it is varying, when it is varying, and how variability is implemented is essential for dealing with these challenges.

In the context of Industry 4.0, variability needs to be understood at multiple levels: at the level of *machines* we are facing an increased demand for customization and machine-specific solutions [40]. Automation solutions are often built based on a core technological platform. Individual solutions are then derived by selecting subsystems, components, and elements from this platform, followed by configuring, adapting and extending this initial solution [21]. Eventually, every machine is individually adapted and tuned for its particular application purpose, either by an original equipment manufacturer (OEM) or directly by the end-user customer. Obviously, this results in high demands for the variability and extensibility of solutions and solution methods. At the level of *production systems* (often also referred to as shop floors), variability is relevant for the choice of machines and their specific customization. Systematically managing variability is thus essential for defining the shop floor layout and production process. At the level of the *products*, variability is fundamental for providing customizable machine designs, individual machine parts, etc. during production. These different levels are obviously highly interdependent and the question of how to produce a variable product goes hand in hand with the question of how to create a flexible production system.

Variability is often the result of new or changing customer requirements, which may lead to re-engineering a machine or its part by adding new or adapting existing features, or by reconfiguring the machine. Variability is also often caused by technological changes: it may be necessary, for instance, to replace a no-longer-produced part, which may affect machine behaviour and ultimately also the production process. Similar effects can be expected when integrating newly-invented machines or parts in an existing production system.

The chapter provides the following contributions: we discuss challenges of dealing with variability in products and production systems. We give an overview about the state-of-the-art in variability engineering. We further illustrate selected methods for variability modeling, variability realization, variability-aware code analysis, and variability evolution using the running example of an injection molding machine. We conclude the paper with a discussion of open research challenges in variability for products and production systems. The rest of the chapter is organized as follows: Sect. 2 discusses variability challenges in industrial automation. Sect. 3 introduces the injection molding machine example. Sect. 4 gives an overview of existing methods for modeling, implementing, and realizing variability, and discusses approaches for verifying, validating, and evolving variable systems. Sect. 5 discusses adoption models for migrating towards product lines and challenges in evolving product lines. Sect. 6 discusses open research challenges. Sect. 7 concludes the chapter.

2 Variability Challenges in Automation

Supporting variability in automation systems ensures the ability of a product to change, thereby addressing different customer wishes. It also ensures the ability of a production process to change, thereby addressing different production needs. Indeed, facilitating such changes is one of the most essential aspects of Industry 4.0 as it denotes the flexibility a product or a production system needs to have. The principles of variability for products and productions are not unlike. After all, a product is a complex system comprising of parts— possibly from different vendors. These parts could change during design time (through engineering) or during runtime (through configuration). Similarly, a production system is a complex system comprising machines and "workers" (its parts abstractly speaking)—again possibly from different vendors or educational backgrounds. Nonetheless, there are also differences, which we discuss in more detail. We illustrate this at the example of a car assembly line (cf. Fig. 1).

2.1 Variability of Different Production Levels

In Industry 4.0, we need to distinguish product and production variability to account for both variable products and variable manufacturing situations.

Product variability concerns the variability in terms of commonalities and differences of different product variants being produced. For example, a car may come with a wide variety of options. These options and their effect need to be understood and designed for.

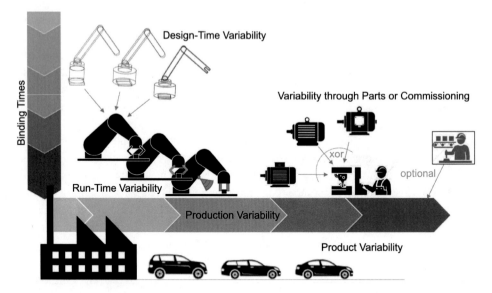

Fig. 1 The assembly line can produce different kinds of cars (product variability) and therefore needs to be reconfigured accordingly (production variability). The variability of different production levels relies on different variability binding time

Product variability comprises

- the commonalities and differences of different product variants,
- definitions of the parts the product is composed of, and
- descriptions of the features provided (or not provided) by the product.

Product variability can be expressed through

- the inclusion/exclusion of parts (e.g., a car with/without an A/C unit),
- the configuration of parts (e.g., pre-definable A/C cooling settings), or
- the variation of parts (e.g., different builds of the A/C for different power consumption needs).

For a convertible car, it does not make sense to allow the addition of a roof rack due to an obvious physical restriction. However, there can be logical restrictions as well. For a convertible car, it probably does make sense to allow the addition of an air conditioning (A/C) unit, but its use should be disabled while the roof is down. The A/C unit or roof rack are features a customer may choose. The restrictions are dependencies. How can these features and their dependencies be expressed?

Production variability concerns the ability of a production system to adjust to different and changing production needs. For example, depending on which features a car includes, the associated assembly process will need to vary. Clearly, the assembly process differs between a convertible car versus a non-convertible car. Production variability is thus essential in addressing product variability. Yet, production variability allows for more than product variability. It also allows for production changes due to environmental issues or labor situations. Production variability needs to be understood and designed for like product variability. Yet, in practical terms, a key difference is the presence of human workers in production systems. Like product variability, production variability relies on the variability of the "parts" that make up production: humans, who obviously are quite capable of change, and machines, which tend to be less flexible. Today, much of production variability is achieved through humans, who are able and willing to change their roles and responsibilities during production.

Production variability comprises

- the variability in the production process,
- definitions of the various steps of the production process (performed by humans and/or machines), and
- descriptions of features the production provides (or does not provide).

Production variability can be expressed through

- the inclusion or exclusion of workers or machines,
- the retraining of workers or the reconfiguration of machines (e.g., attaching a different tool to a machine or changing the settings of a machine), and
- the use of different or additional workers or machines.

Naturally, the variability of the production process needs to accommodate product variability. However, production variability has additional goals: *Production optimization* aims to avoid unnecessary waiting times for workers and machines. This can be achieved, in part, by optimizing the order of products to be produced. For example, assembling a car with an A/C unit might slightly overwhelms an assembly team, perhaps at a certain assembly stage. This problem could be compensated by next assembling a car without an A/C unit if doing so would slightly relieve that assembly team. *Production adaptation* aims to avoid or circumvent environmental influences such as the unavailability of parts or unforeseen events such as sickness of workers. This can be achieved by arranging the order of products to be produced. For example, if A/C units are currently unavailable due to a delay in shipment, then cars without A/C units should be scheduled for production until the A/C units become available again.

2.2   Variability Binding Times

Managing the variability of different production levels relies on dealing with different variability binding times (cf. Fig. 1). Essentially, cyber-physical production systems need to express variability at three stages in the life cycle:

Design-time variability concerns the ability to readily change systems (e.g., machines, products, or production processes) during the design phase. In what ways is a 12V A/C unit different from a 24V A/C unit? In what ways are they the same or similar? Rather than designing two different products, we may consider them as the same product and focus on the differences in their design but also their production. Then, during production, either a 12V or 24V A/C will be produced. However, design-time variability implies that, once produced, it can no longer change to the other.

Configuration-time variability provides the ability for the system to be changed as part of its commissioning, perhaps even its startup. As such, an A/C unit may be usable with both 12V and 24V, however, not at the same time. During commissioning (e.g., assembly), the unit could be configured depending on the situation at hand. Such configuration could be enabled by hardware (a physical switch or a 12V wire connection different from a 24V wire connection) or by software (database entries defining pre-configured settings). Configuration-time variability requires all elements for 12V/24V usage of the A/C unit to be present such that the commissioning decides which voltage to use.

Run-time variability is the ability to readily change systems during operation. An A/C unit that is specifically built for 24V cannot readily be changed during runtime. However, if the A/C unit is both designed and built with the ability to accommodate 12V and 24V, it is possible to change the usage of the A/C unit after construction. Naturally, runtime variability is not only more expensive to build, but also more complex to design. Like with configuration-time variability, the latter requires all elements for 12V/24V to be present and functioning. These elements need to co-exist, which could be a problem in case of feature interactions [41].

2.3   Re-Configurable Production

Production variability requires the variability of workers and machines. Human workers are intrinsically flexible and production variability is today mostly achieved through humans. But what about machinery? In traditional shop floors machinery is largely static. It is bought and commissioned. Changes are mostly limited to changing parts such as the tools the machines use. Today, production variability for machines is mostly about whether or not to use a machine, but not about changing the machine. With Industry 4.0 this is changing: machines, much like their human users, need to support variability to allow for shop floor changes. Herein, we discover one of the most essential problems. How can we change machines during production?

Reconfiguration may happen monthly, weekly, or even daily. Reconfiguration will not always be predictable, but may arise instantaneously, e.g., as a result of delivery shortage, worker illness, or machine failure. The key challenge of Industry 4.0 is: how does a company instantaneously assure the correct production behavior (and implicitly the correct behavior of a shop floor), while the production process changes continuously with little to no advance notice? Traditionally, companies would plan and design the production processes, and plan and design their changes. Unfortunately, this requires significant effort, which limits the ability to frequently change production processes.

Industry 4.0 implies readily configurable production processes
- for changing product needs,
- for changing production needs,
- for changing worker needs, and
- for changing environmental needs.

2.4 Verifying and Validating Variable Products and Production Systems

Unfortunately, the solution to this problem is not only in workers becoming even more flexible to change, but rather also in re-configurable machines supporting flexible production processes. We encounter an even bigger challenge: much like any cyber-physical system, the correct behavior of a production cannot be validated and verified by only considering the correct behavior of its various parts (e.g., machines and "workers"). For example, if a shop floor requires adaptation due to production changes and this adaptation requires changes to machines, then the correct behavior of the shop floor needs to be assured. Several methods are widely used:

Production variability modeling describes how products and machines were designed for change. Today, it is becoming common practice to model the variability of systems using variability modeling languages. A detailed example is provided in Sect. 4. This principle of variability modeling can be applied to production, encompassing the variability models of the various machines. Production variability is analyzed a-priori (during design or commissioning), so that production changes are guaranteed correct for as long as they fit within the analyzed production variability model.

Testing is considered the backbone for assuring correct system behavior (verification and validation). However, testing is expensive and time consuming if not fully automated. And testing is disruptive if changes to the production require testing before production is to resume. In the context of Industry 4.0, testing needs to be shifted to the testing of virtual machines and digital twins of shop floors—hence requiring variability-enabled digital twins of machines from their vendors to validate and verify production behavior before changing it in real life.

Monitoring is useful in assuring that the production system does conform to the expected system behavior [32]. Monitoring does not assure correct system behavior, but ideally provides instant feedback to workers and supervisors, so that corrective measures can be introduced as quickly as possible if deviations are encountered. Some of these corrective measure might be automatable also.

Static analysis is useful for ensuring properties that are deemed essential. It is typically applied in safety-critical domains, usually on source code or formal specifications. As source code for variable machines is likely unavailable (due to their proprietary nature), machine vendors need to provide a (formal) specification of the logical functioning of their machines (in addition to the physical specifications which they already do provide). Machines, their interactions, and the effects of their variation may then be assessed through static analyses of these specifications.

3   Injection Molding Machine Example

We use an injection molding machine for manufacturing plastic parts as a running example for this chapter. Today, injection molding machines are the dominant technology for producing products made of plastic, which range from such simple parts like caps for plastic bottles to complex plastic components of cars. During the injection molding process, the machine heats granulated plastic material and then injects the viscous material into the closed mold chamber with high pressure. After cooling, the mold is opened, and the plastic part gets ejected. Injection molding machines consist of several mechanical components, e.g., the heating system, the injection component, and a complex hydraulic mechanism, as well as complex hardware and software systems for their automation. Such machines rely on the ability to change for varying products, but also play a role in the greater context of production processes, and thus need the ability to change depending on production needs. Given the variety of plastic products, injection molding machines need to vary regarding their capabilities, performance, and size. The machines exist in many different variants and can be further customized with configuration options. It is custom that each individual machine is specifically configured, adapted, and extended to meet the special requirements of its field of application. Often, this includes the development of specific hardware and software components. Needless to say, this overall results in high challenges for managing machine variability.

Injection molding machines thus often provide advanced configuration tools and end-user programming capabilities for customization. For instance, Keba AG (www.keba.com) provides KePlast [22], a hardware and software platform for the automation of injection molding machines. It comprises an industrial PC-based hardware solution with a real-time operating system, a run-time environment, and a development platform based on the IEC-61131 standard [19], a configurable control software framework, a visualization system written in Java, and a configuration tool supporting the interactive customization of solutions based on existing components. The platform exists in several variants, e.g., one is providing specific features for the Chinese market. The platform also comprises many software components for supporting a high number of optional machine components. Thus, the machine configuration also determines the software configuration.

KePlast supports a multi-stage process for developing customized automation solutions [31]: sales and customer-support personnel elicit requirements from the customer. Based on the generic domain solution, engineers then develop concrete automation solutions to meet these requirements. They use a configuration tool that guides the selection and customization of features. The tool automatically derives an initial solution [22], which is then adapted and extended by adding new features or adapting components to meet the specific requirements. During commissioning, engineers fine-tune machines by calibrating the properties of features. The automation solution provides further mechanisms to customize machine behavior during start-up and operation.

The term *feature* [4] is widely used in this process to communicate with customers during development, independent of the specific methods and technologies used. Obviously, the role-specific perspectives and needs define what constitutes a feature: for instance, sales people identify the needs of potential customers in terms of system features, while product managers drive the development of different KePlast product variants by defining features addressing market needs. They use problem-space features to define the scope of products from a market and customer perspective in feature maps comparing different product variants. These spreadsheets comprise high-level system features, feature associations, available hardware options, and references to order numbers used by sales people. At the technical level, software engineers work with solution-space features, i.e., the code implementing a specific functionality denoted by a feature. Features at this level are often cross-cutting, spanning multiple components and sub-systems, and even implemented in multiple programming languages. An example of such a feature is the MoldCavityPressureSensor, which provides a quality index of an injection-molded part by monitoring the cavity pressure. First, the feature comprises the actual sensor component, which can be integrated into the mold chamber. In addition to this machine component, it comprises a control component and a visualization component for monitoring and displaying the pressure curve. The feature is visible to the customer as an extra option. It appears in spreadsheet documents used by sales people and exists in different product variants. It further appears as an option in the software configuration tool. Further, its software components provide various means for customization. Finally, the user interface elements of the feature allow customizing its functionality, such as alarm levels, during operation.

Engineers use a wide range of mechanisms to implement feature variability. For example, they define interfaces to hook in new functionality, they use the software framework's capabilities for adding, exchanging, or reloading modules, and they exploit a mechanisms for checking the presence of machining components, based on which parts of the software can be enabled or disabled at run time. Moreover, the solution framework supports different binding times for variability. The custom configuration tool resolves design-time variability. However, the software solutions also use configuration files, which can be changed in a setup phase before start-up (configuration-time variability). Service personnel usually changes these configuration files for reconfiguration. Further, the automation software provides specific input masks allowing to fine-tune the operations of the machines by operators, often even while the machine is running [30].

4   Variability Engineering

Product line engineering methods and techniques [28] are widely used to develop and manage families of systems. Product line engineering distinguishes two life cycles: in domain engineering, the commonalities among members of families are identified and reusable and customizable components are developed to account for the required variability (development for reuse). In application engineering, the reusable and configurable components then provide the foundation for deriving and developing new family members (development with reuse). These activities rely on methods supporting the systematic engineering of variability during modeling, implementation, verification and validation, as well as evolution. In particular, *variability modeling* (cf. Subsect. 4.1.) is essential in domain engineering for variability analysis and design. Variability models help to understand the commonalities and variability of variants, thereby also defining the scope of a product line. During application engineering, variability models support the derivation of new variants in a product line by using variability models to guide and automate the creation of new products. Methods for *variability realization* (cf. Subsect. 4.2.) guide the implementation. The three basic variability realization mechanisms are annotations, compositions, and transformations. They can be applied to arbitrary variable artifacts including source code and models (e.g., process models or architecture models). Methods for *variability-aware validation and verification* (cf. Subsect. 4.3.) follow different strategies for analyzing the high number of possible product variants. Finally, developers face major challenges when evolving variable systems. Variation control systems uniformly managing both revisions and variants aim to support *variability evolution* (cf. Subsect. 4.4.).

4.1   Variability Modeling

Variability modeling is widely used for managing software product lines. It has been shown that in many important real-world systems such as the Linux kernel and the embedded operating system eCos variability modeling is the only form of modeling used [5], thereby bringing modeling concepts into otherwise mostly code-driven projects. A wide range of techniques for variability modeling exists today [13, 34], which use different abstractions to denote the *units of variability*, i.e., the key concepts used to model variability [13]. Feature modeling, the most prominent technique, uses *features* to define commonalities and variability. In fact, customers, product managers, and engineers nowadays often denote product characteristics as features, as discussed above. As an example, we present the feature model of an injection molding machine's mold component, originally described in [31]. We use this example to define important concepts of feature models based on existing work [12, 24]. The legend in Fig. 2 explains the graphical symbols used to define the key concepts:

Mandatory features are common features that must be present in all products of a product line. This is the case for the feature HydraulicCylinder in our example. *Optional features* in contrast exist only in some variants of a product. For instance, the feature MoldCavityPressure-Sensor in the mold component is an optional subfeature of the mandatory feature Diagnostics. *Hierarchical features* are used to express decomposition or parent-child configuration constraints, thus resulting in a feature tree. For instance, the general feature MoldProtection has two subfeatures DetectionMethod and Reaction. Another example is Diagnostics with its subfeatures ClampingForceMeasurement and MoldCavityPressureSensor. While the feature ClampingForceMeasurement is mandatory, MoldCavityPressureSensor is an optional subfeature of the Diagnostics feature. *Alternative features* are needed in many practical cases to define different variants. For instance, the alternatives for the MoldProportionalValve are either PositionClosed or OpenLoop, while the feature DetectionMethod, a subfeature of MoldProtection, has two alternatives Timeout and DataRecorder. *Or-features* allow the selection of several features within a group of features. For example, the two variants MoldProportialValve

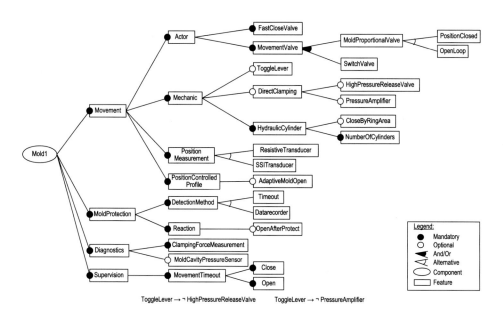

Fig. 2 Feature model of the mold component of an injection molding machine

and SwitchValve are defined for the feature MovementValve, as the system can be equipped with one or both of them. *Cross-tree constraints* are needed for expressing dependencies that exist in the domain or its implementation (or both). For example, *requires* constraints allow defining that one feature relies on another one, while *excludes* constraints are used to indicate that one feature disallows another one. Such constraints are often motivated by implementation dependencies[14]. Configuration constraints restrict the allowed configurations of features, while implementation constraints indicate technical dependencies. For example, the feature ToogleLever excludes the features HighPressureReleaseValve and PressureApplifier (cf. the constraints in the bottom of Fig. 2). *Modules* or *components* have been proposed by some approaches to structure feature models for large-scale systems [31]. Single monolithic feature models are inadequate to deal with the complexity of industrial systems, which has led to the development of multi-product line approaches that support modularizing feature models in various ways [18]. For instance, in our example Mold is the root of the feature, and represents a component within a larger feature model defining the variability of injection modeling machines.

It has been argued that properly decomposing a product line into features, and correctly using features in all engineering phases, is core to the immediate and long-term success of a product line. However, the term feature is a very abstract concept, making it sometimes hard to apply in practice. For instance, a common definition describes a feature as "a distinguishable characteristic of a concept (system, component, etc.) that is relevant to some stakeholder of the concept" [12]. The question of what of constitutes a feature thus depends on the application context and the domain of interest. Yet, defining the purpose, scope, and granularity of features remains hard, specifically when modeling large-scale industrial software systems [31]. For instance, the study by Berger et al. [4] reports empirical results of studying the actual feature usage in industrial product lines of three large companies, including a detailed and contextualized analysis of 23 features. An important aspect of features is to understand the purpose of why they are modelled. Researchers have distinguished *problem space features* generally referring to systems' specifications established during domain analysis and requirements engineering; *configuration space features* defined for easing the derivation of products by resolving variability; and *solution space features* referring to concrete implementations of systems, often by defining mappings of the features to code. In an exploratory study on developing feature models for two large-scale software systems in the domain of industrial automation, Rabiser et al. [31] investigated the purpose, scope, and granularity of feature models of large-scale industrial software systems, also including our KePlast running example.

4.2   Variability Realization

Defining variability occuring on the level of development artifacts (e.g., source code, design models, or documentations) also needs specific mechanisms. At a coarse grain, variability can be realized on a per-variant basis, which essentially amounts to cloning. One variant is copied and modified to fit the requirements for another variant. This technique is called clone-and-own. While it is easy and quick to apply, it leads to problems for maintainability and evolvability in the long run, as all knowledge about the applied changes is lost. Hence, there are more fine-grained variability realization techniques to connect the variability in the problem space (cf. Subsect. 4.1), e.g., in terms of features, to variation in the solution space (artifact level). This allows to automatically derive specific variants on the artifact level from a given configuration. In principle, we can distinguish three different categories of variability realization mechanisms [34] for design-time variability:

- *annotative* (or negative) variability modeling, which considers a superimposition of all possible artifacts and removes specific parts of the superimposed artifacts dependent on the specific variant that should be derived. Typically artifacts are annotated with so called presence conditions in terms of features in order to determine which artifacts are to be used in which variant. A prominent example are C preprocessor annotations.
- *compositional* (or positive) variability modeling, where artifacts are composed depending on the variant that should be derived. Prominent examples are plugin systems or feature-oriented programming [3].
- *transformational* variability modeling, where a core variant is changed by a sequence of transformations until a specific other variant is obtained. Prominent examples are the Common Variability Language (CVL) [16] or delta modelling [9].

For reconfiguration, when variants should be changeable during the runtime of a program, a variation of the annotative variability modeling approach called *variability encoding* [33] can be used: a superimposition of all needed variants is deployed, where features are encoded by runtime flags, which are combined in presence conditions. Those runtime flags and, thus, presence conditions are evaluated in runtime control structures, such as if-statements. If a presence condition evaluates to true the code belonging to those features is executed. Changing the values of the runtime flags then allows switching between different variants.

Regarding variability realization, the KePlast platform discussed above employs different mechanisms: the custom-built configuration tool allows selecting a consistent set of optional and alternative features. Based on the selection, the software solution is built using a combination of compositional and transformational techniques, i.e., the product solution is created by a composition of source files and by the generation of various configuration files. The created software system then comprises a magnitude of configuration settings, which are used in presence conditions for enabling or disabling source code parts at runtime. An interesting hybrid method combines the compositional approach and runtime checks with presence conditions: in the solution framework, a special method allows checking the presence of hardware components; only when a specific hardware component is present, the software part is enabled at runtime. For example, the control software for the MoldCavity-PressureSensor is enabled by checking if the respective hardware endpoint for the sensor is configured. However, when it comes to adapting and extending the solution to its specific application needs, mainly a clone-and-own approach is used. Engineers often rely on previous solutions which they clone and adapt to their needs (cf. Subsect. 5.2.), an approach with potential problems and limitations [20].

Fig. 3 depicts a variability engineering approach for feature-oriented development based on [14, 17], which links problem space models with artifacts and models on the solution space. Models of the implementation artifacts are built by parsing the source code and building a code model in form of an abstract syntax tree. In automation engineering usually

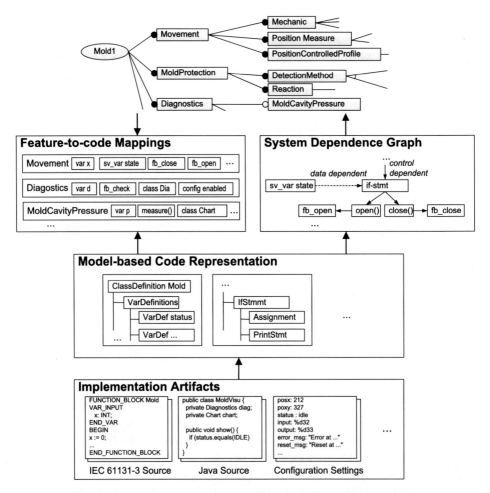

Fig. 3 Variability engineering relies on defining and linking variability information in both the problem space and the solution space. In our running example, the system is implemented in diverse implementation artifacts, including code in multiple programming languages. A model-based representation provides a common view and allows further analyses to compute feature-to-code mappings and a system dependence graph encoding code-level dependencies

many different kinds of artifacts have to be considered, e.g., the KePlast system comprises IEC 61131-3 control code, Java visualization code, and different types of configuration files, thus requiring a abstract syntax tree model covering different languages and technologies. Feature-to-code mappings then map the implementation artifacts to elements in the feature model, thereby providing the foundation for resolving variability and composing products based on a feature selection. Additionally, a system dependence graph, as implemented in [1, 2], can globally represent data and control flow dependencies of a program and enable system-wide dependency analyses at code level. Making such dependency analyses variability-aware allows considering configurations of systems (cf. Subsect. 4.3). Tools integrating feature-to-code mappings and dependency analyses can lift code-level dependencies analyses to the level of features, to then identify inconsistencies or visualize areas of evolution (cf. Subsect. 4.4).

4.3 Variability-Aware V&V

In general, there are four different strategies for post-hoc validation and verification of variability-intensive systems [39]. *Product-based analyses* consider each variant in isolation. This allows to reuse existing analyses without modifications, but is in general not feasible due to the large number of possible variants. It is possible to reduce the analysis effort by exploiting commonality between variants along the lines of regression testing, but the approach essentially stays a product-by-product analysis. In *sampling-based approaches* a representative subset of variants is determined that can then be analysed, e.g., following a product-based approach. Sampling increases the feasibility of the analyses as only a smaller subset of the whole variant space needs to be considered, but cannot ensure the desired properties of every possible variant. *Family-based approaches* analyze a joint representation of all possible variants in one run. While they usually scale very well for large variant spaces, they require a truthful encoding of all possible variants into the family representation, and in some cases also changes to the analyses to make them variability-aware. Furthermore, they are not robust to change, meaning that in case of variant evolution the complete analysis needs to be redone. *Feature-based analyses* are based on a compositional reasoning strategy analysing the components of the variants in isolation and then combining the analysis results to obtain a result for the complete variants from the analysis results of the parts. While highly scalable and robust to change, they first require modular structures in the construction of the variants and second properties that are amenable to compositional analysis, which is not the case for a large number of meaningful properties. Hence, they usually need to be combined with a product- or family-based analysis step. Besides, also a number of combinations of the four main analysis strategies exist. Recent work [6] proposes an orthogonal approach to the four main post-hoc analyses strategies for variant-rich systems by extending correctness-by-construction to software product lines, which allows to derive provably correct variants based on a formal refinement framework.

Static Code Analysis has proven to be an effective software quality assurance technique complementing testing. It aims at detecting problematic code constructs and potential defects in early development stages [29]. Static code analysis methods are well-known and explored for the analysis of single systems [26], i.e., product-based analysis, but are difficult to achieve for family-based analyses, which are thus still rare (cf. [8]).

Angerer et al. [1, 2] have developed methods for configuration-aware static analysis of automation programs. The approach relies on a system dependence graph (SDG), a representation of system-wide dependencies in a program as depicted in Fig. 3. In distinction to other approaches, the SDG uses presence conditions for encoding the variability of a system. In this form, it can also consider run-time variability. For example, it allows identifying relevant code for a concrete product variant and marking code as inactive, if it cannot be executed in a product variant, which is determined by a load-time configuration [2]. This is achieved by first building the SDG, then extracting presence conditions from conditional statements testing configuration settings, and finally identifying inactive code based on control dependencies in the SDG and the concrete product configuration settings. For example, in KePlast an IS_LINKED function can be used for testing if certain hardware endpoints are configured in the current product. Thus, the conditional statement in Fig. 4 tests if the input point di_ImpulseInput is used. As this not the case in the current variant, the respective control code can be greyed out or hidden in an editor, thereby simplifying maintenance.

When developing and maintaining a product line developers also need to determine the impact of changes on all possible variants. Change impact analysis is an important technique in software maintenance to identify the potential consequences of a change, or to estimate what needs to be modified to accomplish a change. However, conventional approaches fail for variable systems, as they do not consider all possible variants. Angerer et al. [1] further presented a change impact analysis technique based on the configuration-aware system dependence graph, which also considers the system configuration to improve the precision of family-based analyses. This allows to assess the impact of a feature update in

```
IF IS_LINKED(di_ImpulseInput) THEN
      // Start and stop the paMeasure algo depending on
      // sv_bDeviceReady flag (postupdate var for this algo),
      // which is false everytime the movement is active
      IF NOT sv_bDeviceReady THEN
            // Reset the stand still detection and the impulse
            // counter and start the measuring algo paMeasure
            fbTStandStill(IN := FALSE);
            sv_iThisStepsActImpulses := 0;
            START_PROCESS_ALGORITHM (paMeasureImpulses);
      ELSE
            STOP_PROCESS_ALGORITHM (paMeasureImpulses);
      END_IF;
END_IF;
```

Fig. 4 A variability-aware editor greying out inactive code, which cannot be executed in the current configuration

an existing system, or to explore what might be missing when transferring a feature from one development branch to another.

4.4   Variability Evolution

Product lines need to evolve continuously to stay current and to meet new requirements. This results in different *versions* which need to be managed during development. In particular, it is challenging to maintain the consistency of features with their implementation artifacts. It is also difficult to manage both revisions and variants created over time.

Feature-to-code consistency When systems evolve, engineers need to extend and adapt feature models to reflect the changes. However, engineers require deep knowledge about the domain and the implementation to avoid inconsistencies between a feature model and its implementation. Ensuring consistency is challenging due to the complexity of both implementation-level artifact dependencies and feature-to-code traces. Feichtinger et al. [14] present an approach to lift code-level dependencies to the level of features (cf. the example shown in Fig. 3). Static code analysis methods based on a system dependence graph are used at the code level for detecting code dependencies. Then, the feature-to-code mappings, i.e., knowledge which code parts implement which features, are used to lift these dependencies to the level of features. This helps engineers to detect and resolve inconsistencies in the feature model, but can also detect implementation deficiencies such as unintended dependencies between feature implementations. Although such family-based analysis approaches recently progressed, they are not yet generally available in widespread IDEs.

As shown in Fig. 3, features at the level of the problem space are commonly implemented in multiple source code artifacts, resulting in complex dependencies at the code level. As developers add and evolve features frequently, it is challenging to keep feature models consistent with their implementation. Combining feature-to-code mappings and code dependency analyses is useful to make engineers aware of possible inconsistencies. Figure 5, for instance, shows a heatmap visualizing the dependency changes of all features in of the KePlast product line. The changes in version 53 of our case study [14] focus on the features Injection, Inject and Plast. The changes introduced new status variables, which cause new dependencies and interactions between the features Inject and Plast.

Uniformly managing both revisions and variants A version can either be a revision or a variant: a *revision* is used to express evolution over time. Revisions are sequential versions, with newer revisions superseding previous ones. A *variant* is used to express concurrent versions. Variants co-exist with other variants at the same point in time. Two versioning approaches can be used to manage revisions and variants [11]: *extensional versioning* explicitly enumerates all existing versions. It then allows to retrieve all versions that were explicitly created before. Examples of such tools are Git or Subversion. *Intensional versioning* expresses a

Fig. 5 A heatmap visualizing code-level dependencies at the level of features for our KePlast example. Darker colors indicate stronger dependency increases

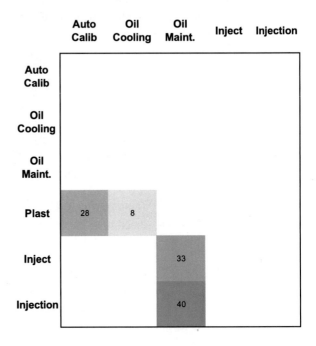

large version space using concepts such as features, configurations, constraints, and construction rules. It then allows the generation of arbitrary versions based on a configuration mechanism that follows these construction rules. Examples of such tools are the variation control systems [24] ECCO [15, 23] or SuperMod [37, 38].

Specifically, the widely used version control systems Subversion and Git help keeping track of the development history by assigning revisions to states of a system over time. However, in practice, evolution over time is rarely just a linear sequence of steps. Branching is a common mechanism for dealing with variants. For instance, short-term *feature branches* exist for as long as it takes to develop a new feature in isolation. Once the new feature is finished, it is merged back with the original code, at which point it becomes inseparable from the system. Another use of branching is the creation of long-term branches with the purpose of creating clones or variants of a system, for example, for different customers with slightly varying requirements. However, neither feature branches nor variant branches are ideal, as the former lead to a loss of variability after the merge and the latter lead to maintenance problems with increasing numbers of variants.

Therefore, *variation control systems* have been conceived. They support intensional versioning and emphasize variant management. For this purpose, they provide capabilities for decomposing software systems into finer-grained features, together with support to automatically manage this variability [24]. For example, the variation control system ECCO provides feature-oriented operations such as commit, checkout, push and pull. It is based on an incremental feature location approach to automate the computation of feature-to-code-mappings. For instance, the mappings show in Fig. 3 are created and refined when committing new or modified features to a repository. ECCO maps additional source code of newly added features (e.g., Diagnostics) or it refines existing mappings to existing features (e.g., MoldProtection).

5 Product Line Adoption and Evolution

Clements distinguished three adoption models for introducing systematic variability management and product lines in organisations [10]: the *proactive* adoption model assumes that all required product variants can be designed and implemented up front. This waterfall-like approach can only work if the requirements and their evolution are very well understood, which may be the case in mature and stable domains. In the *extractive* model, a product line's initial baseline is created by using already existing product variants, which have often been created in a clone-and-own manner. This approach is regarded as effective for organizations transitioning from conventional single-system engineering to software product line engineering. Finally, in the *reactive* adoption model, an existing product line is extended by another product variant incrementally. This model is useful if the future requirements cannot be easily predicted or if only few resources are available for the transition. It is also useful to deal with the flexibility needed in distributed software ecosystems (cf. Subsect. 5.2). The proactive adoption strategy can only be successful in stable and well-known domains. As requirements are highly volatile in many domains, we now discuss the extractive and reactive adoption strategies.

5.1 Extractive Adoption

Product lines are often not developed from scratch. Rather, companies with successful products will start migrating towards a product line approach, if maintaining a family of largely similar products becomes economically infeasible. A common scenario is that variants are first created in an ad-hoc manner by cloning and adapting existing ones. At a certain point, creating further variants is no longer efficient due to difficult reuse and maintenance, caused by the necessary propagation of changes across many variants. The migration or consolidation of the existing set of cloned variants into a managed product line is referred to as extractive product line adoption [10] or reengineering-driven product line development [36].

Product line scoping is an important activity for extractive adoption. In this highly collaborative process stakeholders define what shall be part of the product line [27]. Product line scoping has been described as a three-stage process comprising product mapping, domain potential analysis, and reuse infrastructure scoping [35]. More specifically, essential tasks in this process are the identification of features existing in the different variants, the location of features in the implementation artifacts, the synthesis of a variability model showing how the features can be combined in valid configurations, and the choice of a suitable variability mechanism (e.g., a preprocessor for textual implementation artifacts), to enable the composition of features during product derivation.

5.2   Reactive Adoption

In this adoption model, an existing product line is incrementally extended with another product variant. A common approach is to derive the product variant from the product line that best fits the new requirements and then adapt it accordingly (cf. our running example). This strategy is useful in the domain of industrial products and production, as many companies nowadays need to serve a mass market while at the same time customers demand individual, customer-specific solutions. After deriving an initial product, features are added and adapted to satisfy individual customer requirements, possibly followed by merging back these changes into the original product line. Merging new features into the product line should happen without much delay, as otherwise, the independently maintained product variants will pile up and the product line may gradually degrade into ad hoc clone-and-own reuse.

In many domains, a single company often cannot develop all features requested by customers. To handle this problem, development is frequently organized in software ecosystems (SECOs), i.e., interrelated software product lines involving internal and external developers [7]. In the context of SECOs, companies need to share new or updated features by transferring them to other product lines in the ecosystem. This is, for instance, useful when a feature developed in an individual customer project becomes relevant for another market segment or when updates of features need to be transferred to related products in the ecosystem. The KePlast example shows that managing the evolution in SECOs is challenging, as developers continuously and independently evolve features of the core product line, the cloned product lines, and individual customer products. In particular, it remains difficult to track and understand evolution at the level of features, if not supported with analyses and visualization.

It has been shown that the size and scattering of features are of particular interest for maintenance and evolution [17]. The size of a feature is measured by the magnitude of artifacts mapped to a particular feature (e.g., the number of program elements in source code or the number of data elements in configuration files). The *feature size evolution* is then determined by the relative change of the artifact size between two different points in time. Additionally, features are typically realized in multiple artifacts and locations. The number of contiguous locations of a feature's implementation determines its scattering. For example, if a feature is mapped to all source files within one directory, the directory represents the location of the implementation and the scattering is 1. However, scattering is 5 if the feature is mapped to five independent source files. *Feature scattering evolution* can be computed by the relative change of feature scattering between two different points.

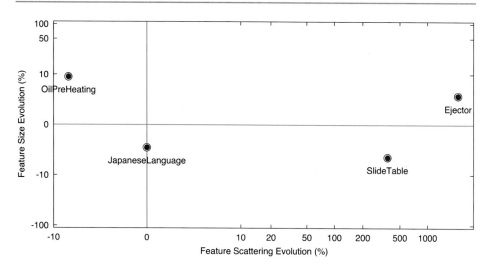

Fig. 6 Feature size evolution and feature scattering evolution of four features of a KePlast industry project. The values in the portfolio are percentages shown in a logarithmic scale.

These metrics can be used to compute a feature evolution diagram as shown in Fig. 6. The portfolio shows the size evolution and scattering evolution of four selected features of an industry project conducted based on the KePlast platform.

Such a visualization can help to identify features leading to high maintenance risks when being adopted. For instance, the scattering evolution of the feature *Ejector* increased by 2150 percent while its size increased by just about 2.7 percent. The high value was caused by user interface code added in many locations. A different example is the feature *JapaneseLanguage*, which was added as a new feature. After the feature was initially added, the feature was only slightly modified. In particular, the deletion of the translation for some features reduced the size while leaving the scattering unchanged. A similar example is the newly added feature *OilPreHeating*, which shows only a small size increase but reduced scattering, caused by the deletion of some no-longer-needed code fragments. The feature *SlideTable* is an example with artifacts added and removed during the project, overall leading to a small reduction in size. Obviously, the newly added code significantly increased the scattering of this feature.

6   Research Challenges

We discuss open research challenges to advance variability management for Industry 4.0 and industrial automation.

Configurable binding times and binding time switching As discussed in Sect. 2, variability can be bound statically at compile time or dynamically at run time. There are advantages and disadvantages of each strategy, with trade-offs that need to be considered. This is especially the case in automation systems: dynamically-bound variability may not be suited for embedded devices with strict memory constraints, or if the unauthorized activation of features by customers must be avoided. Statically-bound variability on the other hand prevents adapting to changes at run time. Ideally, the binding time could be configured per feature during product derivation. However, most variability mechanisms support only one binding time. To achieve different binding times per feature, at least two different variability mechanisms must be used: one for statically bound features and another one for dynamically bound features. Still, the choice which features should be bound statically and which should be bound dynamically needs to be made in advance, as changing variability mechanism is infeasible.

Legacy migration and maintenance in the field Industrial automation software is often tailed to the specific customer or site where it is deployed. This customization often happens on site and is usually not well documented. This leads to a number of challenges in the long run. As the number of custom-tailored variants increases, they become hard to maintain. At some point, it is beneficial to migrate the individual legacy variants towards a structured and systematic reuse approach (cf. Sect. 5). To achieve this, it must be known what variants exist, what features they implement, and how they diverged from other variants over time. Especially the latter is a challenge, as industrial automation systems often remain in the field for decades and change over time as adaptations are made, updates are applied, or fixes are performed. Over time, variants with originally the same features may no longer be identical. While they may still offer the same feature set, the features may have evolved differently in some variants. Furthermore, some features may have been implemented multiple times by engineers unaware of other existing implementations. Over time, the knowledge of what variants have been deployed and what features they support has been lost. This causes problems if such variants ever need an upgrade. For example, when replacing a broken hardware component, the new component may require an update of the control software. It is difficult to guarantee that the update will work in the context of legacy features.

Optimization of non-functional properties with dynamic product lines We showed that variability is often the result of new customer requirements, which may lead to adding new or adapting existing machine features. However, variability is also fundamental for optimizing non-functional properties (e.g., performance, throughput, or energy consumption) of products and production systems [25]. This means that both design-time and operation-time aspects need to be considered. In contrast to product lines with variation points bound at design time, dynamic software product lines allow to bind features at runtime [30]. Monitoring systems at runtime can provide context information to trigger reconfigurations, to provide feedback to feature modeling and feature configuration, or to continuously adapt products and production systems to meet certain constraints (e.g., finding the most eco-friendly configuration during production).

Feature-based evolution A general challenge during the evolution of variable systems is the co-evolution of the modeled variability and the implemented variability (cf. Subsect. 4.4). Another challenge is related to revision management. Traditional version control systems keep track of evolution at system level and not at the level of individual features. The latter would be valuable, for example, to distinguish stable from volatile features, thereby indicating potential design problems. More specific to industrial automation, but also very common in open-source development, are challenges that result from distributed development, as in SECOs (cf. Subsect. 5.2). OEMs, on-site engineers, or even customers adapt products before deployment, meaning that even a well-designed platform can degrade into a set of cloned variants. It is therefore important to keep track of changes and support distributed feature-oriented development. This is beneficial for integrating features developed downstream into the originating platform, when propagating fixes performed upstream in the platform to affected derivations downstream, or for keeping track of feature distribution to avoid redundant development. Finally, similar to the co-evolution of problem and solution space, the co-evolution of different artifact types is especially challenging for variable systems, as each kind of artifact potentially uses its own, custom-tailored variability mechanism. These mechanisms need to be synchronized manually during evolution. A universal variability mechanism would alleviate this problem and allow the uniform evolution of variability across different types of artifacts.

Variability across engineering artifacts While many companies nowadays use variability modeling, it is important to understand that its acceptance is mostly limited to the software within product or production systems. However, neither products nor productions are software systems and, today, it is little understood how software variability relates to, affects, and changes with the hardware it is part of. This is a severe disconnect, often caused by the fact that software engineering is mostly disjoint from hardware engineering. Recent research in mechatronic design suggests an increasing recognition of this problem. It is realized that the ultimate goal must be a comprehensive understanding of features and feature dependencies; and an understanding of the relationships between features/dependencies and the kinds of engineering artifacts they affect (mechanical drawings, computations, electrical planning, software control, etc.). So, if a feature is adapted, it may have hardware and software implications—both of which need to be understood. Understanding the relationship between features in software and features in hardware will thus be a critical part of variability modeling in the next years. Not only does such comprehensive variability modeling benefit the overall understanding, planning, and sale of variable products and production systems, but it also improves the necessary cooperation among the various engineering disciplines involved.

7   Conclusions

Variability is an essential part of Industry 4.0—both in context of product and production systems. Hence, the ability to provide individualized products is essential but so is the ability to customize production continuously for optimization. This book chapter discussed the need for distinguishing product variability and production variability where the latter not only needs to support the former, but also needs to provide room for optimization. Key here are variability binding times. If products are physically different due to design-time variability, then production needs to support their varying production processes. If a product can be reconfigured during run time, then the products become physically similar and so does their production. Designing for run-time variability can thus simplify production variability. However, run-time variability tends to shift focus onto software (to account for differences in behavior). Moreover, products supporting run-time variability are also more expensive because parts may need to be present that are only needed for certain uses. Hence, the more variability is shifted from design time to configuration time or run time, the more stable production processes become, but the more complex the product becomes—a trade-off. Also, run-time variability is not possible for all products.

This book chapter chose an injection molding machine as an illustrative case study, because such machines are products, but at the same time essential parts during production. In discussing its variability, we not only covered essential parts of product variability, but also provided support for production variability. Machines typically exhibit a wide range of variability due to their many different uses during production. A machine's ability to change during production (e.g., its run-time variability) are a key ingredient to production variability. Simply said, production variability requires machine variability. Clearly, the variability binding times of an injection molding machine is an essential part to the variability binding times of a production system. An injection molding machine not designed for supporting run-time variability cannot play an essential role in (and even restricts) production variability.

Variability is a broad topic. This chapter provided an overview of technologies useful for their variability engineering and evolution. The illustrative example demonstrates some of those aspects in more detail. While designing systems for/with variability is essential, doing so is complex due to the combinatorical explosion of how features may interact. Product and production variability is thus not only a modeling problem but also a challenge affecting all phases of a system's life cycle.

Acknowledgements We gratefully acknowledge the financial support by the Austrian Federal Ministry for Digital and Economic Affairs, the National Foundation for Research, Technology and Development, and Keba AG, Austria. Furthermore, we like to acknowledge funding by the Austrian COMET K1-Centre Pro2Future and the Austrian COMET Center K2 program of the Linz Center of Mechatronics of the Austrian Research Promotion Agency (FFG) with funding from the Austrian ministries

BMVIT and BMDW, and the Province of Upper Austria. Finally, we would like to acknowledge funding by the JKU LIT Secure and Correct System Lab (co-funded by the Province of Upper Austria).

References

1. Angerer, F., Grimmer, A., Prähofer, H., Grünbacher, P.: Change impact analysis for maintenance and evolution of variable software systems. Automated Software Engineering **26**, 417–461 (2019)
2. Angerer, F., Prähofer, H., Lettner, D., Grimmer, A., Grünbacher, P.: Identifying inactive code in product lines with configuration-aware system dependence graphs. In: Proceedings 18th International Software Product Line Conference. pp. 52–61. SPLC'14, ACM, New York, NY, USA (2014)
3. Apel, S., Batory, D.S., Kästner, C., Saake, G.: Feature-Oriented Software Product Lines – Concepts and Implementation. Springer (2013)
4. Berger, T., Lettner, D., Rubin, J., Grünbacher, P., Silva, A., Becker, M., Chechik, M., Czarnecki, K.: What is a feature? a qualitative study of features in industrial software product lines. In: Proceedings 19th International Software Product Line Conference. pp. 16–25. SPLC'15, ACM (2015)
5. Berger, T., She, S., Lotufo, R., Wąsowski, A., Czarnecki, K.: Variability modeling in the real: a perspective from the operating systems domain. In: Pecheur, C., Andrews, J., Nitto, E.D. (eds.) Proceedings 25th IEEE/ACM International Conference on Automated Software Engineering. pp. 73–82. ASE 2010, ACM (2010)
6. Bordis, T., Runge, T., Schaefer, I.: Correctness-by-construction for feature-oriented software product lines. In: Proceedings 19th ACM SIGPLAN International Conference on Generative Programming: Concepts and Experiences. pp. 22–34. GPCE'20, ACM (2020)
7. Bosch, J.: Software ecosystems: Taking software development beyond the boundaries of the organization. Journal of Systems and Software **85**(7), 1453–1454 (2012)
8. Brabrand, C., Ribeiro, M., Tolêdo, T., Winther, J., Borba, P.: Intraprocedural dataflow analysis for software product lines. Transactions on Aspect-Oriented Software Development **10**, 73–108 (2013)
9. Clarke, D., Helvensteijn, M., Schaefer, I.: Abstract delta modelling. Mathematical Structures in Computer Science **25**(3), 482–527 (2015)
10. Clements, P.: Being proactive pays off. IEEE Software **19**(4), 28–30 (2002)
11. Conradi, R., Westfechtel, B.: Version models for software configuration management. ACM Computing Surveys **30**(2), 232–282 (1998)
12. Czarnecki, K., Eisenecker, U.: Generative Programming: Methods, Tools, and Applications. Addison-Wesley, Boston, MA (2000)
13. Czarnecki, K., Grünbacher, P., Rabiser, R., Schmid, K., Wąsowski, A.: Cool features and tough decisions: a comparison of variability modeling approaches. In: Eisenecker, U.W., Apel, S., Gnesi, S. (eds.) Proceedings 6th International Workshop on Variability Modelling of Software-Intensive Systems. pp. 173–182. ACM (2012)
14. Feichtinger, K., Hinterreiter, D., Linsbauer, L., Prähofer, H., Grünbacher, P.: Guiding feature model evolution by lifting code-level dependencies. Journal of Computer Languages **63**, 101034 (2021). 10.1016/j.cola.2021.101034
15. Fischer, S., Linsbauer, L., Lopez-Herrejon, R.E., Egyed, A.: Enhancing clone-and-own with systematic reuse for developing software variants. In: Proceedings 30th IEEE International Conference on Software Maintenance and Evolution. pp. 391–400. (ICSME'14) (2014)

16. Haugen, Ø., Wąsowski, A., Czarnecki, K.: CVL: common variability language. In: Proceedings 17th International Software Product Line Conference. SPLC'13, ACM (2013)
17. Hinterreiter, D., Linsbauer, L., Feichtinger, K., Prähofer, H., Grünbacher, P.: Supporting feature-oriented evolution in industrial automation product lines. Concurrent Engineering: Research and Applications 28(4) (2020)
18. Holl, G., Grünbacher, P., Rabiser, R.: A systematic review and an expert survey on capabilities supporting multi product lines. Information and Software Technology 54(8), 828–852 (2012)
19. IEC: IEC 61331-3, programmable controllers – part 3: Programming languages (2003), http://www.iec.ch/
20. Lettner, D., Angerer, F., Grünbacher, P., Prähofer, H.: Software evolution in an industrial automation ecosystem: An exploratory study. In: Proceedings International Euromicro Conference on Software Engineering and Advanced Applications. pp. 336–343. SEAA'14 (2014)
21. Lettner, D., Angerer, F., Prähofer, H., Grünbacher, P.: A case study on software ecosystem characteristics in industrial automation software. In: Proceedings International Conference on Software and Systems Process. pp. 40–49. ICSSP'14, ACM, New York, NY, USA (2014)
22. Lettner, D., Petruzelka, M., Rabiser, R., Angerer, F., Prähofer, H., Grünbacher, P.: Custom-developed vs. model-based configuration tools: Experiences from an industrial automation ecosystem. In: Proceedings MAPLE/SCALE Workshop at the 17th International Software Product Line Conference. pp. 52–58 (2013)
23. Linsbauer, L., Lopez-Herrejon, R.E., Egyed, A.: Variability extraction and modeling for product variants. Software and System Modeling 16(4), 1179–1199 (2017)
24. Linsbauer, L., Schwägerl, F., Berger, T., Grünbacher, P.: Concepts of variation control systems. Journal of Systems and Software 171, 110796 (2021)
25. Munoz, D., Montenegro, J.A., Pinto, M., Fuentes, L.: Energy-aware environments for the development of green applications for cyber-physical systems. Future Gener. Comput. Syst. 91, 536–554 (2019)
26. Nielson, F., Nielson, H.R., Hankin, C.: Principles of Program Analysis. Springer (1999)
27. Noor, M.A., Rabiser, R., Grünbacher, P.: Agile product line planning: A collaborative approach and a case study. Journal of Systems and Software 81(6), 868–882 (2008)
28. Pohl, K., Böckle, G., van der Linden, F.J.: Software Product Line Engineering: Foundations, Principles, and Techniques. Springer (2005)
29. Prähofer, H., Angerer, F., Ramler, R., Grillenberger, F.: Static code analysis of IEC 61131-3 programs: Comprehensive tool support and experiences from large-scale industrial application. IEEE Transactions on Industrial Informatics 13(1), 37–47 (2017)
30. Quinton, C., Vierhauser, M., Rabiser, R., Baresi, L., Grünbacher, P., Schumayer, C.: Evolution in dynamic software product lines. Journal of Software: Evolution and Process 33, 1–25 (2021). 10.1002/smr.2293
31. Rabiser, D., Prähofer, H., Grünbacher, P., Petruzelka, M., Eder, K., Angerer, F., Kromoser, M., Grimmer, A.: Multi-purpose, multi-level feature modeling of large-scale industrial software systems. Software and Systems Modeling 17, 913–938 (2018)
32. Rabiser, R., Schmid, K., Eichelberger, H., Vierhauser, M., Guinea, S., Grünbacher, P.: A domain analysis of resource and requirements monitoring: Towards a comprehensive model of the software monitoring domain. Information and Software Technology 111, 86–109 (2019)
33. von Rhein, A., Thüm, T., Schaefer, I., Liebig, J., Apel, S.: Variability encoding: From compile-time to load-time variability. Journal of Logical and Algebraic Methods in Programming 85(1), 125–145 (2016)
34. Schaefer, I., Rabiser, R., Clarke, D., Bettini, L., Benavides, D., Botterweck, G., Pathak, A., Trujillo, S., Villela, K.: Software diversity: state of the art and perspectives. International Journal on Software Tools for Technology Transfer 14(5), 477–495 (2012)

35. Schmid, K.: A comprehensive product line scoping approach and its validation. In: Tracz, W., Young, M., Magee, J. (eds.) Proceedings 24th International Conference on Software Engineering. pp. 593–603. ICSE'02, ACM (2002)

36. Schmid, K., Verlage, M.: The economic impact of product line adoption and evolution. IEEE Software **19**(4), 50–57 (2002)

37. Schwägerl, F., Westfechtel, B.: Collaborative and distributed management of versioned model-driven software product lines. In: Proceedings of the 11th International Joint Conference on Software Technologies. (ICSOFT)'16), vol. 2, pp. 83–94 (2016)

38. Schwägerl, F., Westfechtel, B.: SuperMod: tool support for collaborative filtered model-driven software product line engineering. In: Proceedings 31st IEEE/ACM International Conference on Automated Software Engineering (ASE). pp. 822–827 (2016)

39. Thüm, T., Apel, S., Kästner, C., Schaefer, I., Saake, G.: A classification and survey of analysis strategies for software product lines. ACM Computing Surveys **47**(1), 6:1–6:45 (2014)

40. Vogel-Heuser, B., Fay, A., Schaefer, I., Tichy, M.: Evolution of software in automated production systems: Challenges and research directions. Journal of Systems and Software **110**, 54–84 (2015)

41. Zave, P.: Feature interactions and formal specifications in telecommunications. Computer **26**(8), 20–30 (1993)

Digital Infrastructures

Reference Architectures for Closing the IT/OT Gap

Patrick Denzler and Wolfgang Kastner

Abstract

The Internet of Things (IoT) is an allegory for the concept of seamlessly connecting intelligent devices. Its application in the industrial domain envisions a next-generation manufacturing industry. Initiatives such as Industry 4.0 promise higher flexibility, improved quality and productivity. Nonetheless, the enhancements cause an increased complexity in a factory and its organisation as they require a seamless collaboration between all involved units, technological systems and individuals. One way of coping with the extended additional complexity is by utilising Architectural Reference Models (ARMs). State-of-the-art architectures combine different perspectives with a standard model to accommodate design choices, remove knowledge barriers and link the physical and virtual realm. This chapter introduces the basic concepts behind architectural designs and points out historical connections and differences between current ARMs. Moreover, it addresses the needs of converging the historically separated Information Technology (IT) and Operational Technology (OT) and exemplifies in a use case how ARMs can assist in closing the gap. Finally, the chapter serves as a foundation for the following chapters, introducing architectural concepts like cloud, fog and edge computing.

P. Denzler · W. Kastner (✉)
Computer Engineering/Automation Systems, TU Wien, Vienna, Germany
e-mail: k@auto.tuwien.ac.at

P. Denzler
e-mail: patrick.denzler@tuwien.ac.at

B. Vogel-Heuser and M. Wimmer (eds.), *Digital Transformation*,
https://doi.org/10.1007/978-3-662-65004-2_4

1   Introduction

When Kevin Ashton, in 1999, coined the term Internet of Things (IoT), he created an allegory for a concept of seamlessly connecting intelligent devices. Although researchers have made technological advances to realise the ubiquitous computer vision since the 1980s, the past few years have seen accelerated progress in various domains, e.g., home and building automation, smart grids, and e-health applications [1].

IoT applied in the industrial sector (Industrial Internet of Things (IIoT)) promises increased innovation, efficiency and quality by making things *"work"* smartly. At the same time, initiatives such as Industry 4.0 [2] aim towards next-generation manufacturing to *"make"* things smartly. Regardless of whether IIoT or Industry 4.0, the intensified interconnection of sensors, actuators, applications, and humans creates an unprecedented step towards the IoT. However, implementing IIoT in an organisation remains a complicated task.

Since it is still common for departments to specialise in one area, current staff members do not have the combined knowledge for integrating IoT vertically and horizontally in all levels of an organisation. For example, manufacturing focuses on the processes involved in producing goods [3]. Information technology (IT) experts prefer a broad and distant perspective and usually ignore the interplay between products, customers and the IT infrastructure. The technical department concentrates on system development and sees IT as something that never really works, and logistics is only concerned about the supply chain [4].

In an industrial setting, next to social and knowledge barriers, there is a historical, technological barrier between IT and operational technology (OT) [5]. Traditionally, OT emphasises efficiency, utilisation, consistency, continuity and safety, while IT drives agility and speed, flexibility, cost reduction, business insight and security. Converging IT and OT is essential for the IIoT.

One solution for organisations to overcome such barriers is to apply reference model architectures to design the intended implementations. Such norms can combine different perspectives with a standard model to support the understanding between heterogeneous communities and bridge technological gaps [4]. By representing each stakeholder's views, e.g., on software interfaces, reliability, security, maintainability and sustainability, reference architectures help employees to master the challenges involved implementing IIoT or Industry 4.0 [1, 2].

The following chapter aims to provide a starting point for readers interested in IoT architectures or needing an overview. The chapter starts with some basic concepts behind architecture design and points out historical connections and differences between current architectural reference models (ARMs). To illustrate the use of architectures, one use case focusing on the IT/OT gap finalises the chapter.

2   Architectural Reference Models

The creation of an ARM is a time consuming and tedious endeavour. It includes identifying and abstracting technologies, functionalities, mechanisms, and protocols relevant to the target domain (e.g., home and building automation, smart grids, e-health, industry) and specific needs, viewpoints of all involved stakeholders [6]. An ARM serves as an overall guideline, and some parts might not be relevant within every particular domain, requiring a very generic formulation process. Moreover, an ARM is a matrix containing models, guidelines, views, perspectives, and design choices that allow derivations into concrete domain-specific architectures. The architecture enables a systems architect to pick the building blocks depending on the targeted systems w.r.t. various dimensions (e.g., security, real-time, semantics or distribution) to form a concrete system. Standards such as the *'ISO/IEC/IEEE 42010:2011 Systems and Software Engineering-Architecture Description* codify conventions, ontologies and standard practices for describing architectures. However, there is no common consensus on building an ARM except that there is a trend in architectures such as Industry 4.0 [2] or IoT [1] to divide an ARM into *Reference Models* and a *Reference Architecture* [6]. Figure 1 visualises the segmentation of the ARM and further breakdowns explained in Sects. 2.1 and 2.2. The division into a reference architecture and models originates from accommodating various information types and relations. Some are best-represented in models, while others fit better in architectures [7].

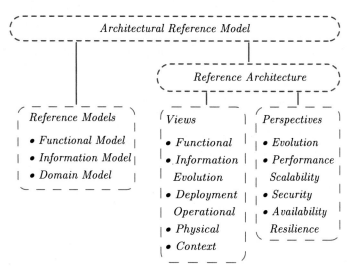

Fig. 1 Elements of an Architectural Reference Model (ARM)

2.1   Reference Architecture, Views and Perspectives

The reference architecture builds upon domain-specific use cases and incorporates functional and non-functional aspects and the collective domain agnostic elements. A common way to further break down system design choices is to introduce architectural views that accommodate a standardised method to structure architectural descriptions [8]. According to Rozanski [9], architectural views provide an intuitive representation of each aspect addressed. The author indicates some commonly used views also depicted in Fig. 1:

- The functional view describes all functional building blocks of the architecture. There is nearly an infinite amount of functional components available, each linked to the chosen domain.
- All static information structures and flows through the system and between components are described within the information view. This view helps to define how relevant information is to be represented in a system.
- The deployment and operation view addresses how a system is realised by selecting technologies and making them communicate and operate comprehensively. It provides a guideline through different design choices for the actual implementation, i.e., how to get from the service description and the identification of the functional elements to the selection of available technologies, to reach the intended system behaviour.
- The physical view represents any physical object for what the system is intended to do, e.g., if the system provides environmental monitoring and control in a building, the physical entities are the people working in the monitored building.
- The context view describes the relationships, dependencies, and interactions between the system and its environment. In detail, the view describes what the system does, where the boundaries are to the outside world and how it interacts with other systems, people, and other external entities [10]. The physical and context views are interconnected.

However, views do not consider that some requirements are qualitative, not only technical and can span over several views. Especially in the case of non-functional or quality properties, views are not sufficient [10]. A potential solution is to introduce architectural perspectives, as suggested by Woods and Rozanski [7]. Perspectives cut across the views and provide a collection of activities, tactics,[1] and guidelines to ensure that the system behaves according to a particular set of quality properties.[2] In some sense, perspectives provide another abstraction layer above the views to consider requirements beyond one view [7]. Modern architectures, later introduced in Sect. 4, such as the Reference Architectural Model Industrie 4.0 (RAMI 4.0), utilise perspectives to a large extent. Woods and Rozanski [7] introduce four possible architectural perspectives (cf. Fig. 1).

---

[1] Tactics are state-of-the-art methodologies used in today's system architectures.

[2] Quality properties are externally visible non-functional properties of a system such as performance, security, or scalability [10].

- The evolution and interoperability perspective addresses that requirements and technologies change over time and create the need for compatibility with newer technologies. The ability of a system to cope with changes after the deployment needs to be balanced against the cost of providing such flexibility.
- The performance and scalability perspective describes two non-functional quality properties. Both are externally visible and difficult to quantify in distributed systems. In other words, the system needs to execute within the given performance profile and adjust flexibly to increasing process volumes.
- Trust, security, and privacy are basic properties for user interaction. They are interrelated and also non-functional. Trust is a complex quality as it is subjective, i.e., the user expects a system to be dependable in all aspects it is built for and behaves accordingly. Security enforces confidentiality, integrity and service access policies and detects and recovers from failure in these security mechanisms. Privacy aims to limit the collection of personally-identifying information.
- Availability and resilience address the system's ability to stay fully or partly operational when things go wrong and how effectively it handles failures that could affect the system's availability.

2.2   Reference Models

Another relevant part of the ARM is the reference model that provides the basic concepts, models, relationships and definitions required to build the previously described architectures. The left side of Fig. 1 depicts three specific models representing the reference model's main parts. Each model provides or receives information from the reference architecture views (shown on the right side of Fig. 2) or the other models. The domain model specifies the specific use case's language, concepts, and entities and their relation [11]. This sub-model creates a shared understanding of all aspects of the chosen domain and is crucial for developing interoperable architectures and systems. In other words, the domain model is the basis for building the other models and relates to the physical, context and deployment views. The information model considers information aspects in a system and is closely connected with the information view that details the information flows. A specific role falls to the functional model; it links concepts and entities introduced by the domain model with current functionalities in the architecture. As with the information model, the functional model correlates with the functional view that provides more details about the required building blocks. Some architectures introduce further groupings as communication and security models that add system security and reliability [12].

The following Fig. 2 aims to visualise the above-described interdependencies between models, views and perspectives. On the right side, the figure shows the relations between the models and the reference architecture as a whole. Moreover, the figure depicts the previously described crosscutting quality of the perspectives over the views that form a grid on the left

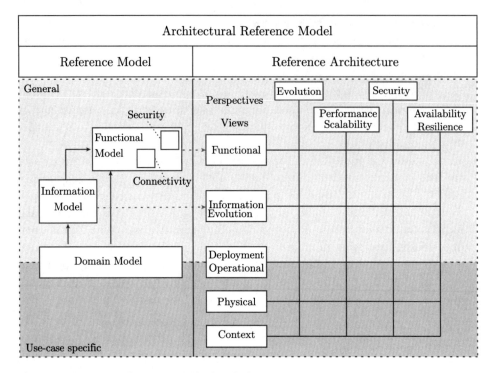

Fig. 2 Architectural Reference Model basic principle

side. Besides, all elements in the dark grey area are use case-specific. In contrast, elements placed in the light grey area have a broader or general applicability, i.e., those elements may apply to several use cases. It needs to be mentioned that Fig. 2 does not claim to be complete; instead, it provides an idea about the interdependencies between the elements of an ARM. Additionally, these interdependencies require an iterative approach while creating the different ARM elements, as a change in a view might affect a model or vice versa.

3   Architectures Before IoT

From a historical perspective, ARMs are never built from scratch; instead, they reuse established and proven concepts from previous architectures. A perfect example for this is the Internet. Without the Internet's established capabilities, the IoT would require an entirely new network architecture. Similarly, in the industrial context, more unique architectures such as RAMI [13] (See Sect. 4) align with older architectures such as the Purdue Enterprise Reference Architecture (PERA) [14]. The following section is a brief introduction to architectures created before the introduction of IoT to understand the relations between ARMs presented in Sect. 4. Those architectures are the backbone of almost any IIoT architecture.

3.1   Internet Protocol Suite and Open Systems Interconnection (OSI) Model

As indicated beforehand, the Internet is an essential element of modern IoT architectures. The Internet's underlying architecture specified in RFC-1122 and RFC-1123 loosely defines a four-layer model where each layer has a specific task (Fig. 3). RFC-1122 covers the communications protocol layers: link, Internet Protocol (IP), and transport and while RFC-1123 describes the application layer and its protocols. Similarly, the OSI model defines seven layers for the same purpose. Both architectures represent elements described as views in the previous section, focusing on technical implementation. The lower levels determine how to access the physical network, while IP is a fundamental feature of the Internet architecture. All Internet transport protocols use IP to carry data from the source to the destination host. The transportation layer, on the other hand, implements end-to-end communication services for applications. Two main transport protocols are the Transmission Control Protocol (TCP) and User Datagram Protocol (UDP). The Internet protocol suite does not further divide the top application layer like the OSI model, although some application layer protocols contain internal sub-layering. There are several standardised appliation layer protocols, but relevant for IIoT general use applications are Message Queuing Telemetry Transport (MQTT), Constrained Application Layer Protocol (CoAP) and Hypertext Transfer Protocol (HTTP) in combination with TCP/UDP over IP [15].

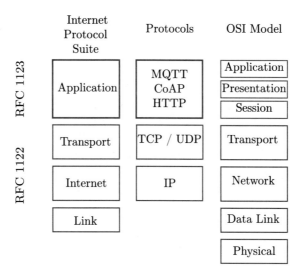

Fig. 3 Internet Protocol Suite and the OSI Model

3.1.1   TCP/UDP over IP

IP plays an essential role on the Internet. In combination with UDP and TCP, IP provides the foundation for all data transport and connectivity. UDP transport implements a connectionless and best-effort delivery service. If a transmission failure occurs, the affected messages will not be resent. The best-effort approach in UDP will send messages as quickly as possible; thus, UDP is suitable for low-latency applications and shorter message sizes. A side effect of best-effort is that messages might not arrive in the same order as sent. Connectivity or frameworks on top of UDP need to be able to deal with message fragmentation. Compared to UDP, TCP is connection-oriented and provides reliable and ordered delivery quality of service. If a transmission failure happens, the affected message will be retransmitted. Resending messages can cause delays for high priority, time-critical messages and block communication channels. For this reason, the latency in TCP varies, which can generate jitter. TCP is suitable for applications that use larger message sizes and require a delivery quality of service.

3.1.2   Application Layer Protocols

HTTP is an application layer protocol designed for the IP suite and represents the foundation for the World Wide Web.[3] Hypertext documents contain hyperlinks to other content on the Internet using Uniform Resource Locators (URLs) and the Uniform Resource Identifiers (URIs) schemes. HTTP uses a stateless request-response (client-server) model as a message pattern. In detail, the client submits an HTTP request message to the server, and the server either provides resources such as HTML files or performs actions specified by the client. The interaction ends with a response message from the server. HTTP requires a reliable transport layer protocol such as TCP.

CoAP is a specialised lightweight web transfer protocol evolved from HTTP and is defined in IETF RFC 7252. The protocol is advantageous for constrained nodes and low bandwidth networks. CoAP supports next to UDP, TCP and Web Sockets several other means of transport as the standards foremost application domain is machine-to-machine (M2M) communication. Like HTTP, CoAP is a client-server protocol where the client makes a request, and the server responds. The methods used by CoAP are the same used by HTTP. CoAP can interact with HTTP and Representational State Transfer (REST)ful Web services and is suitable for device-to-device queries.

MQTT is a connectivity transport standard for lightweight M2M messaging. The standard uses a centralised broker and supports a publish-subscribe communications pattern running on top of TCP/IP. MQTT's primary utilisation lies in device data collection without stiff latency requirements. The goal is to collect data from many devices and transport it to IT systems for further processing, e.g., monitoring applications (many-to-one data collection). MQTT is suitable for large networks and devices with restricted resources. The reader might

---

[3] The RFC 7230 family describes the HTTP protocol in detail.

consider looking into Naik, 2017 [15] for a detailed comparison between application layer protocols.

3.2   The Service-Oriented Architecture (SOA)

Another early architecture related to IoT is SOA. Despite its various definitions, SOA describes a style of software design to provide services through a communication protocol over a network [16]. In SOA, a service is an encapsulated functional unit that remotely offers its services to other units. The intended independence or loose coupling between entities would allow the exchange/update of each unit without interfering with the other components. The idea behind SOA is to provide clean abstraction levels, enabling a modular systems design and ensuring the interoperability of a heterogeneous set of devices. Further, the architecture aims to increase the reliability and scalability of the whole system [17].

From a user's point of view, SOA reduces complexity by facilitating services into standard sets (i.e. like a catalogue of available services). This allows the user to build complex services within a system by combining various function units (modular composability). Moreover, the mapping of services to different protocols and layers is abstracted for the user. SOA is, thus, an essential part to deal with the heterogeneous nature of IoT, where different entities (e.g., sensors and actuators with their possibly different protocols), need to interact over the whole network to execute a common task [18]. Because of its extensive use in Web services, the SOA style found its continuation in modern IoT Reference Model Architectures [19, 20].

In practice, there are several possibilities for how to build a service in a SOA style. The underlying pattern, however, is always more or less the same:

- The service provider exposes its endpoints and describes the available actions at each endpoint.
- The service consumer issues requests and consumes the responses.
- The service provider generates messages to handle requests.

Web services are a representation of such a pattern. While the service interfaces are described by the Web Service Description Language (WSDL), the Simple Object Access Protocol (SOAP) is used for message transfer. Another possibility to implement such patterns is the architectural paradigm REST. In REST, Web services are stateless, and the most common protocol for their implementation is HTTP [21]. Modern middlewares, such as Data Distribution Service (DDS) or OPC Unified Architecture (OPC UA), provide similar patterns. More about communication technologies relevant for the IIoT will follow in Sect. 4.4.

3.3 ISA-95: An Early Reference Architecture for Industry

In a historical context, industrial systems follow the PERA, developed by Williams at Purdue University [14]. PERA became famous in enterprise integration as it arranges different functions in a pyramid structure over several layers. PERA paved the way for standards such as the ISA-95 (IEC 62264:2013) that became an industry standard and proposed an Equipment and Functional Hierarchy Model.

3.3.1 ISA-95 Equipment Hierarchy Model

The Equipment Model of ISA-95 has a hierarchical component structure and exhibits the relations between the different layers. Each layer has areas of responsibility with a focus on manufacturing yet similarly applicable to other industrial domains.

As visualised in Fig. 4, the two layers *Enterprise* and *Site*, are responsible for high-level, strategic decisions and the underlying layers' (physical, geographical and logical grouping). The same applies to the layer *Area* that groups different manufacturing cells/units/lines and zones. Each vertical line represents a specific type of manufacturing (batch, continuous and discrete), including storage. The lower levels describe available resources, e.g., manufacturing equipment or material handling.

To represent object/information models, ISA-95 incorporates the Universal Modeling Language (UML). Information integration also extends to Supervisory Control and

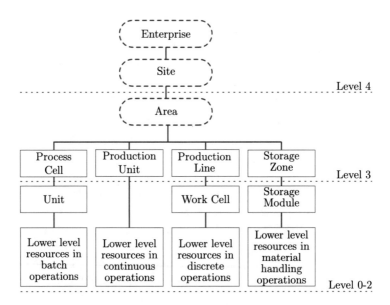

Fig. 4 ISA-95 (Role-Based Equipment Hierarchy Model)

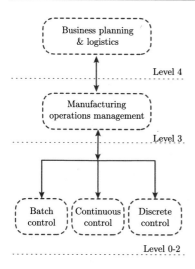

- Level 4. Plant production planning and scheduling, engineering design, purchasing customer order handling.

- Level 3. Production, maintenance, quality, testing, inventory and material flow management.

- Level 2. Process monitoring functions such as supervisory/automated control of processes.

- Level 1. Sensing and process manipulation.

- Level 0. Physical processes.

Fig. 5 ISA-95 (Functional Hierarchy Model)

Data Acquisition (SCADA) and industrial communication technologies (e.g., Modbus, PROFINET, Common Industrial Protocol, EtherCAT).

3.3.2   ISA-95 Functional Hierarchy Model

The ISA-95 functional hierarchy model (Fig. 5) consists of five levels and represents the respective organisational levels in a manufacturing company. The functions span cover business planning, logistic, manufacturing and shop-floor control.

There is a clear relation between the ISA-95 Functional Hierarchy Model and PERA. Levels 3 and 4 contain the high-level enterprise functions and the lower levels the operation and communication functionalities and represent the centralised and monolithic nature of decision-making in ISA-95.

4   Architectural Reference Models for IoT and IIoT

Since the appearance of the IoT idea in 1990, the demand for a reference architecture model has increased steadily. The main drivers were the industry and politics that saw IoT as a possibility to gain economic advantage and increase productivity. During the last decade, several groups formulated architectures to cover the diverse demands of different domains [4]. The following sections introduce the central initiatives for IIoT, the differences between them and their relations to previous architectures.

4.1   Requirements for an IIoT Architecture

In general, IoT aims towards a ubiquitous connected network of things and creates consequently a vast amount of possible implementations spreading over several domains (e.g., home and building automation, smart grids, e-health, industry). A commonly accepted ARM would need to cover an endless amount of cases. Nevertheless, the research on IoT related questions and industry practices provided applicable requirements that help to specify an ARM for industrial applications more accurately. Concretely, several conditions for IIoT appear repeatedly in the context of research and industry use-cases.

- *Collecting, aggregating and analysing data.* Requirements that are relevant for gaining insights and knowledge of processes and offering services.
- *Scalability and Fragmentation.* How to handle dynamically different sizes of systems and fragmented applications (hardware and software).
- *Flexible Device Management.* Significant when adding, re-configuring or removing devices and informing other entities in the network.
- *Security and Privacy.* Important to ensure user privacy and avoid intruders in IoT applications.
- *Safety.* Relevant when humans are part of the system.
- *Communication and Connectivity.* Requirements related to how entities communicate (e.g., unicast, multicast and anycast) and the used technologies.

In addition, a potential ARM for IIoT should include views, perspectives, and models previously mentioned to allow the description of single entities their interactions with other involved entities (e.g., humans, devices and servers) and describe and map the technologies to the specific domain use cases.

4.2   Three ARMs for IIoT

The focus of the most notable ARMs developed in recent years was on interoperability, simplifying development, and implementation. Technological advances and more comprehensive ambitions such as Industry 4.0 provided the foundation for three ARMs relevant for IIoT.

- RAMI 4.0 goes beyond the IoT, adding manufacturing and logistics details [13]. Supported by the acatech initiative,[4] Industry 4.0 started in Germany but is supported by major companies and organisations relevant in the industry.

---

[4] https://www.acatech.de/projekt/forschungsbeirat-industrie-4-0/

- Industrial Internet Reference Architecture (IIRA) has a strong industry focus (founded by AT&T, Cisco, General Electric, IBM, and Intel) [22]. Recently the OpenFog Architecture was incorporated in IIRA.[5]
- The Internet of Things-Architecture (IoT-A) provides a detailed architecture and model from the functional and information perspectives. It includes a detailed analysis of the IoT's information technology requirements [1].

The IoT-A architecture is the oldest of the three ARMs. IoT-A includes several sets of models, guidelines, views, perspectives, and design choices to build fully interoperable domain-specific IoT systems. The creators of the IoT-A ARM use a tree depicted in Fig. 6 to visualise how to apply the architecture in designing a system. Simplified, the data/information created in the tree's roots are adjusted by the trunk (IoT-A guidelines) to fit the specific requirements of a leave (domain). It needs to be pointed out that IoT-A only provides suggestions for the roots and leaves. For example, the choice of communication (e.g., WIFI, IPv6) or device technologies (e.g., sensors, actuators) suitable for a domain (e.g., logistic, transport) lies in the hand of the systems architect. In other words, the trunk only provides the IoT-A specific ARM [1] similar as described above.

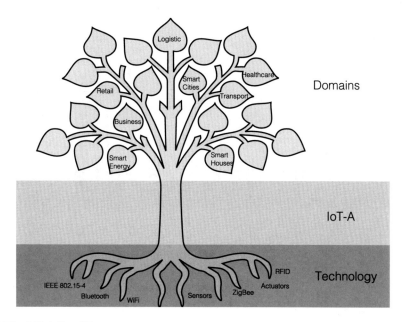

Fig. 6 The IoT-A Tree [1]

---

[5] https://www.iiconsortium.org/press-room/01-31-19.htm

In contrast, the IIRA and RAMI 4.0 are reference architectures containing concepts and methods for establishing concrete systems architectures but not for specific systems. IIRA aims for broad industry applicability by providing a framework to design IIoT systems without making specific recommendations for standards or technologies that comprise these systems. Specific to IIRA are the various perspectives (business and technical) described as viewpoints for identifying and addressing architectural matters as visualised in Fig. 7 [22].

The three-dimensional cube of the RAMI 4.0, as shown in Fig. 8, combines services and data in the sense of Industry 4.0. The hierarchy level addresses the ISA-95 levels while adding products (below level 0) and the connected world (above level 4). The second axis of the cube refers to life-cycle management covering development, production and maintenance. The Layers are the last dimension and span across the (enriched) ISA-95 levels and tackle different levels of interoperability (i.e., business, functional, information, communication, integration, and asset) borrowed from the Smart Grid Reference Architecture Model (SGAM) [13]. Both IIRA and RAMI 4.0 grew upon existing standards such as the ISA-95 and PERA [14] to establish a connection to the industrial domain and reuse established practices. Moreover, the architectures intend to bridge the physical and digital world and

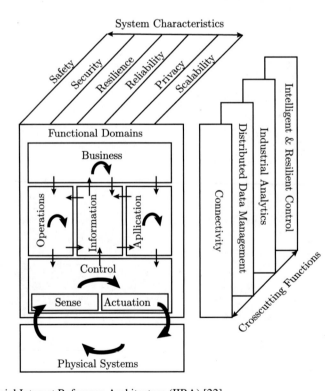

Fig. 7 Industrial Internet Reference Architecture (IIRA) [22]

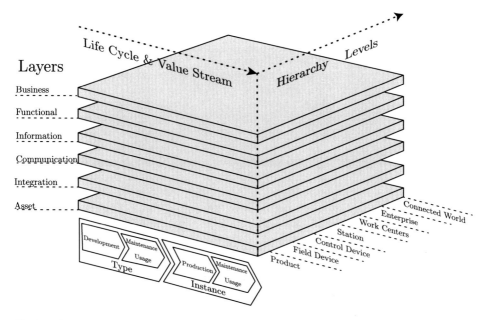

Fig. 8 Reference Architectural Model Industrie 4.0 (RAMI 4.0) [13]

enable the convergence of IT and OT.[6] Establishing IIoT, which focuses on connectivity, data analytics, optimisation, and intelligent operations, requires merging IT and OT [5].

In the shade of the three more prominent initiatives, additional standardisation is ongoing. Table 1 gives an overview of other architectures developed in recent years for IoT and M2M.

4.3 Differences between the Three Architectural Reference Models

The three ARMs are built on thoroughly prepared research in academia and industry to support building IIoT systems. Next to the structural differences, there are other important variations. The previously presented general requirements on an ARM for IoT provide the structure for the elaboration.

- The first difference is how the architectures deal with *data collection, aggregation and information interpretation* to generate valuable business knowledge. In the case of IoT-A, the architecture does not present suggestions for data semantics, rather generic aspects of data and information processing. The IIRA alternatively focuses on industrial functionality, e.g., business, operations (monitoring, optimisation, prognostics), user interfaces, application logic and rules, and information for data analytics and management of big

---

[6] More about the IT/OT gap follows in Sect. 5.

Table 1 Other architectures relevant for IIoT

Category	Initiative	Description
IoT reference architecture models	OpenFog Reference Architecture for Fog Computing [23]	OpenFog Consortium was founded by ARM, Cisco, Dell, Intel, Microsoft and Princeton University in 2015. It aimed to create and maintain the hardware, software and system elements necessary for fog computing and was merged with the Industrial Internet Consortium (IIC) in 2019.
	Standard for an Architectural Framework for the Internet of Things (IoT) [24]	The IEEE P2413 project focuses on the IoT's architectural framework, highlighting protection, security, privacy, and safety issues.
	Arrowhead Framework [25]	This initiative enables collaborative automation by open-networked embedded devices. It represents a major EU project to deliver best practices for cooperative automation.
Further activities	European Research Cluster on the Internet of Things (IREC) [26]	The IREC is involved in various IoT-related issues, including connected objects, the Web of Things, and the future of the Internet.

data in focus. RAMI 4.0 is domain-specific, it adds further aspects such as life cycle and value streams in manufacturing to the IIRA views. It is worth noting that RAMI 4.0 follows a similar matrix philosophy as IoT-A with additional dimensions in the functional layer and hierarchy levels.[7]

- Each architecture includes *scalability*. In IIRA, scalability is a characteristic of a system, while in RAMI 4.0, the functional decomposition in the layers provides the means for a scalable systems design. In IoT-A, scalability is a perspective crossing the different functional views.

---

[7] For further details, see International Electrotechnical Commission [IEC] standards 62890, 62264, and 61512.

- *Device Management, Safety and Security.* In IIRA, IoT-A and RAMI 4.0, the "things" such as sensors, actuators, and tags have a crucial role. All initiatives follow a classic bottom-up approach, mostly known from the automation industry, where data sources (tangible objects) and information needs are the primary references. The architectures provide structures and models on how to implement the interaction between devices and humans. The interactions include patterns for M2M communication standards and safety, security and privacy. In that sense, all architectures provide the required mechanisms across all layers, just with diverse perspectives and granularities.

Recapitulating the comparison, it appears that RAMI 4.0 and the IIRA reference model architecture concentrate on the industrial context and provide more detailed insights. The older IoT-A is more generally applicable to other domains outside of the manufacturing field. Another crucial aspect of IoT systems is communication and connectivity dealt with in the upcoming subsection.

4.4   Connectivity: A Crosscutting Function in IIoT

Ubiquitous communication builds the foundation of IoT and all architectural reference models. Connectivity crosscuts all functional domains and enables data exchange amongst participating components in a system, e.g., sensor data, events, alarms, status changes, commands, and configuration updates [27]. Moreover, connectivity standards assure interoperability between IIoT systems and existing networks and devices with legacy connectivity technologies. The three architectures approach communication and connectivity from different positions.

4.4.1   Communication in IoT-A, IIRA and RAMI 4.0

IoT-A extensively covers the modelling and structuring of IoT abstractly at the communication level (e.g., middleware services and cloud data management). IoT-A encompasses all types of IoT services, business processes and viewpoints from the functional, information and domain-specific use-cases. However, the server architecture (cloud) and realisation on devices are subject to specific implementation. IoT-A is not specifying in detail how connectivity should be solved but provides suggestions for suitable technologies. IIRA and RAMI 4.0, in comparison, focus on different aspects of communication.

IIRA centres on practically oriented use-cases and business processes. To provide further details on establishing communication in IIRA, the IIC published the Industrial Internet Connectivity Framework (IICF). The IIRA is a stack model and defines an open connectivity reference architecture [22], as depicted in Fig. 9 on the left. IIRA has a strong focus on the transport and framework layers and provides a connectivity model to allow device-to-device, device-to application and application-to-application interoperability.[8]

In RAMI 4.0, the communication layer describes a unified Industry 4.0 transmission mechanism, component discovery and data format. The information layer takes care of the transmitted service descriptions and the data model. Figure 9 shows the expanded communication layer on the right side. Moreover, the layer comprises both the life-cycle and the functional hierarchical level axes. The IICF and the RAMI 4.0 communication layer follow the OSI layers and integrate the same standards. Figure 9 visualises the relations between IICF, the RAMI 4.0 communication layer and the OSI model, further detailed in the following subsections.

4.4.2   OSI Layers 5–7 in RAMI 4.0 and IICF

RAMI 4.0 and the IICF (IIRA) use an adapted OSI model for describing communication. The OSI layers 5–7 in Industry 4.0 centre around production and mainly propose OPC UA [28] for connecting manufacturing equipment and process software. Other areas, such as cloud-to-cloud and enterprise-to-cloud communication, remain undefined. IICF promotes other standards such as DDS Web Services and oneM2M besides OPC UA.

OPC Unified Architecture (OPC UA) is an industrial communication architecture for platform-independent, high performance, secure, reliable, and semantic interoperability between field devices, controllers, and applications [28]. It covers the shop floor and the enterprise level. Message transportation and data modelling are the two core components of OPC UA. Considering transport, OPC UA supports multiple types of transport, such as a TCP-based binary protocol for efficient communication and data encoding, and mapping for Web Services, XML, and SOAP over HTTP. The exchange of data in OPC UA primarily follows the client/server model but the publish/subscribe communication pattern is also supported.

Contrary to its predecessor, Open Platform Communications (OPC), which only provided possibilities to represent basic runtime (process) data, OPC UA supports mechanisms to enrich data with additional semantics. Following object-oriented principles, information is modelled in nodes carrying attributes with references linking them while using type hierarchies and inheritance. These information models and their data reside on OPC UA servers and can be discovered, queried and manipulated by OPC UA clients. Based on basic modelling constructs and rules to model data, new models can represent devices of a specific level of the automation pyramid or a dedicated industry domain.

Data Distribution Service (DDS) is an open connectivity framework standard targeted explicitly at IIoT applications and maintained by the Object Management Group (OMG). The main application area of DDS is in control, application, information, operations domains. DDS connects components, gateways or applications and can be deployed in platforms

---

[8] In the following paragraphs, the differences between IIRA and RAMI 4.0 are given priority as IoT-A and IIRA do not specify communication technologies.

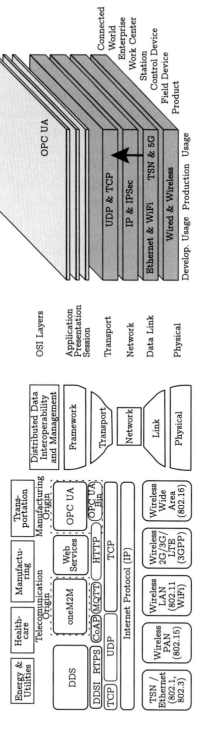

Fig. 9 OSI layers in IICF and RAMI 4.0. Adjusted from [13, 22]

ranging from low-footprint devices to the cloud. All components interact with a shared data space and not directly with each other; hence DDS is a data-centric middleware that enables flexible real-time system integration [29].

oneM2M has its roots in the telecommunications industry and aims for large scale consumer IoT applications [30]. Typical usage of oneM2M includes registration and subscription of devices and applications, service charging and accounting, management of applications and devices, and monitoring. The standard provides a common service layer between applications and connectivity transport. Within this layer, RESTful APIs allow access to services within IoT systems across different industry segments. The so-called Common Service Elements deployment is possible on hosts at the network edge or cloud to improve scalability. At its core, the oneM2M is a horizontal platform architecture that fits within a three-layer model comprising applications, middleware services and networks. The included connectivity standard enables secure and efficient communication between applications hosted on connected machines and devices, enterprise systems and mobile devices. The connectivity services connect application endpoints and native Quality of Service (QoS) and interworking mechanisms that adjust the QoS of the underlying network (e.g., mobile, wireless). The oneM2M service layer supports HTTP, CoAP, MQTT and WebSockets for connectivity and message transport [30].

4.4.3   OSI Layers 0–4 in RAMI 4.0 and IICF

All ARMs transfer the responsibility for M2M communication to the OSI stack's lower layers, considering the aspects of networking, transport, and data links. For instance, the IICF proposes several options to implement the network layer (e.g., IPv6) or for the network and transport layers the UDP and CoAP, respectively. Another option is using MQTT in combination with TCP/IP instead of HTTP. RAMI 4.0 specifies TCP/UDP/IP. In M2M communication, other efforts are employing efficient, scalable, and secure communication stacks. An overview of these efforts and a short description is provided in Table 2.

Table 2 Other M2M communication efforts

Category	Initiative	Description
Machine- to-machine (M2M) standards relevant to the IoT	European Telecommunications Standards Institute Technical Committee (ETSI TC) for M2M	The TC provides IoT communication standards.
	International Telecommunication Union Telecommunication Standardization Sector (ITU-T)	The ITU-T has coordination activities on aspects of identification systems for M2M.

On the link and physical layers, RAMI 4.0 and the IICF propose technologies (wired and wireless), e.g., time-sensitive networking (TSN), 5G. However, none deals with legacy protocols encountered in OT.

5   Combining Information Technology with Operational Technology

A specific challenge for applying IIoT and Industry 4.0 is closing the IT/OT gap and dealing with legacy systems [5]. Factories are complex technical environments built upon software and hardware agents from the domains of IT and OT. Since the 1970s, IT and OT in industrial automation have formed a hierarchical automation pyramid [14] with several layers, as shown in the left part of Fig. 10.

The lower levels of the automation pyramid, close to the factory floor, represent OT devices such as programmable logic controllers (PLCs) with industrial communication systems, e.g., EtherCAT or Profibus. OT representing complex control loops must fulfil strict real-time requirements to guarantee a timely processing of sensor values and a safe operation of actuators, valves, and electrical motors [31]. On a higher level, the control loops are being monitored employing SCADA systems and other industrial applications [32]. Applications in the third OT layer do not have to meet strict real-time requirements; instead, their main focus is high data throughput paired with computational power and internal connectivity. The two top layers contain Manufacturing Execution System (MES), plant management, business, and Enterprise-Resource-Planning (ERP). They utilise commercial off-the-shelf (COTS) IT, such as servers and desktop PCs that interconnect via standard IT communication systems (Ethernet).

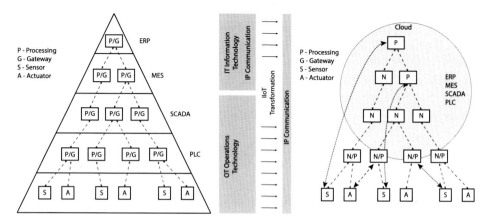

Fig. 10 Automation pyramid transformation towards a flat IIoT architecture. Adjusted from [31]

The actual gap between IT and OT reveals itself when applying the IIoT idea of seamlessly interconnected devices to a flat IP based environment as visualised in the right part of Fig. 10. As the communication systems in OT are optimised for deterministic low latency, tight synchronisation, and low jitter [31], they are difficult to bridge into a standard IT network commonly deployed as Ethernet infrastructures. Similarly, the IT communication networks and systems cannot cope with the deterministic OT requirements. Those differences result in device isolation on the factory floor from the computational resources and the connectivity available in the IT levels of a factory. They, therefore, hinder seamless vertical communication between all devices. Other challenges to close the gap are related to specific OT functionalities, safety and security and the use of legacy communication systems and interfaces [5].

5.1 Legacy Systems and Industrial Communication Technologies

Current OT communication technologies (i.e., industrial communication systems) are well developed and extensively deployed in many industries. Most originated as vendor-specific and special-purpose hardware and protocols covering parts of the connectivity transport and framework functions [27]. Well-known Ethernet-based technologies are Modbus/TCP without any specific real-time capabilities or PROFINET RT with a virtualised LAN mechanism to prioritise real-time communication. For instance, EtherCAT and PROFINET IRT are fully-fledged real-time protocols with their medium access control mechanism and extensions for hard real-time communication. Other functionalities include device management (status, update, configuration) and operational processing that maintains the system's operational integrity. Interoperability between different industrial communication variants is, in general, poor and requires additional gateway solutions. The same applies to syntactic or higher levels of interoperability.

Another challenge in this context is data accessibility between different devices. Transferred data needs to be available across all types of platforms, which requires standardised data modelling methods and a high level of network-independence to enable interoperability. IIRA and RAMI 4.0 propose middleware technologies such as OPC UA and DDS that allow defining domain-specific information models or data-centric approaches for solving this issue [28, 29].

5.2 Fog Computing

IIRA and RAMI 4.0 aim to overcome the strict separation of IT and OT, but they lack clear technological guidance. A favoured solution in this context are generic Fog or Edge Computing architectures [33], that place computes nodes between end devices and the cloud to bridge the gap between IT and OT. Fog computing assumes resourceful nodes close to the "things", such as sensors actuators on the shop floor and acting as aggregation points. Depending on their computational resources, the fog nodes can offer real-time processing of shop floor data close to where it is generated. The architecture aims for an interconnected hierarchy of nodes where higher nodes have larger pools of resources at the cost of increased latency [34]. The last step in the architecture constitutes the cloud offering a large pool of resources at low cost without any latency guarantees. Several initiatives, such as FORA,[9] develop Fog Computing Platforms (FCP) to accommodate the fog computing vision [35].

6 Applying ARM: An Industrial use Case

The following section exemplifies how the introduced ARMs could be applied in a real-world industrial use case. It is neither the intention to present a step by step guide, nor are the technical details of much importance. A legacy system needs to be consolidated on a fog computing platform and updated with other features such as firewalls, novel industrial applications, and resource management in the use-case. In other words, the system update closes the IT/OT gap. The use-case originates from a more extensive study published in [36] and is simplified for this section. For further technical details, the interested reader is referred to this paper.

6.1 Legacy System

The legacy system to be updated builds upon an industrial communication infrastructure (SCADA system [37]) as depicted in Fig. 11. In essence, the SCADA system encompasses several industrial PCs and PLCs that control the production processes and allow configuration and monitoring tasks of the connected machinery. OPC is responsible for data transfer to relevant clients and servers on the IT level. One specific server stores collected data from the shop floor in a database for further data analysis. The gateway shown in the figure allows remote access to the network for maintenance activities and data transfer. The IT network uses standard Ethernet, while EtherCAT ensures real-time communication and timely execution of control loops on the OT level. In short, the system is a perfect example of a system that follows the automation pyramid architecture.

6.2 Objectives and Suitable Reference Architecture

As indicated beforehand, the objective of the use case is to consolidate current functionality and add additional features such as firewalls, novel industrial applications, and resource management. Implementing the objective using the simplified ARM (Fig. 2) makes appar-

---

[9] FORA—Fog Computing for Robotics and Industrial Automation: http://www.fora-etn.eu/.

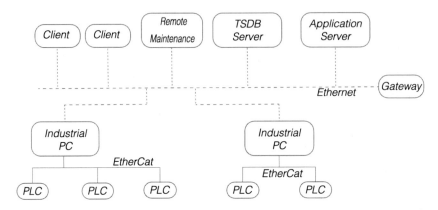

Fig. 11 The legacy system built upon a SCADA system

ent that the use case includes the lower views such as context (*industrial*), physical and operational (*components and their behaviour*). Information related to the other two views, *functional and information*, are not specified in the aim. However, it can be assumed that data flows between the components and that they represent the functional blocks of the system. Similarly not defined are the perspectives. However, an industrial system might need the be secure and have a specific performance. The systems architect gains a better overview by systematically filling the given information and requirements into the views and perspectives. Adding the domain, functional and connectivity models extends the understanding of the interdependencies between the elements of the system.

After the more general analysis of the systems requirements, the systems architect can choose the most suitable ARM. For this use case, IIRA and RAMI 4.0 are reasonable choices as both are meant to be used in an industrial setting. Both architectures provide connectivity models, in IIRA represented by the IICF. However, as the use case requires a fog computing platform and focuses on communication and consolidation of functionality, it makes sense to use a fog computing architecture as depicted in Fig. 12 on the left. The IICF is a suitable addition to a fog computing architecture, as it proposes possible communication technologies as introduced in Sect. 4.4.1. A real-use case would undoubtedly require a more in-depth analysis of all requirements to choose the right ARM.

6.3   Technical Implementation

After choosing an ARM, there are subsequent choices, for example, which technologies are suitable for the use case requirements. This use-case aimed to use a fog computing platform, which encompasses fog nodes (FNs) and a cloud. FNs are nodes with computing resources [35]. A specific feature of an FN is its capability to host virtual machines (VMs) with real-time capabilities and connect to TSN or other OT related communication proto-

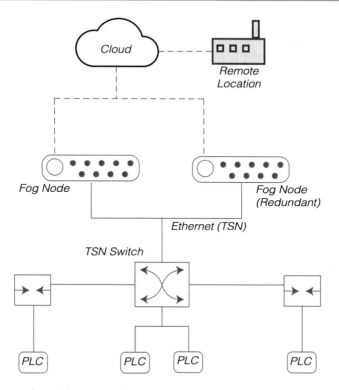

Fig. 12 Fog computing platform network topology

cols. Those capabilities enable FNs to consolidate the functionalities of the legacy system components and run additional functionalities such as firewalls, novel industrial applications, and resource management. As shown in Fig. 12, the cloud and two FNs replace the industrial PCs and dedicated servers. The choice of communication technologies was done based on the IICF and resulted in OPC UA substituting the functionality of OPC and TSN, succeeding the EtherCat network. The TSN enabled switches connect the PLCs.

As iindicated above, the FNs and the cloud host the new and old functionalities. Figure 13 presents an overview of the FN configuration and cloud components required to handle the previous and new functionalities of the system. Below is a summary of the single elements depicted in Fig. 13 based on [36].

- On the cloud level, the graphical user interfaces (GUIs) allow remote maintenance activities. In addition, the DDS middleware allows transferring and receiving data to and from remote locations for further data processing (e.g., archiving, trending, data analysis).
- The primary FN hosts one real-time VM running an OPC UA Pub/Sub entity that receives real-time data from the field devices (PLC) over the TSN network. The Pub/Sub entity transmits the data to a Time Series Data Base (TSDB) on a separate VM containing a cloud connector to forward data to the cloud for further processing.

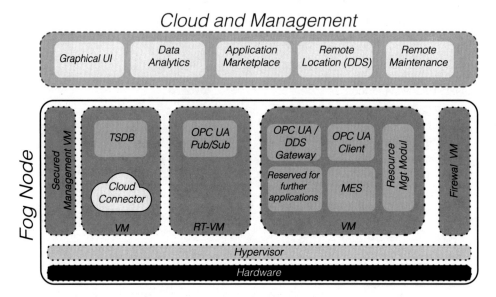

Fig. 13 Fog Node configuration and cloud components

- In another VM, the FN hosts a MES system and an OPC UA client for orchestrating the PLCs. The additional OPC UA/DDS gateway links the local OPC UA middleware with DDS in the cloud.
- Other components, as the secured management VM, protect the system from unauthorised access. The embedded firewall VM defends the running applications.
- The fog computing platform further enables automatic application deployment and dynamic migration of applications between the redundant FNs in case of over utilisation of one FN. A user can choose an application in the application marketplace, and the resource management module deploys the application to one of the FNs with enough resources.

6.4 Summary

While the use-case is simplified, the idea of using an ARM becomes apparent. After the initial analysis of the requirements and splitting them into views, perspectives and models, a system architect gets a clear overview of the interdependencies of the elements of the system. This understanding allows choosing a suitable ARM for the use case. The chosen ARM then supports the system architect building the system and making decisions such as technologies or network topologies. Specific to this use-case, choosing a fog computing architecture combined with the IICF allowed a relatively straightforward legacy system update.

7   Conclusion

The IoT idea is rich in approaches, concepts, and structures and carries the promise of enormous improvements when applied in different domains. Various initiatives have already delivered ARMs and tools to implement IoT in the industry. The presented architectures incorporate older Internet models and architectures from the automation industry and aim for closing the IT/OT gap.

By applying ARMs, the system architects can choose the aspects relevant to their system and ensure that needs of all stakeholders are covered. Architectures such as IoT-A, IIRA and RAMI 4.0 aim for the same target yet have a slightly different focus due to their origin or application domain. The chapter gives a basic understanding of those differences and their historical background that help decide which model is more appropriate for a specific system.

The presented simplified use-case exemplified the application of ARMs on a legacy system to close the IT/OT gap and add further functionalities.

Acknowledgements This work has received funding from the European Union's Horizon 2020 research and innovation programme under the Marie Sklodowska-Curie grant agreement No. 764785, FORA—Fog Computing for Robotics and Industrial Automation.

References

1.   IoT-A, "IoT-A Internet of Things Architecture. https://cordis.europa.eu/project/id/257521," VDI/VDE INNOVATION + TECHNIK GMBH, 2012.
2.   H. Lasi, P. Fettke, H. G. Kemper, T. Feld, and M. Hoffmann, "Industry 4.0," *Business and Information Systems Engineering*, vol. 6, no. 4, pp. 239–242, 2014.
3.   A. Drejer, "Integrating product and technology development," *International Journal of Technology Management*, vol. 24, no. 2–3, pp. 124–142, 2002. https://doi.org/10.1504/IJTM.2002. 003048.
4.   M. Weyrich and C. Ebert, "Reference architectures for the Internet of Things," *IEEE Software*, vol. 33, no. 1, pp. 112–116, 2016.
5.   Harp, Derek R and Gregory-Brown, Bengt, "IT / OT Convergence Bridging the Divide," *NexDefense*, p. 23, 2015.
6.   M. Amadeo, C. Campolo, A. Iera, and A. Molinaro, "Named data networking for IoT: An architectural perspective," in *2014 European Conference on Networks and Communications (EuCNC)*, 2014, pp. 1–5.
7.   E. Woods and N. Rozanski, "Using Architectural Perspectives," in *5th Working IEEE/IFIP Conference on Software Architecture (WICSA'05)*, 2005, pp. 25–35.
8.   P. Shames and T. Yamada, "Reference Architecture for Space Data Systems,"
9.   N. Rozanski and E. Woods, "Applying viewpoints and views to software architecture," *Open University White Paper*, 2005.
10.   N. Rozanski and E. Woods, *Software systems architecture: working with stakeholders using viewpoints and perspectives*. Addison-Wesley, 2012.

11. A. Serbanati, C. M. Medaglia, and U. B. Ceipidor, "Building Blocks of the Internet of Things: State of the Art and Beyond," in *Deploying RFID-Challenges, Solutions, and Open Issues*, IntechOpen, 2011.

12. C. M. MacKenzie, K. Laskey, F. McCabe, P. F. Brown, R. Metz, and B. A. Hamilton, "Reference model for service oriented architecture 1.0," *OASIS standard*, vol. 12, no. S 18, 2006.

13. DIN SPEC 91345:2016-04, "Reference Architecture Model Industrie 4.0 (RAMI4.0)," *DIN Deutsches Institut für Normung*, 2016.

14. T. J. Williams, "The Purdue enterprise reference architecture," *Computers in Industry*, vol. 24, no. 2–3, pp. 141–158, 1994.

15. N. Naik, "Choice of effective messaging protocols for IoT systems: MQTT, CoAP, AMQP and HTTP," in *2017 IEEE International Systems Engineering Symposium (ISSE)*, 2017, pp. 1–7. https://doi.org/10.1109/SysEng.2017.8088251.

16. D. Guinard, V. Trifa, S. Karnouskos, P. Spiess, and D. Savio, "Interacting with the SOA-based Internet of Things: Discovery, query, selection, and on-demand provisioning of Web Services," *IEEE Transactions on Services Computing*, vol. 3, no. 3, pp. 223–235, 2010.

17. B. Li and J. Yu, "Research and application on the smart home based on component technologies and Internet of Things," *Procedia Engineering*, vol. 15, pp. 2087–2092, 2011.

18. M. H. Valipour, B. Amirzafari, K. N. Maleki, and N. Daneshpour, "A brief survey of software architecture concepts and service oriented architecture," *Proceedings - 2009 2nd IEEE International Conference on Computer Science and Information Technology, ICCSIT 2009*, no. April 2014, pp. 34–38, 2009.

19. A. P. Castellani, N. Bui, P. Casari, M. Rossi, Z. Shelby, and M. Zorzi, "Architecture and protocols for the Internet of Things: A case study," *2010 8th IEEE International Conference on Pervasive Computing and Communications Workshops, PERCOM Workshops 2010*, no. June 2014, pp. 678–683, 2010.

20. I. Ishaq, J. Hoebeke, J. Rossey, E. D. Poorter, I. Moerman, and P. De-meester, "Enabling the Web of Things: facilitating deployment, discovery and resource access to IoT objects using embedded web services," *International Journal of Web and Grid Services*, vol. 10, no. 2/3, p. 218, 2014.

21. V. Stirbu, "Towards a RESTful Plug and Play Experience in the Web of Things," in *2008 IEEE International Conference on Semantic Computing*, IEEE, Aug. 2008, pp. 512–517.

22. "Industrial Internet Reference Architecture (IIRA)," *Industrial Internet Consortium*, 2015.

23. OpenFog Consortium, *OpenFog Consortium*, 2015. [Online]. Available: https://www.openfogconsortium.org (visited on 09/07/2018).

24. O. Logvinov, "IEEE 2413-2019 – IEEE Standard for an Architectural Framework for the Internet of Things (IoT)," IEEE SA Standards Association, Standard, 2019.

25. P. Varga, F. Blomstedt, L. L. Ferreira, J. Eliasson, M. Johansson, J. Delsing, and I. M. de Soria, "Making system of systems interoperable–The core components of the Arrowhead framework," *Journal of Network and Computer Applications*, vol. 81, pp. 85–95, 2017.

26. European Research Cluster on the Internet of Things (IREC), *European Research Cluster on the Internet of Things (IREC)*, 2021. [Online]. Available: http://www.internet-of-things-research.eu/index.html.

27. T. Sauter, S. Soucek, W. Kastner, and D. Dietrich, "The Evolution of Factory and Building Automation," *IEEE Industrial Electronics Magazine*, vol. 5, no. 3, pp. 35–48, 2011.

28. W. Mahnke, S.-H. Leitner, and M. Damm, *OPC Unified Architecture*. Springer Science & Business Media, 2009.

29. G. Pardo-Castellote, "OMG data-distribution service: Architectural overview," *Distributed Computing Systems Workshops, 2003. Proceedings. 23rd International Conference on (2003) 200–206*, pp. 200–206, 2003.

30. J. Swetina, G. Lu, P. Jacobs, F. Ennesser, and J. Song, "Toward a standardized common M2M service layer platform: Introduction to onem2m," *IEEE Wireless Communications*, vol. 21, no. 3, pp. 20–26, 2014. https://doi.org/10.1109/MWC.2014.6845045.

31. S. Schriegel, T. Kobzan, and J. Jasperneite, "Investigation on a distributed SDN control plane architecture for heterogeneous time sensitive networks," in *2018 14th IEEE International Workshop on Factory Communication Systems (WFCS)*, Jun. 2018, pp. 1–10.

32. M. Wollschlaeger, T. Sauter, and J. Jasperneite, "The Future of Industrial Communication: Automation Networks in the Era of the Internet of Things and Industry 4.0," *IEEE Industrial Electronics Magazine*, vol. 11, no. 1, pp. 17–27.

33. F. Bonomi, R. Milito, P. Natarajan, and J. Zhu, "Fog Computing: A Platform for Internet of Things and Analytics," in *Big Data and Internet of Things: A Roadmap for Smart Environments, Studies in Computational Intelligence*, vol. 546, Springer International Publishing, 2014, pp. 169–186.

34. A. Al-Fuqaha, M. Guizani, M. Mohammadi, M. Aledhari, and M. Ayyash, "Internet of Things: A Survey on Enabling Technologies, Protocols, and Applications," *IEEE Communications Surveys Tutorials*, vol. 17, no. 4, pp. 2347–2376.

35. P. Pop, B. Zarrin, M. Barzegaran, S. Schulte, S. Punnekkat, J. Ruh, and W. Steiner, "The FORA fog computing platform for industrial IoT," *Information Systems*, vol. 98, p. 101–727, 2021, issn: 0306-4379. https://doi.org/10.1016/j.is.2021.101727. [Online]. Available: https://www.sciencedirect.com/science/article/pii/S0306437921000053.

36. P. Denzler, J. Ruh, M. Kadar, C. Avasalcai, and W. Kastner, "Towards consolidating industrial use cases on a common fog computing platform," in *2020 25th IEEE International Conference on Emerging Technologies and Factory Automation (ETFA)*, vol. 1, 2020, pp. 172–179. https://doi.org/10.1109/ETFA46521.2020.9211885.

37. S. A. Boyer, *SCADA Supervisory Control and Data Acquisition*. USA: International Society of Automation, 2010.

Edge Computing: Use Cases and Research Challenges

Cosmin Avasalcai◉ and Schahram Dustdar◉

Abstract

The continuum increase of connected devices and the rise of new emergent applications with fast response times, higher privacy, and security, push the horizon to a new industrial revolution. As a result, the impact of optimizing production and product transactions manifests a fierce necessity of developing new concepts like the Industry 4.0. The combination of traditional manufacturing and industrial practices with the increasingly large-scale machine-to-machine and the Internet of Things deployments, helps manufacturers and consumers to better communication and monitoring, along with new levels of analysis, providing a truly productive future. Edge computing represents an integral part of Industry 4.0, having the purpose of enabling computational resources closer to the edge of the network. In this chapter, we describe in detail this paradigm by looking at its advantages and disadvantages as well as some representative use cases. Finally, we present and discuss the research challenges found in edge computing, mostly focusing on resource management.

Keywords

Edge Computing • Industry 4.0 • Internet of Things

C. Avasalcai
Technology, Siemens AG, Vienna, Austria
e-mail: cosmin.avasalcai@siemens.com

S. Dustdar (✉)
Distributed Systems Group, TU Wien, Vienna, Austria
e-mail: dustdar@dsg.tuwien.ac.at

B. Vogel-Heuser and M. Wimmer (eds.), *Digital Transformation*,
https://doi.org/10.1007/978-3-662-65004-2_5

125

1   Introduction

The Internet of Things (IoT) devices have seen tremendous technological improvements in their capabilities over the last couple of years. A trend that contributes to the appearance of new use cases such as smart city, smart manufacturing, and smart home, with the power of transforming our daily lives and work environment. However, with the increasing adoption of connected devices, the amount of generated data grows making the current cloud-centric solutions face challenges in meeting the stringent requirements of IoT applications; applications that require low latency as well as better privacy and security.

As a solution to these challenges, researchers proposed a new paradigm, i.e., edge computing, to extend the cloud capabilities closer to the edge of the network. Edge computing enables more computational resources (i.e., processing power, memory, and storage) in the proximity of IoT devices, allowing to process data closer to its origin [27]. Benefits that can transform and optimize the workflow of many different industries like Automotive, Healthcare, and Manufacturing.

For example, by employing edge computing to improve manufacturing efficiencies, smart manufacturing aims to merge the digital (IT) and analog (OT) worlds (building connectivity and orchestration to enable flexibility in physical processes to address a dynamic and global market). Additionally, it seeks to respond in a short time to meet changing demands and conditions in the factory, in the supply network, and to fully integrate manufacturing systems—a key focus of the Industrial 4.0. Clearly, a test of such responsiveness is the capability of customized mass production. Moreover, the manufacturing sector is being fundamentally reshaped by the unstoppable progress of the 4th Industrial Revolution, powered by the IoT and edge computing. Therefore, Industry 4.0 can be seen as the initiators of the smart manufacturing era.

Industry 4.0, like so many new technologies, is not a hot topic as many belief; it is more a rebirth of an older concept that is utilizing newly developed technology. Industry 4.0 is essentially a revision to smart manufacturing that makes use of the latest technological inventions and innovations [13]. Industry 4.0 was coined by the German government initiative and it aims to safeguard a sustainable competitive advantage for the manufacturing base, focusing on connecting the IT and OT using edge devices. The technology identifies itself as the industry that characterizes this century.

In this chapter, we focus on describing edge computing by examining its advantages and disadvantages. Additionally, to further understand the benefits and the challenges of this paradigm, we present an example use case scenario in which edge computing is adopted to build a smart factory. Finally, we examine the research challenges associated with the adoption of edge computing from the point of view of resource management, network communication, and security and privacy issues.

The remainder of the chapter is structured as follows: Sect. 2 defines the edge computing paradigm by describing its architectural features. Next, Sect. 3 describes a smart factory illustrative use case for edge computing, while in Sect. 4 we discuss the challenges that must be overcome to fully integrate edge computing in our society. Finally, Sect. 5 presents our final remarks on edge computing and its challenges.

2 Edge Computing

Edge computing facilitates the operation of computing, storage, and networking services closer to the edge of the network [15] creating a bridge, by adding an additional layer of nodes, between IoT devices (i.e., sensors and actuators) and cloud [26]. The edge layer consists of distributed edge devices with different capabilities, e.g., cloudlets [25], portable edge computers [23], and edge-cloud [11], enabling the deployment of applications in remote locations. An edge device is characterized by (i) heterogeneity, (ii) mobility, and (iii) limited computational resources. Figure 1 presents an overview of an extended cloud-centric architecture, with the addition of edge computing.

Multiple definitions of edge computing are found in the research literature, however, in our opinion, the most relevant is presented in [27]. The authors define edge computing as an enabler for technologies to process data near end-users, i.e., on downstream data for cloud services and upstream data for IoT services. Considering the growing adoption of IoT devices and the stringent requirements of emerging IoT applications, it is clear that processing data closer to the end-users is important. Due to its nature, edge computing has many characteristics [20] described below:

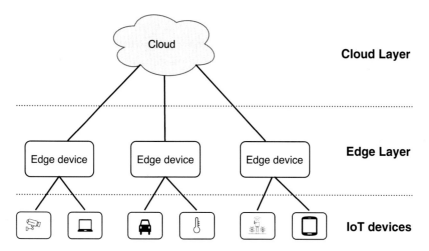

Fig. 1 Edge computing: a bridge between Cloud and IoT devices

1. **Proximity** Since computational resources are available in the proximity of end-users, both cloud and IoT devices can benefit from allocating applications to the edge nodes.
2. **Low latency** The placement of computational resources near the end-user enables deployed applications to provide responses in a short amount of time. A characteristic that aids the cloud in meeting the stringent requirements of latency-sensitive IoT applications.
3. **Increase availability** Maintaining the deployed application operational, even in the absence of a stable connection to the cloud, represents another important characteristic of edge computing. Since there is an extra layer, where deployed application may reside, the applications can work properly independent of the connections to other upper layers (i.e., cloud or any layer which contains more powerful devices).
4. **Device mobility** Edge computing supports device mobility, meaning that enables new IoT devices to connect and use the nearby edge nodes. However, when the IoT devices leave that location, the IoT device interrupts the connection with the old edge device and a new connection forms with the new closest edge node. A behavior that introduces uncertainty into the network; uncertainty that must be considered when deploying and maintaining an application at the edge.
5. **Device heterogeneity** Edge computing consists of distributed edge nodes, communication technologies ensuring the connection between these devices, and different infrastructures. Heterogeneity comes from each edge element, e.g., there may be a variety of differences between edge devices like software, hardware, and technology. Combining all these differences results in the appearance of an interoperability problem. As a result, device heterogeneity represents a challenge and must be considered when deploying an application.
6. **Context-Awareness** Since devices may enter and leave the network at any time without offering any information, context data enables the application coordinator to recover from a bad state, introduced by device mobility, by understanding the environment where the application is deployed. Context data consists of knowledge of device location, environmental characteristics (e.g., temperature sensor, video, and images, etc.), and network information.

Edge computing capabilities shine when converging with IoT and cloud creating novel techniques for IoT systems. Edge computing allows customers to develop and deploy new IoT applications on edge devices, taking advantage of lower latency and increased privacy and security, processing data at the edge without the need of transferring it to a remote location like a cloud. As a result, we consider edge computing as an extension of cloud, helping cloud to meet the stringent requirements of IoT applications, e.g., smart connected vehicles or augmented reality which requires low latency and fast response times, sensors networks that requires location awareness, and smart grids which require large-scale distributed systems.

Many devices can fill the role of an edge device, ranging from resource-constrained devices like smartphones or single-board computers to server-class data centers. Due to this diversity, in the research literature similar paradigms were proposed such as mobile edge computing (MEC) [5] and fog computing [6], that have the same objective—to move computational resources closer to the edge of the network [10]. MEC considers that an edge device is a micro data server placed at a telecommunication relay station and aid resource-constrained devices (i.e., a smartphone) to compute high computational microservices. In contrast, fog computing consists of distributed highly virtualized fog nodes, i.e., cloudlets, that shares the same characteristics as the cloud. We observe that in both cases, the underlying principle is the same, i.e., to extend the cloud and allow the deployment of IoT applications closer to the end-user.

In conclusion, edge computing fills the technological gap found in cloud-centric IoT systems by collaborating with the cloud to create a more scalable and reliable system where IoT applications may be deployed. From this collaboration, new possibilities to deploy the application appears, letting the developer choose if an application should be placed in the cloud or on the edge layer, depending on the application's requirements. An action that can be done manually by the developer or automatic using resource management techniques.

As we can observe, edge computing transformed the current cloud-centric architecture, bringing many advantages to ensure the correct deployment of emergent applications. However, adopting this paradigm in any industry is not a trivial task. Edge computing brings many challenges that makes the development, deployment, and management of IoT applications more difficult—we transition from a target architecture where an application is deployed in a central location, i.e., in the cloud, to a distributed system where a single edge device may not be capable to host an entire application.

To take advantage of the distributed available resources found in an edge computing architecture, we must change the application model and develop novel resource management techniques. For the former, the application model must change from a monolithic to a microservice-based architecture—the developer must divide the application's functionality into multiple microservices [2]. Developing the new application model represents a challenge in itself since the developer must correctly define different requirements for each microservice as well as creating the application's communication flow. For the latter, the current resource management techniques used to deploy an application to the cloud cannot ensure the correct deployment in an edge computing architecture. Deploying an application in an edge computing architecture is not a trivial task since edge devices are heterogeneous, mobile, spatially distributed, and prone to failure. As a result, we require novel resource management techniques to find a deployment strategy for our application and manage the deployed application at the edge of the network. The management of deployed applications, at runtime, is a difficult challenge by itself—we require techniques to recover an application from a faulty state when the target edge architecture has changed (i.e., edge nodes have failed or left the network).

3   Use Cases

Edge computing, in collaboration with cloud computing, can transform the current function-alities of industries by enabling the deployment of emergent IoT applications. For example, the current city infrastructure evolves into a smart city infrastructure by adopting the edge computing architecture providing advantages to people as well as companies; it can create new work environment boosting productivity with smart buildings, can improve the living conditions of a family by creating a smart home environment, and can create smart fac-tories by bringing together the IT and OT resulting in production optimizations and cost reductions. In this section, we focus on the smart factory scenario where edge computing is adopted, presenting and discussing the inherent advantages of edge computing and the research challenges that appear in this context.

3.1   Smart Manufacturing Scenario

Typically, an industrial factory is isolated from the Internet, keeping the IT and OT separated from each other by using private industrial networks. However, this approach was designed for static networks with a limited number of nodes—a scenario that is not true anymore for the modern industrial factories. These factories have seen an increase in the number of par-ticipating nodes and require support for dynamic changes and online reconfigurations [16]. Industry 4.0 and the adoption of edge computing aims at connecting the cloud computing to the manufacturing systems using the Internet; an approach that brings many advantages for industrial factories like, better connectivity, interoperability, and scalability [14].

To better understand the impact of edge computing in a smart factory environment we discuss the following IoT-based manufacturing scenario presented in [8]. The proposed edge computing architecture consists of three different layers, an edge layer, the IoT devices, and the cloud layer. However, in such an architecture the engineer must consider the impact of individual domains when developing it; four different domains are identified, i.e., the device domain, the network domain, the data domain, and the application domain (see Fig. 2).

We present each individual domain below:

1. **The device domain** A domain located either on the IoT devices layer or in the proximity of such devices like sensors, actuators, and robots. The main purpose of the device domain is to provide flexible communication, by using different standardized communication protocols; it enables the capabilities of edge nodes to collect and process data received from IoT devices, based on which it can optimize the control of the industrial machinery.
2. **The network domain** A domain that creates a bridge between the IoT devices and edge nodes. Moreover, it enables a separation between network transmission and the control of industrial machinery with the help of software-defined networking (SDN). Finally, since we are in a environment where applications require low latency, in the deployment

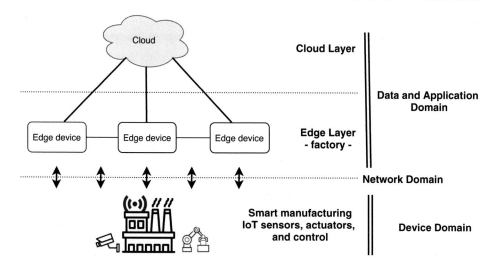

Fig. 2 Smart manufacturing edge computing architecture

of an IoT application we need to control not only the execution of a microservice on a host node but also the transmission of messages between dependent microservices; a solution for the scheduling of messages in the network is offered by the time-sensitive network (TSN).

3. **The data domain** The edge layer is where the data domain resides. In this layer, microservices can be deployed to perform different actions, e.g., take decisions based on the received sensor data or pre-process sensor data for storing in the cloud.

4. **The application domain** This domain offers the possibility to migrate applications from the cloud on the edge devices. A practice that improves the overall manufacturing capabilities of our smart factory and reduces the operational costs.

Considering the edge computing architecture presented in Fig. 2, we continue by presenting two real-world industrial use-cases [9], i.e., (i) accessing and using machine data at runtime and (ii) deploying machine software updates.

The first use case aims at accessing and using the machine data at runtime, gathered from sensors installed on the shop floor. These sensors generate several megabytes of data that provide valuable information regarding the state of the process and each machine – information that can be further used for applications, like predictive maintenance and throughput optimization. However, the current isolated industrial infrastructure lacks the resources to handle this vast amount of generated data—a problem that is solved by adopting edge computing. In this case, edge computing allows the implementation of additional functionalities, enabling data analysis techniques to be performed on-site and transferring vast amounts of data to the cloud for further processing.

The second use case aims at deploying software updates to machines found on the shop floor with a small or no downtime required. To deploy any updates to a industrial factory, many months of prior preparation are required to minimize as much as possible the required downtime. Typically, to install these updates on the physical programmable logic controllers (PLCs), an operator has to do this manually [9]. However, because the production process must stop during this procedure, the operator must perform the updates in a timely manner— an approach that is prone to introduce more errors, resulting in further disruptions. These challenges can be mitigated by using an edge device to host a virtual PLC—PLCs that can be updated automatically from the cloud.

One could say that the aforementioned use cases tackle only the first two challenges discussed briefly in Sect. 2, i.e., developing and deploying of applications. Indeed, in the second use case, we present only the deployment of application updates—applications that are already operational on the edge architecture. However, in reality, in the second use case, the management of the deployed applications is a fundamental part of the use case. As previously mentioned, deploying an application on the target edge computing architecture is not enough—we must ensure the correct functionality of the deployed application throughout its entire life cycle. Therefore, the management of applications is an integral part of any use case that targets application deployment.

In conclusion, adopting edge computing in a manufacturing scenario can bring many advantages and help converge the IT and OT domains to achieve better control and optimization of the manufacturing environment. Furthermore, we can add to a industrial factory many more industrial IoT applications that will improve the overall workflow of a factory. However, many challenges must be solved before edge computing can be adopted safely.

Considering the smart factory scenario, we have identified three research challenges, i.e., security, network communication, and resource management. First, we need to ensure that each edge device is secure enough to withstand any security attack initiated from different sources, e.g., external agents, third-party service providers, and malicious operators [9]. Besides security, ensuring that edge devices can communicate properly represents the second challenge that we must address. Finally, the last and most important challenge is to migrate the applications from cloud-centric architecture to edge computing architecture. As mentioned in Sect. 2, providing resource management in an edge computing architecture is not a trivial task since available resources are distributed among edge devices. We discuss in detail the three challenges in the next section, where we present an introduction to each as well as their research challenges.

4   Research Challenges

One of the main challenges is the integration of OT with IT for the existing factories in the world—a slow and difficult process that may represent a challenge for the next decades. In Sect. 3, we highlight that most of the existing factories use industrial networks that are isolated from the outside world. When transforming a typical industrial factory into a smart factory we must integrate the existing legacy systems into the edge computing architecture—a legacy system is part of the OT and represents the backbone of the factory, e.g., supervisory control and data acquisition (SCADA) is used to build the industrial network that connects multiple industrial PCs and PLCs [7]. More details on how to achieve the IT and OT convergence is presented in [22], while more specific use cases that target predictive maintenance and security are presented in details in [1, 28] respectively. In the use-cases presented in Sect. 3, we can observe that the integration already took place by consolidating the functionality of PLCs, industrial PCs, and gateways into an edge node. However, many challenges must be addressed before adopting edge computing.

In this section, we provide an overview of the three most important challenges that must be solved when adopting the edge computing paradigm, identified in the previous section, i.e., (i) resource management, (ii) security and privacy, and (iii) network management. In a typical Industry 4.0 edge computing architecture, there is a forth challenge, i.e., data management, that will not be discussed in this chapter but plays an equal role in the successful transformation of industrial factories.

4.1   Resource Management

With the introduction of the edge computing paradigm, the demand for novel resource management techniques, to deploy and maintain an application on the IoT network, has increased. Resource management stays at the core of this paradigm as a prime technique to efficiently utilize the available computational resources distributed near the end user. Therefore, resource management plays an important role in the successful adoption of edge computing.

Resource management does not refer only to the deployment of an application but represents an optimal combination of different groups to deploy and manage the application at the edge such that it satisfies all application's requirements [30]. There are four different groups, i.e., resource discovery, resource allocation, resource migration, and resource sharing; each with its own goal.

Resource Discovery Techniques for discovering resources in a distributed network are well covered in the research literature. However, these techniques cannot be leveraged at the edge of the network, due to device heterogeneity, modern workloads (e.g., machine learning applications), and rapidity to discover the available resources [32]. Resource discovery aims at keeping the pool of available resources updated, by discovering resources already deployed at the edge. Hence, seamless discovery and removal of nodes without introducing extra latency and communication overhead is desired. An example of a resource discovery technique deployed in a smart city scenario is presented in [21]. In this paper, the authors propose an edge-to-edge metadata replication framework that uses the kademlia as a communication protocol and elastic search to enable nodes to store and search data in a fast manner. To conclude, the resource discovery plays a more critical role in a smart city scenario, where we target volatile edge computing architectures defined by high uncertainty. In contrast, in a smart factory, there is less uncertainty and typically the participating nodes are know – rendering the need for a resource discovery technique less important.

Resource Sharing By employing resource sharing techniques, we ensure that edge nodes are willing to share resources and collaborate to achieve a common goal. A very important aspect that stays at the core of resource management; without collaboration between nodes, the nodes cannot host applications at the edge of the network. Resource sharing is using incentive techniques to ensure collaboration between nodes. As a consequence, an edge node is motivated to share as many available resources as possible, since sharing more resources will give in return more incentives. Similar to resource discovery, resource sharing is less important in the context of Industry 4.0, where edge nodes are sharing resources willingly without the need of an incentive—all edge nodes belong to the owner.

Resource Allocation After we knows the available resources shared between edge nodes and the microservices' resource requirements as well as the application's objectives, the deployment of the application can start. Resource allocation refers to the process of mapping microservices to edge nodes such that the resource requirements of each microservice and the overall application's quality of service (QoS) are satisfied. Therefore, we employ resource allocation techniques to find a deployment strategy to efficiently use the available resources found in the target edge computing architecture. From the definition of edge computing (Sect. 2), we identify two possibilities to use resource allocation, i.e., microservice offloading and microservice allocation. For the former, the aim is to help the user devices (e.g., a smartphone) to execute applications by offloading high computational microservices to nearby edge nodes; a technique that helps these resource-constrained devices to optimize the utilization of resources. The later has the purpose of extending cloud capabilities in satisfying the application's requirements by allocating a part of the microservices on edge devices where the collected data can be processed.

Resource allocations techniques play an important role in Industry 4.0 since edge computing brings together both IT and OT by acting as a bridge between the two. The consolidation of legacy systems into edge nodes makes the deployment of latency-sensitive applications more challenging. In this setting, on a single edge node, both control applications and latency-sensitive application must co-exist. In our case, a latency-sensitive application represents a non-critical application—such an application may miss a deadline without having high repercussions. Therefore, during the deployment stage, we need to make sure that any new application added to an edge node does not has an impact on the already deployed control applications. To achieve this, we must combine resource allocation techniques to devise a microservice allocation to edge nodes and a scheduling techniques to schedule both control and latency sensitive applications on the local edge node such that the control applications correct functionality is preserved. For example, we can use the decentralized resource allocation technique, presented in [3], to find a satisfiable deployment strategy for a latency-sensitive application. This technique allows each edge node to decide what microservices to host, by using a set of different decision strategies, considering the current available resources and hosted application. Therefore, we can use the scheduling technique presented in [4]—a scheduling technique that ensures the correct functionality of control application when new latency-sensitive application are added to the host node—as a decision strategy to determine if a certain microservice can be mapped on a node.

Resource Migration Considering the dynamic nature of edge computing architectures, the deployment of an application does not ensure the correct functionality over the entire application's lifespan; the network topology used at deployment time is not static, i.e., it is more likely to change multiple times before the application finishes its execution. As a result, we need to develop resource migration mechanisms to recover the application from a bad state—a state where the application's requirements are not satisfied. However, performing a migration of microservices between edge nodes requires more communication overhead. For example, migrating a 35 MB microservice to another node takes 35 s, assuming a 1 MB communication link. We can mitigate the communication overhead by devising a migration technique based on microservice replication—we simply change the communication link to another available replica of that particular microservice, instead of moving a microservice from a failed node to another. In contrast to the first approach, using replication lowers the communication overhead but required more available resources in the target edge computing architecture—each microservice replica will consume a part of the total available resource even when the replicas are not used. As a result, the developer of an edge system must consider the advantages and disadvantages of the two approaches and choose the one that fits best with the system's needs.

In conclusion, creating a resource management technique, to ensure the optimal utilization of the available computational resources shared between multiple resource-constrained devices, represents the first step in the adoption of edge computing. Moreover, it is important to mention that in the context of smart manufacturing and Industry 4.0, some of the groups are not so critical or not important at all. For example, resource sharing and resource discovery are easier to adopt since in a manufacturing context there is a controlled environment, where all nodes belong to the same administrative entity. As a result, there is no need for incentives mechanism or techniques for resource discovery, because the network does not have so many uncertainties compared to a smart city scenario.

4.2   Network Management

Edge computing consists of interconnected devices distributed at the edge of the network. Thus, network management has an important role to fill in the overall network architecture, ensuring that edge devices can communicate to transfer process data from a microservice to another dependent microservice found on a different node. By communicating, nodes can collaborate and share resources to host different deployed applications. Besides the communication path, the network management must provide extra functionality i.e., (i) assists the resource management technique in finding satisfiable deployment strategies, (ii) ensures seamless connectivity for new devices, and (iii) provide a deterministic network for control applications.

Assisting Resource Management When deploying latency-sensitive IoT applications, one of the objectives is to satisfy the end-to-end delay (e2e delay) of the application's communication flow. The e2e delay is the result of the combination of the worst-case execution time (WCET) of a microservice on a node and the communication latency between two dependent microservices and their host nodes. This is particularly important when we deploy applications on an edge computing architecture, since different microservices may reside on different nodes—a mapping that has a high impact on the overall e2e delay of the deployed applications. Therefore, we must introduce two new mechanisms to (i) monitor the latency in the network, at runtime, without introducing extra communication overhead and (ii) find the WCET of microservices. It is important to mention that both latency and WCET are dependent on the microservices' location on the target edge architecture.

Seamless Network Connectivity Considering the uncertainty found in an edge architecture, providing a seamless connectivity mechanism is imperative in a network where both stationary and mobile devices exist. These mechanisms should ensure that new devices can join the network automatically as well as leave it; a job that must be performed by the device requiring from the user a limited technical background. A characteristic that increases the adoption of new edge devices, extending the capabilities of a network in a smart city scenario. An example of a self-organizing approach to seamlessly add new nodes to the system [19]. In this setting, the edge nodes are capable to self-organize either in a hierarchical or a peer-to-peer manner such that the objectives of the system are satisfied, e.g., fault tolerance or proximity awareness. In the context of Industry 4.0, this challenge may not be an issue for the current industrial factories where there is a more or less static environment. However, in the future, this will play an important role, since in a smart manufacturing floor a robot may leave one network and join another to start performing different activities.

Deterministic Network For most IoT applications deployed in a smart city scenario, satisfying a QoS in terms of availability is enough for ensuring great service. Usually, in these cases, the time when a message arrives is not important. However, for control applications that have hard deadlines to meet, knowing the time of arrival for all messages sent between microservices is critical. These types of applications can be found in smart factories, where the objective is to control robots and other factory equipment. As a result, a new communication technology that creates a deterministic network is proposed called a time-sensitive network (TSN) [17]. In TSN there are three types of messages, i.e., time-triggered (TT), audio-video bridging (AVB), and best effort (BE), where the biggest priority is assigned to the TT messages used by control applications. It is interesting, that in a deterministic network it is possible to schedule the messages as well, offering the possibility to ensure a communication flow for the application where the delays are known.

Many other new emerging technologies have been proposed in the research literature, such as software-defined networks, network function virtualization, and network slicing to implement the network, increasing the scalability while reducing the cost [29].

4.3   Security and Privacy

Edge computing enables the digitalization of our environment and everyday lives. However, with an increase in digitalization, we expose our private life to malicious users, as such engineers must develop and enforce new privacy and security techniques on the edge network. Not only that each device has its privacy and security challenges, but these devices also inherit them from the cloud. For example, a burglar can monitor the activity of a home by accessing the edge devices placed in that home, from which it can learn the behavior of the family and when the house is empty, giving the possibility to plan a heist accordingly. As a consequence, upon the adoption of edge devices, enforcing privacy and security is a crucial task.

To identify the privacy and security issues that an edge architecture faces, an engineer can apply the confidentiality, integrity, and availability (CIA) triad model [12], which represents the most three important rules that each architecture must abide. In this model, the privacy is evaluated using the confidentiality and integrity components, while for the security the engineer can use the availability component. Furthermore, according to to [34], four challenges must be overcome when placing devices in the proximity of end-users, i.e., authentication, access control, intrusion attack, and privacy.

Security If resource management is one of the main challenges when adopting the edge computing paradigm from the perspective of hosting applications in the edge ecosystem, security represents the greatest challenge when creating the edge computing ecosystem. There are multiple reasons for this identified in [24]. First, the ecosystem inherits all security issues each component has, hence, it is not enough to protect each component of the system but to create a security mechanism that protects the ecosystem and takes into account the collaboration between all components. This task is not trivial since the mechanism must coordinate with the local security techniques deployed on the other components.

Second, edge devices are resource-constrained devices meaning that applying the security mechanisms deployed in the cloud is not feasible. New techniques must be developed that are not centralized and are autonomous because there are possible scenarios where there is no central control system and at the same time the technique must have a small impact on the edge ecosystem.

Finally, the impact of a successful security attack on an edge architecture has big repercussions for the industries where edge computing is applied since all information about the user will be vulnerable to malicious users. An edge ecosystem represents a big challenge for security since it has multiple layers of technologies that must be protected, resulting in a larger attack surface. In conclusion, if security mechanisms are not developed to protect the architecture, the benefits of adopting edge computing can be outweighed by malicious attacks. This is especially true for Industry 4.0, where ensuring the security of an industrial factory is critical—managing to exploit the security vulnerabilities found in smart manufacturing could result in increased damage to the factory system or even produce catastrophic events. For example, in 2010 there was a malware called Stuxnet that targeted industrial process systems [18]. This malware managed to use the IT resources to manipulate specific processes by monitoring certain variables and change the control commands without being noticed by the system operator. This action leads to material damages, i.e., the product does not meet the required specifications and even permanent damages to the production system.

An extensive study of the security threat model for edge computing paradigm is presented in [24]. The authors identify five different attack points, i.e., network infrastructure, edge node, core infrastructure, virtualization infrastructure, and user devices, each representing the components of an edge ecosystem. For every identified component, a set of possible attacks is discussed. For example, a man in the middle attack, in which an attacker can take control of a part of the network from where eavesdropping or traffic injection attacks can be launched (an example is presented in [33]), is a vulnerability of the network infrastructure component. In contrast, some attacks can target multiple components, e.g., a rogue component can be easily inserted in the network since edge computing by definition consists of many interconnected devices owned by different administrative entities; a rogue component attack can target both the network and core infrastructures as well as the edge node component.

Privacy Privacy represents the process of protecting the user's private data from malicious adversary while in transit [35]. With the current cloud-centric architecture privacy is most vulnerable since all data from the IoT devices is sent to the cloud for further processing. A problem that is diminished by the edge computing paradigm which enables the processing of data closer to its origin. However, the problems of privacy remains since data is still sent between devices and new privacy challenges appear inherited from the edge paradigm such as (i) privacy concerns due to user awareness of privacy rules, e.g., there are more than 80% of WiFi users still use their default password and (ii) the absence of privacy tools for resource-constrained devices [27].

Privacy is not only a concern found in smart city scenarios but also for Industry 4.0 where personal data does not refer to a person but the smart factory as a whole. Next, we present a set of privacy concerns that may apply to Industry 4.0. There are many privacy concerns regarding an edge computing ecosystem that considers the entire data path starting from collecting the data by edge devices, processing it locally on the edge or in the cloud, and disseminating it, if necessary, back to the edge. According to [31], there are five different privacy concerns, i.e., (i) data collection and identification ensuring that data is not only processed locally but it follows the user preferences and legal or administrative frameworks (e.g., the EU General Data Protection Regulation), (ii) aggregation and inference refers to the process of combining data and connecting it to users to whom it belongs, (iii) secondary use, insecurity, and exclusion concern the manipulation and storage of collected data, (iv) decisions and boundaries protect the user from invasive acts from the deployed applications, and (v) appropriation and distortion which targets the protection of user's privacy after the data are collected.

5   Conclusion

With the introduction of edge computing, a shift in the industries has appears where the overall desire is to migrate the execution of latency-sensitive applications from the cloud closer to the edge of the network. The core idea behind edge computing is to enable more computational resources in the proximity of end-users, facilitating the deployment of IoT applications that have stringent requirements like low latency and increased security and privacy; requirements that are harder to satisfy with the current cloud-centric approach. Hence, industries that adopt this paradigm can observe many benefits for their users and production as well. For example, in the automotive domain, with the development of a smart factory, the convergence of IT and OT is possible, resulting in a new architecture that integrates all devices across the entire network, from cloud to devices found on the factory floor; a change that allows for optimization and a reduction of operational costs.

Edge computing has many benefits, however, it introduces a series of challenges that must be solved upon its adoption. We group these challenges into three categories, i.e., resource management, network management, and security and privacy. Each category focus on a particular part of the edge ecosystem, i.e., resource management refers to challenges found when hosting an application on the edge, considering both the deployment and maintenance aspects, while network management looks at the connectivity challenges having the purpose of ensuring seamless integration of new devices; finally, the last category focuses on ensuring the security and privacy of the users.

In conclusion, in this chapter, we aim at introducing the edge computing paradigm and discuss its advantages and disadvantages. Furthermore, to increase the understanding of this paradigm, we present multiple use cases to exemplify the deployment of a latency-sensitive application in this ecosystem. Finally, we identify and debate the most important challenges that must be overcome before the adoption of edge computing.

Acknowledgements The research leading to these results has received funding from the European Union's Horizon 2020 research and innovation programme under the Marie Skłodowska-Curie grant agreement No. 764785, FORA–Fog Computing for Robotics and Industrial Automation.

References

1. Al-Hawawreh, M., Sitnikova, E.: Developing a security testbed for industrial internet of things. IEEE Internet of Things Journal **8**(7), 5558–5573 (2021). https://doi.org/10.1109/JIOT.2020.3032093
2. Avasalcai, C., Murturi, I., Dustdar, S.: Edge and Fog: A Survey, Use Cases, and Future Challenges, chap. 2, pp. 43–65. John Wiley & Sons, Ltd (2020). https://doi.org/10.1002/9781119551713.ch2, https://onlinelibrary.wiley.com/doi/abs/10.1002/9781119551713.ch2
3. Avasalcai, C., Tsigkanos, C., Dustdar, S.: Decentralized resource auctioning for latency-sensitive edge computing. In: IEEE International Conference on Edge Computing (EDGE). IEEE (2019)
4. Barzegaran, M., Karagiannis, V., Avasalcai, C., Pop, P., Schulte, S., Dustdar, S.: Towards extensibility-aware scheduling of industrial applications on fog nodes. In: IEEE International Conference on Edge Computing (EDGE). pp. 67–75. IEEE (2020)
5. Beck, M.T., Werner, M., Feld, S., Schimper, S.: Mobile edge computing: A taxonomy. Citeseer
6. Bonomi, F., Milito, R., Zhu, J., Addepalli, S.: Fog computing and its role in the internet of things. 1st ACM Mobile Cloud Computing Workshop pp. 13–15 (2012)
7. Boyer, S.A.: Scada: Supervisory Control And Data Acquisition. International Society of Automation, Research Triangle Park, NC, USA, 4th edn. (2009)
8. Chen, B., Wan, J., Celesti, A., Li, D., Abbas, H., Zhang, Q.: Edge computing in iot-based manufacturing. IEEE Communications Magazine **56**(9), 103–109 (2018)
9. Denzler, P., Ruh, J., Kadar, M., Avasalcai, C., Kastner, W.: Towards consolidating industrial use cases on a common fog computing platform. In: 2020 25th IEEE International Conference on Emerging Technologies and Factory Automation (ETFA). vol. 1, pp. 172–179 (2020). https://doi.org/10.1109/ETFA46521.2020.9211885
10. Dustdar, S., Avasalcai, C., Murturi, I.: Invited paper: Edge and fog computing: Vision and research challenges. In: 2019 IEEE International Conference on Service-Oriented System Engineering (SOSE). pp. 96–9609 (April 2019). https://doi.org/10.1109/SOSE.2019.00023
11. Elias, A.R., Golubovic, N., Krintz, C., Wolski, R.: Where's the bear?—automating wildlife image processing using iot and edge cloud systems. In: 2017 IEEE/ACM Second International Conference on Internet-of-Things Design and Implementation (IoTDI). pp. 247–258 (April 2017)
12. Farooq, M.U., Waseem, M., Khairi, A., Mazhar, S.: A critical analysis on the security concerns of internet of things (iot). International Journal of Computer Applications **111**(7) (2015)
13. Gilchrist, A.: Introducing industry 4.0. In: Industry 4.0, pp. 195–215. Springer (2016)
14. Givehchi, O., Landsdorf, K., Simoens, P., Colombo, A.W.: Interoperability for industrial cyber-physical systems: An approach for legacy systems. IEEE Transactions on Industrial Informatics **13**(6), 3370–3378 (2017)
15. Gusev, M., Dustdar, S.: Going back to the roots-the evolution of edge computing, an iot perspective. IEEE Internet Computing **22**(2), 5–15 (2018)
16. Gutiérrez, M., Ademaj, A., Steiner, W., Dobrin, R., Punnekkat, S.: Self-configuration of IEEE 802.1 TSN networks. In: IEEE International Conference on Emerging Technologies and Factory Automation (ETFA). pp. 1–8 (2017)

17. Gutiérrez, M., Ademaj, A., Steiner, W., Dobrin, R., Punnekkat, S.: Self-configuration of IEEE 802.1 TSN networks. IEEE International Conference on Emerging Technologies and Factory Automation, ETFA pp. 1–8 (2018)
18. for Information Security (BSI), T.F.O.: Improving it security. shorturl.at/zQZ18 (2021 (accessed March 14, 2021))
19. Karagiannis, V., Schulte, S.: Distributed algorithms based on proximity for self-organizing fog computing systems. Pervasive and Mobile Computing **71**, 101316 (2021)
20. Khan, W.Z., Ahmed, E., Hakak, S., Yaqoob, I., Ahmed, A.: Edge computing: A survey. Future Generation Computer Systems **97**, 219–235 (2019). https://doi.org/10.1016/j.future.2019.02. 050, http://www.sciencedirect.com/science/article/pii/S0167739X18319903
21. Murturi, I., Avasalcai, C., Tsigkanos, C., Dustdar, S.: Edge-to-edge resource discovery using metadata replication. In: IEEE 3rd International Conference on Fog and Edge Computing (ICFEC). pp. 1–6. IEEE (2019). https://doi.org/10.1109/CFEC.2019.8733149
22. Pop, P., Zarrin, B., Barzegaran, M., Schulte, S., Punnekkat, S., Ruh, J., Steiner, W.: The fora fog computing platform for industrial iot. Information Systems **98**, 101727 (2021). https://doi.org/10.1016/j.is.2021.101727, https://www.sciencedirect.com/science/article/pii/ S0306437921000053
23. Rausch, T., Avasalcai, C., Dustdar, S.: Portable energy-aware cluster-based edge computers. In: 2018 IEEE/ACM Symposium on Edge Computing (SEC). pp. 260–272 (Oct 2018). https://doi. org/10.1109/SEC.2018.00026
24. Roman, R., Lopez, J., Mambo, M.: Mobile edge computing, fog et al.: A survey and analysis of security threats and challenges. Future Generation Computer Systems **78**, 680–698 (2018). https://doi.org/10.1016/j.future.2016.11.009, http://www.sciencedirect.com/science/article/pii/ S0167739X16305635
25. Satyanarayanan, M., Bahl, P., Caceres, R., Davies, N.: The Case for VM-Based Cloudlets in Mobile Computing. IEEE Pervasive Computing **8**(4), 14–23 (Oct 2009). https://doi.org/10.1109/ MPRV.2009.82, http://ieeexplore.ieee.org/document/5280678/
26. Shi, W., Dustdar, S.: The promise of edge computing. Computer **49**(5), 78–81 (May 2016). https://doi.org/10.1109/MC.2016.145
27. Shi, W., Cao, J., Zhang, Q., Li, Y., Xu, L.: Edge computing: Vision and challenges. IEEE Internet of Things Journal **3**(5), 637–646 (2016)
28. Strauß, P., Schmitz, M., Wöstmann, R., Deuse, J.: Enabling of predictive maintenance in the brownfield through low-cost sensors, an iiot-architecture and machine learning. In: 2018 IEEE International Conference on Big Data (Big Data). pp. 1474–1483 (2018). https://doi.org/10. 1109/BigData.2018.8622076
29. Taleb, T., Samdanis, K., Mada, B., Flinck, H., Dutta, S., Sabella, D.: On multi-access edge computing: A survey of the emerging 5g network edge cloud architecture and orchestration. IEEE Communications Surveys Tutorials **19**(3), 1657–1681 (thirdquarter 2017). https://doi.org/ 10.1109/COMST.2017.2705720
30. Toczé, K., Nadjm-Tehrani, S.: A taxonomy for management and optimization of multiple resources in edge computing. CoRR **abs/1801.05610** (2018), http://arxiv.org/abs/1801.05610
31. Tsigkanos, C., Avasalcai, C., Dustdar, S.: Architectural considerations for privacy on the edge. IEEE Internet Computing **23**(4), 76–83 (2019)
32. Varghese, B., Wang, N., Barbhuiya, S., Kilpatrick, P., Nikolopoulos, D.S.: Challenges and opportunities in edge computing. In: 2016 IEEE International Conference on Smart Cloud (Smart-Cloud). pp. 20–26 (2016)
33. Wang, Y., Uehara, T., Sasaki, R.: Fog computing: Issues and challenges in security and forensics. In: 2015 IEEE 39th Annual Computer Software and Applications Conference. vol. 3, pp. 53–59 (July 2015). https://doi.org/10.1109/COMPSAC.2015.173

34. Yi, S., Li, C., Li, Q.: A survey of fog computing: concepts, applications and issues. In: Proceedings of the 2015 workshop on mobile big data. pp. 37–42. ACM (2015)
35. Zhou, M., Zhang, R., Xie, W., Qian, W., Zhou, A.: Security and privacy in cloud computing: A survey. In: 2010 Sixth International Conference on Semantics, Knowledge and Grids. pp. 105–112 (Nov 2010). https://doi.org/10.1109/SKG.2010.19

Dynamic Access Control in Industry 4.0 Systems

Robert Heinrich⊙, Stephan Seifermann⊙, Maximilian Walter⊙,
Sebastian Hahner⊙, Ralf Reussner⊙, Tomáš Bureš⊙, Petr Hnětynka⊙
and Jan Pacovský

Abstract

Industry 4.0 enacts ad-hoc cooperation between machines, humans, and organizations in supply and production chains. The cooperation goes beyond rigid hierarchical process structures and increases the levels of efficiency, customization, and individualisation of end-products. Efficient processing and cooperation requires exploiting various sensor

R. Heinrich (✉) · S. Seifermann · M. Walter · S. Hahner · R. Reussner
KASTEL – Institute of Information Security and Dependability, Dependability of
Software-Intensive Systems Group, Karlsruhe Institute of Technology, Karlsruhe, Germany
e-mail: robert.heinrich@kit.edu

S. Seifermann
e-mail: stephan.seifermann@kit.edu

M. Walter
e-mail: maximilian.walter@kit.edu

S. Hahner
e-mail: sebastian.hahner@kit.edu

R. Reussner
e-mail: ralf.reussner@kit.edu

T. Bureš · P. Hnětynka · J. Pacovský
Faculty of Mathematics and Physics, Department of Distributed and Dependable Systems, Charles
University, Prague, Czech Republic
e-mail: bures@d3s.mff.cuni.cz

P. Hnětynka
e-mail: hnetynka@d3s.mff.cuni.cz

J. Pacovský
e-mail: pacovsky@d3s.mff.cuni.cz

© The Author(s), under exclusive license to Springer-Verlag
GmbH, DE, part of Springer Nature 2023
B. Vogel-Heuser and M. Wimmer (eds.), *Digital Transformation*,
https://doi.org/10.1007/978-3-662-65004-2_6

and process data and sharing them across various entities including computer systems, machines, mobile devices, humans, and organisations. Access control is a common security mechanism to control data sharing between involved parties. However, access control to virtual resources is not sufficient in presence of Industry 4.0 because physical access has a considerable effect on the protection of information and systems. In addition, access control mechanisms have to become capable of handling dynamically changing situations arising from ad-hoc horizontal cooperation or changes in the environment of Industry 4.0 systems. Established access control mechanisms do not consider dynamic changes and the combination with physical access control yet. Approaches trying to address these shortcomings exist but often do not consider how to get information such as the sensitivity of exchanged information. This chapter proposes a novel approach to control physical and virtual access tied to the dynamics of custom product engineering, hence, establishing confidentiality in ad-hoc horizontal processes. The approach combines static design-time analyses to discover data properties with a dynamic runtime access control approach that evaluates policies protecting virtual and physical assets. The runtime part uses data properties derived from the static design-time analysis, as well as the environment or system status to decide about access.

1   Introduction

Industry 4.0 combines many different areas such as Digital Manufacturing, Internet of Things, or Cyber-physical Systems [10]. In contrast to classic software systems, the interaction with the real world via sensors and actors is a core feature and enabler for many expected benefits. However, this connection between virtual and physical world also imposes threats. Industrial systems are valuable and frequent subjects to security attacks [11, 25], which can lead to high monetary loss because of stopped production or lost business secrets. Controlling access to resources is one of the most fundamental security requirement [12] that makes attacks more challenging.

In presence of industrial systems, it is not sufficient to focus only on virtual resources such as software systems as part of access control. Physical access control such as limited access to workplaces is crucial as well to protect intellectual property. Employing separate solutions for virtual and physical systems to protect resources is possible. However, combining these access control mechanisms can strengthen security even more or make defining access control policies simpler. For instance, virtual access control can limit access to the user interfaces of a production system, i.e. the software system controlling machines, and physical access control limits access to the production hall in addition, which strengthens the protection. In addition, access control policies become more precise when considering dynamically changing properties of the environment instead of only focusing on subjects and objects. For instance, it is not necessary to give all employees access to the production hall but only those that are assigned to a shift taking place in the very same production hall.

Established access control models [14] usually can express policies in an appropriate way but corresponding mechanisms do not consider dynamically changing properties and provide no means to react to them except for rewriting or at least adjusting policies. Such mechanisms do not scale along with frequent changes. On the other side, dynamic approaches proposed in related work such as the ones discussed in Sect. 6 do not provide means for exploiting design-time information and therefore leave open the question where to get such information from.

To bridge this gap, we propose an approach to formulate access control policies taking into account dynamically changing properties of the environment, the accessing subject and the accessed object as well as a policy evaluation process during runtime. The policies specify situations involving possibly dynamic properties and provide a reaction to this situation, i.e. access is allowed or denied. During runtime, these policies also consider the sensitivity of data that might be accessed. To determine this sensitivity, we propose a static analysis during design-time. The design-time analysis lowers the computational effort during runtime because the results can be gathered before the execution.

We demonstrate our approach by applying it to a realistic scenario covering physical and virtual access control. One dynamic property of the scenario are the locations of factory workers, which influence other properties such as the assignments of workers to shifts. Another property is the status of the worker, i.e. if he/she has collected protective gear that is mandatory for accessing the workspaces. The approach was capable of making appropriate access control decisions for this dynamically changing scenario.

The chapter is organized as follows. Section 2 introduces a running example we will use throughout the chapter. The static data flow analysis to reason about data available in early designs are discussed in Sect. 3. Access Control in highly dynamic environments is discussed in Sect. 4. Section 5 describes application scenarios of our approach. Section 6 discusses related work. The chapter concludes with a summary and outlook in Sect. 7.

2   Running Example

Our running example is an extended Industry 4.0 scenario from [3]. This running example focuses on dynamic physical and virtual access control during a production shift in a factory. Figure 1 illustrates the floor plan of the factory. It consists of a main gate for entering the factory, one dispenser for storing safety gear, three workplaces with gates, and multiple machines within the workplaces. The users in the system are workers and shift foremen. Each worker and foremen is assigned a working shift and a working place, where they work. Additionally, a shift also contains information about possible replacement workers in case a worker is unavailable. The access to the factory is granted for each worker about 30 min before their assigned shift. Then they are also allowed to open the dispenser to retrieve their safety gear. The access to the workplace is granted only with their safety gear and based on their assigned shift. Within their workplace, they can access sensitive information from

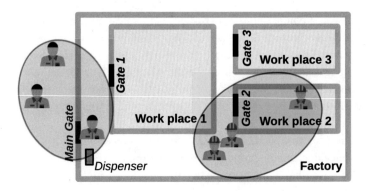

Fig. 1 Floor plan of the running example with different workers [3]

the machines such as the precise temperature. For instance, outside their workplace, they only can access aggregated values, containing less sensitive information. Additionally, the foremen can track the whole process in case of an incident, such as a late worker. In case workers are late for their shift, the foremen can access the workers' phone numbers and send them notifications. Also, in case some workers have not arrived till 15 min before the shift, the system automatically revokes their access rights and selects replacements workers. These replacements are then automatically assigned access rights for the shift and their workplace. Figure 1 shows the scenario before a shift. Two groups of workers are moving to their workplaces. The ones near the main gate are without their safety gear and the ones near workplace 2 have already gotten their safety gear.

3 Static Data Flow Analysis

The static data flow analysis provides the dynamic access control system with the sensitivity of data available at all places in the software design. In order to do that, designers have to model the data processing of the system and run a data flow analysis that yields the sensitivity of data. The used modeling and analysis approach [27, 29] is capable of determining other information about processed data as well, but we focus on sensitivity for the sake of simplicity here. In the following, we give an overview on the modeling language and the analysis results. More details on the tooling are available in Sect. 5.

We use the Palladio modeling language [23] to describe the system design. Modeling the software architecture in Palladio is done by four different roles: component developer, system architect, system deployer, and domain expert. All roles edit their corresponding models in Palladio separately. Figure 2 illustrates the relation between roles and models in Palladio. The component specification stores descriptions of software components and the internal behavior of components. The assembly model stores information about the wiring of component instances. The allocation model stores information about the deployment of component instances. The usage model contains an abstract description of the user behavior and the workload. In general, there is a large amount of modeling languages that we could have used. Especially, there are many languages focused on access control that often also support further security properties (see Sect. 6 for more details). We have chosen Palladio for several reasons: i) Palladio is a domain specific modeling language for describing software architectures. Therefore, it avoids ambiguities often introduced by generic modeling languages that would require using imprecise heuristics. ii) Palladio supports expressing classic aspects of system structures that can be found in other modeling languages as well. This includes service signatures, components providing services and calls between services. Therefore, we assume that the resulting modeling language is not overly complicated to use and that our approach can also be applied to other design-time modeling languages. iii) Palladio has been proven to support various quality properties including performance [6], reliability [7] or maintainability [24], so it is interesting to see if it can also support analyzing access control. Besides these reasons, we would like to stress that we decouple the analysis from the particular modeling language as already described. Therefore, we do not see our decision as a critical or limiting point.

As can be seen from the illustration of the metamodel in Fig. 3, there is only a limited amount of extensions shown by grey elements. These extensions overcome the limitations of the Palladio core language to express behavior in a data-oriented way. The first and most important extension is the introduction of *Data* to the modeling language. To foster integration into the existing modeling language, we do not enforce dedicated data interfaces but reuse the existing operational interfaces consisting of call and return signatures. We assume that data is always exchanged via parameters (which includes return values) in such systems. Therefore, we require that every parameter transports at least one data item. Thus, we can represent that multiple different data items are contained within one parameter without changing the service signature. This enables refining the transported data with low effort. Otherwise, either the service signature would have to be adjusted or a more complex data type modeling would be necessary. Nevertheless, a data item always has an associated data type.

Palladio already provides means for specifying system and user behavior in terms of actions. System behavior is encapsulated in a so-called service effect specification (SEFF). User behavior is encapsulated in a usage description. Roughly said, the actions contained in these behavior descriptions are either i) internal actions taking place within a service or user behavior or are ii) call actions that trigger the behavior of another service. All of these actions can have an effect on the processed data. For instance, an internal action selecting only certain parts of available information might reduce the privacy level from *highly confidential* to *publicly available*. In our running example, the data record of a worker may contain highly sensitive data such as a history of sick days. When selecting only the phone number of the worker, the data might still be sensitive but not as much as the whole data record. To specify this effect on data, we attach a *Data Processing Specification* to the actions. These specifications contain a list of *Data Operations*. These operations consume

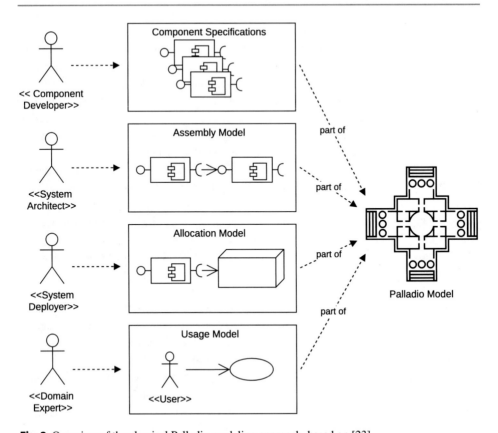

Fig. 2 Overview of the classical Palladio modeling approach, based on [23]

data, such as the data specified via parameters, and yield data. Other operations can use such yielded data to produce new data and yield this produced data. These operations build a processing chain that eventually yields data that is passed to other services via parameters or back to the caller via return values. We provide a set of predefined data operations that can also be extended by further operations depending on the particular domain.

There are five categories of data processing operations as shown in Table 1. Operations in the source category do not consume data but only yield data. These operations are the start of a data flow. This either covers creating completely new data (*CreateData*) or loading data from data stores (*LoadData/LoadAllData*). Obviously, when only yielding data, the sensitivity of data has to be specified explicitly. Operations in the sink category only consume but do not yield data. This covers storing the data (*StoreData*) but also discarding data in various forms (remaining operations). These operations are the end of a data flow. Transmission operations bridge the gap between control flow and data flow by attaching data to parameters and taking data from returns (*PerformDataTransmission*) as well as returning data (*ReturnData*). These operations do not change the sensitivity of data. Relational operations perform operations

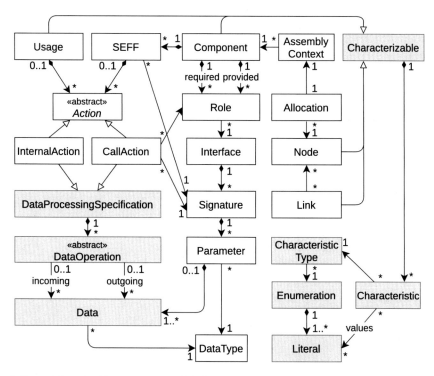

Fig. 3 Data-Driven Architecture metamodel given as UML class diagram. Grey elements are extensions to the Palladio core language given by white elements

of relational algebra on data. The effect of operations on data sensitivity depends on the particular data type and has to be specified per data type. Joining (*JoinData*) builds new data from consumed data parts or fragments. Merging data sets (*UnionData*) builds a new data set by merging two data sets. Extracting a data part or fragment (*ProjectData*) builds new data based on incoming data. Selecting data from a data set (*SelectData*) filters data sets based on a selection parameter. The last operation in a dedicated category is *TransformData*. The operation represents a generic data transformation to be specified by the designer. The operation allows the designer to explicitly state the effect of data processing on data properties such as the sensitivity. This is different to previous operations because it does not require general applicable rules to derive the sensitivity anymore but allows to define these rules in particular for one dedicated operation.

We use characteristics to describe properties, such as sensitivity, of data or system parts. As illustrated in Fig. 3, a characteristic always has a corresponding type. The type refers to a set of possible values. Because this set is a finite set of discrete values, we use enumerations to describe this value set. With respect to our running example, we specify a characteristic type *privacy level* with the values *public*, *internal use*, *sensitive* and *highly sensitive* with increasing sensitivity. A characteristic selects an arbitrary number of values, which means

Table 1 Predefined data processing operations

Category	Operation	In	Out
Source	CreateData	0	1
	LoadData	0	1
	LoadAllData	0	1
Sink	StoreData	1	0
	DeleteData	1	0
	UserReadData	1	0
	SystemDiscardData	1	0
Transmission	PerformDataTransmission	n	m
	ReturnData	1	0
Relational	JoinData	n	1
	UnionData	n	1
	ProjectData	1	1
	SelectData	1(+n)	1
Characteristics	TransformData	1	1

that these values are available in this particular characteristic. In our running example, it is only useful to have exactly one privacy level, so every characteristic only holds one value. Several structural elements within the software architecture are characterizable, which means that they can hold characteristics. In our running example, this is not necessary but in other scenarios, it might be useful to specify roles, clearance levels or criticality of system parts. The characterizable system parts are the following: A usage represents a user, so it is useful to make it characterizable to represent properties of users. A component represents a service provider, for which it is also useful to specify properties. Component instances, i.e., assembly contexts, are allocated to a node, which can also have relevant properties such as a geographical location or an owner. Nodes communicate via links and depending on the scenario it might be necessary to consider the properties of the network connection between nodes.

In our running example, the system service providing the foreman with the phone number of a worker is specified roughly as sketched in Fig. 4. Initially, the foreman sends the worker name to a service of his/her *Management Component*. The action-based specification of the service is given in the lower part of the management component. To provide the phone number, the component calls the *DB Component* and extracts it. To specify the data processing more precisely, the designer extends the actions by data specifications. The specification of the call action simply contains a *PerformDataTransmission* operation that receives a list of workers from the database component. The internal action is specified by two actions. The first action selects a particular worker based on the name of the worker from the list of workers. The second action extracts the phone number of that worker. The data specifi-

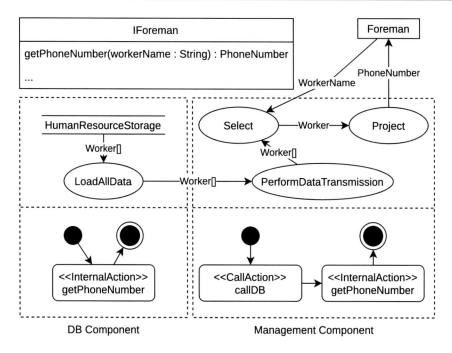

Fig. 4 Excerpt of the Palladio model representing the phone number provider used by the foreman

cation of the internal action of the database component simply loads all workers from the database *HumanResourceStorage*. All shown operations except for the *Project* operation do not change the privacy level but simply forward it. The *Project* operation effectively lowers the privacy level because it only selects the phone number that is considered less sensitive compared to all information about a worker. The effect of data processing, which is the effect on the privacy level in our running example, is defined for every triple of data processing operation type, input data type and output data type. Because the characteristic types are usually case-specific, the definition of data processing effects are also specified individually per case. However, reusing the effect in cases using the same characteristic types is possible.

The data flow analysis traces data including its properties through the system and determines which actor has access to which data. A label propagation algorithm carries out this analysis. The propagated labels are the literals of characteristic types and the propagation rules stem from the data processing effect of the used operators. Afterwards, the raw result of the analysis is transformed into a text file specifying a) the subject that accesses data b) the type of access, which is always reading in our case c) the object that is accessed d) the determined privacy level. The line for the phone number access, for example, looks like this: `foreman;read(phoneNumber);worker;internal-use`. More details on the data flow analysis are given in Sect. 5.

4   Dynamic Access Control

To specify access control in a highly dynamic environment of Industry 4.0, we base the approach on our previous work [4], where we defined the concept of autonomic *ensembles*. An ensemble is a dynamically formed group of components (they can be both software and hardware components but also, importantly, directly uncontrollable components like people, etc.) that is formed to achieve joint goal or perform a coordinated activity. Components within a single ensemble are selected dynamically at runtime, based on a set of predicates defined in the ensemble.

Using ensembles for specifying access control in a system, an ensemble definition represents a particular situation in a system and defines components taking part in the situation and access rules for these components. The ensembles are established at the moment the particular situation occurs in the system and, in the same way, it ceases its existence when the situation disappears.

Compared to usual component-based approaches, we do not require direct control of the components. Thus, as mentioned above, components can be of any kind—software and hardware but also people or even places.

In both cases of components and ensembles, our approach distinguishes between definitions (types) and instances of components and ensembles. I.e., for components we can have a single type defining a worker or a shift (in the running example) and then multiple instances of them. Similarly for ensembles, we can have an ensemble type defining access to workplace and then as many instances as we have shifts at a particular time.

The ensembles' definitions in our approach represent an access control rule. In more detail, the ensemble definition specifies a situation, identifies components taking roles in the situations and grants/denies access to particular resources.

In the rest of this section, we describe the above shown concepts on the running example and further describe the semantics.

4.1   Specification of the Running Example

For easy creation of access control specifications and rapid testing of our approach, we have created a Scala-based domain specific language (DSL). Listing 1 shows an excerpt of the specification of the running example (described in Sect. 2). Being defined as an Scala internal DSL, its usage requires at least basic knowledge of the Scala language; on the other hand, the necessary Scala concepts are more-or-less the same as in any other modern programming language.

Scala is an ideal language for rapid creation of DSL, as it has a variable and flexible syntax (compared to more traditional languages like, e.g., Java). Thanks to this feature, it is easy to create new "keywords", which from the implementation perspective are regular methods. When calling a method with a single argument, the round parentheses around arguments can be replaced with curly braces. This is ideal when a particular method takes as its argument another method (or function) and thus its call can be seen as a block of code prefixed by a keyword (in reality, it is a method call with an anonymous function passed as its argument). A particular example in our DSL is, e.g., the "keyword" `situation`, which actually is a call of a method which accepts a boolean function as its argument. These method calls, which are placed directly in the class body, are executed during the class instantiation, i.e., they are part of the default constructor. Another nice feature of Scala is that methods can be used as infix operators. E.g., the `isAfter` operator on line 45 is just a method call on the object `now`.

In our DSL, both the component and ensemble types are modeled as classes while their instances are the particular classes' instances. The particular classes have to extend the `Component` class or `Ensemble` class respectively. In the example, there are six components (lines 3–25), which represent the physical components in the running example—namely doors, head-gear dispensers, workers, work places and factories. Each of the components has its attributes (called a *component knowledge* in the terminology of ensembles), however, as we do not directly control the components, we can only observe their values. Here, for example, all of the components have the attribute `id` and most of them have the `position` attribute.

The ensembles can be hierarchically nested (an ensemble can contain other ensembles). This means that components, which are members of an ensemble, have to be also members of the parent ensemble. Thus, a top-level ensemble (in our DSL it has to extend the `RootEnsemble` class) describes a goal of the system as a whole while the sub-ensembles decompose the system into sub-goals, which are easily manageable. A component can be member of many ensembles at the same time (even directly unrelated ensembles) reflecting a common requirement that a single component can be simultaneously in many different situations.

In the example, there are four ensemble types. The top-level ensemble—`FactoryTeams` (starting at line 27)—represents all the individual `ShiftTeam` ensembles (line 116). As shifts are specified externally, the `FactoryTeams` ensemble only models the teams. The `FactoryTeams` ensemble declares a global constraint for the system that standby workers need to be assigned to the shifts where required and a standby worker cannot be shared among several shifts (lines 117–120). This is necessary, since the selection of standby workers and giving them the respective access control rights is performed by the ensemble system. The `FactoryTeam` ensemble is instantiated for every factory in the example (line 122).

The `ShiftTeam` ensemble (lines 28–114) models most of the activities in the system. It is parameterized by the Shift component instance (which defines the shift). The individual access control rules are represented by sub-ensembles. The ensemble declares 5 lists (lines 29–32) which aid identification of workers during their selection by sub-ensembles. In general, the `ShiftTeam`'s sub-ensembles can be divided into ensembles, which (i) assign permission to individual workers and which (ii) notify workers about selection for or removal from a shift. The former ones are the ensembles `AccessToFactory`, `AccessToDispenser`, `AccessToWorkplace`,   while   the   latter   ones   are

`CancellationOfWorkersThatAreLate` and `AssignmentOfStandbys`. The `NotificationAboutWorkersThatArePotentiallyLate` ensemble performs both functions. All of them have the same structure, which is as follows. First, there is a definition of the *situation*, which defines a spatial and temporal condition under which the ensemble is formed. For the `AccessToFactory`, the current time has to be in the interval of start of the shift minus 30 min and end of the shift plus 30 min. It is similar for `AccessToDispenser`, but there is a different time interval. In the case of the `AccessToWorkplace`, there is an extra condition (in addition to the situation) that the workers must have a headgear from the dispenser expressed as a selection of the shift workers with the headgear (line 52). All these three ensembles assign (lines 39, 48, and 57) the particular permissions (to enter the factory, use the dispenser, enter the workplace) to the workers selected by the conditions.

The `NotificationAboutWorkersThatArePotentiallyLate` ensemble detects workers assigned to the shift but not present in the factory (line 61) 20 min before start of the shift (line 63). For these workers, the ensemble notifies the particular foreman that they are late and allows the foreman to see the workers' phone numbers (the foreman can call them to "hurry up"—line 65) and their distance[1] from the factory (to see whether there is a chance to come in time yet—line 66).

The `CancellationOfWorkersThatAreLate` ensemble is similar to the previous one, but it detects workers, which are late even 15 min before start of the shift (line 73), and notifies them that they are canceled from the shift (line 75).

Finally, the `AssignmentOfStandbys` ensemble selects and notifies the standby workers that replace the canceled late workers. To do so, there is defined the sub-ensemble `StandbyAssignment` (line 79), which selects a suitable standby worker for a particular canceled worker (the selected standby has to have the same capabilities as the canceled one). The sub-ensemble is instantiated for each canceled worker (lines 85 and 86). The constraint (line 91) requires that a single standby worker is not used as a replacement for several workers. Within the given time interval (the situation definition at line 88), the ensemble notifies the selected workers to come (line 93) and the foreman (line 94) of the particular shift.

The ensemble `NoAccessToPersonalDataExceptForLateWorkers` expresses access control assertions (i.e., forbidden situations). Such assertions serve as safe-guards to detect potential inconsistencies in the specification. At runtime, they are used to verify that access control rules determined by ensembles. In detail, it is described and discussed in the following section.

From a technical point of view, a part of the ensembles in the example is declared as classes while other ensembles are declared as objects. This is an exploitation of another feature of the Scala language—the object is a class definition with a singleton instance. The name after the object keyword refers to the instance. The ensembles that are needed in a single instance only are therefore defined as objects.

---

[1] As the exact position of a worker outside the factory can be potentially very sensitive information, in our implementation we are using abstracted values like *close, far*, etc.

To summarize our DSL, there are two predefined classes (`Ensemble` and `Component`) for extensions. Plus, there are six new "keywords" (actually methods) that are `situation`, `constraints`, `rules`, `allow`, `deny`, and `notify`. The semantics of these keywords will be explained in the following section.

```scala
 1  class TestScenario(scenarioParams: TestScenarioSpec) extends Model with ModelGenerator {
 2    ...
 3    class Door(val id: String, val position: Position) extends Component
 4    class Dispenser(val id: String, val position: Position) extends Component
 5    class Worker(
 6      val id: String, var position: Position,
 7      val capabilities: Set[String], var hasHeadGear: Boolean
 8    ) extends Component {
 9      def isAt(room: Room) = room.positions.contains(position)
10    }
11    class WorkPlace(
12      id: String, positions: List[Position], entryDoor: Door
13    ) extends Room(id, positions, entryDoor) {
14      var factory: Factory = _
15    }
16    class Factory(
17      id: String, positions: List[Position], entryDoor: Door,
18      val dispenser: Dispenser, val workPlaces: List[WorkPlace]
19    ) extends Room(id, positions, entryDoor)
20    class Shift(
21      val id: String, val startTime: LocalDateTime,
22      val endTime: LocalDateTime, val workPlace: WorkPlace,
23      val foreman: Worker, val workers: List[Worker],
24      val standbys: List[Worker], val assignments: Map[Worker, String]
25    ) extends Component
26
27    class FactoryTeam(factory: Factory) extends RootEnsemble {
28      class ShiftTeam(shift: Shift) extends Ensemble {
29        val canceledWorkers = shift.workers.filter(wrk => wrk notified AssignmentCanceledNotification(shift))
30        val calledInStandbys = shift.standbys.filter(wrk => wrk notified CallStandbyNotification(shift))
31        val availableStandbys = shift.standbys diff calledInStandbys
32        val assignedWorkers = (shift.workers union calledInStandbys) diff canceledWorkers
33
34        object AccessToFactory extends Ensemble {
35          situation {
36            (now isAfter (shift.startTime minusMinutes 30)) &&
37              (now isBefore (shift.endTime plusMinutes 30))
38          }
39          allow(shift.foreman, "enter", shift.workPlace.factory)
40          allow(assignedWorkers, "enter", shift.workPlace.factory)
41        }
42
43        object AccessToDispenser extends Ensemble {
44          situation {
45            (now isAfter (shift.startTime minusMinutes 15)) &&
46              (now isBefore shift.endTime)
47          }
48          allow(assignedWorkers, "use", shift.workPlace.factory.dispenser)
49        }
50
51        object AccessToWorkplace extends Ensemble {
52          val workersWithHeadGear = (shift.foreman :: assignedWorkers).filter(wrk => wrk.hasHeadGear)
53          situation {
54            (now isAfter (shift.startTime minusMinutes 30)) &&
55              (now isBefore (shift.endTime plusMinutes 30))
56          }
57          allow(workersWithHeadGear, "enter", shift.workPlace)
58        }
59
60        object NotificationAboutWorkersThatArePotentiallyLate extends Ensemble {
61          val workersThatAreLate = assignedWorkers.filter(wrk => !(wrk isAt shift.workPlace.factory))
```

```
62      situation {
63         now isAfter (shift.startTime minusMinutes 20)
64      }
65      workersThatAreLate.foreach(wrk => notify(shift.foreman, WorkerPotentiallyLateNotification(shift, wrk)))
66      allow(shift.foreman, "read.personalData.phoneNo", workersThatAreLate)
67      allow(shift.foreman, "read.distanceToWorkPlace", workersThatAreLate)
68    }
69
70    object CancellationOfWorkersThatAreLate extends Ensemble {
71      val workersThatAreLate = assignedWorkers.filter(wrk => !(wrk isAt shift.workPlace.factory))
72      situation {
73         now isAfter (shift.startTime minusMinutes 15)
74      }
75      notify(workersThatAreLate, AssignmentCanceledNotification(shift))
76    }
77
78    object AssignmentOfStandbys extends Ensemble {
79      class StandbyAssignment(canceledWorker: Worker) extends Ensemble {
80         val standby = oneOf(availableStandbys)
81         constraints {
82            standby.all(_.capabilities contains shift.assignments(canceledWorker))
83         }
84      }
85      val standbyAssignments = rules(canceledWorkersWithoutStandby.map(wrk => new
              StandbyAssignment(wrk)))
86      val selectedStandbys = unionOf(standbyAssignments.map(_.standby))
87      situation {
88         (now isAfter (shift.startTime minusMinutes 15)) && (now isBefore shift.endTime)
89      }
90      constraints {
91         standbyAssignments.map(_.standby).allDisjoint
92      }
93      notify(selectedStandbys.selectedMembers, StandbyNotification(shift))
94      canceledWorkersWithoutStandby.foreach(wrk => notify(shift.foreman, WorkerReplacedNotification(shift,
              wrk)))
95    }
96
97    object NoAccessToPersonalDataExceptForLateWorkers extends Ensemble {
98      val workersPotentiallyLate =
99         if ((now isAfter (shift.startTime minusMinutes 20)) && (now isBefore shift.startTime))
100           assignedWorkers.filter(wrk => !(wrk isAt shift.workPlace.factory))
101        else Nil
102     val workers = shift.workers diff workersPotentiallyLate
103     deny(shift.foreman, "read.personalData", workers, PrivacyLevel.ANY)
104     deny(shift.foreman, "read.personalData", workersPotentiallyLate, PrivacyLevel.SENSITIVE)
105   }
106   rules(
107     // Grants
108     AccessToFactory, AccessToDispenser, AccessToWorkplace,
109     NotificationAboutWorkersThatArePotentiallyLate,
110     CancellationOfWorkersThatAreLate, AssignmentOfStandbys,
111     // Assertions
112     NoAccessToPersonalDataExceptForLateWorkers
113   )
114  }
115
116  val shiftTeams = rules(shiftsMap.values.filter(shift => shift.workPlace.factory == factory).map(shift => new
            ShiftTeam(shift)))
117  constraints {
118    shiftTeams.map(shift => shift.AssignmentOfStandbys
119      .selectedStandbys).allDisjoint
120  }
121  }
122  val factoryTeams = factoriesMap.values.map(factory => root(new FactoryTeam(factory)))
123  }
```

Listing 1 Access control specification

4.2   Semantics

Formally, the specification describes a constraint optimization problem. A solution to the problem determines which ensemble instances are to be formed and which are their members. Every ensemble can be instantiated multiple times (once for every instance of the situation it reflects). To connect the instance of an ensemble with a particular instance of the situation, ensembles are typically parameterized (using constructor arguments). Only in case an ensemble is a singleton, the parameters are missing (e.g., in case of `CancellationOfWorkersThatAreLate`). An ensemble instance is created only if the situation it reflects happens. Formally, it is created if the condition specified in the `situation` block is true.

Each ensemble identifies its members components either directly (e.g., in the `AccessToFactory` and `AccessToDispenser` ensembles) or by listing potential members and formulating constraints governing their selection (e.g., in the`AssignmentOfStandbys` and `StandbyAssignment`). The identification of potential members is done through functions `oneOf` and `unionOf`. The result is a set variable that represents a selection from the set of components given to these functions as a parameter. The valuation of these variables is constrained by the `constrains` blocks, which make it possible to express the constraints using common logical and set operators.

When ensembles are nested, the parent ensemble identifies all potential sub-ensembles. This is done with the `rules` statement. The sub-ensembles are however only instantiated if their `situation` condition holds. Ensembles are properly nested, thus a sub-ensemble instance can be only instantiated if the parent ensemble is instantiated, too. Formally, the `rules` statement creates a condition whose existence of a sub-ensemble instance implies the existence of the containing ensemble.

Nested ensembles form a tree structure. The root is identified using the `root` statement.

If an ensemble is instantiated, its `allow` statements determine the permissions. Each allow permission is formed as a triple `<subject,verb,object>`. Our access control model denies every access request, unless it is explicitly allowed. Thus, deny rules by themselves are not needed. Nevertheless, we use them in our approach as runtime assertions. We say that a specification is consistent if there are no conflicting allow and deny rules. Each deny permission is formed as a triple `<subject,verb,object>` or as a quadruple `<subject,verb,object,privacy_level>`. The form with the privacy relates to what we presented in Sect. 3. It makes it possible to restrict the deny rule only to data at a particular level of sensitivity. This is useful to protect data privacy by means of access control because it allows us to dynamically assign and restrict access to data based on changing situations in the environment.

An instantiated ensemble may also perform notifications (specified using `notify`) statement. This is needed to let a user know about permissions that have been dynamically assigned and revoked. The notify statement has the form `<target,message>`. The semantics is that every such pair is notified only once. All subsequent notifications for the same pair are ignored.

5 Application Scenarios

We have already presented the models used to describe the static and dynamic part of our approach in the previous sections, but have only roughly described how these models were used to make access control decisions. Therefore, we briefly give an overview on how both parts of the approach are combined to decide about access to information or locations in Sect. 5.1. Afterwards, we give insights into the corresponding tool support that we used to realize our running example. We will start with the tooling for static analyses in Sect. 5.2 and continue with tooling for dynamic access control enforcement in Sect. 5.3.

5.1 Overview of the Combined Approach

The overall goal of the combined approach is to decide about virtual or physical access requests during runtime. To do so, there is a decision point that receives access requests from various services or locations that the decision point answers. This decision point is the last action in the overview given in Fig. 5. To properly decide about requests, the decision point needs the set of applicable rules, i.e., the policy that specifies allowed and denied access requests.

Which rules are applicable depends on the current system state and system context. The ensembles of the runtime approach consider these dynamically changing situations to filter all defined rules for the applicable ones in a solving process. The solving process needs the ensembles, the system or context state and the privacy levels of processed data. The state information originates from a set of various probes that collect state information. The ensembles contain the access rules and criteria for when the rules shall be applied during design-time. The privacy levels are the result of a sensitivity analysis executed on the software architecture during design-time. The privacy levels serve as additional source of information for deciding on the applicability of a rule as we will show later.

As can be seen in Fig. 5, the only manual steps in the process sketched before are the definition of the software architecture and the ensembles. Both steps have to be done during design-time. During runtime, the results of these manual steps are used.

There are two roles involved in defining both artifacts: A software architect is responsible for defining the system architecture including the data processing and turning access control policies into ensembles (in cooperation with an access control expert). The software architect has to know the architectural description language Palladio and its extension for data processing as well as the internal DSL for specification of ensembles. An access control expert supports the software architect by defining the access control policies as well

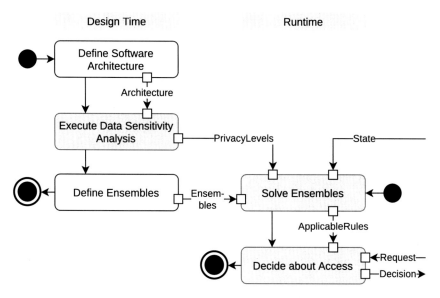

Fig. 5 Overview on combination of static design-time and dynamic runtime parts of approach given as UML activity diagram. White elements are executed manually, grey elements are executed automatically

as additional data processing operations and data properties. Access control experts also have the possibility to extend or to define new design-time analyses. In our scenario, the sensitivity analysis is predefined, so the access control expert does not have to define that particular analysis anymore. The analysis of the static design-time analysis is specified in a logic programming language that the access control expert has to know. The analysis definition uses the processing operations and data properties to determine the privacy levels. In the running example, the privacy levels are the data properties. The combination of data is an example of a data processing operation. The effect of such a combination is that the highest privacy level available on inputs is applied to the output. It is reasonable to have a dedicated access control expert defining these artifacts because a software architect might not have the required expertise to do so. The software architect can assume that the access control expert covers all important properties and processing operations that the architect can reuse afterwards.

5.2   Palladio Design-Time Tooling

The overall process to determine data sensitivity during design time is shown in Fig. 6. We start with a conventional system design created using the non-extended version of Palladio. After modeling the architecture, we expect the architect to extend it with data processing specifications that include specifying data, data exchange between services and data processing steps by using the modeling language presented in Sect. 3. The modeling language is available as metamodel[2] and can be used by a prototypical graphical editor.[3] To decouple our analysis approach from the particular modeling language, we defined a dedicated analysis model[4] that is tailored to our analysis and only includes the essential aspects of the design. We later describe this model and how it uses the effects of data processing for every operation that we discussed in Sect. 3. The mapping of the architecture to the analysis model is done automatically by a model-to-model transformation.[5] The actual analysis is carried out by a logic program. Again, the transformation into the logic program is done automatically by a model-to-text transformation.[6] Architects now can execute the data sensitivity analysis within the logic program. Internally, an interpreter takes the logic program and executes a query for the sensitivity. In the last step, the sensitivities are extracted in form of privacy levels from the analysis result. This requires mapping system elements from the logic program back to the architecture. This is done automatically by looking up the elements in the traces of the model transformations. Eventually, the privacy levels are written into a file to be used during runtime by the dynamic access control analysis.

The analysis metamodel as shown in Fig. 7 is a description of the system design tailored to data flow analyses. A system consists of system usages and operations. System usages are calls to the system by external actors. Operations are processing steps of the system that consume and produce data. The interface of such an operation is given by variables. There are variables representing parameters, returns and states. Operation calls from external actors to operations, as well as calls between operations transport data by defining assignments to target states or parameters. The operations specify their effect by assignments to returns. Assignments can assign constant values but can also use parameters or states to derive a value. A value represents one particular characteristic such as the privacy level *sensitive*. In terms of label propagation, these values are labels and the assignments are the propagation functions. For a sake of simplicity, the elements representing the value types and the terms for the assignments are not shown in Fig. 7.

The automated mapping from the software architecture to the analysis model is as follows. There is always exactly one system for each modeled software architecture. Every usage, i.e., every user, is mapped to a system usage. Every data processing operation is mapped to an operation. All data assigned to parameters in the architecture become parameter variables of the corresponding operation. All data assigned to returns in the architecture become return variables of the corresponding operation. The variable assignments of return definitions of an operation are given by the processing effects of the data processing operations in the

---

[2] https://github.com/Trust40-Project/Palladio-Addons-DataProcessing-MetaModel

[3] https://github.com/Trust40-Project/Palladio-Addons-DataProcessing-Editor

[4] https://github.com/Trust40-Project/Palladio-Addons-DataProcessing-PrologModel

[5] https://github.com/Trust40-Project/Palladio-Addons-DataProcessing-AnalysisTransformation

[6] https://github.com/Trust40-Project/Palladio-Addons-DataProcessing-PrologModel/tree/master/bundles/org.palladiosimulator.pcm.dataprocessing.prolog.transformation

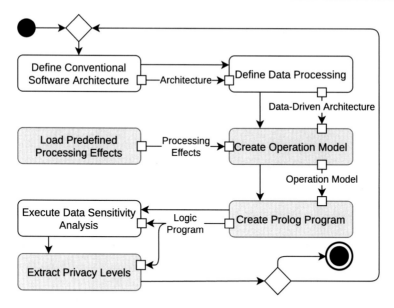

Fig. 6 Process for deriving privacy levels during design-time given as UML activity diagram. White elements are executed manually, grey elements are executed automatically

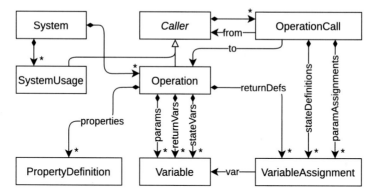

Fig. 7 Analysis metamodel tailored to data flow analyses

software architecture. For instance, the projection operation separating the phone number from a worker data record has the effect of reducing the sensitivity. The variable assignments of parameter definitions are always copy assignments that just pass variables as they are to the next operation. The next operation is determined by the data dependencies in the software architecture. If a data processing operation requires a certain data item that another operation emits, there is a data dependency. This data dependency is mapped to an operation call from the emitting operation to the receiving operation. There are variables and property definitions for every possible value of all possible characteristic types. For instance, there is a variable for every privacy level for the return value of the phone number getter. An assignment of a truth value to such a variable indicates whether the privacy level is available.

An automated model-to-text transformation carries out the mapping to the logic program. Most parts of the transformation simply map elements one by one to logical facts. The remainder of the transformations adds logical rules to perform the label propagation. Because both steps are mostly straight forward, we omit a detailed discussion of the transformation to save space. Instead, we focus on the query for the generated logic program and explain relevant parts of the logic program while explaining the query. The goal of the query shown in Listing 2 is to find the privacy level VAL of an output VAR of an operation OP with data type T. Please note that the resolution algorithm of Prolog is capable of finding all possible solutions for a query, which means that all possible bindings for the given variables are found. This is exactly what we are aiming for because we want to get all privacy levels for all outputs of all operations in the system. To do so, we define that the interesting characteristic type is the privacy level in line 1. In line 2, we find a return variable VAR with the data type T for the operation OP. Again, this line automatically considers all returns of all operations. After that line, all of these variables are bound, i.e. have a value. Lines 3–4 ensure that we have a valid call stack S, which is just a list of called operations and operation calls. The call stack has to have the current operation OP on the top, which means that the previous call has been made to this operation and there is no call after that call. In line 5, we determine the privacy level VAL for the return variable VAR for the given call stack S. The call stack is important here because operations can be called from various places, which can have an effect on the privacy level. For instance, consider a simple echo operation that just returns what it received. If the operation is called with highly sensitive data, it will return highly sensitive data. Therefore, it is crucial to know the call stack for determining the privacy level. Under the hood, the `returnValue` rule goes back the call stack until it can find a privacy level by applying the label propagations we mentioned before as part of the analysis metamodel. One out of many results is given in lines 7 to 13. It just reports all bindings to all variables mentioned in the query. In line 12, we can see that the privacy level of the phone number is now *internal-use*.

```
1   ?- ATTR = 'PrivacyLevel',
2   operationReturnValueType(OP, VAR, T),
3   S = [OP|_],
4   stackValid(S),
5   returnValue(S, VAR, ATTR, VAL).
6
7   ATTR = 'PrivacyLevel',
8   OP = 'getPhoneNumber',
9   VAR = 'RETURN',
10  T = 'PhoneNumber',
11  S = ['getPhoneNumber'|...],
12  VAL = 'internal-use';
13  ...
```

Listing 2 Queries to logic program for determining privacy levels of yielded data

For the running example, we can reuse the shown query, so no manual definitions are necessary. However, access control experts can define further queries or extend the existing query by using a set of predefined logic predicates. This certainly requires some effort in learning the predicates and also requires some limited knowledge in logic programming. However, supporting customize queries provides huge flexibility and allows using the analysis approach for other purposes and scenarios. As we have shown in another publication [28], such flexible query definitions can not only be used to collect information such as privacy levels but can also look for design issues violating confidentiality requirements in a software architecture. Alternatively, a domain specific language (DSL) can be used to define data flow constraints which requires no knowledge about the analysis process [15].

In the last step, we create a CSV file based on the returned results. The CSV file consists of the columns subject, action, object and privacy level. In the chosen example, the subject is the foreman. The action is reading the phone number. The object is a worker. The privacy level is *internal-use*. The created CSV file is passed to the tooling considered with dynamic access control during runtime that we describe in the following.

5.3   Runtime Decision Making

For runtime decision making (also described in [3]), we base our implementation on the standard MAPE-K loop [20]. The loop phases work as follows. a) Monitoring: data about the current situation are collected (in the running example, a position of all the workers, data about the shifts, etc. b) Analysis: Ensembles are instantiated according to the observed situation. Then the specification of ensembles is translated into a constraint satisfaction problem (CSP). A CSP solver is then applied to find a model for the logical theory described by the CSP and thus the ensemble instances are determined. c) Planning: Determined ensemble instances provides particular access grants. d) Execution: The access grants are applied to the system and thus system is updated to conform with the current situation observed in phase (a).

Given the fact that we employ a CSP solver to determine the ensemble instances and their member components, a question naturally arises about the scalability of such an approach as the CSP solving has inherently exponential complexity. To showcase how our approach scales, we have conducted a series of experiments of different scenario sizes and two points of time in the scenario: a) 17 min before the shift and b) 13 min before the shift.

The different scenario sizes are achieved by varying the number of workers in a shift (from 50 to 500) and the percentage of late workers that are canceled from the shift (from 5% to 20%). There are 3 shifts running at the same time. The shifts compete for the same standby workers. The number of standby workers available is equals to $no\_of\_workers\_in\_shift * percentage\_of\_late\_workers * 5$.

These two times were selected because they mark two major cases in the scenario. Case (a) corresponds to the time when multiple ensembles are instantiated, but all of these perform selection of member components directly. Thus no complex constraint optimization is needed. Case (b), on the other hand, involves the assignment of standbys—i.e., one unique suitable standby for each worker that is late. This assignment represents an optimization problem which is inherently NP hard—it is essentially a scheduling problem.

From the implementation perspective, we deal with both these two cases the same way—we translate the specification to a constraint solving problem (CSP) and use a CSP solver to figure out, which ensemble instances should exist and which components belong to which ensemble instance. Having this in place, the ensemble instances determine the allow, deny and notify rules. In case of the direct selection in case (a), the constraint problem contains essentially no alternatives to select from, thus, the CSP solving amount to traversing the constraint graph and grounding the variables to their only permissible value. Thus, even thought it is processed by the CSP solver, the computation time is essentially linear to the number of components and potential ensemble instances. In case (b), there are multiple mutually exclusive options that the CPS solver has to traverse. This is, as expected, exponential, but only in the dimension that determines variants—in our case the number of workers that have been cancelled (and thus with the number of standby workers shared between the three shifts).

We conducted the scalability experiments on Intel(R) Xeon(R) CPU E5-2660, running on 2.20GHz. In the case of #1, we performed a warmup of 1000 computations and collected 10,000 measurements for each size of the scenario. In the case of #2, we performed a warmup of 10 computations and collected 100 measurements for each scenario size. We excluded computations which exceeded 60 s. The results are shown in Fig. 8a and b.

In case (a), the time needed to resolve the ensemble instances scales linearly with the number of workers in a shift. Given the fact that 3 shifts are evaluated together, we can easily compute the access rules for 1500 workers in approx. 3.5 ms. Case (b) exhibits exponential

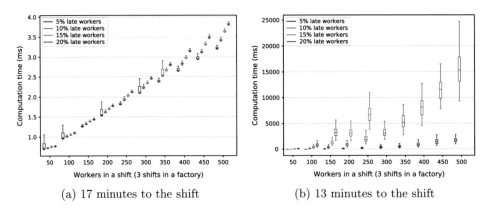

(a) 17 minutes to the shift

(b) 13 minutes to the shift

Fig. 8 Evaluation results

growth in time with the increasing percentage of late workers. Nevertheless, even for 10% of late workers, it is possible to assign access grants to 1500 persons in 15 s.

The same assignment process is performed for each factory in our running example. As each factory has its own pool of shared standby workers, the ensembles (and consequently the access rules) can be evaluated independently (as it is so in our test case) and in parallel. Thus, the number of factories does not have a significant impact on the computation time, which makes it possible to scale our use-case to arbitrarily large instances—assuming that the size of a shift remains below 500 workers and the percentage of workers that do not come to the shift is below 10%, which we believe is a realistic assumption.

5.3.1 Visualization

To allow for rapid development of access control specifications and immediate observations of results, we have developed a visualization, which provides not only a live view of the simulations, but can also be used for visualization of actual real-life data. Figure 9 shows a screenshot from the visualization of the running example where most of the workers in the factory were simulated and the highlighted worker showed an actual person requiring interaction with actual devices (access card readers, etc.).

The visualization has been developed using our tool IVIS [8], which provides a framework for easy creation of IoT related visualization.

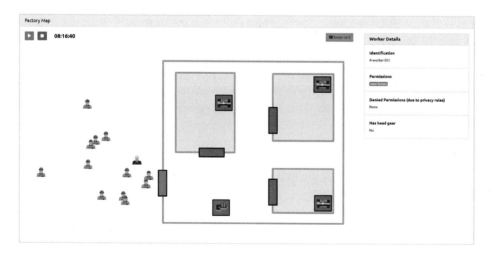

Fig. 9 Demo visualization

6   Related Work

In this section we discuss work in several areas of research related to our approach.

Access Control Types Access control is one way to protect the resources of the system. It regulates who can access protected resources and therefore increase the confidentiality of the system. One established way of access control is Role-based Access Control RBAC [13]. It groups users to roles and abstracts from the underlying user, which increases the comprehensibility of access policies. However, the sole focus on roles does not allow to easily model fine-grained access control policies that depend on the environment, like in our running example, the period for entering the factory. Organisation-based Access Control (OrBAC) [18] considers more contextual information, by explicitly modeling context information [9]. Originally, it could not support multiple organizations. However, newer approaches [32, 35] exist for handling multiple organizations, as they might exist in Industry 4.0. In contrast to our approach, their analysis does not support data flow definitions or analysis based on the software architecture. Another approach considering the context for access control is Attribute-based Access Control (ABAC) [16]. Here different Boolean values can form logical conjunctions to model access policies.

Self-Adaptive Access Control Policies Self-adaptive RBAC (saRBAC) [30] models a system by using Markov chains to determine unusual user behavior. On discovery, it automatically adapts the access policies to mitigate potential attacks. In contrast to our approach, the system's architecture is not considered, and only the user behavior is relevant for the adaption. Verma et al. [34] describe an approach, which provides a policy generation for dynamically established coalitions. This is similar to our dynamic approach. However, the coalitions consist only of people with the same shared goal. Additionally, they are not dynamically described as in our approach. Bailey et al. [5] introduce an approach for the management of dynamic policy adaption and optimization using a MAPE-K loop. While this is similar to our dynamic analysis, we use a unified modeling approach for components under direct control and those beyond direct control, such as humans.

Model-Driven Approaches for Confidentiality Nguyen et al. [22] provide an overview of different model-driven security analysis approaches. They conclude that most model-driven approaches analyze confidentiality or access control, similar to our approach. The design-time sensitivity analysis basically is a confidentiality analyses that contributes to the overall access control analysis provided by the complete approach. In the following, we only discuss closely related approaches as examples. UMLSec [17] is an UML profile extensions for annotating security properties. These properties then can be analyzed with the CARiSMA [1] tool. In contrast to our approach, they work on the control flow and have no coupling to dynamic runtime policy management. Another UML extension is SecureUML [21]. It uses internally an RBAC approach, which can be extended with OCL statements to support dynamic access control. It also provides an export function to use the policies during runtime. However, they provide no analysis based on the data flow. Also, since their export functions generate specific Java code, it cannot be reused easily in non-Java environments. The iFlow [19] approach is an information flow analysis, which also uses UML profiles. Another confidentiality analysis is SecDFD [33]. Similar to our static analysis, they analyze the system based on the data flow. However, in contrast to our approach, there is no coupling to a dynamic runtime analysis. R-PRIS [26] investigates changes during runtime, which might introduce confidentiality issues. It uses a runtime model, which is compared to a set of privacy rules. However, R-PRIS [26] only considers the location to determine the confidentiality.

Code-Based Data Flow Analysis Similar to our static analysis, which analyzes the data flow based on the architectural model, there are approaches that analyze them on source code. Joana [31] and KeY [2] are two typically representatives for this. While they do not need architectural models for the analysis, they need the full source code and are specific to one language. Especially in Industry 4.0 environments, where multiple different software systems interact, it might be complicated to apply them.

7 Conclusion and Outlook

We presented an approach to realize dynamic access control policies with a focus on Industry 4.0 systems that covers virtual and physical access control. The combination of both access control types has the potential to strengthen the overall protection of data and physical entities. The approach consists of two parts. First, there is a static design-time analysis to determine the sensitivity of exchanged data in the system. Second, a runtime policy decision point continuously evaluates the dynamic access control policies using the system state, information about actors as well as precalculated data sensitivity. We demonstrated that our approach was capable of making appropriate access control decision in presence of a dynamically changing execution context by applying the approach to a realistic scenario in Industry 4.0.

In the future, we will consider not only dynamic changes in the system or environment but also dynamic changes in the policies required to handle new situations. Adjusting the policies might be necessary because a new situation arises and there needs to be a policy change in order to keep the system and its processes at least partially running. This requires detecting upcoming situations and reacting to them. We will address this challenge as part of the FluidTrust[7] project.

Acknowledgements This work was funded by the DFG (German Research Foundation)—project number 432576552, HE8596/1-1 (FluidTrust). It was also supported by funding from the topic Engi-

---

[7] https://fluidtrust.ipd.kit.edu

neering Secure Systems of the Helmholtz Association (HGF) and by KASTEL Security Research Labs (46.23.03). The work was also partially supported by the Czech Science Foundation project 20-24814J and partially supported by Charles University institutional funding SVV 260451.

References

1. Amir Shayan Ahmadian et al. "Model-Based Privacy Analysis in Industrial Ecosystems". In: *Modelling Foundations and Applications - 13th European Conference, ECMFA@STAF 2017, Marburg, Germany, July 19–20, 2017, Proceedings.* Ed. by Anthony Anjorin and Huáscar Espinoza. Vol. 10376. Lecture Notes in Computer Science. Springer, 2017, pp. 215–231. https://doi.org/10.1007/978-3-319-61482-3_13.
2. Wolfgang Ahrendt et al., eds. *Deductive Software Verification - The KeY Book - From Theory to Practice.* Vol. 10001. Lecture Notes in Computer Science. Springer, 2016. ISBN: 978-3-319-49811-9. https://doi.org/10.1007/978-3-319-49812-6.
3. Rima Al Ali et al. "Dynamic security rules for legacy systems". In: *Proceedings of the 13th European Conference on Software Architecture, ECSA 2019, Paris, France, September 9–13, 2019, Companion Proceedings (Proceedings Volume 2)*, ed. by Laurence Duchien et al. ACM, 2019, pp. 277–284. https://doi.org/10.1145/3344948.3344974.
4. Rima Al Ali et al. "Toward autonomically composable and context-dependent access control specification through ensembles". In: *Int. J. Softw. Tools Technol. Transf.* 22.4 (2020), pp. 511–522. https://doi.org/10.1007/s10009-020-00556-1.
5. Christopher Bailey, David W. Chadwick, and Rogério de Lemos. "Selfadaptive federated authorization infrastructures". In: *J. Comput. Syst. Sci.* 80.5 (2014), pp. 935–952. https://doi.org/10.1016/j.jcss.2014.02.003.
6. Steffen Becker, Heiko Koziolek, and Ralf H. Reussner. "The Palladio component model for model-driven performance prediction". In: *J. Syst. Softw.* 82.1 (2009), pp. 3–22. https://doi.org/10.1016/j.jss.2008.03.066.
7. Franz Brosch et al. "Architecture-Based Reliability Prediction with the Palladio Component Model". In: *IEEE Trans. Software Eng.* 38.6 (2012), pp. 1319–1339. https://doi.org/10.1109/TSE.2011.94.
8. Lubomír Bulej et al. "IVIS: Highly customizable framework for visualization and processing of IoT data". In: *46th Euromicro Conference on Software Engineering and Advanced Applications, SEAA 2020, Portoroz, Slovenia, August 26–28, 2020.* IEEE, 2020, pp. 585–588. https://doi.org/10.1109/SEAA51224.2020.00095.
9. Frédéric Cuppens and Alexandre Miège. "Modelling Contexts in the Or- BAC Model". In: *19th Annual Computer Security Applications Conference (ACSAC 2003), 8–12 December 2003*, Las Vegas, NV, USA. IEEE Computer Society, 2003, pp. 416–425. https://doi.org/10.1109/CSAC.2003.1254346.
10. Mohammad Dastbaz. "Industry 4.0 (i4.0): The Hype, the Reality, and the Challenges Ahead". In: *Industry 4.0 and Engineering for a Sustainable Future.* Ed. by Mohammad Dastbaz and Peter Cochrane. Cham: Springer International Publishing, 2019, pp. 1–11. ISBN: 978-3-030-12953-8. https://doi.org/10.1007/978-3-030-12953-8_1.
11. Jyoti Deogirikar and Amarsinh Vidhate. "Security attacks in IoT: A survey". In: *2017 International Conference on I-SMAC (IoT in Social, Mobile, Analytics and Cloud) (I-SMAC).* 2017, pp. 32–37. https://doi.org/10.1109/I-SMAC.2017.8058363.

12. David F. Ferraiolo et al. *Policy Machine: Features, Architecture, and Specification*. en. Tech. rep. NIST IR 7987r1. National Institute of Standards and Technology, Oct. 2015, NIST IR 7987r1. https://doi.org/10.6028/NIST.IR.7987r1.

13. David F. Ferraiolo et al. "Proposed NIST standard for role-based access control". In: *ACM Trans. Inf. Syst. Secur.* 4.3 (2001), pp. 224–274. https://doi.org/10.1145/501978.501980.

14. Steven Furnell, ed. *Securing information and communications systems: principles, technologies, and applications*. en. Artech House computer security series. Boston: Artech House, 2008. ISBN: 978-1-59693-228-9.

15. Sebastian Hahner et al. "Modeling Data Flow Constraints for Design-Time Confidentiality Analysis". In: *18th IEEE International Conference on Software Architecture Companion, ICSA Companion 2021*, Stuttgart, Germany, March 22–26, 2021. IEEE, 2021, pp. 15–21. https://doi.org/10.1109/ICSA-C52384.2021.00009.

16. Vincent C. Hu, D. Richard Kuhn, and David F. Ferraiolo. "Attribute- Based Access Control". In: *Computer* 48.2 (2015), pp. 85–88. https://doi.org/10.1109/MC.2015.33.

17. Jan Jürjens. Secure systems development with UML. Springer, 2005. ISBN: 978-3-540-00701-2. https://doi.org/10.1007/b137706.

18. Anas Abou El Kalam et al. "Organization based access contro". In: *4th IEEE International Workshop on Policies for Distributed Systems and Networks (POLICY 2003), 4–6 June 2003, Lake Como, Italy*. IEEE Computer Society, 2003, p. 120. https://doi.org/10.1109/POLICY.2003.1206966.

19. Kuzman Katkalov et al. "Model-Driven Development of Information Flow- Secure Systems with IFlow". In: *International Conference on Social Computing (SocialCom'13)*. IEEE Computer Society, 2013, pp. 51–56. https://doi.org/10.1109/SocialCom.2013.14.

20. Jeffrey O. Kephart and David M. Chess. "The Vision of Autonomic Computing". In: *Computer* 36.1 (2003), pp. 41–50. https://doi.org/10.1109/MC.2003.1160055.

21. Torsten Lodderstedt, David A. Basin, and Jürgen Doser. "SecureUML: A UML-Based Modeling Language for Model-Driven Security". In: *UML 2002 - The Unified Modeling Language, 5th International Conference, Dresden, Germany, September 30–October 4, 2002, Proceedings*. Ed. by Jean-Marc Jézéquel, Heinrich Hußmann, and Stephen Cook. Vol. 2460. Lecture Notes in Computer Science. Springer, 2002, pp. 426–441. https://doi.org/10.1007/3-540-45800-X_33.

22. Phu Hong Nguyen et al. "An extensive systematic review on the Model- Driven Development of secure systems". In: *Inf. Softw. Technol.* 68 (2015), pp. 62–81. https://doi.org/10.1016/j.infsof.2015.08.006.26.

23. Ralf H. Reussner et al. *Modeling and Simulating Software Architectures: The Palladio Approach*. The MIT Press, 2016. ISBN: 026203476X.

24. Kiana Rostami et al. "Architecture-based Assessment and Planning of Change Requests". In: *Proceedings of the 11th International ACM SIG-SOFT Conference on Quality of Software Architectures, QoSA'15 (part of CompArch 2015)*. ACM, 2015, pp. 21–30. https://doi.org/10.1145/2737182.2737198.

25. Ahmad-Reza Sadeghi, Christian Wachsmann, and Michael Waidner. "Security and privacy challenges in industrial internet of things". In: *Proceedings of the 52nd Annual Design Automation Conference, San Fran-cisco, CA, USA, June 7–11, 2015*. ACM, 2015, 54:1–54:6. https://doi.org/10.1145/2744769.2747942.

26. Eric Schmieders, Andreas Metzger, and Klaus Pohl. "Runtime Model- Based Privacy Checks of Big Data Cloud Services". In: *Service-Oriented Computing - 13th International Conference, ICSOC 2015, Goa, India, November 16–19, 2015, Proceedings*. Ed. by Alistair Barros et al. Vol. 9435. Lecture Notes in Computer Science. Springer, 2015, pp. 71–86. https://doi.org/10.1007/978-3-662-48616-0_5.

27. Stephan Seifermann, Robert Heinrich, and Ralf H. Reussner. "Data-Driven Software Architecture for Analyzing Confidentiality". In: *IEEE International Conference on Software Architecture, ICSA 2019, Hamburg, Germany, March 25–29, 2019*. IEEE, 2019, pp. 1–10. https://doi.org/10.1109/ICSA.2019.00009.

28. Stephan Seifermann et al. "A Unified Model to Detect Information Flow and Access Control Violations in Software Architectures". In: *Proceedings of the 18th International Conference on Security and Cryptography, SE-CRYPT 2021, July 6–8, 2021*. Ed. by Sabrina De Capitani di Vimercati and Pierangela Samarati. SCITEPRESS, 2021, pp. 26–37. https://doi.org/10.5220/0010515300260037.

29. Stephan Seifermann et al. "Detecting Violations of Access Control and Information Flow Policies in Data Flow Diagrams". In: *The Journal of Systems and Software* (2022). accepted, to appear.

30. Carlos Eduardo da Silva et al. "Self-Adaptive Role-Based Access Control for Business Processes". In: *12th IEEE/ACM International Symposium on Software Engineering for Adaptive and Self-Managing Systems, SEAMS@ICSE 2017, Buenos Aires, Argentina, May 22–23, 2017*. IEEE Computer Society, 2017, pp. 193–203. https://doi.org/10.1109/SEAMS.2017.13.

31. Gregor Snelting et al. "Checking probabilistic noninterference using JOANA". In: *it Inf. Technol.* 56.6 (2014), pp. 280–287. https://doi.org/10.1515/itit-2014-1051.

32. Khalifa Toumi, César Andrés, and Ana R. Cavalli. "Trust-orBAC: A Trust Access Control Model in Multi-Organization Environments". Ed. by Venkat N. Venkatakrishnan and Diganta Goswami. Vol. 7671. Lecture Notes in Computer Science. Springer, 2012, pp. 89–103. https://doi.org/10.1007/978-3-642-35130-3_7.

33. Katja Tuma, Riccardo Scandariato, and Musard Balliu. "Flaws in Flows: Unveiling Design Flaws via Information Flow Analysis". In: *IEEE International Conference on Software Architecture, ICSA 2019, Hamburg, Germany, March 25–29, 2019. IEEE, 2019*, pp. 191–200. https://doi.org/10.1109/ICSA.2019.00028.

34. Dinesh C. Verma et al. "Generative policy model for autonomic management". In: *2017 IEEE SmartWorld, Ubiquitous Intelligence & Computing, Advanced & Trusted Computed, Scalable Computing & Communications, Cloud & Big Data Computing, Internet of People and Smart City Innovation, SmartWorld/SCALCOM/UIC/ATC/CBDCom/IOP/SCI 2017, San Francisco, CA, USA, August 4–8, 2017*. IEEE, 2017, pp. 1–6. https://doi.org/10.1109/UIC-ATC.2017.8397410.

35. Zeineb Ben Yahya, Farah Barika Ktata, and Khaled Ghédira. "Multiorganizational Access Control Model Based on Mobile Agents for Cloud Computing". In: *2016 IEEE/WIC/ACM International Conference on Web Intelligence, WI 2016, Omaha, NE, USA, October 13–16, 2016*. IEEE Computer Society, 2016, pp. 656–659. https://doi.org/10.1109/WI.2016.0116.

Challenges in OT Security and Their Impacts on Safety-Related Cyber-Physical Production Systems

Siegfried Hollerer⬤, Bernhard Brenner, Pushparaj Rajaram Bhosale, Clara Fischer, Ali Mohammad Hosseini, Sofia Maragkou, Maximilian Papa, Sebastian Schlund, Thilo Sauter and Wolfgang Kastner⬤

Abstract

In Cyber-Physical Production Systems (CPPS), integrity and availability of hardware and software components are necessary to ensure product quality and the safety of employees and customers, while the confidentiality of engineering artifacts and product details must be kept to hide company secrets. At the same time, an increasing number of Internet connected control systems causes the presence of new attack vectors. As a result, unauthorized hardware/software modifications of CPPS components through cyber attacks

S. Hollerer · P. R. Bhosale · W. Kastner (✉)
Computer Engineering/Automation Systems, TU Wien, Vienna, Austria
e-mail: k@auto.tuwien.ac.at

S. Hollerer
e-mail: siegfried.hollerer@tuwien.ac.at

P. R. Bhosale
e-mail: pushparaj.bhosale@tuwien.ac.at

B. Brenner
Telecommunications/Networks, TU Wien, Vienna, Austria
e-mail: bernhard.brenner@tuwien.ac.at

C. Fischer · M. Papa · S. Schlund
Managementwissenschaften/Mensch-Maschine-Interaktion, TU Wien, Vienna, Austria
e-mail: clara.fischer@tuwien.ac.at

M. Papa
e-mail: maximilian.papa@tuwien.ac.at

S. Schlund
e-mail: sebastian.schlund@tuwien.ac.at

B. Vogel-Heuser and M. Wimmer (eds.), *Digital Transformation*,
https://doi.org/10.1007/978-3-662-65004-2_7

become more prevalent. This development raises the demand for proper protection measures significantly, not only to ensure product quality and security but also the safety of people working with the machinery. In this chapter, we describe vulnerable assets of Operational Technology (OT) and identify information security requirements for these assets. Based on this assessment, possible attack vectors and threat models are discussed. Furthermore, measures against the mentioned threats and security relevant differences between OT and Information Technology (IT) systems are outlined. To manage a CPPS and its related threats, risk management will be addressed in more detail. Although safety and security should no longer be viewed as isolated, there are several challenges of integrating safety and security, which can lead to struggles and trade-offs. For this reason, the "Safety and Security Lab in Industry" currently investigates different aspects of future integrated solutions covering both safety and security. Challenges of such integrated solutions are outlined at the end of the chapter.

1   Introduction

Today's intelligent production environments employ a plethora of communication networks, ranging from classical fieldbus systems and (industrial) Ethernet solutions to traditional office networks and, more recently, an increasing number of wireless technologies [1]. The implementation of this communication structure, however, requires expertise and standards to address challenges with security impacts on vulnerable human-machine safety [2]. In order to deal with this topic in more detail, we first briefly define the current network architectures and possible communication actors.

1.1   Production Network: The Automation Pyramid

Networks used in industrial production systems have a wide variety of requirements that are not the same at every level of the company. Historically, such systems were structured as an automation pyramid like the one shown in Fig. 1. When moving from the top to down, the number of components increases, though the number of data decreases. For example, the highest layer aggregates data from the bottom layers consisting of multiple sensors, actuators, and controllers. This kind of data flow makes data density higher in the upper

A. M. Hosseini · S. Maragkou · T. Sauter
Computertechnik/Software-Intensive Systems, TU Wien, Vienna, Austria
e-mail: ali.hosseini@tuwien.ac.at

S. Maragkou
e-mail: sofia.maragkou@tuwien.ac.at

T. Sauter
e-mail: thilo.sauter@tuwien.ac.at

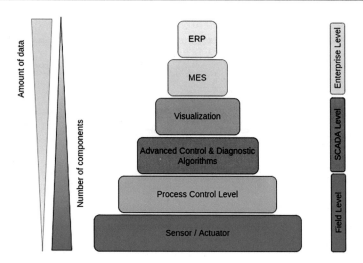

Fig. 1 Automation pyramid [5]

layers. The automation pyramid dates back to the 1970s and is no uniform way of describing production systems. Actually, there is a large variety of such hierarchical models, and recent developments around Industry 4.0 have further increased system complexity [3].

A contemporary way of looking at the automation hierarchy is to distinguish between Information Technology (IT) and Operation Technology (OT). Here, OT provides availability and safety in monitoring and controlling technical processes. In contrast, IT focuses on computer systems, networks, and software to process and distribute data. In recent years, however, the IT and OT boundaries are disappearing, and the borders between the automation pyramid layers are not that accurate anymore [4].

1.2   Cyber-Physical Systems

A Cyber-Physical System (CPS) combines two components, a physical entity, and a cyber unit. Figure 2 shows an abstract depiction of a CPS and its interfaces, where hardware components (i.e., mechanical, electric, and electronic parts) are enhanced with software. This enables the system to interact with its environment, receive signals, data and commands, as well as deliver them [6]. The seamless integration of physical computation and network components results in a complex nature of two orthogonal components (cyber and physical) which makes protecting such systems go beyond securing each of them in an isolated manner [7].

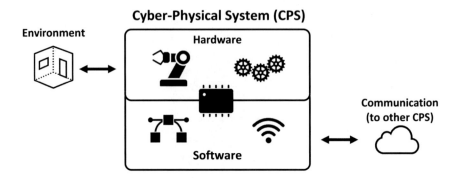

Fig. 2 Abstract depiction of a CPS

1.3   Cyber-Physical Production System (CPPS)

If one or more CPSs are integrated into a production environment, a CPPS is created. Here, CPSs are combined and communicate with each other, and use the information of the others to optimize the production process [8].

Figure 3 illustrates an example of a CPPS built of multiple CPSs. This system includes a collaborative robot (cobot), which is a stationary lightweight production robot without the need of any safety fences [9]. A mobile robot can also accept jobs from the control system and then perform transport tasks as CPS [10]. As an extension, a mobile robot with an attached industrial robot, also called a mobile manipulator, enables more complex and flexible work tasks [11].

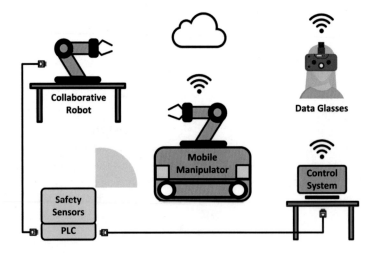

Fig. 3 CPS in a production environment forming a CPPS

To ensure that these tasks are carried out safely, the entire environment can be monitored by various sensors and Industrial Control Systems (ICS), for example, in the form of Programmable Logic Controllers (PLCs) [12]. The required data can then be exchanged either in a wireless or wired fashion. Finally, this data can be used for various algorithms such as predictive maintenance or presented directly to the user through data glasses [13]. Owing to their complexity, robots, even more so in collaborative settings, are good examples of CPS that process production jobs [14].

1.4 Motivation

In summary, CPSs provide data accessing and processing functionality. Furthermore, the introduction of CP(P)S into production system engineering brings, on one hand, many benefits, such as a gain of efficiency for manufacturers. On the other hand, the reliance on complex and all-time connected software introduces new attack vectors. A major risk lies in the increasing capabilities of the CPS components: As devices become more intelligent and interconnected, the risk of cyber-attacks is increasing. Besides data theft, espionage, and the associated economic damage, attacks on, e.g., production robots can lead to human injury. Awareness is slowly increasing, but still, more than 48% of the surveyed German companies reported that they see a low risk of a cyber-attack on their own company [15].

Therefore, this work will focus on the challenges in OT security and their impacts on safety-related CPPS. In particular, we identify the current state of the art and especially its limits and challenges in the information security of modern CPPSs based on a holistic view, including mechanical properties, control systems, risk management, and network communication.

2 Vulnerable Assets of a CPPS

In Sect. 1, a CPPS and examples of possible components are presented (cf. Fig. 3). In case of an attack on this system, all components of a CPPS can be affected. But what are the exact assets of a CPPS that have to be protected against attacks?

An overview including vulnerable assets of a CPPS is presented in Fig. 4, to illustrate the answer to this question.

On one hand, CPPSs consist of tangible assets. These can be seen as components of a single CPS and can be physical, cyber, or both. On the other hand, a CPPS has intangible assets, which can be seen as safety and security requirements every system should comply with.

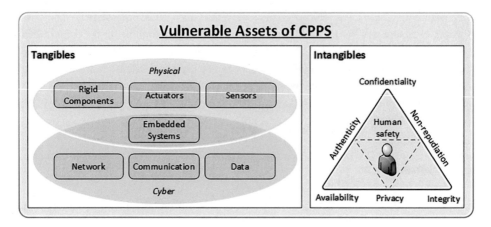

Fig. 4 Graphical overview of the tangible and intangible assets of a CPPS (based on [6, 16])

2.1   Tangible Assets

The tangible assets assigned to a CPPS represent the technologies used. They can be considered as the baseline of understanding the following terms and concepts since each technology can be investigated separately under the concept of safety and security. Therefore, these assets should continue to be considered as parts of a CPPS (cf. Fig. 3 in Sect. 1) [6].

Furthermore, these assets are also categorized by physical and cyber components. The physical ones are the rigid components, like actuators and sensors. Embedded systems consist of both physical and cyber components. Cyber assets are networks, communication systems, and data. These are now described in more detail below.

Rigid Components are the parts including bodies, joints, and various force elements. They represent the mechanical parts of a CPS. From the security point of view, the rigid components are an asset because:

- they are needed for the proper function of the whole CPPS and thus for the CPPS's availability,
- they can be attacked physically,
- they must be protected, for example, by means of physical access control.

Actuators are the elements of a CPPS that enable the active, direct interaction with the physical environment and the desired impact on the product. Examples for actuators are motors executing certain movements, heating/cooling, optical or acoustic emissions. Actuators are assets because they:

- can cause damage or injuries if operated by unauthorized subjects,
- are relevant for the proper operation of the CPPS.

Sensors serve the purpose of detecting and quantifying the physical aspects of the environment. We consider the set of sensors an asset because:

- the integrity of proper sensor values is crucial to product quality and human's safety,
- their availability is crucial to the reliable operation of the CPPS.

Embedded Systems are a combination of computer hardware and software designed for a specific function [17]. They can be seen as the brain of the system, and we consider them as assets because they:

- control the actuators and the sensors which are assets themselves,
- log data where integrity is of relevance for the proper function of the CPPS and regular product parameters,
- implement their specific control logic (e.g., open and closed-loop control).

Networks include wired and wireless connections to enable the CPS to communicate with other CPSs as well as with other network nodes, such as engineering workstations, control hosts, or log servers. We also consider the network components as an asset because:

- the confidentiality, integrity, and authenticity (three of the intangible assets) of messages depend on the network,
- the availability of the network itself and of sufficient bandwidth depends also on the network component.

Communication itself refers to the information exchanged between devices within the system. While "network" focuses on the connection and all measures necessary to function properly, like addressing and administrative traffic, "communication" is the set of actual payload exchanged. We define the communication component as an asset since:

- the confidentiality, integrity, and authenticity (three of the intangible assets) of exchanged information depends on this component as well,
- the component itself uses a network but may build its own secure channel to exchange information in an authenticated and confidential way.

Data include all the information that can be stored on a storage medium. Data are created or processed in control units and may be distributed among subsystems. The data component is also defined as an asset because:

- the confidentiality is relevant for the company's competitive advantage,
- the integrity of data is relevant for product quality,
- the integrity of programs, artifacts, and orders is crucial for proper production operation and production safety.

2.2   Intangible Assets

The second group of vulnerable assets of a CPPS (cf. Fig. 3 in Sect. 1) are the intangible ones. Apart from the tangible assets, these intangible assets of a system, the mandatory safety and security requirements, have to be protected as well.

Human safety is at least from the ethics viewpoint the most important intangible asset. For this reason, this asset is placed in the middle of the triangle of Fig. 4. Its aim is that no matter what is going to happen with any kind of system, a human worker shall never get harmed by any machine or its environment. A basic requirement to guarantee this would be the machine directive as national law. Furthermore, every CPS has to follow specific standards and guidelines for human safety, as every system is different.

The basic security requirements to prevent a system of an attack are represented by the so-called "CIA triad" (see Fig. 4). In the standard form, it consists of confidentiality, integrity, and availability [18].

Confidentiality means that unauthorized persons do not have access to any data or functions of the system [19]. It is considered as an asset because:

- lack of confidentially of sensitive data, such as product design, may lead to serious economic loss and loss of market position,
- unauthorized access to control signals, like control switches, can cause an unwanted interrupt in the production line or a false alarm,
- unauthorized control access can be the entry point for a multitude of attacks.

Availability implies that data and functions of the system are available whenever desired by authorized personnel [20]. Availability is one of the assets of a CPPS because:

- in a production system, non-availability can result in delayed delivery,
- non-availability of safety-related infrastructure can cause injuries,
- non-availability of information can cause a competitive disadvantage for the CPPS operator.

Integrity refers to the property of accuracy and completeness [20]. Here, it means the accuracy and completeness of data, messages exchanged within the CPPS, and also control circuitry. This implies that every change of these assets is only performed by authorized personnel. For the following reasons, integrity is considered as an asset since:

- integrity of data, like product parameters, is crucial to ensure the products functional and safety requirements as intended,
- it is crucial for program logic, since machinery may only be given authorized, intended instructions,
- integrity of integrated circuitry is important to prevent hardware trojans,
- it is crucial for exchanged messages (including machine-to-machine communication), for proper and intended operation of the CPPS.

The classical CIA triad can be extended with three additional assets: authenticity, privacy, and non-repudiation [16].

Authenticity is defined as the property that every entity within the system is what it claims to be [20]. The requirements set by this property can be seen as a subset of the requirements of "integrity". Breaking the authenticity of a CPPS would mean injecting or modifying exchanged messages or hardware—for example, on the supply chain. We consider it an asset because:

- broken authenticity may result in unexpected behavior of the CPPS, faulty products, or even hardware trojans within the CPPS's components itself.

Privacy relates to an individual's wish to manage who has access to personal information like age or address. In the context of a CPPS, personal information can also include data that are created during a production process [21]. Besides personal interests, also for ethical reasons, not all of these should be accessible to everyone. Therefore, privacy can be considered as an asset because:

- the leakage of employees' information in a company such as a name, date of birth, age, phone number, and address might endanger their life, especially in a sensitive organization or company,
- some CPPS technologies, like tracking modules, track and monitor not only the machines but also the human workers.

Non-repudiation is opposed to the authentication process and aims to identify malicious people who have already proven their identity through an authentication process [22]. We consider it as an asset because:

- authenticated people may act as a Trojan to plunder other intangible assets, like privacy and confidentiality.

Though clearly distinct in theory, tangible and intangible assets are interrelated and cannot be treated separately. Tangible assets can be attacked in a deliberate manner, which may also have an unintentional effect on intangible assets. For example, if the sensors are attacked, it can have an impact on human safety. Moreover, if the data of a process are attacked, confidentiality and integrity can no longer be guaranteed.

3   Threat Modeling and Attack Vectors

Threat modeling is referred to as a structured approach that allows a systematic identification and rating of all the security-related threats that are most likely to affect a system under consideration [7]. There are three main types of threat modeling, each of them focusing on another aspect: asset modeling, attacker modeling, and software modeling [23]. Within the context of this work, we focus on the assets presented above and attacker modeling.

3.1   Complexity of CPPS Attacker Modeling

In [24], an attacker model for CPS (cf. Fig. 3 in Sect. 1) is proposed in which CPSs are divided into two layers: a physical layer that consists of sensors and actuators interacting with given properties of the environment of interest (such as temperature or humidity) and the cyber layer that transforms the perceived data into information, exploited for driving the environment towards a given goal—possibly, but not necessarily, by means of the actuators. The physical layer inevitably requires physical access to be attacked, whereas the cyber layer requires some form of a direct or indirect communication channel to be attacked. The author further mentions that, as opposed to an IT system, a cyber-physical system can be influenced only by affecting its environment. However, there are possibilities to influence the physical layer also without physical access, if access to the cyber layer is possible in advance. This is in particular feasible if hardware relies on a system on a chip (SoC) since many manufacturers depend on third-parties in design as well as in manufacturing [25]. Due to the need to outsource the design and fabrication procedure and the use of third-party intellectual property (3PIP) cores, SoCs are becoming vulnerable to malicious modifications. A hardware Trojan inserted by electronic design automation (EDA) tools or through 3PIP cores can cause not only denial of service (DoS) attacks but can also leak important information or even change the original functionality of the circuitry.

3.2   CPS-Specific Threat Modeling

Attacker models from the IT world are no longer sufficient for the area of CPSs (cf. Fig. 3 in Sect. 1) for two reasons. First, there exist differences in requirements and maintainability of OT components which are discussed in more detail in Sect. 4.1. Second, attacker models from IT, such as the Dolev-Yao model widely used in protocol security and distributed systems, neglect the existence of additional attack vectors of the physical space [26]. This additional attack surface requires extended models [7, 24, 27].

In [27], a taxonomy of CPS-specific attacker models is presented, and a framework based on the proposed taxonomy is derived. This taxonomy is illustrated in Table 1. Beyond that, they propose a set of attacker profiles, covering most attacker types found in the observed set of publications. In their framework, they distinguish six types of attackers, classified by properties such as their capabilities (i.e., qualification and resources), interests, and levels of required stealth. Their categorization starts with the "basic user" who is assumed to have neither high attack knowledge nor a lot of resources and ends with the "nation-state attacker" who is assumed to have high knowledge and high resources. For the context of our work, we refer to their attacker model.

3.3   Threats Against CPPS Assets

In Sect. 2 and with Fig. 4, we listed possible assets, that is, components of a CPPS that are worth protecting. Next, we address possible attacks, i.e. threats, to these assets. However, we do not provide a comprehensive taxonomy of network attacks. It would be beyond the scope of this work for the IT as well as for the OT case [28, 29]. In general, we want to structure the possible threats into the four dimensions as depicted in Fig. 5. A threat either concerns logical entities or physical components and is carried out either in a remote or onsite (local) manner. A combination of threats w.r.t. remote-logical, local-physical, etc. is possible. The two-dimensional matrix lists a few examples that fall into these categories.

Physical Threats
A physical threat is caused by potential attacks having a physical impact on the CPPS. This physical impact can either be destruction, removal (including theft) or an influence on systems component with the aim to cause irritation—e.g., by influencing sensors maliciously. Obviously, the physical components of the CPPS are prone to this threat. Nevertheless, physical attacks may also impact logical assets: data, for example, are endangered by the possibility that the physical data carriers are stolen or destroyed. Examples of physical attacks include theft and vandalism, but also the installation of additional hardware for the conduction of logical attacks for a later point in time. Putting paint or other substances on a sensor to disable it is just another example of a physical attack.

Logical Threats
Logical threats are threats that address the cyber components of the CPPS. In particular, logical threats may endanger either communication payload, data, or the availability of the cyber layer. Successful logical attacks may have consequences for the physical layer as well,

Table 1 Attacker Types [27]

Name	Description
Basic user	This is the basic unstructured hacker, cracker, or hobbyist and someone who uses established and potentially also automated techniques to attack a system. The adversaries have average access to hardware, software, and Internet connectivity purchasable with average personal funds or theft from their employers
Insider	The insiders are basic users with the difference of the employment position inside the company. The privileges they own tightly correlate to their employment position (user, administrator, supervisor, etc.). This type is of high importance for systems where the main protection is an air gap between the systems network and Internet which is often the case in current CPS
Hacktivist	The hacktivists (a word combining the words hacker and activist) use their hacking abilities to promote a political agenda. Their intentions are often related to freedom of information
Terrorist	The terrorist or cyber-terrorist is a politically motivated attacker who uses computers and IT to cause severe disruption or widespread fear
Cybercriminal	This is the "black hat" type of hacker, i.e., an attacker with high knowledge and skills but criminal intentions. This category of attackers takes advantage of known vulnerabilities and may even find zero-day vulnerabilities on their own. Their goals can range from blackmailing and espionage (industrial forebrain) to sabotage
Nation-state attacker	This type of attacker is sponsored by a government. They are possibly belonging to a state organization for carrying out offensive cyber operations. Typical targets are general intelligence, but also public infrastructure systems, traffic management, and power or water systems

e.g., control code or hardware designs can be influenced through logical attacks. Examples for logical attacks can be taken from IT networks and reach from Denial of Service attacks, or malware and information exfiltration, to taking control over a target CPS.

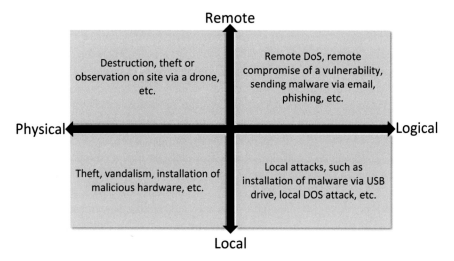

Fig. 5 Categorization of Threats

Local Threats Versus Remote Threats

We further divide threats into a second dimension since attacks can be conducted locally, that is, within the premises of the company, or remotely. Physical attacks are in most cases local but can also be conducted remotely—for example, by steering a drone towards the premises of a company for destruction, theft, or observation purposes. Although logical attacks are in principle carried out in a remote manner, the ability to access network nodes physically opens a variety of additional attack vectors (cf. Table 2).

3.4   Attack Vectors

An attack vector (also called *threat vector*) is an entry point on any accessible asset [30]. Each attack vector aims at one or more of the basic security requirements (the CIA triad). Successful attacks can lead to loss of control, loss of insight into the process, or loss of data. For example, this can include situations where users note unusual behavior but cannot make changes to the system, or the information about the underlying system is flawed and the user is not aware of the manipulation [31].

An attack surface is any asset or information which could be misused by an attacker. This includes system information reachable from the Internet, such as product data sheets or exposed sensitive information about system components which could aid hackers, up to posts of employees in social media. Large attack surfaces are, for instance, systems and products directly coupled to the Internet. Isolated computers in internal networks, like lab computers, are considered small attack surfaces [30].

Attack vectors are basically methods of how a hacker will attempt to attack assets. Those vectors include how assets handle data on communication paths. Communication paths are important since they can change the potential impact of an attack. For instance, if an Internet-facing asset with a low criticality rating can be used to access other internal (more critical) systems, it should also get attention. Table 2 shows common attack vectors.

Exploiting vulnerabilities through the mentioned attack vectors can cause unwanted scenarios such as malware infection. For example, the ICS malware *Stuxnet* was discovered in 2010 and targeted PLCs (Programmable Logic Controllers). Stuxnet was able to attack Siemens PCS 7, S7 PLC, and WinCC systems. This malware exploited four vulnerabilities which led to remote code execution. At the end of 2010, approximately 100,000 hosts were infected around the globe. Stuxnet traditionally got into an environment via a USB flash drive or external hard drive. After plugging in the infected external drive, the malware looked for systems that had Siemens Step7 software running to execute the malicious payloads via peer-to-peer network communication. [30] Mapping this behavior to the attack vectors in Table 2, *physical access* via USB ports opened the doors for the malware to the internal network. Before doing any harm and starting the infection, the malicious USB device had to be placed on a system in the internal network by someone. Therefore, the root cause was, in this case, *users*, who plugged in the compromised USB device on a suitable system where the infection had its origin. Afterwards, it spread through the *network* to *ICS systems and devices*. Table 2 lists further common attack vectors.

4 Measures Against Threats

Consortia from academia and industry have defined different security standards, guidelines, and best practices that provide recommendations or requirements. Besides standards focusing on general information security standards, there are industry-specific security standards as well [30].

There are significant differences when dealing with security in the ICS domain compared to the IT domain. These differences demonstrate the need for a different approach regarding security measures in ICS environments (cf. Fig. 3 in Sect. 1).

4.1 Security Relevant Differences Between IT and OT Systems

The performance and reliability requirements of ICSs differ from those of IT systems. Typically, communication systems, operating systems, and applications are used that may be considered unconventional in IT environments. Originally, IT systems were isolated from ICS, where proprietary control protocols using specialized hardware and software were running. There is a trend that (industrial) Ethernet and Internet Protocol (IP) devices are replacing legacy solutions which increases the attack surface of ICS. IT solutions are adopted to enable

Table 2 Common Attack Vectors [30]

Access Vector	Possible Attack Vectors
Network	Adjacent internal networks (wired/Ethernet) (e.g., business network or DMZ)
	Compromised dual-homed device from an adjacent device (e.g., engineering workstation)
	Compromised devices on the local network
	Internet (including cloud and multi-tenant environments)
	WiFi networks
	Split tunneling (often via VPN used by staff at insecure remote locations)
	Other radio connections such as WirelessHART, ZigBee
ICS and devices	CPS, sensors, actuators, controllers, industrial PCs, PLCs, SCADA systems, ...
Applications (residing on workstations, servers and devices)	Network service ports which includes industrial protocols like Modbus (default TCP/IP port 502), SSH (default TCP/IP port 22) and remote desktop (default TCP/IP port 3389), etc.
	File input/insertion
	User input (includes local applications and web interfaces)
	Data input such as libraries and DLLs
Physical access	USB ports (including USB sticks, keyboards or smartphones charged via USB)
	Serial and parallel ports
	Other data ports at the device like SATA/eSATA, HDMI, Display Port, PS/2 Keyboard and mouse input
Users (social engineering)	Direct social interaction via phone or in person
	Email and social media
	Office applications such as PDF readers, email clients and Internet browsers
Supply chain	Chip/hardware modification
	Application/firmware code modification

corporate connectivity and remote access, which leads to less isolation for ICS from both company internal networks (e.g., office and enterprise IT) and the external networks (e.g., Internet or remote support connection for system integrators). In such scenarios, however, defense-in-depth approaches have been proposed many years ago [32] and are now a proven security strategy, together with appropriate access control [33].

The following aspects are relevant for security considerations in ICSs [34]:

Timeliness and Performance Requirements ICSs are typically time-critical due to the need for reliable and deterministic responses. Therefore, acceptable delay and jitter are defined by the individual installation. Throughput is secondary in ICS environments. In IT, it is the other way around: Throughput is prioritized, while delay and jitter are not critical. Automated response time or response to human interaction is crucial in some cases (e.g., the usage of collaborative robots in CP(P)Ss).

Availability Requirements ICS processes are typically continuous, and unexpected outages cannot be tolerated. Outages (e.g., due to maintenance) have to be scheduled days or weeks in advance. To meet the high availability requirement (ICSs often cannot be simply restarted without affecting production), exhaustive testing before deploying is necessary. Since even just a temporal outage is often not an option, ICSs can have redundant components installed to ensure availability during outages.

Risk Management Requirements This topic is discussed in more detail in Sect. 5.

Physical Effects ICS devices (e.g., PLCs, operator stations) control and interact with physical processes, which can lead to physical events, whereas IT does not interact with the physical environment in a comparable manner.

System Operation ICS operating systems and networks can be quite different from IT and require different skill sets, experience, and expertise. Therefore, industrial communication networks are often managed by control engineers instead of IT administrators.

Resource Constraints ICSs typically use resource-constrained systems (CPS), which lack the possibility to include state-of-the-art IT security capabilities (e.g., resource intensive encryption, error logging, password protection). Furthermore, using exactly the same IT security practices at ICSs can lead to availability and timing issues. Adding resources, computing power, or features to meet current security capabilities may not be possible.

Communications Communication protocols and media may be proprietary and different from most IT environments.

Change Management Using outdated software can be a severe vulnerability on both IT systems and ICSs. In IT environments, software updates (including security patches and fixes) are typically deployed regularly in a timely fashion based on security policies and procedures which can be automated. In the ICS environment, software updates have to be tested exhaustively by both the vendor and the end-user prior to deployment. This process also has to be scheduled days or weeks in advance. Beyond that, ICSs may require revalidation during the update process and can use older, unsupported versions of operating systems which can lead to the situation that patches, updates, or fixes are no longer available. Besides, the software change management process can also be applied for hardware and firmware, which also requires careful assessments by experts.

Managed Support IT systems usually allow diversified support, whereas ICS support is sometimes only possible through a single vendor. Sometimes ICS vendor license and service agreements prevent the usage of third-party security solutions leading to the loss of service support when doing so without vendor acknowledgment or approval.

Component Lifetime The lifetime of usual IT components is 3 to 5 years because of the speed of technology evolution in this area. In the ICS environment, the typical lifetime is in the order of decades [12].

Component Location Usually, IT components and some ICSs are located in facilities accessible by local transportation. Largely distributed ICS components can be installed at far distances and generate a higher transport effort. Thus, the component location also has to be considered for physical security measures.

These mentioned differences in IT environments also lead to additional or different attack vectors and show the need for specific threat modeling approaches. To protect against those attack vectors, governments and industry organizations have defined different security standards, guidelines, and best practices that provide recommendations or requirements tailored to ICSs [30]. The standards, *IEC 62443* and *NIST Special Publication 800-82* are examples for such security standards tailored for ICSs.

Weak security can lead to situations where operators of machines could get harmed, as the following example shows: The machine should enter a safe state once the light grid has been activated. Due to weak or missing security measures, the system gets manipulated and data of the light barrier can not be evaluated anymore. Thus, the safety function is deactivated and the machine continues to operate also when operators are next to it. Such safety-related communication in ICSs is addressed by the *IEC 61784* standard.

4.2   IEC 62443

IEC 62443 consists of a series of standards that provide a framework to address and mitigate security vulnerabilities in Industrial Automation and Control Systems (IACS). It can be used by asset owners, system integrators, product suppliers, service providers, and compliance authorities (including government agencies and regulators to verify compliance with laws). Product suppliers, service providers, and system integrators can use it to validate if their products and services can provide the security capability to meet the asset owner's security requirements.

The previously mentioned security requirements are based on the following seven security related topics, called foundational requirements in IEC 62443:

- Identification and authentication control
- Use control
- System integrity
- Data confidentiality
- Restricted data flow
- Timely response to events
- Resource availability

Figure 6 illustrates the structure of the IEC 62443 standard series. It is designed for several layers of an ICS to address a comprehensive set of potential threats and security vulnerabilities. The layer *general* clarifies the used terminology, definitions, and vocabulary, which leads to a common understanding throughout the whole standard series. The part *policies & procedures* deals with organizational topics, *system* addresses technical architectures as

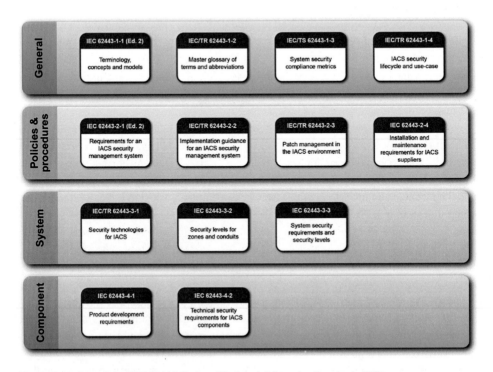

Fig. 6 Overview of the IEC 62443 Series of Industrial Security Standards [35]

a whole and *component* focuses on single components of a system (e.g., a robot). Different parts of the standard series are still in draft or reviewing state at the time of writing.

4.3   NIST Special Publication 800-82

The latest version of this standard, revision 2, was published in 2015 and includes recommendations for security, management, operational, and technical security controls in ICS environments [30]. NIST Special Publication 800-82 rev. 2 provides an overview of ICSs, typical system topologies and architectures, known threats and vulnerabilities and suggests security countermeasures to decrease corresponding risks. The suggested countermeasures are described on a conceptional, high-level view (e.g., doing network segmentation) as well as on a more technical, detail-oriented view (e.g., recommendations for common services, like DNS, HTTP, (T)FTP, DHCP, SSH, SMTP, SNMP, DCOM). It also advises how ICS security programs can be developed and deployed. Another part of the standard deals with the application and implementation of security controls to ICS environments and has a broad intended audience:

- Control engineers, integrators, and architects who design or implement secure ICS
- System administrators, engineers, and other IT professionals who administer, patch, or secure ICS
- Security consultants who perform security assessments and penetration testing of ICS
- Managers who are responsible for ICS
- Senior management who are trying to understand implications and consequences as they justify and apply an ICS security program to reduce impacts on business functionality.
- Researchers and analysts who try to understand the security needs of ICS
- Vendors that develop products that will be deployed as part of an ICS

4.4   IEC 61784

The standard IEC 61784 relates to industrial communication and defines a group of protocols specific communication profiles in line with the IEC 61508 series. The third part of the standard addresses safety and security aspects and demonstrates the principles of functional safety-related communication, taking into account issues raised by IEC 61508. Functional safety is a part of overall safety that can ensure the correct operation of machinery and systems.

Detection of a potentially threatening situation is a primary part of functional safety. Subsequently, the activation of a protective or corrective device or mechanism is required to prevent hazardous events. In IEC 61508, the safety communication layers provide the necessary confidence in the transportation of information transmission in the underlying communication system. The IEC 61784-3 introduces safety communication layers that enable, e.g., a fieldbus to work in the application areas that demand functional safety and provides details about:

- Fundamental principles for implementing the requirements of the IEC 61508 series for safety-related data communication, including potential communication faults, corrective measures, and considerations influencing data integrity.
- Functional safety communication profiles for various communication profile families in IEC 61784-1 and IEC 61784-2, including safety layer extensions to the communication service and protocols sections of the IEC 61158 series.

5   Risk Management

Risk management provides an important insight into the safety and security implementation for industry and also makes it cost-effective [36]. The primary focus is to:

- provide safety and security to personnel, critical infrastructure, information, and assets,
- prevent damage to environment and image (reputation) of an organization, and
- provide stability and maintenance of efficiency of CPS for a safe and secure operation of a CPPS.

As CPPSs have physical and cyber assets, which impact the risk assessment, the approach to risk management must take into account the primary focus mentioned above and provide convenient security solutions to the critical and non-critical components of the constituting CPSs. Relevant terms defined for the risk management process include [37],

- Risk: effect of uncertainty on objectives (goals or outcomes),
- Residual risk: risk that remains after risk treatment,
- Acceptable risk: informed decision of the organization to go ahead with the risk, based on analysis,
- Tolerable Risk: organization's readiness to bear the risk after the risk treatment in order to achieve its objectives.

Risk management is a continuous cycle of identification, evaluation, and prioritization of risks followed by coordinated and economical application of resources to minimize, monitor, and control the probability or impact of unfortunate events or to maximize the realization of opportunities [38]. The risk management process provides a necessary framework for the actions to be taken. These are the influential steps applied by the management to estimate and control risks.

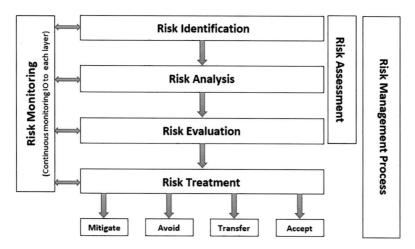

Fig. 7 Risk management process

The objective is to have an understanding of the economic impact of one or more compromised safety and security features of CPSs and the overall disruption in the functionality of a CPPS. The first step is to identify the risk, then to analyze and evaluate it. Later, this risk is introduced to the treatment steps. The last step is not a separate action per se, but to monitor the result of the previous step continuously and to calculate the residual risk. This residual risk should be within the acceptable level provided by the organization. The documentation of these steps helps in future analysis. Figure 7 illustrates the steps involved in the risk management process.

Risk Identification

Risk identification is the process of finding, recognizing, and describing risks [37]. It defines a risk profile by identifying the sources and events of risks. It also describes the causes and potential consequences of the identified risk. Historical data provided by the organization, stakeholder analysis, expert opinions, and theoretical analysis are used [38]. Risk identification for a CPS can be performed by following the threat modeling and attack vector analysis mentioned in Sect. 3.

Risk Analysis

Risk analysis determines the nature of risks and also the level of risks [37]. It marks the basis for risk evaluation and the future decisions that will follow for the risk treatment. One of the prerequisites to risk analysis is threat analysis. There are two methods that can be used [39]:

- Qualitative analysis is a non-statistical approach and an in-depth examination of an asset, threat, and vulnerabilities. They are scaled from "High" to "Low" or 1 to 5, based on

Table 3 Risk assessment methods based on [40]

Safety assessment methods	Security assessment methods	Integrated methods
Fault Tree Analysis (FTA)	Attack Tree Analysis (ATA)	Boolean logic-driven Markov processes (BDMP)
Failure modes and effects analysis (FMEA)	Cyber Physical Security (CPSec)	Non-functional requirements (NFRs)
Systematic theoretic process analysis (STPA)	Systematic theoretic process analysis Security (STPA-sec)	Systematic theoretic process analysis—safety and security (STPA-SafeSec)
Model based Engineering (MBE)	Bayesian Network approach	Bayesian belief network (BBN)
Goal Tree-success tree and master logic diagram (GTST-MLD)	Traditional security technology	Six-step model (SSM) and information flow diagram (IFD) integration approach
Hazard and operability methodology (HAZOP)		

their impact, consequence, and likelihood. By this approach, the risks are prioritized for further investigation.

- Quantitative analysis is a statistical approach to obtain the loss in financial terms. It depends highly on historical data. As risk analysis is a measurement of events that may occur, quantitative analysis is a method to determine estimates and not exact values.

Risk Evaluation

Risk evaluation is a process of comparing the results of risk analysis with risk criteria to determine whether the risk and/or its magnitude is acceptable or tolerable [37]. It assists in the decision about risk treatment. Threats and attacks, when evaluated, would possess high risk if appropriate measures are not in place.

Risk Assessment

Risk assessment is an overall process of a combination of risk identification, risk analysis, and risk evaluation. A number of risk assessment methods are available today. Some practically implemented risk assessment methods are mentioned in Table 3. The table is divided into safety-related, security-related, and integrated approaches for risk assessment of CPS [40].

Risk assessment for CPPSs need to cope with a dynamic threat landscape and attack vectors. This can only be achieved by using risk assessment techniques which are dynamic in nature. Most of the methods applied for assessment in ICSs are static in nature. Such methods will become irrelevant when applied to a dynamic production process because CPS safety and security aspects would vary w.r.t. the application.

Fault Tree Analysis (FTA) is used to model complex failure modes of the system w.r.t. component failure modes. It is a qualitative and quantitative failure mode impact analysis. The values of reliability and safety are computed based on the system information. FTA is a top-down and deductive method. Drawbacks of this method are its implementation costs and difficulty to find all possible accidental paths [41]. Failure Modes and Effects and Analysis (FMEA) is a bottom-up, inductive and iterative analysis approach, which is based on reliability theory. Even though it reduces the timing and resource cost of system development considerably, it spends time on tracking failures. It does not consider the relationship between different failure components and also is limited to analyzing a single cause of impact [42]. Hazard and operability methodology (HAZOP) conducts analysis and review for industrial processes to determine any deviation from design conditions. This method mainly depends on an expert's analysis and fails to highlight the issues. It is incapable of handling the numerous and complex interactions within CPSs [43]. Goal Tree Success Tree-Master Logic Diagram (GTST-MLD) is a goal-oriented method that uses prior knowledge to identify hazards. It represents various elements of the system hierarchically. GTST-MLD ignores undesired interactions and hence cannot adequately analyze CPS vulnerabilities, failures, and errors [44].

Attack Tree Analysis (ATA) is an independent module widely used for security assessments. It presents the steps in an attack process in the form of a graph. There are no standards for attack tree construction. The limitation of ATA as a quantitative analysis method is that it depends on known attacks from historical data and intuitive expert opinion for attacks. A standard attack library is available for some CPS applications. System Theoretic Process Analysis (STPA) is a technique used for hazard analysis based on the system theoretic accident model and processes (STAMP). STPA aims to identify the causes of hazards and accident scenarios along with the entire accident process [45]. STPA-sec is an extension of STPA, considering security issues. STPA-SafeSec considers safety and security equally and performs integrated analysis to get the optimum results. The influencing factors for safety and security issues are taken into consideration in a single framework in STPA-SafeSec. Although STPA-SafeSec can be used to analyze the interdependencies of safety and security of a CPS, it cannot quantify the risk [46].

Boolean logic-driven Markov Processes (BDMP) obtain quantitative and qualitative analysis for safety and security risk assessment. BDMP is a dynamic process that satisfies the real-time characteristics of CPSs. BDMP's mathematical modeling makes it clearer and more intuitive. The actual situation and assumption of network structure have a great impact on the results. Some wrong assumptions of network components will make the analysis inaccurate [47]. Non-functional requirements (NFR) are used in qualitative evaluation methods, e.g., to answer questions related to the fulfillment of safety and security. NFRs need to be repeated after every system update, which may not be economical [48]. Bayesian Belief Network (BBN) is a probabilistic graphical model that utilizes Bayesian inference for conditional dependency. Bayesian inference helps to assess the risk under unknown threats and hazards. The obvious challenges in building a BBN model are lack of historical data and prior knowledge of a large number of attacks [49].

A standard that can be of use for the risk assessment process in the machine life cycle is ISO 12100. This document's scope is the general design framework of machine development for the safety of people, property, and the function of a machine, with least amount of risk. Principles of risk assessment and risk reduction are specified to help designers. Guidelines for identifying hazards and estimating and evaluating risks during relevant phases of the machine life cycle, and for the elimination of hazards or the provision of sufficient risk reduction, are described too [50].

Risk Treatment

Risk treatment aims to control and manage risks. The management implements one of the following options [34]:

- Mitigate: control measures to reduce risk and/or its impacts,
- Transfer (or share): some or all the risks are transferred to some external entity, such as insurance company or business partner,
- Avoid: management terminates the activity which provides the potential risk,
- Accept: management goes ahead with the activity despite the risk in place.

A problem-solving approach should be ensured to achieve an optimum mitigation solution. Applying all the security measures provided by a security professional can be complex and costly. Risk management can provide an optimal solution for required security solutions which is agreed upon by all the decision-makers.

6 Challenges of Integrating Safety and Security

By combining OT and IT, a system is created that has to fulfill safety and security standards at the same time. Changes in physical behavior can lead to state changes in the system. On the other hand, changes in the system can cause modified physical behavior [18]. Therefore, it is obvious that safety and security go hand in hand in the context of CPSs.

Furthermore, implementing an increasing amount of CPSs leads to more systems that must be supplied with updates. This fact, however, contradicts the basic idea of an easy and fast integration to a reliable CPPS because these independently deployed updates can bring uncertainties into the system. A concept for solving this problem can be found within the DevOps approach. This approach adapts the updates regarding the operating status of the machines included in the whole system.

Finally, it can be assumed that there is often no clear separation between OT and IT in the production environment of a CP(P)S for reasons of simplicity. Both safety and security in combination, therefore, play an important role in the implementation. Consequently, rethinking is needed, and OT networks should be secured according to state-of-the-art knowledge and standards (e.g., IEC 62443).

6.1 Current Status and Objectives

Currently, the focus of systems is rather on safety or on security only. Looking at physical systems, the emphasis has been more on safety, especially when a new machine is integrated. Nevertheless, a safe implementation is not easy and differs in every application. For example, a risk analysis must be carried out anew for each implementation, even if only a small adjustment has been made. It could happen that this new scenario leads to new dangers, which must be avoided. However, most importantly, a safe system does not make the entire application safe. Due to the mutual integration of safety and security, security must also be considered from the very beginning since the inseparable interaction can create new hazards under cyber attacks.

According to ISO 12100, a standard for risk assessment and risk reduction, risks must first be reduced by coherently safe designs, then by technical protective measures, and finally by user information. In each step, both safety and security can be addressed. Safe construction can help to protect the human worker against workplace accidents. However, interfaces vulnerable to IT security threats can also be protected by safely constructed enclosures. When using technical measures, attention must also be paid to whether they can be cyber attacked or open up new vulnerabilities. And finally, in showing the right usage for a safe operation, employees can also be trained for increased awareness of social engineering attacks.

However, companies affected do not yet have a high level of awareness about the significance of security [51], and therefore mostly safety aspects are addressed at present. Therefore, the main objective is to create awareness of a combined safety and security implementation of CPPSs. This would probably also require standards and guidelines to be revised or created. Fortunately, with laws such as the European Union network and information system security directive[1] or the Austrian network and information system security act,[2] a trend in the right direction can already be seen.

6.2 Challenges Relevant for the Physical Layer

Securing a device is challenging, especially when the device is part of a CPPS, where the interoperability of the system is very complex. Also, security can come with a cost since

---

[1] https://www.enisa.europa.eu/topics/nis-directive

[2] https://www.ris.bka.gv.at/GeltendeFassung.wxe?Abfrage=Bundesnormen& Gesetzesnummer=20010536

it can have an impact on other aspects of the system [52]. At the same time, human safety should always be the first priority when designing a CPPS.

Security Challenges

According to [52], the general system security challenges to tackle are reduced performance, high power consumption, delays, high cost, and compatibility issues. More specifically, at the device level, two main factors should be taken into account: shielding the device and securing the design of the device.

Shielding a device can be an important countermeasure for a physical invasive attack. The invasive case can be coped with pure physical anti-invasive or invasion detection countermeasures. For very sensitive hardware that stores keys, it can be a practical solution to cast them into pure plastic or metal, ideally within a unique crystalline or polymer structure that can be preshared by the vendor [53]. Since maintainability must, in many cases, be given, it is not practical to apply such anti-invasive techniques for every device. However, in this case, one can still build cases so that unauthorized openings can be detected. Such techniques include sealed packaging and sealed device cases [54], painted screws or plumbs, and more advanced techniques such as a tamper sensing mesh [55].

In addition, changes to hardware and the device's design could be detected by electrical or other methods of aggregated measures that store certain values (electric field emission, electric resistance, size, weight, etc.) In this case, the main challenge is applying proper and sufficient measures so that unauthorized hardware changes result in significant differences. To countermeasure unauthorized changes in integrated circuits, modern hardware Trojan detection techniques are applicable [56]. Also, security challenges related to the design of electronic hardware can be raised from the requirements of modern devices. According to [25], the design complexity requested from CPPS is a pressing point to SoC design houses. The design and the production of electronic circuitry are so cost and time-demandingthat it is almost impossible for a single SoC design house to manufacture its own SoC without outsourcing help. Most of the SoC design houses are integrating their own intellectual property (IP) with third-party intellectual property (3PIP) and outsource at third-party companies the fabrication and packaging. This way, a lot of untrusted entities are created or integrated into the desired design. Malicious functionality, such as a hardware Trojan, can have been inserted at any part of the design and at any time during the production process of the device. Even though there are some methods to detect such malicious circuitry like logic testing or side-channel analysis [57], the complexity of the attacks will increase in the next years. With that, the detection methods should evolve, too.

Integration Challenges

When it comes to safety w.r.t. operational and personnel safety, there has been a lot of good practice principles in place. Some safety measures in use today can seem close to the security measures needed.

Developers face challenges in the key parameters, namely power, performance, and real-time fault detection. Combining safety and security increases system complexity, runtime errors, delays in operation, and costs [58]. An understanding of both safety and security concepts and best practice implementation in the design cycle can help to integrate and implement these concepts to hardware properly.

6.3   Software-Related Challenges

Challenges to the security of modern CPPS stem especially from the differences between IT and OT systems and their respective requirements.

Updates Versus Availability
Software testers can only prove the presence of vulnerabilities and not their absence, which is why often security vulnerabilities in software products are found after shipment. The common solution is to fix software vulnerabilities by offering security updates to the clients. In fact, it can be assumed that regular updates are one of the most important practical measures to ensure a software products' security [59].

At the same time, regular software updates are difficult to implement in the OT area due to the high requirements regarding availability and operational safety. Software changes, such as updates, can have unforeseen consequences on the functionality of the software. This is why operators tend to test software updates for weeks prior to implementation at the production site [60]. As a consequence, the delay and the cost to close certain vulnerabilities are raised significantly.

Lifetime Versus Changes
The software world is characterized by relatively short-lived technologies. Furthermore, software products can rely upon the fact that re-deployments do typically not require physical efforts. This short lifetime enables software engineers to learn from conceptual mistakes and fosters an evolution towards conceptual security. For a CPPS, where typical operation time reaches up to 50 years [12] and major software changes often go hand in hand with physical replacements, a quick change to another technology is not an option. Instead, security vulnerabilities must be dealt with through workarounds and by adding measures to the existing technology.

There exist approaches that aim to implement cyber security measures at the design phase of a CPPS [12, 61]. However, the approaches do not solve the problem of insecure legacy systems.

Deployment and Dependencies
In many cases, software security measures rely upon quick deployability of changes, such as automated (over-the-air) updates and security-related bug fixes. Such approaches signif-

icantly reduce reaction time to close detected vulnerabilities and, likewise, help the vendor to keep the number of vulnerable devices low. However, these advantages do not exist in the OT world, where CPSs with their own firmware interact in a coordinated manner within the CPPS and where requirements regarding availability prohibit concepts that automatically change a system. This leaves the operators with the challenge to manually update many independent systems from several independent manufacturers. Keeping all of these systems up-to-date is often difficult: not only is the effort significantly higher, but it results in a dependency on many vendors at a time.

7   Conclusion

This chapter focused on the explanation of the state of the art of OT security in CPPS and outlining existing challenges. We explained conceptional differences between IT and OT and the components of a CPS. We also identified possible attacks and attackers and listed relevant standards that, when implemented, should defend against cyber attacks. Also, the importance of risk management w.r.t. CPPSs was addressed.

The transition from classical control electronics to smart devices and complex software as well as the introduction of technology that has its origins in the IT area, such as IP-based communication protocols, IEEE 802.11-based wireless communication technology, cloud data storage, and the radical rise in sensory capabilities of these systems brought significant gains in efficiency and versatility into production plants. However, this transition introduced many novel security issues into OT that have previously been almost exclusive to IT networks, such as malware and botnets or information exfiltration.

The situation is made more difficult by the fact that production plants may pose serious safety threats if behaving unexpectedly. As a consequence, the demand for security measures in the OT area rose significantly. During our research, we observed that necessary security countermeasures have to fit the very specific requirements of production systems, which differ significantly from those of IT environments. Examples are low computational resources and constrained network bandwidth, very high reliability, safety constraints, the long operation time of equipment, and software and cost-intensive deployment of changes.

Additionally, some safety requirements (for example, an emergency stop function) prohibit the use of some known classical IT security measures such as long passwords and other interactive and effort-taking authentication methods. These requirements pose several major obstacles to OT security, and as a consequence, many security concepts that have been successfully developed and prove to work well for the IT area cannot be implemented, at least not in the same way, for many OT systems. Furthermore, weaknesses in the security of an ICS can lead to safety impacts, which makes security and safety interconnected.

Regarding the history of safety and security, these two concepts were independent, and each of them has particular goals, tools, and methodology. However, over the past few years, as systems have become more complex, the borders between safety and security are beginning to vanish. Safety can influence security and vice versa. For instance, strengthening one of them may result in weakening another. Therefore, the weak integration of these two concepts might lead to an ineffective and inefficient system.

Successful integration requires some preliminaries, and scientists should consider the priorities during integration. The first step is to identify the common need for the safety and security communities. Some standards regulated safety and security separately. For example, IEC 61508 consists of safety-related measures, and IEC 62443 addresses security issues. Recently, ISA (International Society of Automation) distinguished a need for integration between safety and security and organized a working group named "Work Group 7 - Safety and Security" to examine alignment and common security and safety issues [62]. Next, identifying the interdependencies is critical since there is a mutual relation between safety and security. Finally, standardization can help to achieve efficient interoperation, harmonization, and integrated safety and security.

References

1. S. Vitturi, C. Zunino, and T. Sauter, "Industrial communication systems and their future challenges: Next-generation Ethernet, IIoT, and 5G," *Proceedings of the IEEE*, vol. 107, no. 6, pp. 944–961, 2019.
2. I. Reithner, M. Papa, B. Lueger, M. Cato, S. Hollerer, and R. Seemann, "Development and Implementation of a Secure Production Network," *Proceedings of the 31st DAAAM International Symposium*, pp. 736–745, 2020.
3. J. Jasperneite, T. Sauter, and M. Wollschlaeger, "Why we need automation models: Handling complexity in Industry 4.0 and the Internet of Things," *IEEE Industrial Electronics Magazine*, vol. 14, no. 1, pp. 29–40, 2020.
4. E. J. Colbert and A. Kott, *Cyber-security of SCADA and other industrial control systems*. Springer, 2016, vol. 66.
5. M. Bajer, "Control systems integration using OPC standard," *AGH Master Thesis, W. Grega-Supervisor, Krakow & Antwerp*, 2008.
6. E. Geisberger and M. Broy, *Integrierte Forschungsagenda Cyber-Physical Systems: acatech STUDIE*. Deutschland: acatech, 2012.
7. G. Martins, S. Bhatia, X. Koutsoukos, K. Stouffer, C. Tang, and R. Candell, "Towards a systematic threat modeling approach for cyber-physical systems," in *2015 Resilience Week (RWS)*. IEEE, 2015, pp. 1–6.
8. R. E. Petruse, I. Bondrea, and I. C. Nicolae, "Main requirements of a cyber physical production system demonstrator," *Acta Universitatis Cibiniensis. Technical Series*, vol. 71, no. 1, pp. 76–80, 2019.
9. International Organization for Standardization (ISO), "Robots and robotic devices - Collaborative robots," Geneva, CH, Feb. 2016.
10. R. Siegwart, I. R. Nourbakhsh, and D. Scaramuzza, *Introduction to Autonomous Mobile Robots*, 2nd ed. Cambridge, Massachusetts: The MIT Press, 2004.
11. O. Khatib, "Mobile manipulators: Expanding the frontiers of robot applications," in *Field and Service Robotics*, A. Zelinsky, Ed. Springer, 1998, pp. 6–11.
12. B. Vogel-Heuser, T. Bauernhansl, and M. ten Hompel, Eds., *Handbuch Industrie 4.0 Bd. 2: Automatisierung*, 2nd ed., ser. Springer Reference Technik. Berlin: Springer, 2017.

13. Y. Ro, A. Brem, and P. Rauschnabel, *Augmented Reality Smart Glasses: Definition, Concepts and Impact on Firm Value Creation*. Gewerbestrasse 11, 6330 Cham, Switzerland: Springer International Publishing AG, 2017, ch. 12, pp. 169–181.

14. A. Grau, M. Indri, L. L. Bello, and T. Sauter, "Industrial robotics in factory automation: From the early stage to the Internet of Things," in *IECON 2017 - 43rd Annual Conference of the IEEE Industrial Electronics Society*, 2017, pp. 6159–6164.

15. Statista, "Wie hoch schätzen Sie das Risiko für Ihr Unternehmen ein, Opfer von Cyberangriffen/Datenklau zu werden?," 2019, accessed: 2020-10-16. [Online]. Available: https://de.statista.com/statistik/daten/studie/760006/umfrage/wahrgenommenes-risiko-von-cyberangriffen-unter-unternehmen-in-deutschland/.

16. C. Fife, "What's Required To Secure The IoT?" 2015, accessed: 2020-10-23. [Online]. Available: https://www.citrix.com/blogs/2015/04/09/whats-required-to-secure-the-iot/.

17. Barrgroup-Dictionary, "Embedded System," 2020, accessed: 2020-10-14. [Online]. Available: https://barrgroup.com/embedded-systems/glossary-embedded_system.

18. TÜV Austria, Fraunhofer Austria Research GmbH, "Safety & security in der Mensch-Roboter-Kollaboration," 2016. [Online]. Available: https://www.tuv.at/fileadmin/user_upload/docs/group/innovation/tuv-austria-white-paper-deutsch/003_tuv_austria_white_paper_III_einfluss_it_security_sicherheit_in_der_mensch_roboter_kollaboration_fraunhofer_DE_WEB.pdf.

19. M. Kumar, J. Meena, R. Singh, and M. Vardhan, "Data outsourcing: A threat to confidentiality, integrity, and availability," in *2015 International Conference on Green Computing and Internet of Things (ICGCIoT)*. IEEE, 2015, pp. 1496–1501.

20. F. Accerboni and M. Sartor, "ISO/IEC 27001'," *Quality Management: Tools, Methods, and Standards. Emerald Publishing Limited*, pp. 245–264, 2019.

21. Y. Lu and M. Zhu, "A control-theoretic perspective on cyber-physical privacy: Where data privacy meets dynamic systems," *Annual Reviews in Control*, vol. 47, pp. 423–440, 2019.

22. P. Van Aubel, E. Poll, and J. Rijneveld, "Non-repudiation and end-to-end security for electric-vehicle charging," in *2019 IEEE PES Innovative Smart Grid Technologies Europe (ISGT-Europe)*. IEEE, 2019, pp. 1–5.

23. A. Shostack, *Threat modeling: Designing for security*. John Wiley & Sons, 2014.

24. R. Vigo, "The cyber-physical attacker," in *International Conference on Computer Safety, Reliability, and Security*. Springer, 2012, pp. 347–356.

25. M. T. Swarup Bhunia, *Hardware Security: A Hands-on Learning Approach*. Morgan Kaufmann, 2019.

26. D. Dolev and A. Yao, "On the security of public key protocols," *IEEE Transactions on information theory*, vol. 29, no. 2, pp. 198–208, 1983.

27. M. Rocchetto and N. O. Tippenhauer, "On attacker models and profiles for cyber-physical systems," in *European Symposium on Research in Computer Security*. Springer, 2016, pp. 427–449.

28. N. Hoque, M. H. Bhuyan, R. C. Baishya, D. K. Bhattacharyya, and J. K. Kalita, "Network attacks: Taxonomy, tools and systems," *Journal of Network and Computer Applications*, vol. 40, pp. 307–324, 2014.

29. A. Humayed, J. Lin, F. Li, and B. Luo, "Cyber-physical systems security-a survey," *IEEE Internet of Things Journal*, vol. 4, no. 6, pp. 1802–1831, 2017.

30. C. Bodungen, B. Singer, A. Shbeeb, K. Wilhoit, and S. Hilt, *Hacking Exposed Industrial Control Systems: ICS and SCADA Security Secrets & Solutions*, 1st ed. New York: McGraw-Hill Education, 2016. [Online]. Available: https://mhebooklibrary.com/doi/book/10.1036/9781259589720.

31. S. J. Templeton, "Security aspects of cyber-physical device safety in assistive environments," in *Proceedings of the 4th International Conference on PErvasive Technologies Related to Assistive Environments*, ser. PETRA '11. New York, NY, USA: Association for Computing Machinery, 2011. [Online]. Available: https://doi.org/10.1145/2141622.2141685.

32. A. Treytl, T. Sauter, and C. Schwaiger, "Security measures in automation systems-a practice-oriented approach," in *2005 IEEE Conference on Emerging Technologies and Factory Automation*, vol. 2, 2005, pp. 847–855.

33. A. Valenzano, "Industrial cybersecurity: Improving security through access control policy models," *IEEE Industrial Electronics Magazine*, vol. 8, no. 2, pp. 6–17, 2014.

34. K. A. Stouffer, V. Pilitteri, M. Abrams, and A. Hahn, "NIST Special Publication 800-82 Revision 2. Guide to Industrial Control Systems (ICS) Security: Supervisory Control and Data Acquisition (SCADA) Systems, Distributed Control Systems (DCS), and Other Control System Configurations Such as Programmable Logic Controllers (PLC)," Gaithersburg, MD, USA, 2015.

35. "IEC 62443-3-3:2013 Industrial communication networks - Network and system security - Part 3-3: System security requirements and security levels," 2013.

36. D. R. Preiss, *Risk analysis techniques in engineering*. TÜV Austria Akademie GmbH, 2020.

37. International Organization for Standardization (ISO), "ISO/IEC guide 73:2009 - risk management - vocabulary," 2009.

38. D. W. Hubbard, *The Failure of Risk Management: Why It's Broken and How to Fix It*. Wiley, 2009.

39. P. Gregory, *CISA Certified Information Systems Auditor All-in-One Exam Guide, Fourth Edition*. McGraw-Hill, 2019.

40. S.-H. Y. Xiaorong Lyu, Yulong Ding, "Safety and security risk assessment in cyber-physical system," *IET Cyber-Physical Systems: Theory & Applications*, vol. 4–3, pp. 221–232, 2019.

41. E. Ruijters and M. Stoelinga, "Fault tree analysis: A survey of the state-of-the-art in modeling, analysis and tools," *Computer Science Review*, vol. 15–16, pp. 29–62, 2015. [Online]. Available: https://www.sciencedirect.com/science/article/pii/S1574013715000027.

42. L. Grunske, R. Colvin, and K. Winter, "Probabilistic model-checking support for FMEA," pp. 119–128, 10 2007.

43. M. Rausand and S. Haugen, *Hazard Identification*. John Wiley & Sons, Ltd, 2020, ch. 10, pp. 259–337. [Online]. Available: https://onlinelibrary.wiley.com/doi/abs/10.1002/9781119377351.ch10.

44. M. Modarres and S. W. Cheon, "Function-centered modeling of engineering systems using the goal tree-success tree technique and functional primitives," *Reliability Engineering & System Safety*, vol. 64, no. 2, pp. 181–200, 1999. [Online]. Available: https://www.sciencedirect.com/science/article/pii/S0951832098000623.

45. D. Lee, J. Lee, S.-W. Cheon, and J. Yoo, "Application of System-Theoretic Process Analysis to Engineered Safety Features-Component Control System," 2013.

46. I. Friedberg, K. McLaughlin, P. Smith, D. Laverty, and S. Sezer, "STPA-safesec: Safety and security analysis for cyber-physical systems," *Journal of Information Security and Applications*, vol. 34, pp. 183–196, 2017. [Online]. Available: https://www.sciencedirect.com/science/article/pii/S2214212616300850.

47. S. Kriaa, M. Bouissou, L. Piètre-Cambacedes, and Y. Halgand, "A Survey of Approaches Combining Safety and Security for Industrial Control Systems," *Reliability Engineering and System Safety*, vol. 139, pp. 156–178, 02 2015.

48. L. Chung and J. C. S. do Prado Leite, *On Non-Functional Requirements in Software Engineering*. Berlin, Heidelberg: Springer Berlin Heidelberg, 2009, pp. 63–379.

49. A. Kornecki, N. Subramanian, and J. Zalewski, "Studying interrelationships of safety and security for software assurance in cyber-physical systems: Approach based on Bayesian belief networks," pp. 1393–1399, 01 2013.

50. International Organization for Standardization (ISO), "ISO 12100:2010-general principle for design-risk assessment and risk reduction." 2010.

51. Federal Ministry for Climate Action, Environment, Energy, Mobility, Innovation and Technology (BMK) Austria, "Sicherheit für die digitale Transformation der Produktion," 2020, accessed: 2020-10-22. [Online]. Available: https://www.bmk.gv.at/themen/innovation/publikationen/produktion/sigi.html.

52. J.-P. A. Yaacoub, O. Salman, H. N. Noura, N. Kaaniche, A. Chehab, and M. Malli, "Cyber-physical systems security: Limitations, issues and future trends," *Microprocessors and Microsystems*, vol. 77, p. 103201, 2020. [Online]. Available: http://www.sciencedirect.com/science/article/pii/S0141933120303689.

53. S. F. D'amato and D. W. Mallik, "Plastic molding of articles including a hologram or other microstructure," Dec. 10 1991, US Patent 5,071,597.

54. C. A. Cole and J. T. Weber, "Package integrity indicating closure," Apr. 2 2013, US Patent 8,408,792.

55. V. Immler, J. Obermaier, K. K. Ng, F. X. Ke, J. Lee, Y. P. Lim, W. K. Oh, K. H. Wee, and G. Sigl, "Secure physical enclosures from covers with tamper-resistance," *IACR Transactions on Cryptographic Hardware and Embedded Systems*, vol. 2019, no. 1, p. 51-96, Nov. 2018. [Online]. Available: https://tches.iacr.org/index.php/TCHES/article/view/7334.

56. Y. Liu, K. Huang, and Y. Makris, "Hardware trojan detection through golden chip-free statistical side-channel fingerprinting," in *Proceedings of the 51st Annual Design Automation Conference*, 2014, pp. 1–6.

57. M. M. T. Bhunia Swarup, *The Hardware Trojan War*. Springer-Verlag GmbH, 2017. [Online]. Available: https://www.springer.com/de/book/9783319685106.

58. B. Bailey, "Optimization challenges for safety and security," 2019, accessed: 2020-09-25. [Online]. Available: https://semiengineering.com/optimization-challenges-for-safety-and-security/.

59. W. A. Arbaugh, W. L. Fithen, and J. McHugh, "Windows of vulnerability: A case study analysis," *Computer*, vol. 33, no. 12, pp. 52–59, 2000.

60. A. A. Cárdenas, S. Amin, and S. Sastry, "Research challenges for the security of control systems." in *HotSec*, 2008.

61. B. Brenner, E. Weippl, and A. Ekelhart, "Security related technical debt in the cyber-physical production systems engineering process," in *IECON 2019-45th Annual Conference of the IEEE Industrial Electronics Society*, vol. 1. IEEE, 2019, pp. 3012–3017.

62. G. Sabaliauskaite and A. P. Mathur, "Aligning cyber-physical system safety and security," in *Complex Systems Design & Management Asia*. Springer, 2015, pp. 41–53.

Runtime Monitoring for Systems of System

A Closer Look on Opportunities for Manufacturers in the Context of Industry 4.0

Michael Vierhauser and Alexander Egyed

Abstract

Software-intensive systems in general and Cyber-Physical Systems (CPS) in particular have drawn considerable attention from both industry and academia in recent years, with companies increasingly adopting Cyber-Physical Production Systems (CPPS). Regardless of the domain in which these systems are deployed, what they have in common is a shift from traditional software engineering principles towards a development process where software, hardware, and human actors controlling these systems are deeply interwoven and dependent upon each other. To ensure safe operation, it is crucial that an SoS complies with its requirements. Engineers and maintenance personnel must monitor if and how the SoS meets its requirements at runtime, e.g., to correctly verify timing behavior, or measure performance and resource consumption. In this chapter, we provide a brief introduction to SoS and CPPS, and investigate runtime monitoring from two different angles. First, we discuss requirements and challenges from the machine vendor's perspective. Second, we focus on the customer itself, i.e., typically a shop floor owner who has to combine a multitude of different components, both machinery and software systems for her production system. Finally, we discuss potential applications and benefits of runtime monitoring and provide an outlook presenting current research lines in SoS monitoring.

M. Vierhauser (✉)
Secure and Correct Systems Lab, Johannes Kepler University Linz, Linz, Austria
e-mail: michael.vierhauser@jku.at

A. Egyed
Institute for Software Systems Engineering, Johannes Kepler University Linz, Linz, Austria
e-mail: alexander.egyed@jku.at

© The Author(s), under exclusive license to Springer-Verlag GmbH, DE, part of Springer Nature 2023
B. Vogel-Heuser and M. Wimmer (eds.), *Digital Transformation*,
https://doi.org/10.1007/978-3-662-65004-2_8

Keywords

Cyber-Physical Production Systems • Systems of Systems • Runtime Monitoring

1   Introduction

Software-intensive systems in general and Cyber-Physical Systems (CPS) in particular have drawn considerable attention from both industry and academia in recent years. With the steady growth of smart cities, autonomous systems, or Internet of Things (IoT) applications, CPS are becoming pervasive in everyone's daily life. Furthermore, in an industrial context, in order to increase profitability and competitiveness, companies are increasingly adopting Cyber-Physical Production Systems (CPPS) and shop floor automation technologies. Industry 4.0, as the fourth industrial revolution, emphasizes the trend towards automation, the use of "intelligent machines", advanced sensors to monitor the systems at runtime, and smart networks to connect and exchange data between machines and also humans. As an example, the EU has dedicated 1B Euros in research funding for "Factories of the Future" and Smart Manufacturing to "[...] adapt manufacturing to a smart, green and inclusive economy" [5]. Often, these kinds of systems can be seen as Systems of Systems (SoS) which are characterized as large, decentralized, and heterogeneous systems, where different parts (or independent systems) have to work together and interact to achieve the overall goal [34]. Regardless of the domain in which these systems are deployed, what they share and have in common is a shift from traditional systems- and software engineering principles towards a development process where software, hardware, and human actors controlling these systems are deeply interwoven and dependent upon each other. This, in turn, not only affects the way software is used but also impacts the way CPS are designed, developed, maintained, and evolved. Furthermore, for many CPS that involve or interact with humans, these systems must not "only" deliver their required functionality but it must do so in a way that ensures that the system is safe and secure for its intended use. Due to the interplay of a vast number of different components with hardware, software, and human actors, operating these systems the intended way and ensuring their safe behavior at runtime is a non-trivial and challenging task.

In this chapter, we focus on how these systems can be monitored at runtime and how data can be collected, and what purpose this data can serve in order to make educated decisions about the state of the system.

In Sect. 2, we provide a brief introduction to SoS, provide background about their characteristics and challenges developers and engineers phase when putting such systems in operation. Subsequently, in Sect. 3, we analyze two different perspectives in which runtime monitoring and runtime data serve rather distinct purposes that nevertheless share a common goal: The view of the machine vendor on the one hand and the shop flow owner on the other hand. In Sect. 4, we then present possible research directions and examples of runtime monitoring for large-scale SoS and CPS. Finally, in Sect. 5 we revisit ReMinds [46], a framework for requirements-based monitoring of SoS. ReMinds employs a domain-specific language to define constraints and leverages incremental consistency checking to perform constraint checks at runtime.

2   Systems of Systems and Cyber-Physical Production Systems

SoS are large, decentralized, and heterogeneous systems with operational and managerial independence. They integrate "a finite number of constituent systems which are independent and operable, and which are networked together for a period of time to achieve a certain higher goal" [17]. These systems typically come into existence gradually and are often the result of decades of development by different teams. These types of systems are a vital part of the infrastructure of many domains, including defense, industrial production, healthcare, and infrastructure and transportation to name but a few. Examples for such SoS range from industrial automation systems [44], military applications [6], emergency response [25], or Information Systems [39].

Maier [22] and Daham [7] have identified four different types of SoS that vary in terms of their independence, and how they are managed and controlled.

The first type are *Directed SoS*. In this case, the SoS is centrally managed during long-term operation and fulfills the specific purpose or any new emerging purpose. While the systems contributing to the SoS operate independently, their operational mode however is subordinated to a centrally managed purpose.

The second type of SoS are *Collaborative SoS*. Compared to a directed SoS, in a Collaborative SoS systems voluntarily collaborate to fulfill the agreed central purposes without a central authority managing the process and systems.

The third type, *Virtual SoS* lack a central management authority and centrally agreed purposes. Large-scale behavior emerges and may be desirable but the "super-system" must rely upon implicit mechanisms to maintain it.

The fourth and last group are Acknowledged SoS. While these SoS have designated objectives and resources, the constituent systems retain their independent ownership, objectives, and funding. Changes in the constituent systems are based on collaboration between the SoS and the system.

One of the key aspects of such large-scale systems, that need to be taken into consideration when ensuring their correct behavior, is that the different constituent parts of an SoS might have been designed and implemented by different teams (operational and managerial independence) using different technologies, and might have not even been designed to work together in the first place. Different stakeholders are involved in the engineering process and are in charge of the requirements, which may be cross-cutting, overlapping, or even conflicting. SoS comprise various heterogeneous systems which differ regarding their architecture, technologies, or development organization.

These characteristics make coordinated evolution of the entire SoS, or even certain parts of the System of Systems extremely challenging. Different release cycles of the constituent systems result in systems evolving independently, while interoperability needs to be ensured. Therefore, SoS do not necessarily exhibit predictable and dependable behavior and it is difficult to conceive SoS as a single entity resulting in a continuous process of reacting to these changes. Monitoring these systems at runtime and checking predefined constraints to ensure that, for example, interfaces adhere to their predefined requirements plays an important role [44].

According to Boehm and Lane [3], traditional 20th-century acquisition and development processes do not work well on such systems due to their characteristics, i.e., "*not only [...] they integrate multiple, independently developed systems, but also they are very large, dynamically evolving, and unprecedented with emergent requirements and behaviors, and complex socio-technical issues to address*". Similarly, in work on ultra-large-scale systems Sullivan [38] suggests "*to rethink software research and to work o a new synthesis of systems engineering and software engineering that looks beyond traditional engineering*".

In order to address these challenges, a broad body of research exists ranging from improved testing strategies for SoS [26], ensuring interoperability between components [24], or introducing a digital representation, a digital-twin to better understand, analyze and simulate the behavior of such systems. One fundamental aspect when dealing if uncertainty and emerging behavior is to monitor the SoS (and its constituent parts) at runtime to be able to detect, and more importantly react to undesired behavior deviating from predefined requirements. The research area of runtime monitoring has gained traction providing support for monitoring the behavior of such systems [1, 46], their performance [10, 42], or resource consumption. Collecting these indicators from the system at runtime, and more importantly analyzing them in order to perform corrective actions when necessary is a vital part of operating such large-scale, complex, and heterogeneous, systems.

In the following sections, we focus on runtime monitoring support for Systems of Systems, with a focus on Industrial Production Systems and Industry 4.0 applications.

3   Runtime Monitoring of Industry 4.0 Applications—The Two Perspectives

As envisioned almost a decade ago [19] with the new paradigm of Industry 4.0 and the increasing automation and inception of Cyber-Physical Systems in the production process new opportunities and challenges emerge. This not only affects the owner and operator of the shop floor introducing new machinery, hardware, and software with the goal to reduce production costs and increase capacity and efficiency but also on the manufacturer and machine vendor that has to meet new demands regarding configurability and interoperability of their systems within a greater ecosystem. In the remainder of this section, we take a closer look at these two sides of the story in the context of runtime monitoring. First from the viewpoint of the machine producer who produces one (or more) part of a much bigger shop floor and second from the viewpoint of the shop floor owner who is in charge of the entire SoS, changes parts of it, needs to maintain and evolve it over time.

3.1   The Machine Vendor View

A machine vendor produces one part contributing to a much bigger System of Systems ultimately forming the shop floor. The challenges he faces in that scenario is that while he has control over his own hardware, sensors, components, and software there is little to no knowledge on how, the machine will be used, under which condition it will operate, and what unforeseen factors might influence the system at runtime when interacting with other parts. This includes, for example, the environmental variables, which some of them might be unknown, some only partially known, and some might even be inaccurate or incorrect or knowledge about how the machine is meant to interact with other machines or workers. To exacerbate the situation, the latter falls into the area of safety-critical behavior and can have significant impact on the safety of the workers working on the shop floor in general or interacting with the machine in particular. In cases where vendors provide multiple parts of the system (or even the entire system), integration testing can ensure that the different parts work together according to the defined requirements. In cases, however, where this is not the case, techniques such as software-in-the-loop testing allow for simulating other components and environmental impact factors before the actual hardware-in-the-loop testing, e.g., during site acceptance testing (SAT) or during the actual factory acceptance test (FAT) where the system must be validated. Additionally, this information about the deployed systems can be further leveraged for product innovation and for data-driven business models [36].

Collecting information about the status of the machine and then, more importantly, inferring meaningful information has to be inferred from the data has become a crucial task [21]. This requires the machine to be monitored at runtime to collect information about its state, incoming and outgoing information, as as well as its surrounding. This allows the machine vendor to, internally, for example, perform checks of input and sensor values to detect deviations from predefined specifications and even facilitates the possibility to, at runtime adapt to changing (external) factors.

3.2   The Shop Floor Owner View

While the machine vendor has a micro-level view on the system, i.e., the machines he is manufacturing and their immediate interfaces, the shop floor owner purchasing and subsequently operating these machines has to have a more macro-level view on the entire SoS keeping track of all its different constituent parts. Different machines, from different vendors, using different technologies, etc. have to work together on the shop floor to manufacture the final product. While these machines have been purchased in the first place with a specific purpose in mind, given certain specifications and requirements, the shop floor owner might know very little to nothing about how these machines behave and work internally. The machine might conform to certain industry standards, such as OPC UA [28] ensuring a standardized machine-to-machine communication protocol, this however does not necessarily mean that the full behavior of the system (including software and hardware) is obvious. Furthermore, to exacerbate the situation, depending on the type of machine and the terms of purchase, shop floor owners have no or at least only limited ability to update or modify these machines, being dependant upon the machine vendor and service contracts.

While in the past situation was acceptable due to the limited amount of flexibility and autonomy expected by such machines this has drastically changed in the context of Industry 4.0. Machines are expected to exhibit a certain level of autonomy or even capabilities to self-adapt to certain changes in the environment and they may experience some level of automated maintenance which needs to be carefully orchestrated within the context of the entire shop floor. Techniques such as Condition (Based) Monitoring (CM or CBM) and Prognostics and Health Management have gained traction that allow predictive maintenance to detect deterioration of mechanical parts, or predict performance problems in an early stage before the problem occurs. This form of runtime monitoring and data analysis allows the shop floor owner to plan maintenance early and thus avoid costly standstill periods of a machine, parts of a production line, or in the worst case the entire shop floor if problems are not detected leading to an error, crash of the software, or a broken or worn-out piece of hardware.

With the increasing automation of machines, shop floors, and entire factories both shop floor owners and machine vendors have a vital interest in making their machines "more intelligent". While their intentions and desired purposes might differ, what unifies them is the need for being able to collect and subsequently analyze data collected during operation, i.e., while the systems and the machines are operating in their environment. This, in turn, requires new capabilities, such as standardized interfaces and certain challenges to be taken into consideration. In the following, we present a brief overview of potential applications of runtime monitoring in different stages of the lifecycle of the system and for different purposes.

4   Potential Applications of Monitoring

Traditional engineering processes heavily rely on testing the system in various different aspects. Both manual testing approaches (such as code walkthroughs, technical reviews, or code inspections), as well as automated testing approaches are widespread and an active research community has formed around this topic investigating topics such as automated test case generation and prioritization [29, 32], fuzzy testing or regression testing. However, testing does not guarantee the absence of errors and especially in the context of CPS and

SoS, where emergent behavior is one key characteristic, it is simply not possible to cover all potential situations during development.

Monitoring the system during operation (i.e., monitoring the system at runtime), allows going beyond what can be tested and covered by these tests during development. Runtime monitoring is a vital part that helps to ensure a correct, safe, and secure operation of a machine even if the machine is modified or its environment changes.

Similar to testing, a broad research community has emerged exploring various different aspects of runtime monitoring for various different domains [30]. Examples range from runtime verification, requirements-based monitoring, or performance monitoring. In general, a wide variety of non-functional requirements (NFRs) can be monitored and subsequently checked for compliance with the specified requirements. What should be monitored for a certain system or machine and what is important depends, to a large extent, on the type of the machine or system and application area.

Model-based approaches using widely used modeling languages, such as UML and SysML or domain-specific languages (DSL) can also be leveraged for modeling diverse heterogeneous components, such as communication architectures allowing easy collection and data analysis in an industrial automation system [35, 41]. Leveraging domain-agnostic approaches, modeling and describing system components with either SysML or a dedicated DSL furthermore avoids premature commitment to certain technologies and facilitates flexible code generation based on these models.

4.1 Monitoring Safety Properties

On a shop floor, with different types of machines and humans interacting with these machines, safety-related properties could be of particular importance. Image a robot performing drilling or welding tasks where humans require to maintain a minimum safety distance at any time during operation. Additional sensors can be used to ensure that the immediate surrounding of the robot is clear and that nobody has violated the minimum safety distance. Should one sensor report a person stepping into the area the robot can slow down, pause, or even shut down depending on the distance and severity of the violation.

4.2 Condition Monitoring

Another area that has recently gained traction is Condition (Based) Monitoring (CM or CBM) for industrial automation systems. The ISO/TC 108/SC 5 committee attends to "Condition monitoring and diagnostics of machine systems" and the ISO 17359 standard provides guidelines for setting up a condition monitoring program for machines. This included the *"Standardization of the procedures, processes and equipment requirements uniquely related to the technical activity of condition monitoring and diagnostics of machines systems in which selected physical parameters associated with an operating machine system are periodically or continuously sensed, measured and recorded for the interim purpose of reducing, analyzing, comparing and displaying the data and information so obtained and for the ultimate purpose of using this interim result to support decisions related to the operation and maintenance of the machine system"* [16].

Condition Monitoring is related to preventive maintenance where components are routinely and periodically inspected with the goal to detect and fix problems before an error or component failure occurs avoiding costly downtimes. CM buils upon this paradigm by collecting information about the system and/or mine using sensor data, performance data, or execution logs to facilitate (automated) self-diagnostic capabilities [18]. Maintenance decisions can then be made based on condition monitoring. This could for example include data such as vibration data indicating the need for changing a component, power voltages, or a degradation in the storage performance of a software component indicating the need for database maintenance. Examples for condition monitoring being used in industry range from smart factories and socio-cyber-physical systems [14] to power generators [40]. Furthermore, in cyber-physical systems using distributed networks, integrating diverse embedded sensors, and employing a variety of different actuators, SCADA (Supervisory Control and Data Acquisition) systems are often used for gathering real-time data, (remote) monitoring, controlling processes and informing the user about the health and status of the various constituent components of the system. These systems are inherently complex in nature as they span across various different facilities and need to cover multiple different production processes. Based on the SCADA architecture, approaches have emerged to detect and prevent security vulnerabilities, such as intrusion detection and prevention [12, 27] or to monitor and optimize the performance of the system based on the real-time data collected by the SCADA system data [8, 15, 47].

5   Requirements-Based Monitoring for Systems of Systems

As software systems are subject to continuous change and ongoing evolution during their lifetime, requirements describing the software's intended behavior are becoming more important and complex. This is particularly true for SoS where requirements originate in different parts of the system, at different levels of granularity, and need to address cross-cutting concerns and SoS wide behavior. Furthermore, this goes hand in hand with a shift from requirements being statically defined during the elicitation phase, and then forgotten, towards requirements as runtime entities which are ubiquitous throughout the whole lifecycle of the software system [2]. To specifically address these issues, requirements-based monitoring approaches stress the need for continuously checking the adherence of systems to their requirements, particularly during their operation [33]. Software *Monitors* are used to observe the behavior of a software system and, at runtime, check whether it still behaves as intended or if it deviates from its defined requirements.

For example, in an industrial factory, a factory-wide automation system is in charge of managing, tracking, and monitoring materials and goods at the different production stages. It typically comprises both hardware and software developed and manufactured by different vendors and companies sizing up to several million LoC. The systems have heterogeneous architectures, they have been developed using diverse technologies, and they frequently interact, e.g., when exchanging data controlling the production process. Although these different systems are engineered independently, there are manifold dependencies in a production process that need to be considered when planning their joint operation. Furthermore, there may be dependencies between components within one particular system, such as that certain steps that are performed by a robot need to occur in a certain order or within a certain timeframe. Although such requirements and their dependencies are carefully specified and managed during development, it is crucial to monitor them after deployment to detect inaccurate and erroneous behavior at runtime.

This is particularly important after changing or upgrading certain components, or when the manufacturing process and hence the arrangement of machines change on the shop floor. In the following, we describe a typical scenario for runtime monitoring from our previous experience in this domain [44]. The emphasizes the need for runtime monitoring and constraints that can be dynamically (and continuously) checked at runtime.

The scenario (cf. Fig. 1) starts with an issue being reported describing a deviation from the expected system behavior after changing the configuration of one of the machines on the shop floor. Due to the interplay of several machines, systems, and components, there may be a couple of different reasons for this new behavior and both hardware and software issues could have caused the problem. Without any runtime information in place, this would trigger a cumbersome manual process where the different parts need to be inspected one by one to pinpoint the origin of the problem. With a proper runtime monitoring infrastructure in place however, a service engineer in charge of the system can easily retrieve more details about the

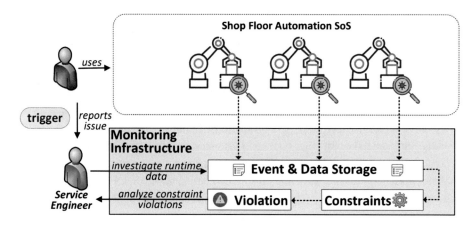

Fig. 1 Scenario

state of the system, e.g., by analyzing recorded event data or performance information. This task is simplified if constraints related to the issue have already been defined and violations have been recorded. If necessary, the service engineer can also add new constraints to the monitoring infrastructure to be made aware if a similar problem should occur in the future. Once the problem has been uncovered and fixed, these newly added constraints can also serve as tests to ensure that the problem has in fact been resolved and does not resurface at a later date.

5.1   Challenges for Monitoring Systems of Systems

When dealing with the complex behavior and structure of SoS, a number of challenges arise when designing a runtime monitoring solution that need to be taken into consideration. Based on a number of case studies [31, 44] and analysis of the domain [30] we have identified key challenges for monitoring SoS from the perspective of engineers and service staff. In the following, we provide a brief overview of these challenges [46].

Level of Granularity of Monitoring Data

Depending on the purpose of the monitoring solution, e.g., for analyzing low-level events, or checking data exchange between systems, different types of data and information needs to be collected and analyzed. Especially in the context of an SoS, this means different sources from where information is provided. For example, in a factory automation system following a traditional SCADA multi-layer architecture, retrieving information about inter-system communication, e.g., between level 2 and level 3 systems requires instrumenting the interfaces provided by these systems. Getting information about inter-process communication, for example, for tracking products across different production stages, requires instrumenting the communication paths of the involved processes. Collecting information about user interactions requires instrumenting the HMI and its communication with the different systems. Additionally, a variety of other sources can be leveraged to collect information about the systems and their components, such as log files, or event data archived in databases.

Runtime Checks Across Different Systems

SoS are based on the principle of interaction and collaboration. Multiple different systems, components, and machines are involved in order to satisfy customer requirements. Furthermore, as external services are used, the systems constituting an SoS do not only communicate internally, but share data with other systems. This in turn results to the situation that it is often not possible to allocate requirements to a specific component or part of the system. Traditional monitoring approaches focusing on single systems, however, are limited with respect to checking such global SoS properties and, furthermore, lack support for instrumenting different systems and adequately capturing their interaction.

Heterogeneous Technologies

As SoS consist of a number of different systems and components, it is very likely that individual parts of an SoS are designed, implemented, and maintained, by different teams within a company, or even by different external or third-party vendors. This means that these systems typically use different technology stacks, with diverse architectural styles. While this is a challenge itself for SoS operation, this has also a major impact when providing monitoring support. System instrumentation and data collection can not be limited to a specific technology or implementation language, and instrumenting interfaces between different systems can pose further challenges. Instrumenting these diverse sources of information requires the development of domain-specific and often application-specific probes to accommodate the different technologies and system architectures.

Diverse and Changing Requirements and Monitors

SoS are subject to continuous maintenance and evolution in order to meet fix bugs, accommodate new hardware, or satisfy new requirements. For monitoring an SoS this means that substantial parts of the monitoring solution also need to change over time and be adapted and need to co-evolve with the system under monitoring. To support these adaptations, runtime monitors need to be customizable to accommodate customer-specific usage scenarios and system variants. Furthermore, depending on the application context, monitors need to be adapted. For example, during simulation more detailed monitoring can be performed, i.e., more data can be collected and analyzed without interfering with the production process. However, during production, on a live system, resources are more critical requiring monitors to be configurable to cope with these different requirements.

Performance and Monitoring Overhead

Since SoS are typically large in size, with a number of different components and subsystems, large amounts of data have to be collected and analyzed. One key requirement of a runtime monitoring solution, however, is that it does not affect the system under monitoring, while collecting and analyzing data at the same time. This means that, especially during system operation, the monitoring solution must not significantly slow down the system during operation and its performance overhead must be carefully controlled. Furthermore, in order to be usable in the context of a CPPS, for example, the monitoring approach must scale to industrial requirements regarding the number of monitors, frequency of events, and amount of data collected (Fig. 2).

5.2   A Requirements Monitoring Model

Providing support for Requirements Monitoring of an SoS requires, on the one hand, the technological capabilities to perform monitoring activities, but additionally, due to the size

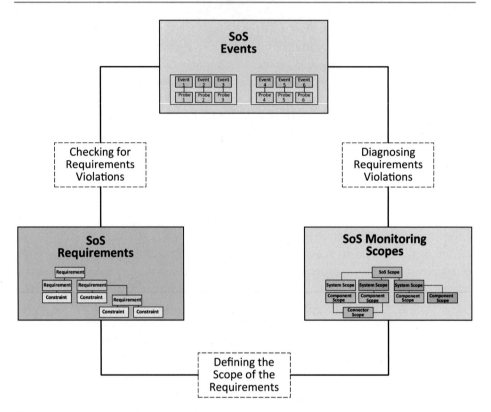

Fig. 2 Requirements Monitoring Model

and complex nature of SoS, requires means for managing, analyzing, and connecting various different elements involved in the monitoring process (i.e., the Requirements, Constraints, and Monitors collecting information). The goal is to reduce the burden of the developer to trace an occurring violation to its origin and ultimately resolve the problem. Based on the previously mentioned challenges, different aspects and their dependencies need to be described, and we have identified three main dimensions that need to be taken into consideration: (i) the Components (i.e., different hardware and software systems, as well as their interfaces) of the SoS, (ii) the actual Requirements that describe the behavior of the SoS, and finally (iii) the Monitors embedded within the system collecting data at runtime. These three dimensions provide the foundation for our Requirements Monitoring Model (cf. Fig. 2) [45].

Dimension I: Requirements

A Requirement describes a functionality, a property, or behavior of the SoS to be monitored at runtime [13]. Due to its sheer size, only a subset of the requirements that are defined for the SoS are considered as relevant to be monitored and checked at runtime and thus need to be included in the Requirements Monitoring Model. Depending on the purpose of monitoring, for example, monitoring performance-related requirements, timing issues, or behavior, different requirements need to be included. Furthermore, depending on the granularity and rigor requirements are selected for the different systems and components. In order to provide a more fine-grained analysis, these may need to be refined and broken down into sub-requirements before they can be used in the context of runtime monitoring. Afterwards, to facilitate runtime checks, to ensure that the SoS adheres to its requirements, these need to be formalized, typically in the form of constraints (cf. Sect. 5.3). Each requirement is then assigned to a dedicated Scope (cf. Dimension III). This connection between Scopes and Requirements enables monitoring specific parts of the SoS and raise alarms for Certain components or Systems if constraint checks fail, meaning that requirements are violated. In an SoS context, however, it might not always be possible to assign Requirements to a single System or Component. Requirements may be cross-cutting, spanning across multiple different systems, or different parts of the SoS.

Dimension II: Events

As the systems part of an SoS are typically quite diverse with regards to technology and architecture, a uniform representation of the collected information is essential when reasoning and checking constraints across the different parts of the SoS. Explicitly describing and modeling the different events and their respective types that are collected by Probes at different locations within the SoS provides allows modelers to define a taxonomy for structuring and filtering information. An Event can be a relevant system operation or interaction with the system that happens at a specific scope at a specific point in time. The events are provided by Probes and can subsequently be used to perform constraint checks at runtime. A Probe, in turn, is a single and encapsulated unit that provides arbitrary information extracted or intercepted from a system under monitoring [23]. Depending on the technology and architecture of the system the way how information is collected differs. Thus Probes largely depend upon the system to be monitored.

Dimension III: Monitoring Scopes

Finally, Monitoring Scopes connect and allocate both Probes instrumenting the system and Requirements relevant for specific components. A Monitoring Scope describes an area of interest to be monitored in the context of the overall SoS architecture. For example, a scope can represent a specific system part of the SoS, one or more components, or a connector (such as interfaces or APIs) between the different parts of an SoS. As part of our Requirements Monitoring Model, we have identified five different types of Scopes. These different types can then be instantiated for a concrete SoS: 1) the type *SoS* represents the root of the hierarchical scope model, i.e., it is related to SoS-wide properties or behavior that can not be strictly assigned to a single system or component in the SoS; 2) the Scope *System* represents a single system part of the SoS. System-level requirements are associated with this Scope; 3) the type *Component* refines the System Scope and is used to describe subsystems and components of a system. It can be further used to logically group capabilities and functionalities to indicate team responsibilities. The granularity of how scopes are modeled for a specific SoS thus largely depends on the SoS architecture, and to a certain extent on the organizational structure of the SoS. Furthermore, two additional Scope Types are denoted to interactions between different SoS parts: 4) the *System Interaction* Scope addresses the communication of systems in the SoS, whereas 5) the *Component Interaction* Scope describes interactions of components. These five different types of Scopes facilitate the creation of a hierarchical representation of different monitoring concerns, and in turn, provide a fine-grained view on the overall SoS and its constituent systems. Figure 3 shows an example of the Requirement Monitoring Model with different scopes visualized in our ReMinds Tool for an industrial automation system [45].

Besides the three dimensions, a vital part of the Requirements Monitoring Model is the connection between them. By creating links between requirements and the respective constraints and the different scopes violated constraints can be directly traced to their origin in the scope and hence in the respective system or component. Furthermore, each constraint is linked to at least one (or more) events. These events, collected at runtime, are used to perform the actual constraint checks and detect violations. Probes, which are responsible for collecting these events are also associated with a specific monitoring scope. This in turn

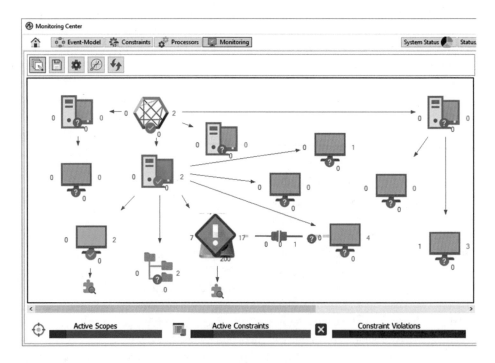

Fig. 3 Example from the ReMinds Monitoring UI [45]

supports fine-grained, event-level diagnosis of each constraint violation and its allocation in the SoS.

5.3   A Domain-Specific Language for SoS Constraint Checking

Domain-specific languages are have recently gained popularity in the domain of software engineering for various different purposes [20], such as variability modeling [11], or machine learning [37]. In contrast to general-purpose languages, DSLs are designed for a specific purpose in a particular domain, or for a particular problem. While various different types of constraint languages, for different purposes, have been proposed, e.g., requirements-level methods [43], UML-based approaches, or formal approaches [4] their support for different types of constraints is often limited. Depending on the monitoring approach, for example for performance properties, temporal properties, or event sequence analysis, different types of constraint checks are required. Additionally, approaches typically are for a particular application domain or technology, such as service-based systems, making it hard to apply to other types of systems or in other domains. Finally, due to the complex nature, and special skills required, for example, temporal logic, many existing constraint languages are deemed inconvenient by industrial end users as Furthermore, most existing approaches do not provide adequate tool support to facilitate easy creation and maintenance of constraints.

In order to provide a simple, and easy to use language for diverse constraints, focusing on usefulness and practical applicability, we have devolved a DSL that allows engineers to define diverse types of constraints, including temporal, structural, and constraints on event data. Furthermore, to reduce overhead and increase performance, the constraint checker allows to incrementally validate constraints at runtime [9], ensuring that constraint violations of certain parts of the SoS can be reported instantly to the user. To facilitate maintenance activities, and to adapt to different usage scenarios, constraints can be deployed and activated/deactivated at runtime.

We assume a stream of events observed at runtime, which are then collected in the runtime instance of the Requirements Monitoring Model. When creating the DSL and refining it based on the feedback received from industrial end users, we collected best practices and lessons learned which we deem useful for researchers and practitioners working in this area:

Follow the YAGNI—"you aren't gonna need it"—Principle when Developing a DSL
Before developing "yet another" constraint language, we investigated, analyzed, and discussed a variety of alternative existing constraint languages. Despite the fact that a number of these approaches would have provided support for the required constraint checks, the reaction of our industry partner was different. This was mainly because these languages provide too many features which are not needed in the concrete application context and defining constraints was regarded as difficult using formal notations. A strong emphasis was on being able to define constraints as close as possible to natural language with only those

concepts that are actually required. A key lesson thus is to *keep a language as simple as possible*, only including concepts necessary for the concrete context. The constraint language should be oriented towards end users and hide the complexity of the underlying constraint engine.

Embrace Iterative Language Design for SoS

As SoS typically contain diverse, heterogeneous systems, with different teams involved in their development developing a new domain-specific language addressing all the diverse needs may be difficult. The development of our constraint DSL started by interviewing different teams responsible for different parts of the SoS. This helped us to identify commonalities and specific requirements for the language. With each new iteration of the DSL, which was discussed with the stakeholders, new features were added and existing ones refined according to the feedback we received. This iterative language design helped to develop a solution that addresses the needs and requirements of the different stakeholders.

Simplify and Provide Automation Support for Extending the DSL

While keeping the constraint language simple is on fundamental key principle, new types of constraints and checks might still be needed as the system evolves. For example, in our context, while in a first iteration support for simple range checks and data constraints with primitive data objects were provided—and deemed sufficient at that time—later on, more complex checks of data objects and their relations were required. To be able to provide a simple language that covers stakeholder requirements, easy extension and refinement of the DSL is key. Using existing tool-support and technologies such as XText and XTend, provided by the Eclipse framework facilitates easy adaptation and provides sophisticated tool support for the end user based on the grammar of the DSL.

Keep the Mapping of the DSL to the Constraint Engine Flexible

While evolving and iteratively refining the DSL we learned that it might be required to switch out the constraint engine, for example, with an increasing number of events and constraint checks, the original engine might not be suitable anymore. From our experience, while our Java-based incremental checker worked well in our application scenario, future applications and requirements might require other constraint engines types of constraints. Therefore, one lesson learned was to *define a clear interface between the language and the engine* to allow replacing and updating them independently. For instance, while keeping our simple DSL, the underlying constraint engine could be replaced with a complex event processing engine, specialized in checking event sequences.

Support Dynamic Constraint Management

Industrial use cases demonstrate the necessity to frequently adapt existing constraints, or add new constraints, even while the system under monitoring and the monitoring infrastructure itself are running. This could for example be used to investigate an unforeseen and emerging issue, or to further diagnose an emerging requirements violation. Providing a dynamic approach, where *constraints can be easily updated and changed without shutting down or restarting the monitoring infrastructure*, therefore helps to keep maintenance efforts to a minimum.

6   Conclusion

This chapter reported on existing approaches and open issues for runtime monitoring of CPS in the Context of Systems of Systems. We discussed monitoring challenges from two different perspectives, namely the machine vendor, responsible for her part of a much bigger shop floor system, and the shop floor owner in charge of the entire shop floor who needs to operate a slew of different machines from different vendors running different kinds of software. Runtime monitoring provides a wide variety of potential application scenarios, depending on the type of SoS and the needs of the system owner, ranging from the support of preemptive maintenance using condition monitoring, to ensuring safe behavior when monitoring safety-critical properties of a system. We focused on one particular research area, requirements-based monitoring for SoS. With our ReMinds framework [46] we provide a holistic approach for SoS runtime Monitoring. It provides a domain-specific language allowing engineers and maintenance personnel to easily define and maintain different types of constraints which are automatically checked at runtime based on events and data collected from the instrumented SoS. This, in turn, facilitates the rapid detection and subsequent diagnosis of deviations from the specified and desired behavior of the system helping to avoid costly maintenance downtimes or critical system failures during production. u

References

1. Baresi, L., Guinea, S.: Event-based multi-level service monitoring. In: 2013 IEEE 20th International Conference on Web Services. pp. 83–90. IEEE (2013)
2. Bencomo, N., Whittle, J., Sawyer, P., Finkelstein, A., Letier, E.: Requirements reflection: requirements as runtime entities. In: Proceedings of the 32nd ACM/IEEE International Conference on Software Engineering-Volume 2. pp. 199–202 (2010)
3. Boehm, B., Lane, J.A.: 21st century processes for acquiring 21st century software-intensive systems of systems. Crosstalk: Journal of Defence Software Engineering **19**(5), 4–9 (2006)
4. Chen, F., d'Amorim, M., Roşu, G.: A formal monitoring-based framework for software development and analysis. In: International Conference on Formal Engineering Methods. pp. 357–372. Springer (2004)
5. Comission, E.: Factories of the Future. https://ec.europa.eu/digital-single-market/en/factories-future (2020), [Online; accessed 20-April-2020]
6. Dahmann, J., Rebovich, G., Lane, J., Lowry, R., Baldwin, K.: An implementers' view of systems engineering for systems of systems. In: 2011 IEEE International Systems Conference. pp. 212–217. IEEE (2011)

7. Dahmann, J.S., Baldwin., K.J.: Understanding the current state of us defense systems of systems and the implications for systems engineering. In: Proc. of the 2nd Annual IEEE Systems Conf. (2008)
8. Dao, P.B., Staszewski, W.J., Barszcz, T., Uhl, T.: Condition monitoring and fault detection in wind turbines based on cointegration analysis of scada data. Renewable Energy **116**, 107–122 (2018)
9. Egyed, A.: Instant consistency checking for the uml. In: Proceedings of the 28th international conference on Software engineering. pp. 381–390 (2006)
10. Ehlers, J., van Hoorn, A., Waller, J., Hasselbring, W.: Self-adaptive software system monitoring for performance anomaly localization. In: Proceedings of the 8th ACM international conference on Autonomic computing. pp. 197–200 (2011)
11. Eichelberger, H., Schmid, K.: Ivml: a dsl for configuration in variability-rich software ecosystems. In: Proceedings of the 19th International Conference on Software Product Line. pp. 365–369 (2015)
12. Fernandez, J.D., Fernandez, A.E.: Scada systems: vulnerabilities and remediation. Journal of Computing Sciences in Colleges **20**(4), 160–168 (2005)
13. Fickas, S., Feather, M.S.: Requirements monitoring in dynamic environments. In: Proceedings of 1995 IEEE International Symposium on Requirements Engineering (RE'95). pp. 140–147. IEEE (1995)
14. Fleischmann, H., Kohl, J., Franke, J.: A reference architecture for the development of socio-cyber-physical condition monitoring systems. In: 2016 11th system of systems engineering conference (SoSE). pp. 1–6. IEEE (2016)
15. Gonzalez, E., Stephen, B., Infield, D., Melero, J.J.: Using high-frequency scada data for wind turbine performance monitoring: A sensitivity study. Renewable energy **131**, 841–853 (2019)
16. ISO—The International Organization for Standardization: ISO/TC 108/SC 5 Condition monitoring and diagnostics of machine systems. https://www.iso.org/committee/51538.html (2020), [Online; accessed 01-January-2020]
17. Jamshidi, M.: Systems of Systems Engineering: Principles and Applications. CRC Press (2017)
18. Jin, T., Mechehoul, M.: Minimize production loss in device testing via condition-based equipment maintenance. IEEE transactions on automation science and engineering **7**(4), 958–963 (2010)
19. Kagermann, H., Lukas, W.D., Wahlster, W.: Industrie 4.0: Mit dem internet der dinge auf dem weg zur 4. industriellen revolution. VDI nachrichten **13**(11), 2 (2011)
20. Kosar, T., Bohra, S., Mernik, M.: Domain-specific languages: A systematic mapping study. Information and Software Technology **71**, 77–91 (2016)
21. Lee, J., Bagheri, B., Kao, H.A.: A cyber-physical systems architecture for industry 4.0-based manufacturing systems. Manufacturing letters **3**, 18–23 (2015)
22. Maier, M.W.: Architecting principles for systems-of-systems. In: Proc. of the INCOSE Int'l Symp. pp. 565–573. Wiley (1996)
23. Mansouri-Samani, M., Sloman, M.: Monitoring distributed systems. IEEE network **7**(6), 20–30 (1993)
24. Mendes, A., Loss, S., Cavalcante, E., Lopes, F., Batista, T.: Mandala: an agent-based platform to support interoperability in systems-of-systems. In: Proceedings of the 6th International Workshop on Software Engineering for Systems-of-Systems. pp. 21–28 (2018)
25. Nielsen, C.B., Larsen, P.G., Fitzgerald, J., Woodcock, J., Peleska, J.: Systems of systems engineering: basic concepts, model-based techniques, and research directions. ACM Computing Surveys (CSUR) **48**(2), 1–41 (2015)
26. de Oliveira Neves, V., Bertolino, A., De Angelis, G., Garcés, L.: Do we need new strategies for testing systems-of-systems? In: 2018 IEEE/ACM 6th International Workshop on Software Engineering for Systems-of-Systems (SESoS). pp. 29–32. IEEE (2018)

27. Oman, P., Phillips, M.: Intrusion detection and event monitoring in scada networks. In: International Conference on Critical Infrastructure Protection. pp. 161–173. Springer (2007)
28. OPC Foundation: OPC Unified Architecture. https://opcfoundation.org/about/opc-technologies/opc-ua (2020), [Online; accessed 01-January-2020]
29. Panichella, A., Kifetew, F.M., Tonella, P.: Automated test case generation as a many-objective optimisation problem with dynamic selection of the targets. IEEE Transactions on Software Engineering **44**(2), 122–158 (2017)
30. Rabiser, R., Schmid, K., Eichelberger, H., Vierhauser, M., Guinea, S., Grünbacher, P.: A domain analysis of resource and requirements monitoring: Towards a comprehensive model of the software monitoring domain. Information and Software Technology **111**, 86–109 (2019)
31. Rabiser, R., Vierhauser, M., Grünbacher, P.: Assessing the usefulness of a requirements monitoring tool: a study involving industrial software engineers. In: 2016 IEEE/ACM 38th International Conference on Software Engineering Companion (ICSE-C). pp. 122–131. IEEE (2016)
32. Ramler, R., Klammer, C., Buchgeher, G.: Applying automated test case generation in industry: a retrospective. In: 2018 IEEE International Conference on Software Testing, Verification and Validation Workshops (ICSTW). pp. 364–369. IEEE (2018)
33. Robinson, W.N.: A requirements monitoring framework for enterprise systems. Requirements engineering **11**(1), 17–41 (2006)
34. Sage, A.P.: System of Systems Engineering: Innovations for the 21st Century, vol. 58. John Wiley & Sons (2011)
35. Sollfrank, M., Trunzer, E., Vogel-Heuser, B.: Graphical modeling of communication architectures in network control systems with traceability to requirements. In: IECON 2019-45th Annual Conference of the IEEE Industrial Electronics Society. vol. 1, pp. 6267–6273. IEEE (2019)
36. Sorescu, A.: Data-driven business model innovation. Journal of Product Innovation Management **34**(5), 691–696 (2017)
37. Sujeeth, A.K., Lee, H., Brown, K.J., Rompf, T., Chafi, H., Wu, M., Atreya, A.R., Odersky, M., Olukotun, K.: Optiml: an implicitly parallel domain-specific language for machine learning. In: ICML (2011)
38. Sullivan, K.: Ultra-large-scale systems (keynote talk). In: Proc. of the 12th International Symposium on Component Based Software Engineering (2009)
39. Teixeira, P.G., Lopes, V.H.L., Dos Santos, R.P., Kassab, M., Neto, V.V.G.: The status quo of systems-of-information systems. In: 2019 IEEE/ACM 7th International Workshop on Software Engineering for Systems-of-Systems (SESoS) and 13th Workshop on Distributed Software Development, Software Ecosystems and Systems-of-Systems (WDES). pp. 34–41. IEEE (2019)
40. Tian, Z., Jin, T., Wu, B., Ding, F.: Condition based maintenance optimization for wind power generation systems under continuous monitoring. Renewable Energy **36**(5), 1502–1509 (2011)
41. Trunzer, E., Prata, P., Vieira, S., Vogel-Heuser, B.: Concept and evaluation of a technology-independent data collection architecture for industrial automation. In: IECON 2019-45th Annual Conference of the IEEE Industrial Electronics Society. vol. 1, pp. 2830–2836. IEEE (2019)
42. Van Hoorn, A., Waller, J., Hasselbring, W.: Kieker: A framework for application performance monitoring and dynamic software analysis. In: Proceedings of the 3rd ACM/SPEC International Conference on Performance Engineering. pp. 247–248 (2012)
43. Van Lamsweerde, A.: Requirements engineering: From system goals to UML models to software, vol. 10. Chichester, UK: John Wiley & Sons (2009)
44. Vierhauser, M., Rabiser, R., Grünbacher, P.: A case study on testing, commissioning, and operation of very-large-scale software systems. In: Companion Proceedings of the 36th International Conference on Software Engineering. pp. 125–134 (2014)

45. Vierhauser, M., Rabiser, R., Grünbacher, P., Aumayr, B.: A requirements monitoring model for systems of systems. In: 2015 IEEE 23rd International Requirements Engineering Conference (RE). pp. 96–105. IEEE (2015)
46. Vierhauser, M., Rabiser, R., Grünbacher, P., Seyerlehner, K., Wallner, S., Zeisel, H.: Reminds: A flexible runtime monitoring framework for systems of systems. Journal of Systems and Software **112**, 123–136 (2016)
47. Yang, W., Jiang, J.: Wind turbine condition monitoring and reliability analysis by scada information. In: 2011 Second International Conference on Mechanic Automation and Control Engineering. pp. 1872–1875. IEEE (2011)

Blockchain Technologies in the Design and Operation of Cyber-Physical Systems

Abel Gómez⊙, Christophe Joubert⊙ and Jordi Cabot⊙

Abstract

A blockchain is an open, distributed ledger that can record transactions between two parties in an efficient, verifiable, and permanent way. Once recorded in a block, the transaction data cannot be altered retroactively. Moreover, smart contracts can be put in place to ensure that any new data added to the blockchain respects the terms of an agreement between the involved parties. As such, the blockchain becomes the single source of truth for all stakeholders in the system.

These characteristics make blockchain technology especially useful in the context of Industry 4.0, distributed in nature, but with important requirements of trust and accountability among the large number of devices involved in the collaboration. In this chapter, we will see concrete scenarios where cyber-physical systems (CPSs) can benefit from blockchain technology, especially focusing on how blockchain works in practice, and which are the design and architectural trade-offs we should keep in mind when adopting this technology both for the design and operation of CPSs.

A. Gómez (✉) · J. Cabot
Internet Interdisciplinary Institute (IN3), Universitat Oberta de Catalunya (UOC), Barcelona, Spain
e-mail: agomezlla@uoc.edu

J. Cabot
e-mail: jordi.cabot@icrea.cat

C. Joubert
Prodevelop SL, Valencia, Spain
e-mail: cjoubert@prodevelop.es

J. Cabot
ICREA, Barcelona, Spain

B. Vogel-Heuser and M. Wimmer (eds.), *Digital Transformation*,
https://doi.org/10.1007/978-3-662-65004-2_9

Keywords
Blockchain • Distributed Ledger • Smart Contracts

1 Introduction

The Fourth Industrial Revolution—or Industry 4.0—is characterized by the need of combining complex engineering systems to move towards the automation and optimization of manufacturing technologies. In any Industry 4.0 initiative, cyber-physical systems (CPSs) that connect and integrate physical and digital elements in a single system are a key aspect. Unlike more traditional embedded systems, a full-fledged CPS is typically designed as a network of interacting elements with physical input and output instead of as standalone devices [18].

One of the major shortcomings in CPSs is their current centralized architecture which will struggle to scale up to meet the demands of future CPSs [5]. In this chapter, we will see why (and how) blockchain can be a key enabler to solve this challenge.

Blockchain is a distributed ledger. A ledger is essentially a record of transaction data. In a distributed ledger, the consensus of replicated, shared, and synchronized digital data is geographically spread across multiple sites, countries, or institutions [30]. Data is internally represented as a growing list of blocks where each block is linked to the previous one via a cryptographic hash [22].

Blockchain does not need a central administrator or a centralized data storage, which makes it ideal as a technology on top of which implementing virtual currencies. This was its very first application with the creation of Bitcoin[1] in 2008. However, since then, blockchain has been used in a large variety of scenarios [27]: education, health, social media, organizations' governance,[2] and even artificial intelligence [15]. We believe the combination of CPSs/IoT and blockchains will disrupt existing processes across a variety of industries. As an example, we will describe in this chapter the impact of blockchain adoption in the supply chain of maritime trade, for example, to register the events associated with actions carried out in the transport chain—such as entry or exit of a container—and the validation of those operations via smart contracts.

But blockchain can also help in the specification and generation of the CPSs systems themselves. Design and monitoring of CPSs are more and more based on model-driven engineering (MDE) [6] techniques that allow working with the CPS at a higher abstraction level and disregard, to a certain extent, low-level technical details of the concrete platforms the CPS is executed upon. Blockchain can be used to complement these methods in at least two critical aspects of the design and operation of a CPS: *runtime traceability* and *design accountability*. As the single source of truth for all stakeholders in the system, blockchains

---

[1] https://bitcoin.org/
[2] https://aragon.org/

can be used to record all changes in the system and make sure every single one of those changes can be traced back to the stakeholder that performed that change. Given that data in a blockchain cannot be altered retroactively, this recorded information can be used at any time to identify the liability and responsibility of everyone involved in the CPS development and monitoring.

The rest of this chapter is structured as follows. Section 2 introduces the key concepts of a blockchain data structure while Sect. 3 describes how that data can be safely manipulated by means of smart contracts. Then, Sect. 4 describes how blockchains can be useful to design and operate CPSs. Finally, Sect. 5 discusses some trade-offs and strategic decisions that must be considered when choosing the right blockchain design for your business, and Sect. 6 gives some final remarks to conclude the chapter.

2 Starting from the Beginning: What is a Blockchain?

A blockchain is a data structure defined as a hash tree of blocks representing transactions, where each block is identified by its hash value and contains the hash value of its predecessor. As its name suggest, although a blockchain is a tree, there is only one agreed upon chain of blocks, being the blocks of diverging branches considered as stale blocks.

Blockchain-like data structures are not new and, in fact, have already been used before in wide-spread technologies like Git.[3] But their popularity exploded with the irruption of Bitcoin and its innovations. Specifically, the innovation introduced by Satoshi Nakamoto [21]— the presumed pseudonymous of the person or persons who developed Bitcoin—was the Proof of Work (PoW) of the mining/consensus algorithm, which ensures that the agreed-upon state of the data is supported by at least 51% of the mining nodes. The algorithm is applied during the creation of new blocks for the blockchain and determines which blocks are to be accepted.

Although a blockchain is a quite simple data structure, several different concepts play an important role for its understanding. Thus, in the next subsections, we first describe the most important concepts in blockchains; and second, we describe what a blockchain looks like and how it is built.

2.1 Blockchain Basic Concepts

A blockchain is a data structure that, by design, allows an efficient and secure verification of its contents thus preventing a malicious actor to tamper with it. This is achieved by putting together different concepts from the computing field. Next, we introduce some of these basic concepts that play an important role:

---

[3] https://git-scm.com/

Blockchain — A blockchain is a distributed ledger that contains a continuously growing list of records, the so-called blocks. As we will see later, a blockchain is designed in such a way that altering its contents becomes extremely difficult in a reasonable time.

Block — A block is a record inside a blockchain containing any kind of data. Besides data, a block must contain, at least, an index denoting its position in the chain, the hash value of all the information it contains (both data and metadata, to be able to verify any modification), and the hash value of its predecessor in the blockchain.

Hashing function — A hashing function is a function able to map information of any size to a fixed-sized *hash* value. In short, a hashing function will comply with the following conditions: *(i)* it is relatively easy to compute; *(ii)* it will always produce the same hash value for the same input data; *(iii)* it is impossible to recover the input data from its hash value; *(iv)* different input data should likely produce different hash values; and *(v)* small changes in input data should lead to big changes in the hash value.

Proof of work (PoW) — A PoW is a mechanism where the requester of a service must solve a relatively complex problem in order to access a given service. The PoW power resides in its asymmetry: it should be relatively complex to solve the problem posed by the service provider in the PoW, but it should be fairly easy to check the solution calculated by the requester.

Block mining — Block mining is the process of validating a given block by solving a PoW. The PoW to solve while mining a block is, typically, being able to find a specific value (the so-called *nonce* value, see below) whose hash function produces a hash value validating some specific conditions. In a blockchain, the PoW is used as a consensus mechanism used to agree which additions to the blockchain are valid.[4]

2.2   A Blockchain Under the Microscope

Figure 1 shows what a fraction of a simple blockchain with minimal (meta-)data would look like. Specifically, the first three blocks of the blockchain are shown. Each block contains the following information: *(i)* an index, shown in the black area at its top; *(ii)* a *prev-hash* value, pointing to the previous block in the chain; *(iii)* a hash value, calculated by applying a hash function to all the information stored in the chain, including both data and metadata, such as *prev-hash* or *nonce*; *(iv)* a *nonce* value, which is an arbitrary number; and *(v)* a block of data, typically limited to a maximum size. Since the first block—*block 0*—has no predecessor, its *prev-hash* value is set to 0. This special block is called *genesis block*.

As it can be observed, all the hash values start with a given number of leading zeros (four zeros). This is because, in the design of our blockchain, we have defined this as a requirement to consider a block as valid: the hash value of a block must be smaller than a

---

[4] Other methods for reaching consensus than PoW have been proposed—Proof of Stake (PoS), Deferred Proof of Stake (DPoS), or Practical Byzantine Fault Tolerance (PBFT), to name a few—but for the sake of simplicity, we will focus on PoW.

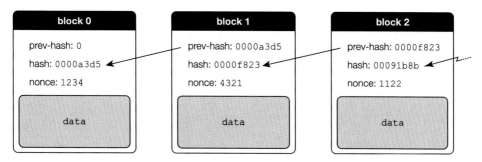

Fig. 1 A simple blockchain

specific *target value* (0000ffff in our example). However, the probability of obtaining a hash value meeting this condition by chance is extremely low. This is where the *nonce* value enters into play, and this is the PoW of our blockchain: we must find the *nonce* value that makes the hashing function applied over a block return a valid hash value (i.e., lower than 0000ffff). As aforementioned, this process of finding the proper *nonce* value is called *mining*.

Now we will illustrate how the chain is populated by explaining how an hypothetical *block 3* is added. At this point, we assume that a network of interconnected peers exists, each one containing a full copy of the blockchain. The process for adding the *block 3* is, simplifying, as follows: a peer decides to add a block, so it creates it and sets a *nonce* of 0 in the block. Then, the peer calculates the hash of the block. If the hash is smaller than the target value, then the block is added to the chain and it is communicated to all its connected peers in the network. If the hash is higher than the *target value*, the peer increments the *nonce* and checks again. This process is repeated until a valid *nonce* value is found and then proceeds as mentioned above.

It could be the case, however, that another peer in the network creates another *block 3* (or even more blocks). In short, when peers intercommunicate and share their versions of the blockchain, a few rules apply: if the chains have diverged, but have the same length, they do nothing (the conflict will be resolved later when the chains grow); if one of the peers has a longer (valid) chain, this one takes precedence (i.e., the shortest chain is discarded and considered stale). Figure 2 describes this scenario: in the case of the blockchain having both valid blocks 3a and 3b, both are maintained as valid (Fig. 2a). When block 4 is added to the chain containing block 3a, the branch of 3b is marked as stale since it becomes the shortest chain (Fig. 2b). It is important to remark that, if the data in the stale blocks is not contained within the valid blocks of the resulting blockchain, they need to be recomputed to be added in a subsequent block. Although there is no guarantee when a stale block will be added to the blockchain, eventually, it will.

At this point, it can be seen that modifying any block would require to recalculate its *nonce* value so that the hash value is still valid. But not only recalculating the modified block

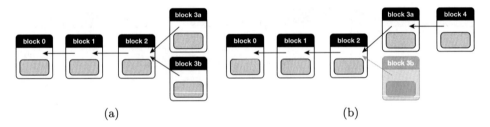

Fig. 2 A blockchain with two branches

is necessary: all subsequent blocks need to be recalculated too, since they use the hash value of the corresponding previous block in the calculation of their hashes. Additionally, this has to be done in such a way that the modified chain is longer than the chain trusted by other peers in the network. This is why tampering with a blockchain is impractical: any minor modification would require both an enormous amount of processing power and controlling a big portion of the peers in the network.

3   Manipulating Data in a Blockchain

We have seen that blocks in a blockchain can contain any arbitrary data, but, how can we ensure that data contained within a blockchain is only modified in a specific valid and secure way? This is the question that *smart contracts* try to answer.

Smart contracts, similarly to blockchains, are not anything new; and also similarly to blockchains, were tightly tied to digital currencies in their origin. The term *smart contract* was initially coined by Nick Szabo—a computer scientist and cryptographer known for his research in digital currencies and their legal aspects—in the mid nineties, and was further developed in *Formalizing and Securing Relationships on Public Networks* [29]. In this work, Szabo discusses the similarities and differences between smart contracts and traditional business procedures based on written contracts, controls, and static forms; and proposes to use cryptography to secure algorithmically specifiable relationships.

Thus, a smart contract is exactly this: instead of specifying the contractual clauses of the relationship among different parties using natural language in a written contract, the clauses are algorithmically specified using—for example—a programming language. Szabo explains how smart contracts work using a vending machine example: the vending machine implements a smart contract between the buyer and the vendor; and when the buyer introduces the coins in the machine, it dispenses—as specified by its program—the desired product and the change according to the displayed price.

Focusing on blockchain technology, smart contracts can be used to ensure that modifications of data in the blockchain are securely done. Taking as an example a blockchain implementing a cryptocurrency—such as Bitcoin—we see that data contained in chain blocks are money transactions done between peers. As it can be guessed, it is important not only that data is not altered once it is recorded—as we have seen in Sect. 2—but also that modifications (i.e., data in new blocks) are done according to a set of rules. Examples of such rules are ensuring that the sender of some amount of money is the owner of some specific funds; or that the receiver of a specific transfer is allowed to spend the received money. These are simple smart contracts that ensure that the integrity of data stored in the blockchain is not compromised.

3.1   Smart Contracts, Languages, and Turing Completeness

In domain specific blockchains—such as Bitcoin—smart contracts like the ones above can be directly encoded in the blockchain protocol itself for the sake of simplicity and security. But that goes at the cost of flexibility and extensibility: if new ways of interacting with data in the blockchain are envisaged, it may be too costly—or even impossible—to add support for them. This is why most blockchain technologies provide their specific languages to define smart contracts.

For example, a cryptocurrency blockchain initially designed to support transactions between two peers may need to support other types of transactions such as co-owned wallets. This type of wallets could require that all—or a configurable majority of—the owners of the wallet sign a transaction in order to authorize it. These types of wallets in fact exist in Bitcoin—and are known as *multisig wallets*—and indeed, the way they work is not directly encoded in the Bitcoin protocol, but it is specified in a language called *Bitcoin Script*.

Bitcoin Script is the language provided by Bitcoin to specify smart contracts, but as a domain-specific blockchain for cryptocurrency, it does not need to be excessively expressive. As a consequence, Bitcoin Script is a purposely simple language, with very few instructions (*operation codes*), and with a very simple stack-based computing paradigm. In fact, Bitcoin Script is so limited that it does not support loops or recursion, making it Turing incomplete. This is in fact a deliberate design decision since the *halting problem* is undecidable for programs written in Turing-complete languages: keeping the language Turing-incomplete serves as a security countermeasure against malicious smart contracts that may run forever thus provoking a denial of service.

It must be taken into account, however, that smart contracts are nothing but simple computer programs, and as such, they do not need to be tied to a specific domain nor they need to have limited power. The power of smart contracts can be unleashed by providing a Turing complete language, with rich semantics, and a high-level concrete syntax. With such a language, a blockchain can be seen as a completely distributed and secure storage and computing platform. An example of such a platform is *Ethereum*[5] thanks to its language for smart contracts definition, *Solidity*.[6]

---

[5] http://ethereum.org/
[6] https://solidity.readthedocs.io/

It must be noted that, although blockchain technologies do not need to necessarily rely on cryptocurrencies, this is not the case for Ethereum: on the Ethereum network, users have access to accounts in *ether*, the main currency used on the Ethereum blockchain. This currency is used to trade with computing power in the blockchain.

There are two types of accounts in Ethereum: *externally owned accounts*, which are controlled by a private key—i.e., an individual—and can send transactions but cannot publish code; and *contract accounts*, which have a state and an associated code—the smart contract—that can be executed by transactions and can call the code of other smart contracts. The smart contracts are stored in the blockchain itself thus being immutable: once a contract is deployed to a certain address on the Ethereum network, it is included in a block and cannot be changed. This is why smart contracts provide a secure way to make a procedure both publicly available as well as unalterable.

Solidity, the main language used for defining smart contracts, is an object-oriented language with a syntax similar to JavaScript that compiles to bytecode. This bytecode is executed by the *Ethereum Virtual Machine* (EVM), which runs on every full Ethereum node. A mining node is an Ethereum client, which downloads and verifies the blockchain and mines new blocks. The reward for successfully mining new blocks is the *transaction fee* as well as a pre-set *mining reward*, both of which are included in the mined block. As aforementioned, Solidity is Turing-complete, thus allowing the execution of any arbitrary code.

3.2   Smart Contracts in Turing-complete Languages: The Case of Ethereum

But, did we not just say that Turing-complete languages are a possible vector of attacks? This is where the concept of *gas* comes into play. The *gas* was developed and implemented in order to prevent system abuse by a malicious actor who attempts to run a denial of service attack on the network by spamming it with transactions.

In Ethereum, executing an actual transaction has a cost in gas, calculated as a fixed fee plus a variable fee that depends on the actual operations executed by the EVM.[7] Thus, when a transaction is sent, a user has to set two values: first, the price that will be paid for each unit of gas consumed during the execution (i.e., mining); and second, the maximum amount of gas that the transaction will be allowed to consume. If a block is successfully mined, all the gas included in the transactions is collected by the miner, which is converted to ether at the price specified by the transaction senders. This is the aforementioned *transaction fee*. But there are several cases in which the transaction is not fully executed or it is even rejected. If a transaction runs for too long, and exceeds the maximum amount of gas, it will stop and any changes done will be reverted. In that case, the miner will get anyway the fee for the gas—computation power—consumed. For example, an endlessly running loop is not possible on Ethereum since, for sure, at some point it will consume the maximum amount of

---

[7] All these gas fees are specified in the Ethereum Yellow Paper [32].

gas. But, if the gas limit for the transaction is set too high—for example, because a malicious attacker wants to ensure that the loop runs at least for enough time to harm the network—it can not even be included into a block since blocks also have a gas limit which is used to control the size of mined blocks.

As it can be seen, the gas prevents the *halting problem* at execution time. However, we can also see that we are limiting the computation power of the EVM. Thus, although Solidity is a Turing complete language, it runs on a *pseudo Turing-complete* virtual machine—the EVM—since it will stop executing a smart contract as soon as it runs out of gas.

3.3 Example of a Smart Contract

Listing 1 shows what an example smart contract[8] looks like in Solidity. The smart contract implements a simple form of a cryptocurrency. The goal of the contract is to ensure that only its creator can create new coins; but anyone with an Ethereum address—i.e., keypair—can send coins to each other.

```
1 // SPDX-License-Identifier: GPL-3.0
2 pragma solidity >=0.5.0 <0.7.0;
3
4 contract Coin {
5       // The keyword "public" makes variables accessible from other contracts
6       // "address" is a 160-bit value not allowing arithmetic operations
7       address public minter;
8       // "mapping" is a kind of hash table that does not allow obtaining the
9       // list of keys nor the list of values
10      mapping (address => uint) public balances;
11
12      // Constructor code is only run when the contract is created
13      constructor() public {
14          minter = msg.sender;
15      }
16
17      // Send an amount of newly created coins to an address
18      // Can only be called by the contract creator
19      function mint(address receiver, uint amount) public {
20          require(msg.sender == minter);
21          balances[receiver] += amount;
22      }
23
24      // Send an amount of existing coins from any caller to an address
25      function send(address receiver, uint amount) public {
26          require(amount <= balances[msg.sender], "Insufficient balance.");
27          balances[msg.sender] -= amount;
28          balances[receiver] += amount;
29      }
30 }
```

Listing 1. Example of a smart contract in *Solidity*

Line 1 specifies the source code license—specifying the license is important since publishing the source code is the default—while line 2 specifies the compliance of the source code. Solidity is still under development, and important changes in the language may happen

---

[8] Adapted   from   https://solidity.readthedocs.io/en/latest/introduction-to-smart-contracts.html# simple-smart-contract.

inadvertently. The *pragma* directive allows us to guarantee that the contract will behave as expected as far as the specified language versions are met.

The actual *Coin* contract is specified between lines 4–30. The contract basically stores two pieces of information: the address of the contract creator—the *minter* (line 7)—and the *balances* of all the *addresses*—i.e., accounts—that have received any funds (line 10). Both *minter* and *balances* are public and can be queried from any other contract; but a big caveat must be done with respect to the latter: since the *balances* are a *mapping*, it is only possible to get the *balance* for a known *address*. Bulk queries, such as getting all the registered addresses and all the stored balances, are not allowed.

The constructor (lines 13–15) is only run once: when the contract is created. When a contract is created, a *contract account* with an Ethereum address is created for it as previously said in Sect. 3.1. The constructor makes use of the *msg* special global object, which allows accessing the blockchain. *msg.sender* is the address where the current function call came from—in this case the contract creator. Thus, line 14 registers the identity of the contract creator at contract creation time. This value will remain unchanged in the contract throughout its whole lifecycle.

Once a contract is created, anybody with an Ethereum address can invoke the functions of the contract. Functions work similarly to contract clauses. For example, the *mint* function (lines 19–22) regulates how new coins can be created. On the one hand, line 20 guarantees that only the contract creator—the *minter*—can execute this operation. If this requirement is not met, the execution fails and no changes are done. On the other hand, line 21 specifies that the *receiver*—any Ethereum *address*—will receive the amount of newly created coins specified by the *minter*.

Finally, the send function (lines 25–29) regulates how the transfer of funds must be done. As aforementioned, anybody with an Ethereum address can invoke this function. In this case, line 26 ensures that the sender has enough funds to transfer; line 27 substracts the funds from the sender's balance; and line 28 adds the funds to the receiver's balance. It is important to remind that contract invocations are transactional, so it is not possible that funds get substracted from an account but are not transferred to another.

3.4   Challenges in Contracts Lifecycle

Software evolves for an unlimited number of reasons: changes in the requirements, bugs, improvements, refactorings, etc. Smart contracts, as a kind of software themselves, are not an exception. But although smart contracts are executable, in fact, they are saved as any other kind of data within the blockchain. That means that smart contracts remain immutable once they are deployed into the blockchain, thus prohibiting to deploy new changes by design. This poses an additional challenge in the design of smart contracts, since, for example, a contract like the one explained above indeed will remain in the blockchain forever once deployed.

Nevertheless, there exist several approaches to deal with evolution in smart contracts, most of them based on well-known *software design patterns* [11, 12, 17]:

Smart contract destruction — It could be the case that the blockchain platform provides a method to destroy contracts. This is the case of Ethereum and the *selfdestruct* operation. However, it is the contract who must destroy itself. That means that this *self-destruction clause* must be encoded in the smart contract from the beginning—of course secured with some kind of identity check like in the *mint(...)* method in Listing 1—to avoid a malicious destruction.

Once a smart contract has been destroyed, a new version can be deployed. This, however, has important drawbacks: *(i)* since the new version will receive a different address, any user of the smart contract must be aware of it in order to use the new version; *(ii)* in case the removed smart contract is invoked—for example by mistake for not updating the address—it will consume the ether sent in the transaction anyway although the contract does not perform any action, thus being that ether lost forever; *(iii)* destruction is permanent, and a destroyed smart contract cannot be restored in the address it was initially deployed; or *(iv)* data stored within the old smart contract must be migrated to the new smart contract.

Smart contract deactivation — To prevent some of the previous issues—ether loss, permanent destruction—smart contracts may be deactivated instead. In this case, a *circuit breaker* pattern can be implemented within the contract. Thus, the contract will keep an enabled/disabled state—which can be toggled by the contract creator—that will enable/disable the contract. This change in the state of the contract can be either temporary or permanent, but in any case, does not imply destroying it. This can be useful in case the contract needs to be disabled while an external service is upgraded, because a bug has been found, or because the contract is no longer needed once a specific state has been reached.

This solution, however, still poses some drawbacks, since in case a new version is needed, it would still require to share the address of the new contract, and again, it would still require a data migration.

Delegation, proxies and registries — In order to solve the previous problems, some other patterns can be applied, such as *delegation*, *proxy* or *registry*, among others. In short, these patterns allow to solve the problem of address change and data migration by using multiple contracts that call each other. For example, using a simple *delegation pattern*, a contract can delegate part of its logic to another contract (so called *delegated contract*). For example, this delegated contract can be in charge of only managing the data. Thus, if a new contract with new logic is deployed, there is no need to do a data migration: the new contract only needs to make use of the delegated contract.

The *proxy pattern* can be used in a similar way: a *proxy contract* is a contract with minimal logic that, once deployed, redirects all the requests it receives to another contract. Thus, the proxy contract acts as the single entry point that always remains unchanged.

Previous patterns can also be combined with the *registry pattern*: a *registry contract* can act as a central storage for other contracts that can be retrieved, for example, by a custom identifier rather than their address. Thus, a contract wanting to delegate or redirect a specific function to another contract, only needs to query the registry to get its latest version. For example, the *Ethereum Name Service*[9] is an example of the registry pattern.

4   Use Cases

Once we have understood the basics of defining and manipulating blockchains, we will see in this section how they can help in the operation and design of CPSs by looking at two specific use cases: supply chains in maritime trade and model-based collaborative design of CPSs.

4.1   Supply Chain in Maritime Trade

More than 80% of global trade by volume is carried by the international shipping industry. Any increased efficiency in the maritime trade sector can have significant effects on global Gross Domestic Product (GDP). Blockchain is one of the key drivers for enhancing the efficiency of maritime trade in the future, and brings many opportunities and benefits, namely:

- better means of sharing, distributing and verifying information and for transferring digital assets;
- a driver for process automation;
- time and cost reductions.

But blockchain-based solutions for the maritime sector are still in their infancy. This is an active research area[10] looking how to best adapt blockchain to the specific characteristics, technology and processes of this domain. For instance, in the maritime trade sector and at its interfaces, there will be a number of blockchain solutions that will need to be interoperable. Working on open international interoperability standards between different blockchain networks is of high importance and several international standard bodies, such as UN/CEFACT [31] are playing an important role in this area. Another aspect to consider is the use of port community systems (PCS), which are strategic assets for process harmonization and integration. Such systems could bring added value to the implementation of blockchain-based business processes in the maritime trade sector. As an example, PCS may be able to bridge different blockchain local/global networks and the different technology adoption levels of users.

---

[9] https://ens.domains

[10] http://www.dataports-project.eu

Recently, a large number of blockchain proof of concept and trial implementation initiatives have been put in place in the maritime trade to start exploring these issues. But a more massive adoption could be delayed due to a number of open challenges. Beyond the interoperability issue above, we also witness a lack of experts with know-how about maritime logistics and the lack of developers with blockchain expertise. Additional challenges to implementing blockchain in maritime trade are: the technology maturity; the long transaction confirmation time; legal issues and regulatory recognition; data ownership, personal privacy and general data protection regulation (GDPR); overlapping between current competing solutions; better knowledge on where to introduce—and where not—blockchain technologies; existence of multiple players with different technology adoption levels; need to change business processes; missing open standards; cybersecurity threats and risks; or the ability of micro, small and medium-sized enterprises to be integrated into blockchain-based systems.

Some of these challenges—such as the access to skilled blockchain developers, a general issue due to the huge demand of blockchain professionals—are common to any attempt to deploy a blockchain strategy in a new domain and will be further discussed in Sect. 5.

Despite these challenges, numerous blockchain maritime applications have appeared. Most of them involve the registration of events associated with actions carried out in the transport chain, such as entry or exit of a container in the different stages of transport, opening actions, control or verification of the containers at a certain stage[11, 12, 13]. The information must have adequate privacy mechanisms to be shared only with the actors involved in the processes, or with whom it is wished to share the information. The consultation operations of the registered events are carried out on the database of the blockchain ledger. The consumer launches the information request to the corresponding smart contract that validates the permission rules, retrieves the events registered in the blockchain and returns them to the consumer.

As a case of on-chain data—where the full data must be stored in the blockchain—we can mention the sharing of verified gross mass (VGM) of containers composed of multiple data entities, such as vehicles, weights, berths and locations.[14] In the case of off-chain data—where the data is stored outside the blockchain (e.g., port database, data provider, international data structure, etc.)—another example could be the sharing of the consignment note together with registering the digital signatures of the shipper and road-haulier, the proof of goods delivery, and the digital hash of the control document. This would ensure the consensus, origin, immutability and finality of these legal documents and data, acting as a digital notary of these documentation[15, 16]. Other initiatives are cryptocurrency for

---

[11] https://www.tradelens.com

[12] https://github.com/blockfreight

[13] https://t-mining.be

[14] International Maritime Organization, MSC.1/Circ.1475, 9 June 2014.

[15] https://ipcsa.international

[16] https://wavebl.com

container liners and their customers to reduce counterparty risk of default of a cargo shipping agreement (booking).[17]

4.2   Collaborative Design of CPSs

Blockchain technology can also play a key role in the development of CPSs and not just during their runtime execution as described in the previous section.

Development and evolution of CPSs is more and more based on MDE principles. In MDE, models are the key element of all system engineering activities. We can use modeling techniques to specify CPSs at a higher abstraction level, disregarding, to a certain extent, low level technical details of the concrete platforms the CPS is executed upon [3, 8, 13].

Due to their essential complexity, all CPSs projects involve a large number of people that discuss and collaborate in the definition of a CPS, many times involving external companies—e.g., outsourcing or offshoring [4] scenarios—and consultants. These project participants may work on separate parts of the project, require a different perspective on the project, various access rights, etc. The MDE community has come with a number of practical solutions for these problems—e.g., see the approaches to generate model views based on a user profile [7].

Nevertheless, there are three major challenges that have not been yet satisfactorily addressed:

Accountability — The need to track the actions of each team member interacting with the system. For instance, this can be needed for contribution attribution, e.g., to perform quality assessment or calculate payments to external parties.

Explainability — The need to explain and justify the behaviour of the CPS in the future by tracking that behaviour to past design decisions.

Intellectual Property (IP) Protection — CPSs models are key assets for a company and as such they must be protected. We need mechanisms to detect and deal with stolen IP, including from our own collaborators.

We believe integrating blockchain technologies in MDE—and in general, in software and systems engineering [2]—could be a significant step forward towards solving these three issues. By tracking all changes on the CPS models—including by *whom*, from *where*, with which access rights, etc.; but also the *why* and reasons for the change—we could leverage the benefits of blockchain to be able to provide *accountability*, *explainability* and *IP protection* to the CPSs design process.

This is still an open research area though some promising results have been presented in the last couple of years [10, 14, 25]. They have as core element a blockchain metamodel as the "glue code" between the CPS models and the blockchain technology.

---

[17] https://www.300cubits.tech

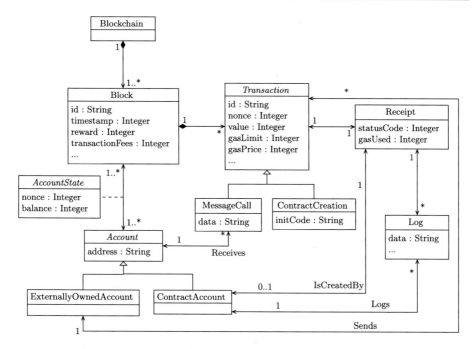

Fig. 3 A blockchain metamodel (extracted and adapted from [23])

This metamodel is used to enrich the model data with all the auxiliary information required to properly store all the model manipulation data in the blockchain for future analysis. An example of such blockchain metamodel is the one proposed by Olivé [23] for Ethereum, as shown in Fig. 3.

Without getting into too much detail, we can note how the metamodel comprises the key metaclasses to represent the modification operations in Ethereum as explained in Sect. 3.[18] Please note that, for simpler scenarios, we could simply just include a taxonomy of model changes [19] and link the type of change to the affected model elements, as done in [16]. This way, we can also describe how the need for that change was decided (*explainability*) linked to the user that executed those operations (*accountability*).

Storing all this data together with the model itself in a blockchain will also fully protect the company's IP. Unfortunately, this is not always possible as models can be large and pretend to save the full model in a block would not scale at all. There are several alternatives to address this issue—see also the discussion in the next section. We could, for instance, store only the changes in the blockchain while keeping the model on an external support. Or, if our main concern is the detection of stolen versions of our model, store only the model hash in the blockchain and not the model itself.

---

[18] A deeper explanation of the concepts and metaclasses can be found in the original publication [23].

Robust hashing algorithms produce the same or very similar hashes for similar inputs. Moreover, they are capable of resisting attacks—i.e., modifications—that change non-essential properties of the asset. Adaptations of robust hashing to models [20] show how they can successfully detect model copies even with a certain degree of modifications.

5   Designing My Blockchain

This section covers some of the socio-technical challenges [9] to consider when adopting a blockchain infrastructure in an organization project.

We first go through the list of aspects to consider one by one. We then cover a number of "packaged" blockchain solutions—favouring each of them a different set of qualities—that can help us accelerate the implantation based on our needs. Some of these solutions have even *as a service—blockchain as a service* (BaaS) [28]—option to speed up even more the deployment of your own blockchain-based process—e.g., see *AWS Blockchain templates*,[19] *Oracle's cloud blockchain platform*[20] or *Huawei's blockchain service*.[21]

5.1   Socio-Technical Challenges

This subsection goes into greater depth on the key aspects you should consider before starting any blockchain project.

Alignment of stakeholder interests — In traditional business sectors, like maritime supply chain, many stakeholders are involved and should change their mind and adjust their corporate culture if a blockchain-based solution is used instead of current electronic exchange format. To be a success, this would require economies of scale and cooperation among potential competitors.

Standards — A key property in many domains is the fluidity of information. This can be achieved by adhering to standard protocols and data definitions and formats. The successful adoption of robust blockchain standards could have the effect of making the number of parties involved irrelevant because the opportunity to corrupt it would be reduced to zero.

Data integrity from source—e.g. traceability — The trustworthiness of the information carried by a blockchain depends on verified inputs occurring as early in the chain as possible. Therefore, it is important in the design of blockchain systems to focus on when and how data is to be verified.

---

[19] https://aws.amazon.com/blogs/aws/get-started-with-blockchain-using-the-new-aws-blockchain-templates/

[20] https://www.oracle.com/application-development/cloud-services/blockchain-platform/

[21] https://www.huaweicloud.com/intl/en-us/product/bcs.html

Data collection — When it comes to automatize data collection, the technology to achieve it usually is based on radio frequency identification (RFID), QR codes and respective scanners, as well as IoT sensors. However, implementation challenges with accurate or incomplete readings are still observed when they are deployed at scale.

Anomaly management — Rules governing a blockchain system should be designed to let a network recognize nodes with the authority to make correcting entries to original data. This also affects the contracts to manipulate the blockchain, as discussed in Sect. 3.

Regulation — The potential for hiding or obfuscating important information is of concern to governments. Without regulation it is possible for entire economies to operate out of sight, thereby avoiding taxes, fees and financial laws such as those on money-laundering.

Costs — An obvious example is the large amount of electricity used in PoW systems. This has been heavily criticized because of its impact. For this reason, many public blockchains that use PoW—such as Ethereum—are moving toward the use of other consensus mechanisms for the verification of data blocks.

Securing the Blockchain — As discussed above, common PoW protocols are very secure but their cost make them unlikely to find favour in sectors like supply-chain. A different trade-off may be required here, for instance, using other protocols such as PoS.

Privacy and liability — Strong permission-based access protocols offer a theoretical level of privacy that should meet the most exacting standards of business and governmental agencies. The issue of liability remains when confidential information may be stolen by a malicious actor or shared with an unauthorized user.

5.2   Enterprise Blockchain Platforms

When a company decides to implement a blockchain solution for its business, like the maritime sector,[22] it has to make a technological choice. Will the solution involve cryptocurrency? Bitcoin, Ether, other? Will it require high transaction rate throughput? Complex smart contracts? Under what licences? …?

To answer this need, a number of blockchain platforms have emerged combining different features and technologies. These platforms differ in many aspects, among which we find their unique purpose, consensus protocol, maturity, performance and scalability, privacy and confidentiality, security and identity, scalability and interoperability, licence, network access, transaction cost/maximum number of transaction per second, smart contract support, cryptocurrency, cloud offerings and ease of use. Table 1 gives an overview of the current four

---

[22] http://www.dataports-project.eu

Table 1 Comparison of four main enterprise blockchain platforms

	HYPERLEDGER FABRIC	ETHEREUM (EEA)	CORDA	QUORUM
CONSORTIUM	Linux foundation	Ethereum foundation	R3 consortium	JP Morgan Chase
LICENSE	Apache 2	GPL	Apache 2	LGPL-3.0
CONSENSUS	Kafka—Solo	Proof of work (POW)	Notary nodes	Raft-based—Istanbul BFT
SMART CONTRACTS	Chaincodes	EVM smart contract	Smart contract	EVM smart contract
STATE	Blockchain and Database (key/value)	Blockchain and Database (key/value)	State Transaction chain	Blockchain and Database (key/value)
TRANSACTION COST	Free	GAS + Mining commission	Free	Free
MAX. TRANSACTIONS/S	3500TPS	25TPS	600TPS	100TPS
CRYPTOCURRENCY	No	Ether	No	No

main enterprise blockchain platforms, namely Hyperledger Fabric,[23] Ethereum (EEA),[24] Corda,[25] and Quorum.[26]

While the Hyperledger Fabric presents the benefit to be general purpose with a wide support of privacy mechanisms, and a global state shared over the network, it is less suitable for bilateral peer-to-peer contracts. Ethereum (EEA) also offers strong privacy mechanism, as well as interoperability with Ethereum-based solutions, but it suffers from poor flexibility for protocol optimization, and specific solutions for restricted purposes. On the contrary, Corda presents a flexible and easily orchestrated transaction processing, in particular for bilateral contracts, and state visibility. However, its on-chain business logic is not well adapted for consensus processes, and it presents poor confidentiality and privacy—notary witnesses transactions, simple cryptographic technologies. Finally, Quorum presents better performance than public Ethereum, with support for private transactions, but at a higher cost, with a small support community and mostly fitted for financial services and applications.

A more complete list of enterprise blockchain platforms can be found in [24].

---

[23] https://www.hyperledger.org/use/fabric

[24] https://ethereum.org/enterprise

[25] https://www.corda.net

[26] https://www.goquorum.com

6   Conclusions

We have covered the main concepts of blockchain technology and its benefits but also the potential issues you will face when starting to adopting blockchain in your organization.

However, blockchain technologies and environments are changing at a fast pace. Thus, we expect that many of those challenges will either disappear or get simplified by new developments and the release of more mature tools in the area.

Some of these new tools and approaches are pushing the limits of blockchain and even exploring alternative distribute ledger architectures, mostly to address scalability concerns. Hedera Hashgraph[27] is one of these initiatives. Based on a hashgraph consensus algorithm [1], it focuses on delivering fast and inexpensive transactions. XuperChain[28] is blockchain-compatible, but it introduces parallel contract execution and verification. IOTA[29] uses a direct acyclic graph (DAG) called the *Tangle* instead of a blockchain to store the ledge. IOTA allows different branches of the DAG to eventually merge, resulting in a much faster overall throughput.

One way or the other, blockhain and distributed ledgers technology are here to stay. While right now, its adoption can be at the *early-adopters/early-maturity* phase [26], we believe blockchain will end up impacting all business domains, especially all those related to Industry 4.0 and CPSs. If you are not yet monitoring this field, we hope this chapter is just the first step towards your blockchain adoption strategy.

Acknowledgements This work has received support from the AIDOaRt project, funded by the ECSEL Joint Undertaking under grant agreement No. 101007350.[30]

References

1. Baird, L.: The swirlds hashgraph consensus algorithm: Fair, fast, byzantine fault tolerance. Swirlds, Inc. Technical Report SWIRLDS-TR-2016 **1** (2016)
2. Beller, M., Hejderup, J.: Blockchain-based software engineering. In: Sarma, A., Murta, L. (eds.) Proceedings of the 41st International Conference on Software Engineering: New Ideas and Emerging Results, ICSE (NIER) 2019, Montreal, QC, Canada, May 29-31, 2019. pp. 53–56. IEEE/ACM (2019). 10.1109/ICSE-NIER.2019.00022
3. Bernardi, S., Domínguez, J.L., Gómez, A., Joubert, C., Merseguer, J., Perez-Palacin, D., Requeno, J.I., Romeu, A.: A systematic approach for performance assessment using process mining - An industrial experience report. Empir. Softw. Eng. **23**(6), 3394–3441 (2018). 10.1007/s10664-018-9606-9
4. Blinder, A.S.: Offshoring: the next industrial revolution? Foreign affairs pp. 113–128 (2006)

---

[27] https://www.hedera.com

[28] https://github.com/xuperchain/xuperchain

[29] https://www.iota.org

[38] This Joint Undertaking receives support from the European Union's Horizon 2020 research and innovation programme and Sweden, Austria, Czech Republic, Finland, France, Italy and Spain.

5. Braeken, A., Liyanage, M., Kanhere, S.S., Dixit, S.: Blockchain and cyberphysical systems. Computer **53**(09), 31–35 (sep 2020). 10.1109/MC.2020.3005112
6. Brambilla, M., Cabot, J., Wimmer, M.: Model-Driven Software Engineering in Practice, Second Edition. Synthesis Lectures on Software Engineering, Morgan & Claypool Publishers (2017). 10.2200/S00751ED2V01Y201701SWE004
7. Brunelière, H., Burger, E., Cabot, J., Wimmer, M.: A feature-based survey of model view approaches. Software and Systems Modeling **18**(3), 1931–1952 (2019). 10.1007/s10270-017-0622-9
8. Casale, G., et al.: DICE: quality-driven development of data-intensive cloud applications. In: Gray, J., Chechik, M., Kulkarni, V., Paige, R.F. (eds.) 7th IEEE/ACM International Workshop on Modeling in Software Engineering, MiSE 2015, Florence, Italy, May 16-17, 2015. pp. 78–83. IEEE Computer Society (2015). 10.1109/MiSE.2015.21
9. Cataldo, M., Herbsleb, J.D., Carley, K.M.: Socio-technical congruence: a framework for assessing the impact of technical and work dependencies on software development productivity. In: Rombach, H.D., Elbaum, S.G., Münch, J. (eds.) Proceedings of the Second International Symposium on Empirical Software Engineering and Measurement, ESEM 2008, October 9-10, 2008, Kaiserslautern, Germany. pp. 2–11. ACM (2008). 10.1145/1414004.1414008
10. Falazi, G., Hahn, M., Breitenbücher, U., Leymann, F.: Modeling and execution of blockchain-aware business processes. SICS Softw.-Intensive Cyber Phys. Syst. **34**(2-3), 105–116 (2019). 10.1007/s00450-019-00399-5
11. Fowler, M.: Analysis Patterns: Reusable Objects Models. Addison-Wesley Longman Publishing Co., Inc., USA (1996)
12. Fowler, M.: Patterns of Enterprise Application Architecture. Addison-Wesley Longman Publishing Co., Inc., USA (2002)
13. Gil, M., Joubert, C., Torres, I.: Model-driven engineering IDE for quality assessment of data-intensive applications. In: Binder, W., Cortellessa, V., Koziolek, A., Smirni, E., Poess, M. (eds.) Companion Proceedings of the 8th ACM/SPEC on International Conference on Performance Engineering, ICPE 2017, L'Aquila, Italy, April 22–26, 2017. pp. 173–174. ACM (2017). 10.1145/3053600.3053633
14. Härer, F., Fill, H.: Decentralized attestation of conceptual models using the ethereum blockchain. In: Becker, J., Novikov, D. (eds.) 21st IEEE Conference on Business Informatics, CBI 2019, Moscow, Russia, July 15–17, 2019, Volume 1 – Research Papers. pp. 104–113. IEEE (2019). 10.1109/CBI.2019.00019
15. Harris, J.D., Waggoner, B.: Decentralized and collaborative ai on blockchain. In: 2019 IEEE International Conference on Blockchain (Blockchain). pp. 368–375. IEEE Computer Society, Los Alamitos, CA, USA (jul 2019). 10.1109/Blockchain.2019.00057
16. Izquierdo, J.L.C., Cabot, J.: Collaboro: a collaborative (meta) modeling tool. PeerJ Comput. Sci. **2**, e84 (2016). 10.7717/peerj-cs.84
17. Larman, C.: Applying UML and Patterns: An Introduction to Object-Oriented Analysis and Design and Iterative Development (3rd Edition). Prentice Hall PTR, USA (2004)
18. Lee, E.A.: Cyber physical systems: Design challenges. In: 2008 11th IEEE International Symposium on Object and Component-Oriented Real-Time Distributed Computing (ISORC). pp. 363–369. IEEE (2008)
19. Lehnert, S., Farooq, Q., Riebisch, M.: A taxonomy of change types and its application in software evolution. In: IEEE 19th International Conference and Workshops on Engineering of Computer-Based Systems (ECBS). pp. 98–107. IEEE Computer Society (2012). 10.1109/ECBS.2012.9
20. Martínez, S., Gérard, S., Cabot, J.: Robust hashing for models. In: Wasowski, A., Paige, R.F., Haugen, Ø. (eds.) Proceedings of the 21th ACM/IEEE International Conference on Model Driven

Engineering Languages and Systems, MODELS 2018, Copenhagen, Denmark, October 14–19, 2018. pp. 312–322. ACM (2018). 10.1145/3239372.3239405

21. Nakamoto, S.: Bitcoin: A Peer-to-Peer Electronic Cash System, https://bitcoin.org/bitcoin.pdf

22. Narayanan, A., Bonneau, J., Felten, E., Miller, A., Goldfeder, S.: Bitcoin and cryptocurrency technologies: a comprehensive introduction. Princeton University Press (2016)

23. Olivé, A.: The conceptual schema of ethereum. In: Dobbie, G., Frank, U., Kappel, G., Liddle, S.W., Mayr, H.C. (eds.) Conceptual Modeling. pp. 418–428. Springer International Publishing, Cham (2020)

24. Pelz-Sharpe, A.: Enterprise blockchain – market forecast & scenarios 2019-2024 (2019), https://www.deep-analysis.net/wp-content/uploads/2019/08/DA-190812-Ent-Blockchain-forecast.pdf

25. Rocha, H., Ducasse, S.: Preliminary steps towards modeling blockchain oriented software. In: 1st IEEE/ACM International Workshop on Emerging Trends in Software Engineering for Blockchain, WETSEB@ICSE 2018, Gothenburg, Sweden, May 27–June 3, 2018. pp. 52–57. ACM (2018), http://ieeexplore.ieee.org/document/8445060

26. Rogers, E.M.: Diffusion of innovations. Simon and Schuster (2010)

27. Ruoti, S., Kaiser, B., Yerukhimovich, A., Clark, J., Cunningham, R.K.: Blockchain technology: what is it good for? Commun. ACM 63(1), 46–53 (2020). 10.1145/3369752

28. Samaniego, M., Jamsrandorj, U., Deters, R.: Blockchain as a service for iot. In: 2016 IEEE International Conference on Internet of Things (iThings) and IEEE Green Computing and Communications (GreenCom) and IEEE Cyber, Physical and Social Computing (CPSCom) and IEEE Smart Data (SmartData). pp. 433–436 (2016)

29. Szabo, N.: Formalizing and securing relationships on public networks. First Monday 2(9) (Sep 1997). 10.5210/fm.v2i9.548

30. UK Government, O.f.S.: Distributed Ledger Technology: beyond block chain. Tech. rep., UK Government (2016)

31. United Nations: Blockchain in Trade Facilitation: Sectoral challenges and examples (ECE/TRADE/C/CEFACT/2019/INF.3) (2019), https://unece.org/fileadmin/DAM/cefact/cf_plenary/2019_plenary/CEFACT_2019_INF03.pdf

32. Wood, G.: Ethereum: A secure decentralised generalised transaction ledger (41c1837 – 2021-02-14) (2021), https://ethereum.github.io/yellowpaper/paper.pdf

Data Management

Big Data Integration for Industry 4.0

Daniel Obraczka⊙, Alieh Saeedi⊙, Victor Christen⊙ and Erhard Rahm⊙

Abstract

The fourth industrial revolution promises a new quality of automation with smart manufacturing devices sharing enormous amounts of data. A crucial step in fulfilling this promise is developing advanced data integration methods that are able to consolidate and combine heterogeneous data from multiple sources. We outline the use of knowledge graphs for data integration and provide an overview of proposed approaches to create and update such knowledge graphs, in particular for schema and ontology matching, data lifting and especially for entity resolution. Furthermore, we present data integration use cases for Industry 4.0 and discuss open problems.

Keywords

Industry 4.0 • Big data • Data integration • Knowledge graph

This work was supported by the German Federal Ministry of Education and Research (BMBF, 01/S18026A-F) by funding the Center for Scalable Data Analytics and Artificial Intelligence (ScaDS.AI) Dresden/Leipzig.

A. Saeedi · V. Christen · E. Rahm (✉)
Institute for Computer Science, Leipzig University, Leipzig, Germany
e-mail: rahm@informatik.uni-leipzig.de

A. Saeedi
e-mail: saeedi@informatik.uni-leipzig.de

V. Christen
e-mail: christen@informatik.uni-leipzig.de

D. Obraczka
ScaDS.AI, Leipzig University, Leipzig, Germany
e-mail: obraczka@informatik.uni-leipzig.de

© The Author(s), under exclusive license to Springer-Verlag GmbH, DE, part of Springer Nature 2023
B. Vogel-Heuser and M. Wimmer (eds.), *Digital Transformation*,
https://doi.org/10.1007/978-3-662-65004-2_10

247

1   Introduction and Related Work

The success of Industry 4.0 is based on the transforming technologies of the last decade: the Internet of Things and Big Data [37]. The Internet of Things enables communication and exchange of data between physical objects (e.g., sensors) to implement certain services and reach autonomous decisions. In the medical domain, for example, the Internet of Things can improve services such as monitoring, diagnostics, and treatment by utilizing interconnected devices that observe the vitality of persons [72]. The idea of Industry 4.0 is similarly based on the close interaction of decentralized systems such as production systems and products, to achieve self-controlled and self-optimizing processes. Big Data comes into play due to the enormous amount of different kinds of data that are continuously generated, exchanged, and to be processed. This data has to be standardized to enable their interpretation and autonomous decisions. Moreover, the different kinds of data can be collected, transformed, and integrated to support a holistic analysis and optimization of the different production processes, production lines, etc. [22].

The challenges of Big Data are usually characterized by the "V" properties of *Volume*, *Velocity*, *Variety* and *Veracity*. These challenges are all relevant for Industry 4.0. In particular, disconnected sources in manufacturing processes generate a massive amount of data (Volume) at a high rate (Velocity) for further processing [22]. Variety refers to the need to process different kinds of heterogeneous data, in particular structured data (such as events or database records), semi-structured data (documents, log files, error reports), and unstructured data (e.g., images, audio files, and videos). Veracity finally asks for providing a high data quality to enable valid analysis results.

Data integration is the task to combine and enrich data from multiple sources for data analysis. *Big Data Integration* is data integration for Big Data that has to address the V challenges, in particular, Variety to deal with heterogeneous data of different kinds and Veracity to achieve high data quality. Additionally, the requirements Volume and Velocity lead to high-performance demand to deal with the massive amount of continuously produced data. The high data quality and performance requirements are best met with so-called physical data integration approaches that bring the data from different sources into a dedicated repository such as a data warehouse or knowledge graph. Such repositories can be maintained and used on a distributed cluster platform with many processors to achieve fast data processing and analysis. Furthermore, such approaches can apply comprehensive data preprocessing to improve data quality, in particular by extracting information from semi- and unstructured sources and for performing transformation and cleaning approaches for data consolidation [32, 75]. Physical data integration such as the creation and continuous update of a data warehouse or a knowledge graph also entails several steps, including the task of entity resolution to identify (match) and fuse different representations of the same real-world entity such as for a product part or customer.

While there is a huge amount of previous research and commercial activities in the area of data integration [13, 74], there is only little work focusing specifically on data integration for Industry 4.0. Some work has been done on the use of dedicated process knowledge repositories for workflow analysis [61], and for the enrichment and maintenance of unstructured documents such as failure and performance reports [52]. Most repositories focus on certain data types, applications or certain phases in the value chain of products. Process knowledge data consists of structured rules, information about data mining models and results as structured data. On the other hand, documents such as failure reports and unstructured data are essential as well. Groeger et al. [23] propose a repository for maintaining these types of data for each manufacturing step.

In the remainder of this chapter, we focus on (Big) data integration with knowledge graphs that can semantically integrate and interrelate many entities of different types for data analysis. Knowledge graphs are more flexible than data warehouses that are built on relational databases with a rather static, predefined schema that prevents the easy addition of new kinds of heterogeneous entities and their relationships. We begin by motivating the topic by outlining selected industrial use cases for data integration in Sect. 2. In Sect. 3, we introduce knowledge graphs and give an overview of the methods for constructing them. The important task of entity resolution is the topic of Sect. 4 that explains the main steps and how its performance can be improved to deal with Big Data. We close with a summary and outlook to open problems.

2   Data Integration Use Cases

Knowledge Graphs (KG) and other semantic technologies have become a viable option for companies to organize complex information in a meaningful manner. The semantic representation of data can improve understandability of complex data making development of new technologies more efficient [18], and improve quality control in manufacturing processes [97]. Not only software giants like Facebook, Google and Microsoft, but also production companies like Siemens [78] or news conglomerates like Thomas-Reuters [91] turn towards semantic representations of their data. Aibel, a service company in the energy sector, has reportedly saved more than 100 million Euros through better representation of their products using ontologies [90].

In the following we will look at some examples, where companies integrated heterogeneous data sources into semantic repositories.

In a Bosch factory[38] Surface Mount Technology is used to mount electrical components directly on circuit boards. Different machines are needed in this process, e.g., to place the electronic parts or inspect the solder joints. To detect failures in the manufacturing process, the integration of several data sources coming from machines of different vendors is necessary. This data integration relies on a domain ontology. An ontology is a semantic data structure, which contains known concepts and relationships and can be used to ensure the consistency in the data integration process. The machine components in the manufacturing pipeline produce log data in the form of JSON files. These are extracted and stored in a

PostgreSQL database which is then manually mapped to the ontology. Through the use of the Ontop[1] framework a Virtual Knowledge Graph is created from the ontology and the mappings to the original data sources. The manufacturing process data can then be analyzed by sending SPARQL (a semantic querying language) queries which are translated to SQL queries to the original data sources. In an evaluation this approach returned results in tens of seconds, which the researchers deemed a reasonable amount of time for their use case. What is still missing is a more comprehensive data analysis that goes beyond the use of queries, e.g., the use of machine learning to identify erroneous processing steps.

Siemens relies on a similar approach to unify multiple data sources in their smart manufacturing process [78]. A common ontology is used and the heterogeneous sources are mapped to this ontology. The resulting KG is used as a basis to integrate dynamically occurring events in their factory into the KG. The researchers present an approach for event-enhanced KG completion using a machine learning approach to jointly learn KG embeddings as well as event sequence data embeddings. In their evaluation they show, that their approach leads to good quality KG completion and can aid in the synchronisation of the physical and digital representations of a smart factory.

Jirkovský et al. [35] investigate the use of semi-automatic ontology matching to integrate an Excel File containing Ford spare part records and the Ford supply chain ontology. They utilize extensive preprocessing to enrich the Excel records with implicit information contained in part numbers and abbreviations. Multiple similarity measures are used for element pairs which are fed into a self-organizing map, which is a type of artificial neural network that can be trained in an unsupervised fashion. The trained model can classify entity pairs and present the user with examples, where it is least confident about its classification.

3   Knowledge Graphs

In this section we first present the foundations of semantic technologies for knowledge graphs. We then present the necessary steps to semantically integrate heterogeneous data sources for creating and evolving such knowledge graphs.

3.1   Knowledge Graph Foundations

In Fig. 1 we can see an example snippet of a KG. We will use this illustration to subsequently introduce RDF, ontologies and finally what a KG is.

RDF The standard that is used to create KGs with their entities and relationships is called RDF (Resource Description Framework), which is a recommendation[2] of the W3C (World

---

[1] https://ontop-vkg.org
[2] https://www.w3.org/TR/rdf-primer/

Wide Web Consortium). An RDF graph is a set of triples. Using such triples we can make statements about entities and their relations. An example of a triple we can see in Fig. 1 is

```
Part123 manufacturedIn ProcessingStep123 .
```

An RDF Graph can have three different kinds of nodes: IRIs (Internationalized Resource identifiers), literals or blank nodes. IRIs are generalizations of URIs and give each resource a unique identifier. To express values such as strings, dates or numbers literals are used. RDF enables the user to also state the datatype and if the literal is a string a language tag can be provided. Blank nodes are anonymous resources, that enable more complex structures.

Ontologies An ontology is a formal description of knowledge using machine-processable specifications. These specifications have well defined meanings and contain known concepts and relationships [30]. For example, in Fig. 1 we express that the entity Part123 belongs to the class Screw with the triple

```
Part123 rdf:type Screw .
```

Ontologies build on description logic, which enables reasoning engines to check logical consistency and correctness. Such reasoning possibilities are advantageous in the Industry

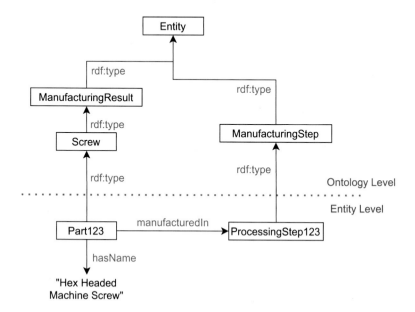

Fig. 1 Example snippet of a KG

4.0 setting to make implicit information explicit. For example, [49] use reasoning as data enrichment step to infer compatibility of parts.

Furthermore, ontologies provide a so-called vocabulary, which is a set if IRIs, that can be used in RDF graphs.[3] In Fig. 1 for example we use the RDF vocabulary, by utilizing `rdf:type` to express that an entity is an instance of a class. Incorporating vocabularies is a common technique to rely on already existing ontologies and makes integration of different semantic systems easier. For the Industry 4.0 context there already exist ontologies like e.g. CORA (Core Ontology for Robotics and Automation) [71] that can be a useful starting point for companies. An overview over other ontologies for Industry 4.0 can be found here [86].

Knowledge Graph The terms ontology and KG are sometimes erroneously used as synonyms. KGs often integrate multiple sources into a single ontology and are able to derive new knowledge through reasoning [16]. While ontologies often focus on the conceptual modeling, knowledge graphs include a large number of entities and relations as instances of concepts and relationships, which introduces the need of instance-level data integration such as entity resolution. In the industry 4.0 context often the more specific term *industrial knowledge graph* (e.g. at Siemens [31]) is used. A notable example of an open-source KG is the *Industry 4.0 Knowledge Graph* [3]. This KG contains information about standards used in smart manufacturing and relations between standards.

3.2   Knowledge Graph Construction

The construction of a KG entails the integration of (heterogeneous) data sources and enriching the data with semantic information. The integration process generally necessitates the following steps:

1. Creation of the KG ontology
2. Mapping of data sources to the KG ontology which requires *schema or ontology matching*
3. Preprocessing of data sources to extract, and clean entities and transform them into the RDF format which is also known as *data lifting*
4. Categorization of entities to assign them to the ontology concepts, e.g. for entities extracted from documents. This task can be addressed with machine learning by utilizing already assigned entities as training data [79]
5. *Entity Resolution* to identify duplicate entities and fuse them together in the knowledge graph.

Bear in mind, that some of these tasks can happen in different order (e.g., integrate the data sources first and then perform data lifting or vice versa) or even overlap (e.g., classification of entities can happen in the data lifting step). Moreover, the knowledge graph has to be

---

[3] https://www.w3.org/TR/rdf11-concepts/#vocabularies

continuously updated to incorporate new data and even new data sources. This asks for incremental methods to evolve the KG ontology and to add entities incrementally.

In the following we will start by presenting schema and ontology matching, followed by the data lifting task. Entity Resolution will be discussed separately in Sect. 4.

Schema and Ontology Matching Smart factories produce a plethora of different data formats from a vast number of sensors, databases, spreadsheets etc. To tackle this *variety* aspect of Big Data, companies have to unify these data collections under a common schema, a task that is referred to as schema matching. Schema matching aims to determine semantic correspondences between metadata, database schemata or in the special case of ontology matching between ontology elements. The high degree of semantic heterogeneity between sources makes this a difficult task, especially since not only one-to-one matches have to be found, but also more complex relationships like e.g., generalizations or part-of relations.

A central element of schema matching systems are matchers, which determine the similarity between concepts/attributes of the given schemata. Different types of matchers exist, namely instance- and metadata-based matchers. Instance-based matchers rely on already known instance matches between data sources and mostly rely on the instance overlap among concepts to determine how similar concepts are. Matchers that rely on metadata can further be divided into element-level and structure-level matchers, where the former use similarity between concept names sometimes utilizing dictionaries and the latter exploit structural information in ontologies e.g., the children or parents of concepts. Matching frameworks typically rely on a combination of different types of matchers to achieve a high quality result [24]. Matchers can be executed sequentially, in parallel or a mixture of both.

To illustrate this let us look at an example from the smart product lifecycle, where products from different vendors generate data, that we need to integrate [88]. Table 1 lists six sample products from five different provider sources such as *www.ebay.com* and *www.buzzilions.com*. The descriptions represent six cameras from two manufacturers *Canon* and *Nikon*. As shown, *entity 1* and *entity 2* as well as *entity 4*, *entity 5*, and *entity 6* represent the same real-world camera. We can see that schemata between data sources vary immensely. This is not only apparent by the different number of properties for the same entities, but also in the very different representation of the same attributes. For example, *entity 1* has an attribute *effective megapixel count* with a value 10.1, as well as an attribute *pixel count* with the value 10 Megapixel, while the matching *entity 2* has an attribute *megapixels* with the value 10.1 MP. All three attributes would have to be determined to be the same. Data preprocessing can alleviate some heterogeneity e.g., replacing common abbreviations like MP for Megapixel. A schema matching approach will first have to classify entities from the given sources. In the example, the entities are all of the type camera, but the data sources might contain e.g., camera cases, which have to be separated from camera entities. Secondly, classification of properties helps to reduce the search space e.g., the property *compatible with macintosh* in *entity 1* should be treated as a Boolean variable rather than a string, and therefore not compared with other string attributes.

Table 1 Example raw data

property	value
entity 1	
"source"	"www.buzzillions.com"
"page title"	"Canon EOS 40D Digital SLR Camera"
"compatible with macintosh"	"Yes"
"depth inches"	"2.9"
"digital slr"	["Body Only", "Body With Lens"]
"effective megapixel count"	"10.1"
"height inches"	"4.2"
"lcd display size inches"	"3"
"lcd viewer"	"3 Inch"
"manufacturers warranty hardware"	"1 Year"
"megapixels"	"10.0"
"optical zoom"	"4x"
"pixel count"	"10 Megapixel"
"shutter speed"	"1/8000-30 second"
"skuprice"	"1299.9900"
"still image resolution max"	"3888 x 2592"
"usb port"	"(1) Mini-B"
"weight pounds"	"1.63"
"width inches"	"5.7"
entity 2	
"source"	"www.ebay.com"
"brand"	"Canon"
"megapixels"	"10.1 MP"
"model"	"40D"
"mpn"	"EOS 40D"
"screen size"	"3"
"type"	"Digital SLR"
entity 3	
"source"	"www.priceme.co.nz"
"page title"	"Canon EOS 400D New Zealand Prices - PriceMe"
"focus adjustment"	"Automatic focus, Manual focus"
"image stabilizer"	"Without Image Stabilizer"
"light sensitivity"	"ISO 100, ISO 1600, ISO 200, ISO 400, ISO 800, Auto"
"optical sensor"	"CMOS"
entity 4	
"source"	"www.gosale.com"
"page title"	"Nikon D3100 14.2MP Digital SLR on sale for $461.20"
"camera type"	"SLR"
"ean13"	"0018208097982"
"manufacturer"	"Nikon"
"megapixels"	"14.2 MP"
"product number mpn"	"D3100 18-55 5"
"retail price"	"$949.00"
"upc"	"018208097982"
entity 5	
"source"	"www.ebay.com"
"page title"	"Nikon D3100
"mpn"	"33858"
"screen size"	"3"
"upc"	"018208254866"
entity 6	
"source"	"www.walmart.com"
"page title"	"Nikon 14.2MP DSLR Camera with VR Lens, 3LCD"
"model no"	"Nikon D3100 Kit"
"shipping weight in pounds"	"3.6"
"walmart no"	"000609532"

Reduction of search space is a general problem in schema matching. Given two schemata, the comparison of every element of one schema with every element of the other schema has quadratic complexity. This large search space has not only detrimental effects with regards to scalability but can also negatively impact match quality given the higher number of error possibilities. The main strategies to narrow the search space are early pruning of dissimilar elements and partitioning of the ontologies [73]. Early pruning means discarding element pairs with low similarity early in the matching process. Especially, in sequential matching workflows this enables early matchers to alleviate the burden of unnecessary comparisons for subsequent matchers. For example, after determining the attribute name similarity of *usb port* in *entity 1* and *brand* in *entity 2* is low, the comparison of these attributes can be omitted in further steps. Peukert et al. [70] employ filters to discard element pairs beneath a certain similarity threshold. The threshold can be predefined or dynamically set depending on already calculated comparisons and mapping results. Partitioning-based approaches divide the ontologies in smaller parts so that only partitions have to be compared. This not only reduces the number of necessary comparisons, but makes these match tasks easily parallelizable.

Several different aspects of the data will have to be considered in order to create a high quality match result. A schema matching workflow will have to incorporate the similarity of attribute names and attribute values. The use of pre-trained word embeddings or synonym dictionaries can be beneficial to match attributes, that are dissimilar on character level, while being close semantically like `brand` and `manufacturer`. LeapME [2] relies on word embeddings and meta-information of property names and property values as input for a dense neural network. The classifier is trained on labeled property pairs and the corresponding feature vectors. The trained model can then be used to obtain matching decisions between unlabeled property pairs and their similarity scores. To integrate data about smart energy grids, Santodomingo and colleagues [87] use background knowledge from a database of electrical terminology. This background knowledge is used to find words with similar meanings to extend the strings of entities in the given ontologies. The authors utilize several matcher components, such as a linguistic module, which reduces words to their root form and filters out stop words, that are uninformative in the matching process (e.g., "the"), as well as threshold-based similarity components to derive matching decisions.

While binary matching approaches, unifying two sources, are most common, schema matching in the industry 4.0 context usually requires more holistic approaches that are able to consolidate multiple sources as shown in the example. Although it is possible to perform this task by sequentially matching two sources until all sources are integrated, specific approaches have been developed that cluster elements of multiple sources directly. Gruetze et al. [26] align large ontologies by clustering concepts by topic. Topical grouping is done by using Wikipedia pages related to concepts which result in category forests, that are a set of Wikipedia category trees. Utilizing the tree overlap alignments are generated. Megdiche et al. [54] model the holistic ontology matching task as maximum-weighted graph matching problem, which they solve within a linear program. Their approach is extensible with different linear constraints, that are used to reduce incoherence in resulting alignments. Roussille et al. [81] extend existing pairwise alignments of multiple sources by creating a graph with entities from ontologies as nodes, and correspondences as edges. They determine graph-cliques to detect the holistic alignment.

For a more general overview over ontology matching we refer the interested reader to this survey [63] and for a more detailed discussion of large-scale ontology and schema matching to [73].

Data Lifting The data in organizations usually has to be semantified, since it resides in formats which contain no machine-readable semantics such as relational databases or spreadsheets or even unstructured formats such as plain text. The necessary conversion process is called *data lifting*, since the data is not only transformed, but also "lifted" to a higher data level which contains semantic information [92].

While schema matching and data lifting both are concerned with mappings between different aspects of data sources, they have a different focus. Schema matching aims to consolidate heterogeneity between data sources and any enrichment of the data consists of implicit information that was scattered among different data sources. Data lifting seeks to transform data into RDF. While the mapping of e.g., a relational database to an existing ontology can be seen as a form of schema/ontology matching, data lifting is mainly concerned with transformation of the data into a different format.

The transformation process can be done manually by using specific mapping languages. The simplest is the *direct mapping*,[4] which performs a quick conversion of a relational database to RDF. The relational database should have well-defined primary and foreign keys and meaningful table and column names. While being simple, the direct mapping approach has the drawback of not being able to reuse existing popular vocabularies. For a more sophisticated conversion the mapping language *R2RML*[5] can be used. It enables the user to have more control over the mapping process. The use of manually created mappings is frequently mentioned in the industry 4.0 context. The German industrial control and automation company Festo describes their struggles with their previous monolithic Java application for data transformation in this paper [49]. They have since moved to use custom R2RML mappings to transform relational data into entities of their KG. Similarly, Kotis and Katasonov [46] propose rule-based mappings in their semantic smart gateway for the Web of Things.

While mapping languages enable powerful transformations, they require domain experts to go through a laborious process of writing many mapping rules, even with tool support. To address this problem learning-based transformation approaches have been devised in a research field called *ontology learning*. In the following we will present some examples from the field. For a more thorough overview over the field of ontology learning we refer the reader to this recent survey [50].

---

[4] http://www.w3.org/TR/rdb-direct-mapping/
[5] https://www.w3.org/TR/r2rml/

Maedche and Staab [51] first conceptualize ontology learning to address the need of simplifying the ontology engineering process by enabling the semi-automatic integration of a wide range of sources including web documents, XML files as well as databases and existing ontologies. They rely on dictionaries to extract concepts and use hierarchical clustering to build a taxonomical structure in their ontology. Using association rule mining with a class hierarchy as background they derive possible relationships that are presented to the user. Modoni et al. [56] present a rule-based approach to automatically transform relational databases to ontologies. Their ontology integration approach uses the mediator pattern, which does not physically integrate the ontologies but rather provides a common interface to distributed data sources. The mediation is done through custom mapping rules. The authors illustrate their approach with a case study of a mould production company, which is faced with integrating their various data sources.

4   Entity Resolution

In the smart product lifecycle and Industry 4.0. in general, a deluge of data from numerous sources is generated [88] requiring Big Data techniques for the collection, integration and analysis of heterogeneous data. Entity Resolution (ER) or data matching is a main step for data integration and the creation/evolution of knowledge graphs. It is the task of identifying entities within or across sources that refer to the same real-world entity. ER for Industry 4.0 requires fast and scalable solutions (Volume) as well as advanced methods to incrementally add new data or even new data sources either in a real-time or evolutionary way (Velocity) [21].

ER is typically implemented by a multistep workflow, as shown in Fig. 2. The input is data from multiple sources that may differ enormously in size and quality, and the output is a set of clusters, each of which contains all matching entities referring to the same real-world entity. The shown preprocessing step has already been discussed and entails data cleaning actions such as handling missing values, smoothing noisy values, and identifying and correcting inconsistent values [9]. Furthermore, schema matching can be applied to identify matching properties that can be used for determining the similarity of entities for ER. To match the cameras shown in Table 1, preprocessing may include transforming values into the same unit, lower casing strings, applying canonical abbreviations to harmonize property values, and assigning the same name to matching properties to facilitate similarity computations.

The blocking step prevents comparing irrelevant entities with each other. For instance, in our running camera example (Table 1), cameras with different manufacturers will be placed in different blocks in order to avoid comparing *Nikon* cameras with *Canon* cameras. Then in the pair-wise matching step, the similarity of candidate pairs are computed by applying a set of similarity methods on the property values of the entities. Finally, the clustering step uses computed similarities to group the same entities in the same cluster. Clustering facilitates fusion of the same entities into one unique representative entity.

Data Sources   Sets of Clusters

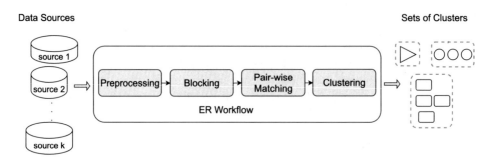

Fig. 2 Entity Resolution Workflow

The main ER steps of blocking, matching and clustering will be discussed in the following subsections, emphasizing on techniques related to Big Data. We will also outline incremental ER solutions to deal with the incremental addition of new entities and even new sources to a knowledge graph. Finally, we briefly discuss some ER prototypes for Big Data.

4.1   Blocking

Blocking aims at improving performance and scalability by avoiding that every entity has to be compared with every other entity for determining matching entity pairs, leading to a quadratic complexity. Therefore, blocking methods intend to restrict the comparisons only to those pairs that are likely to match. Standard Blocking (SB) [19] and Sorted Neighborhood (SN) [28] are two popular blocking methods that both utilize a so-called *blocking key* to group entities. The key is mostly specified by an expert and is the result of a function on one or several property values, e.g. the initial five letters of the manufacturer name or page title property for the camera example (Table 1). Since real data is noisy, generating one blocking key per entity may not allow finding all matches. Hence, it can be necessary to generate multiple blocking keys per entity, leading to multi-pass blocking [29, 44] that can find more matches and thus improve recall over the use of single blocking key. Since determining suitable blocking keys can be a tedious and difficult task, approaches based on both supervised [6, 20] and unsupervised [39] Machine Learning (ML) have been proposed to learn blocking keys. [67] gives a comprehensive overview of blocking techniques.

To further improve runtime and scalability, the blocking methods can be parallelized to utilize multiple machines in a cluster. This is relatively easy to achieve on partitioned input data by utilizing the MapReduce [12] framework or newer frameworks such as Apache Spark [95] that build on MapReduce. Moreover, since the sizes of the output blocks can be skewed, achieving good load balancing is the major challenge for parallel blocking and ER. Kolb et al. propose the load-balanced SB [43] and SN [42] based on the MapReduce framework.

For semi-structured, textual data or in absence of an aligned schema across sources, schema-agnostic token-based blocking approaches have been proposed. The basic Token Blocking (TB) [65] generates a candidate match based on the common tokens of property values of a pair. Like with traditional blocking methods, scalability can be improved by a MapReduce-based implementation [67] and ensuring load balancing [11]. Since the basic TB may create too many candidate pairs, newer schema-agnostic approaches reduce them by pairing tokens from synthetically similar properties, considering only selected properties, or comparing only the entities of the same type [67]. Furthermore, block post-processing approaches such as meta-blocking [66, 89] can largely reduce the number of candidate matches. A very different approach is [62] that totally ignores property values but determines candidate matches based on relations between entities.

4.2   Pair-wise Matching

The decision on whether a pair of entities is a likely match is based on the similarity of the two entities, which is determined by one or multiple similarity functions. These functions mostly determine the similarity of property values depending on the data type (string, numerical, date, geographical coordinates etc.). Typically, several such similarity values need to be combined to derive a match or non-match decision. Traditional approaches such as threshold-based or rule-based methods classify the matching status for each pair independently. In threshold-based classification, a specified threshold considers all pairs with similarity above a certain value as matches. On the other hand, in rule-based classification, a rule specifies a match predicate consisting of property-specific similarity conditions that are combined with logical operations [9]. For the camera example (Table 1), the match decision may be based on the similarity of the properties "page title" and "megapixels" although the latter property is not present for all entities shown.

Another line of research called collective ER [5] uses both property value similarity and relational information for determining the similarity of two entities. Here, the ER process is mostly iterative because changes in similarity or matching status of one pair affects the similarity value of the neighbouring pairs. Such approaches are more difficult to scale than with the standard approaches, where candidate pairs are compared independently. To better scale collective ER, Rastogi et al. [77] propose a generic approach that executes multiple instances of the matching task and constructs the global solution by message passing.

Manually determining the properties to match, similarity functions and similarity thresholds is a complex task, especially for heterogeneous and noisy data. Hence, a better alternative is often to apply supervised ML approaches to find optimal match configurations to determine matching entity pairs. These approaches can utilize traditional ML techniques such as SVM, logistic regression or random forests [40] but also newer approaches based on deep learning. Barlaug et al. [4] provides an overview about ER proposals utilizing deep neural networks including the approaches DeepER [14], DeepMatcher [57] and Hi-EM [96]. These approaches typically utilize embeddings for textual property values by transforming either words or their characters to numerical representations that preserve the semantic similarity between property values. Word embeddings are able to convert a long sequence to a short one, but they can not necessarily cover all possible words for specialized domains. The generation of embeddings can make use of pretrained models such as word2vec [55], GloVe [69] or fastText [7] that are derived from large corpora such as Wikipedia [4].

4.3   Clustering

The matches determined by the pair-wise similarity calculations are often contradicting and therefore only match candidates. The final matches are determined by applying a clustering approach on the set of candidate match pairs that form a similarity graph where matching entities are linked with each other. The baseline approach for entity clustering is to determine the transitive closure or connected components over the match links. Note, that general clustering algorithms like K-means that need a predefined number of clusters are not suitable for ER.

The Connected components algorithm does not consider the strength or similarity of candidate matches, and can thus cluster even weakly similar entities. There is a large spectrum of alternatives some of which, e.g. Stable Marriage [53] and Hungarian algorithm [47] are suited when the input consists of two duplicate-free sources. For deduplicating a single source, Hassanzadeh et al. [27] comparatively analyzed several clustering algorithms. For some of them, such as Correlation Clustering, parallel implementations based on iterative processing and message passing have been proposed [8, 64]. Saeedi et al. [83] comparatively evaluate the effectiveness and scalability of parallel implementations of several clustering schemes from [27] for the case of multiple data sources. Recently, Yan et al. [94] proposed a novel hierarchical clustering approach that avoids so-called hard conflicts inside clusters where the weakest similarity in a cluster is below a critical threshold. This is achieved by not merging candidate cluster pairs if this would lead to such a hard conflict. The approach is used within an industrial ER framework that is applied on billions of customer records on a daily basis.

Another line of research focuses on designing methods and algorithms for clustering entities from multiple duplicate-free sources [59, 84] or clustering entities from combined duplicate-free and dirty (duplicate-containing) data sources [48]. The proposed approaches outperform more general approaches such as correlation clustering.

4.4   Incremental ER

Incremental ER approaches are needed to address the "Velocity" characteristic of Big Data to deal with dynamic or evolving data such as new incoming entities or even new data sources. Incremental approaches generally fall into two categories: 1) real-time approaches that are mostly applied in query processing and deal with individual new entities and 2) evolutionary approaches that deal with the addition of several entities of even a complete new data source in order to update an already existing knowledge graph without repeating the ER process for all data.

1) *Real-time approaches* leverage dynamic blocking and indexing techniques [76] as well as dynamic pair-wise matching methods [1, 33, 93] that support the fast matching of entities at query time.

2) *Evolutionary approaches* focus on updating the knowledge graph. Gruenheid et al. propose a generic greedy approach for such an incremental ER and clustering [25]. This method is extended in [58] to avoid computations on already integrated portions of the data that are unlikely to be affected by the new data. Scalable approaches for incremental entity clustering that also support the addition of new data sources are investigated in [60, 85]. In particular, [60] proposes an incremental entity clustering based on a `Max-Both` strategy that adds a new entity to the maximally similar cluster only if there is no other new entity of the same input source with a higher similarity. [85] proposes a method called `n-depth reclustering` for incremental linking and clustering that is even able to repair existing clusters for improved quality and a reduced dependency on the insert order of new entities.

4.5   ER Prototypes

There are several ER prototypes suitable for Big Data that are surveyed in [10] including Dedoop [41], Magellan [45], FAMER [82], Silk [34], MinoanER [15], and JedAI [68]. Each of them implements the whole ER pipeline in a parallel way and includes novel Big-Data-specific approaches for at least one step of the pipeline. Dedoop is one of the early systems and based on MapReduce [12]; it implements the load balancing techniques discussed in the subsection on blocking. Silk, MinoanER, JedAI and a non-public version of Magellan are implemented on top of Apache Spark[6] while FAMER uses Apache Flink.[7] FAMER additionally supports the incremental addition of new entities and new data sources [85] and can deal with entities from multiple sources (>2), while MinoanER supports schema-agnostic ER methods to deal with heterogeneous and noisy web entities.

---

[6] https://spark.apache.org/
[7] https://flink.apache.org/

5   Conclusion & Open Problems

We presented an overview over the Big Data challenges for data integration posed by the fourth industrial revolution. We advocated the use of knowledge graphs for the integrated and semantically consolidated representation of heterogeneous data as a basis for data analysis and production optimization. Creating and continuously updating knowledge graphs is challenging and we presented approaches for the tasks of schema/ontology matching, data lifting/semantification and especially for entity resolution. We also discussed some published data integration use cases for Industry 4.0.

The current state for Big Data integration using knowledge graphs in Industry 4.0 is still in an early stage and requires too much manual effort. The common use of manual mapping rules for data lifting and/or schema matching can be justifiable for horizontal integration cases with already well structured high quality data. However, more efforts are needed to bridge the gap between the (semi-)automatic data integration tools developed in academia and manual matching efforts that are prevalent in the industry to establish robust methods for integrating the complex data of industrial applications. Especially the increasing interconnection of different domains (e.g. IoT, Smart Factories and Smart Grids) calls for more automated integration concepts, that could enable "plug & play" capabilities of smart machinery [17]. Solving these challenges is not reasonably possible without incremental ER solutions that keep knowledge graphs in sync with the physical realities present in smart factories, within a reasonable time frame. The possibility of integrating increasingly larger data sources asks for scalable solutions. The triple stores used in Semantic Web applications can become a bottleneck, which necessitates alternative solutions [36]. The use of frameworks that rely on property graph models (e.g. Neo4j[8] or Gradoop [80]) can be a viable alternative to triple stores in some use cases.

The interdiscplinary nature of Industry 4.0 necessitates a close cooperation between domain experts of the respective manufacturing domain, ontology engineers and data scientists [38]. We believe, that this is not only true for individual projects in this domain, but for the research in this direction as a whole.

References

1. Altwaijry, H., Kalashnikov, D.V., Mehrotra, S.: Query-driven approach to entity resolution. Proceedings of the VLDB Endowment **6**(14), 1846–1857 (2013)
2. Ayala, D., Hernández, I., Ruiz, D., Rahm, E.: Leapme: Learning-based property matching with embeddings (2020)
3. Bader, S.R., Grangel-González, I., Nanjappa, P., Vidal, M.E., Maleshkova, M.: A knowledge graph for industry 4.0. The Semantic Web **12123**, 465 – 480 (2020)
4. Barlaug, N., Gulla, J.A.: Neural networks for entity matching: A survey. arXiv preprint arXiv:2010.11075 (2020)

---

[8] https://neo4j.com/

5. Bhattacharya, I., Getoor, L.: Collective entity resolution in relational data. ACM Transactions on Knowledge Discovery from Data (TKDD) **1**(1), 5–es (2007)
6. Bilenko, M., Kamath, B., Mooney, R.J.: Adaptive blocking: Learning to scale up record linkage. In: Sixth International Conference on Data Mining (ICDM'06). pp. 87–96. IEEE (2006)
7. Bojanowski, P., Grave, E., Joulin, A., Mikolov, T.: Enriching word vectors with subword information. Transactions of the Association for Computational Linguistics **5**, 135–146 (2017)
8. Chierichetti, F., Dalvi, N., Kumar, R.: Correlation clustering in mapreduce. In: Proceedings of the 20th ACM SIGKDD international conference on Knowledge discovery and data mining. pp. 641–650 (2014)
9. Christen, P.: The data matching process. In: Data Matching, pp. 23–35. Springer (2012)
10. Christophides, V., Efthymiou, V., Palpanas, T., Papadakis, G., Stefanidis, K.: An overview of end-to-end entity resolution for big data. ACM Computing Surveys (2020)
11. Chu, X., Ilyas, I.F., Koutris, P.: Distributed data deduplication. Proceedings of the VLDB Endowment **9**(11), 864–875 (2016)
12. Dean, J., Ghemawat, S.: Mapreduce: simplified data processing on large clusters. Communications of the ACM **51**(1), 107–113 (2008)
13. Dong, X.L., Srivastava, D.: Big data integration. Synthesis Lectures on Data Management **7**(1), 1–198 (2015)
14. Ebraheem, M., Thirumuruganathan, S., Joty, S., Ouzzani, M., Tang, N.: Distributed representations of tuples for entity resolution pp. 1454–1467 (2018)
15. Efthymiou, V., Papadakis, G., Stefanidis, K., Christophides, V.: Minoaner: Schema-agnostic, non-iterative, massively parallel resolution of web entities. arXiv preprint arXiv:1905.06170 (2019)
16. Ehrlinger, L., Wöß, W.: Towards a definition of knowledge graphs. In: SEMANTiCS (Posters, Demos, SuCCESS) (2016)
17. Ekaputra, F.J., Sabou, M., Biffl, S., Einfalt, A., Krammer, L., Kastner, W., Ekaputra, F.J.: Semantics for Cyber-Physical Systems: A cross-domain perspective. Semantic Web **11**(1), 115–124 (2020). https://doi.org/10.3233/SW-190381, https://doi.org/10.3233/SW-190381
18. Elmer, S., Jrad, F., Liebig, T., Ul Mehdi, A., Opitz, M., Stauß, T., Weidig, D.: Ontologies and reasoning to capture product complexity in automation industry. CEUR Workshop Proceedings **1963**, 1–2 (2017)
19. Fellegi, I.P., Sunter, A.B.: A theory for record linkage. Journal of the American Statistical Association **64**(328), 1183–1210 (1969)
20. Giang, P.H.: A machine learning approach to create blocking criteria for record linkage. Health care management science **18**(1), 93–105 (2015)
21. Gölzer, P., Cato, P., Amberg, M.: Data processing requirements of industry 4.0 - use cases for big data applications. In: Becker, J., vom Brocke, J., de Marco, M. (eds.) 23rd European Conference on Information Systems, ECIS 2015, Münster, Germany, May 26-29, 2015 (2015), http://aisel.aisnet.org/ecis2015_rip/61
22. Gröger, C.: Building an industry 4.0 analytics platform - practical challenges, approaches and future research directions. Datenbank-Spektrum **18**(1), 5–14 (2018). https://doi.org/10.1007/s13222-018-0273-1, https://doi.org/10.1007/s13222-018-0273-1
23. Gröger, C., Schwarz, H., Mitschang, B.: The manufacturing knowledge repository - consolidating knowledge to enable holistic process knowledge management in manufacturing. In: Hammoudi, S., Maciaszek, L.A., Cordeiro, J. (eds.) ICEIS 2014 - Proceedings of the 16th International Conference on Enterprise Information Systems, Volume 1, Lisbon, Portugal, 27-30 April, 2014. pp. 39–51. SciTePress (2014). https://doi.org/10.5220/0004891200390051, https://doi.org/10.5220/0004891200390051

24. Gross, A., Hartung, M., Kirsten, T., Rahm, E.: On matching large life science ontologies in parallel. Lecture Notes in Computer Science (including subseries Lecture Notes in Artificial Intelligence and Lecture Notes in Bioinformatics) **6254 LNBI**, 35–49 (2010). https://doi.org/10. 1007/978-3-642-15120-0_4

25. Gruenheid, A., Dong, X.L., Srivastava, D.: Incremental record linkage. Proceedings of the VLDB Endowment **7**(9), 697–708 (2014)

26. Gruetze, T., Böhm, C., Naumann, F.: Holistic and scalable ontology alignment for linked open data. CEUR Workshop Proceedings **937** (2012)

27. Hassanzadeh, O., Chiang, F., Lee, H.C., Miller, R.J.: Framework for evaluating clustering algorithms in duplicate detection. Proceedings of the VLDB Endowment **2**(1), 1282–1293 (2009)

28. Hernández, M.A., Stolfo, S.J.: The merge/purge problem for large databases. ACM Sigmod Record **24**(2), 127–138 (1995)

29. Hernández, M.A., Stolfo, S.J.: Real-world data is dirty: Data cleansing and the merge/purge problem. Data mining and knowledge discovery **2**(1), 9–37 (1998)

30. Hitzler, P., Krötzsch, M., Rudolph, S.: Foundations of Semantic Web Technologies. Chapman & Hall/CRC (2009)

31. Hubauer, T., Lamparter, S., Haase, P., Herzig, D.: Use cases of the industrial knowledge graph at siemens. CEUR Workshop Proceedings **2180** (2018)

32. Ilyas, I.F., Chu, X.: Data cleaning. Morgan & Claypool (2019)

33. Ioannou, E., Nejdl, W., Niederée, C., Velegrakis, Y.: On-the-fly entity-aware query processing in the presence of linkage. Proceedings of the VLDB Endowment **3**(1-2), 429–438 (2010)

34. Isele, R., Bizer, C.: Learning expressive linkage rules using genetic programming. arXiv preprint arXiv:1208.0291 (2012)

35. Jirkovský, V., Kadera, P., Rychtyckyj, N.: Semi-automatic ontology matching approach for integration of various data models in automotive. Lecture Notes in Computer Science (including subseries Lecture Notes in Artificial Intelligence and Lecture Notes in Bioinformatics) **10444 LNAI**(August), 53–65 (2017). https://doi.org/10.1007/978-3-319-64635-0_5

36. Jirkovsky, V., Obitko, M., Marik, V.: Understanding data heterogeneity in the context of cyber-physical systems integration. IEEE Transactions on Industrial Informatics **13**(2) (2017). https://doi.org/10.1109/TII.2016.2596101

37. Kagermann, H., Wahlster, W., Helbig, J.: Recommendations for implementing the strategic initiative industrie 4.0 – securing the future of german manufacturing industry. Final report of the industrie 4.0 working group, acatech – National Academy of Science and Engineering, München (2013), https://en.acatech.de/wp-content/uploads/sites/6/2018/03/Final_report__ Industrie_4.0_accessible.pdf

38. Kalaycı, E.G., Grangel González, I., Lösch, F., Xiao, G., Ul-Mehdi, A., Kharlamov, E., Calvanese, D.: Semantic Integration of Bosch Manufacturing Data Using Virtual Knowledge Graphs, vol. 12507 LNCS. Springer International Publishing (2020). https://doi.org/10.1007/978-3-030-62466-8_29, http://dx.doi.org/10.1007/978-3-030-62466-8_29

39. Kejriwal, M., Miranker, D.P.: An unsupervised algorithm for learning blocking schemes. In: 2013 IEEE 13th International Conference on Data Mining. pp. 340–349. IEEE (2013)

40. Koepcke, H., Thor, A., Rahm, E.: Learning-based approaches for matching web data entities. IEEE Internet Computing **14**(4), 23–31 (2010)

41. Kolb, L., Rahm, E.: Parallel entity resolution with dedoop. Datenbank-Spektrum **13**(1), 23–32 (2013)

42. Kolb, L., Thor, A., Rahm, E.: Parallel sorted neighborhood blocking with mapreduce. arXiv preprint arXiv:1010.3053 (2010)

43. Kolb, L., Thor, A., Rahm, E.: Load balancing for mapreduce-based entity resolution. In: 2012 IEEE 28th international conference on data engineering. pp. 618–629. IEEE (2012)

44. Kolb, L., Thor, A., Rahm, E.: Multi-pass sorted neighborhood blocking with mapreduce. Computer Science-Research and Development **27**(1), 45–63 (2012)
45. Konda, P., Das, S., Suganthan GC, P., Doan, A., Ardalan, A., Ballard, J.R., Li, H., Panahi, F., Zhang, H., Naughton, J., et al.: Magellan: Toward building entity matching management systems. Proceedings of the VLDB Endowment **9**(12), 1197–1208 (2016)
46. Kotis, K., Katasonov, A.: Semantic interoperability on the web of things: The semantic smart gateway framework. In: Barolli, L., Xhafa, F., Vitabile, S., Uehara, M. (eds.) Sixth International Conference on Complex, Intelligent, and Software Intensive Systems, CISIS 2012, Palermo, Italy, July 4-6, 2012. pp. 630–635. IEEE Computer Society (2012). https://doi.org/10.1109/CISIS.2012.200, https://doi.org/10.1109/CISIS.2012.200
47. Kuhn, H.W.: The hungarian method for the assignment problem. Naval research logistics quarterly **2**(1-2), 83–97 (1955)
48. Lerm, S., Saeedi, A., Rahm, E.: Extended affinity propagation clustering for multi-source entity resolution. Datenbank-Spektrum (2021)
49. Liebig, T., Maisenbacher, A., Opitz, M., Seyler, J.R., Sudra, G., Wissmann, J.: Building a knowledge graph for products and solutions in the automation industry. CEUR Workshop Proceedings **2489**, 13–23 (2019)
50. Ma, C., Molnár, B.: Use of Ontology Learning in Information System Integration: A Literature Survey. Communications in Computer and Information Science **1178 CCIS**, 342–353 (2020). https://doi.org/10.1007/978-981-15-3380-8_30
51. Maedche, A., Staab, S.: Ontology learning for the semantic web. IEEE Intell. Syst. **16**(2), 72–79 (2001). https://doi.org/10.1109/5254.920602, https://doi.org/10.1109/5254.920602
52. Mazumdar, S., Varga, A., Lanfranchi, V., Petrelli, D., Ciravegna, F.: A knowledge dashboard for manufacturing industries. In: Garcia-Castro, R., Fensel, D., Antoniou, G. (eds.) The Semantic Web: ESWC 2011 Workshops - ESWC 2011 Workshops, Heraklion, Greece, May 29-30, 2011, Revised Selected Papers. Lecture Notes in Computer Science, vol. 7117, pp. 112–124. Springer (2011). https://doi.org/10.1007/978-3-642-25953-1_10, https://doi.org/10.1007/978-3-642-25953-1_10
53. McVitie, D.G., Wilson, L.B.: Stable marriage assignment for unequal sets. BIT Numerical Mathematics **10**(3), 295–309 (1970)
54. Megdiche, I., Teste, O., dos Santos, C.T.: An extensible linear approach for holistic ontology matching. In: Groth, P., Simperl, E., Gray, A.J.G., Sabou, M., Krötzsch, M., Lécué, F., Flöck, F., Gil, Y. (eds.) The Semantic Web - ISWC 2016 - 15th International Semantic Web Conference, Kobe, Japan, October 17-21, 2016, Proceedings, Part I. Lecture Notes in Computer Science, vol. 9981, pp. 393–410 (2016). https://doi.org/10.1007/978-3-319-46523-4_24, https://doi.org/10.1007/978-3-319-46523-4_24
55. Mikolov, T., Sutskever, I., Chen, K., Corrado, G.S., Dean, J.: Distributed representations of words and phrases and their compositionality. Advances in neural information processing systems **26**, 3111–3119 (2013)
56. Modoni, G.E., Doukas, M., Terkaj, W., Sacco, M., Mourtzis, D.: Enhancing factory data integration through the development of an ontology: from the reference models reuse to the semantic conversion of the legacy models. International Journal of Computer Integrated Manufacturing **30**(10), 1043–1059 (2017). https://doi.org/10.1080/0951192X.2016.1268720, https://doi.org/10.1080/0951192X.2016.1268720
57. Mudgal, S., Li, H., Rekatsinas, T., Doan, A., Park, Y., Krishnan, G., Deep, R., Arcaute, E., Raghavendra, V.: Deep learning for entity matching: A design space exploration. In: Proceedings of the 2018 International Conference on Management of Data. pp. 19–34 (2018)
58. do Nascimento, D.C., Pires, C.E.S., Mestre, D.G.: Heuristic-based approaches for speeding up incremental record linkage. Journal of Systems and Software **137**, 335–354 (2018)

59. Nentwig, M., Groß, A., Rahm, E.: Holistic entity clustering for linked data. In: 2016 IEEE 16th International Conference on Data Mining Workshops (ICDMW). pp. 194–201. IEEE (2016)
60. Nentwig, M., Rahm, E.: Incremental clustering on linked data. In: 2018 IEEE International Conference on Data Mining Workshops (ICDMW). pp. 531–538. IEEE (2018)
61. Niedermann, F., Schwarz, H., Mitschang, B.: Managing insights: A repository for process analytics, optimization and decision support. In: Filipe, J., Liu, K. (eds.) KMIS 2011 - Proceedings of the International Conference on Knowledge Management and Information Sharing, Paris, France, 26-29 October, 2011. pp. 424–429. SciTePress (2011)
62. Nin, J., Muntés-Mulero, V., Martinez-Bazan, N., Larriba-Pey, J.L.: On the use of semantic blocking techniques for data cleansing and integration. In: 11th International Database Engineering and Applications Symposium (IDEAS 2007). pp. 190–198. IEEE (2007)
63. Otero-Cerdeira, L., Rodríguez-Martínez, F.J., Gómez-Rodríguez, A.: Ontology matching: A literature review. Expert Systems with Applications **42**(2) (2015). https://doi.org/10.1016/j.eswa.2014.08.032
64. Pan, X., Papailiopoulos, D., Oymak, S., Recht, B., Ramchandran, K., Jordan, M.I.: Parallel correlation clustering on big graphs. In: Advances in Neural Information Processing Systems. pp. 82–90 (2015)
65. Papadakis, G., Ioannou, E., Palpanas, T., Niederee, C., Nejdl, W.: A blocking framework for entity resolution in highly heterogeneous information spaces. IEEE Transactions on Knowledge and Data Engineering **25**(12), 2665–2682 (2012)
66. Papadakis, G., Papastefanatos, G., Palpanas, T., Koubarakis, M.: Scaling entity resolution to large, heterogeneous data with enhanced meta-blocking. In: EDBT. pp. 221–232 (2016)
67. Papadakis, G., Skoutas, D., Thanos, E., Palpanas, T.: A survey of blocking and filtering techniques for entity resolution. CoRR, abs/1905.06167 (2019)
68. Papadakis, G., Tsekouras, L., Thanos, E., Pittaras, N., Simonini, G., Skoutas, D., Isaris, P., Giannakopoulos, G., Palpanas, T., Koubarakis, M.: Jedai3: beyond batch, blocking-based entity resolution. In: EDBT. pp. 603–606 (2020)
69. Pennington, J., Socher, R., Manning, C.D.: Glove: Global vectors for word representation. In: Proceedings of the 2014 conference on empirical methods in natural language processing (EMNLP). pp. 1532–1543 (2014)
70. Peukert, E., Berthold, H., Rahm, E.: Rewrite techniques for performance optimization of schema matching processes. Advances in Database Technology - EDBT 2010 - 13th International Conference on Extending Database Technology, Proceedings pp. 453–464 (2010). https://doi.org/10.1145/1739041.1739096
71. Prestes, E., Carbonera, J.L., Fiorini, S.R., Jorge, V.A.M., Abel, M., Madhavan, R., Locoro, A., Gonçalves, P.J.S., Barreto, M.E., Habib, M.K., Chibani, A., Gérard, S., Amirat, Y., Schlenoff, C.: Towards a core ontology for robotics and automation. Robotics Auton. Syst. **61**(11), 1193–1204 (2013). 10.1016/j.robot.2013.04.005, https://doi.org/10.1016/j.robot.2013.04.005
72. Qadri, Y.A., Nauman, A., Zikria, Y.B., Vasilakos, A.V., Kim, S.W.: The future of healthcare internet of things: A survey of emerging technologies. IEEE Commun. Surv. Tutorials **22**(2), 1121–1167 (2020). https://doi.org/10.1109/COMST.2020.2973314, https://doi.org/10.1109/COMST.2020.2973314
73. Rahm, E.: Towards Large-Scale Schema and Ontology Matching. Schema Matching and Mapping pp. 3–27 (2011). https://doi.org/10.1007/978-3-642-16518-4_1
74. Rahm, E.: The case for holistic data integration. In: Proc. ADBIS. pp. 11–27. Springer (2016)
75. Rahm, E., Do, H.H.: Data cleaning: Problems and current approaches. IEEE Data Eng. Bull. **23**(4), 3–13 (2000)
76. Ramadan, B., Christen, P., Liang, H., Gayler, R.W.: Dynamic sorted neighborhood indexing for real-time entity resolution. Journal of Data and Information Quality (JDIQ) **6**(4), 1–29 (2015)

77. Rastogi, V., Dalvi, N., Garofalakis, M.: Large-scale collective entity matching. arXiv preprint arXiv:1103.2410 (2011)
78. Ringsquandl, M., Kharlamov, E., Stepanova, D., Lamparter, S., Lepratti, R., Horrocks, I., Kroger, P.: On event-driven knowledge graph completion in digital factories. Proceedings - 2017 IEEE International Conference on Big Data, Big Data 2017 **2018-Janua**, 1676–1681 (2017). https://doi.org/10.1109/BigData.2017.8258105
79. Ristoski, P., Petrovski, P., Mika, P., Paulheim, H.: A machine learning approach for product matching and categorization. Semantic Web **9**(5), 707–728 (2018)
80. Rost, C., Thor, A., Fritzsche, P., Gómez, K., Rahm, E.: Evolution analysis of large graphs with gradoop. In: Cellier, P., Driessens, K. (eds.) Machine Learning and Knowledge Discovery in Databases - International Workshops of ECML PKDD 2019, Würzburg, Germany, September 16-20, 2019, Proceedings, Part I. Communications in Computer and Information Science, vol. 1167, pp. 402–408. Springer (2019). https://doi.org/10.1007/978-3-030-43823-4_33, https://doi.org/10.1007/978-3-030-43823-4_33
81. Roussille, P., Megdiche, I., Teste, O., Trojahn, C.: Boosting holistic ontology matching: Generating graph clique-based relaxed reference alignments for holistic evaluation. Lecture Notes in Computer Science (including subseries Lecture Notes in Artificial Intelligence and Lecture Notes in Bioinformatics) **11313**(November), 355–369 (2018). https://doi.org/10.1007/978-3-030-03667-6_23
82. Saeedi, A., Nentwig, M., Peukert, E., Rahm, E.: Scalable matching and clustering of entities with famer. Complex Systems Informatics and Modeling Quarterly **16**, 61–83 (2018)
83. Saeedi, A., Peukert, E., Rahm, E.: Comparative evaluation of distributed clustering schemes for multi-source entity resolution. In: European Conference on Advances in Databases and Information Systems. pp. 278–293. Springer (2017)
84. Saeedi, A., Peukert, E., Rahm, E.: Using link features for entity clustering in knowledge graphs. In: European Semantic Web Conference. pp. 576–592. Springer (2018)
85. Saeedi, A., Peukert, E., Rahm, E.: Incremental multi-source entity resolution for knowledge graph completion. In: European Semantic Web Conference. pp. 393–408. Springer (2020)
86. Sampath Kumar, V.R., Khamis, A., Fiorini, S., Carbonera, J.L., Alarcos, A.O., Habib, M., Goncalves, P., Howard, L.I., Olszewska, J.I.: Ontologies for industry 4.0. Knowledge Engineering Review **34** (2019). https://doi.org/10.1017/S0269888919000109
87. Santodomingo, R., Rohjans, S., Uslar, M., Rodríguez-Mondéjar, J.A., Sanz-Bobi, M.A.: Ontology matching system for future energy smart grids. Engineering Applications of Artificial Intelligence **32** (2014). https://doi.org/10.1016/j.engappai.2014.02.005
88. Schmidt, M., Galende, M., Saludes, S., Sarris, N., Rodriguez, J., Unal, P., Stojanovic, N., Vidal, I.G.M., Corchero, A., Berre, A., Cattaneo, G., Geogoulias, K., Stojanovic, L., Decubber, C.: Big data challenges in smart manufacturing: A discussion paper on big data challenges for bdva and effra research & innovation roadmaps alignment. Tech. rep., Big Data Value Association (2018), https://bdva.eu/sites/default/files/BDVA_SMI_Discussion_Paper_Web_Version.pdf
89. Simonini, G., Bergamaschi, S., Jagadish, H.: Blast: a loosely schema-aware meta-blocking approach for entity resolution. pvldb 9, 12 (2016), 1173–1184 (2016)
90. Skjæveland, M.G., Gjerver, A., Hansen, C.M., Klüwer, J.W., Strand, M.R., Waaler, A., Øverli, P.Ø.: Semantic material master data management at AibEL. CEUR Workshop Proceedings **2180**, 4–5 (2018)
91. Song, D., Schilder, F., Hertz, S., Saltini, G., Smiley, C., Nivarthi, P., Hazai, O., Landau, D., Zaharkin, M., Zielund, T., Molina-Salgado, H., Brew, C., Bennett, D.: Building and Querying an Enterprise Knowledge Graph. IEEE Transactions on Services Computing **12**(3), 356–369 (2019). https://doi.org/10.1109/TSC.2017.2711600

92. Villazon-Terrazas, B., Garcia-Santa, N., Ren, Y., Faraotti, A., Wu, H., Zhao, Y., Vetere, G., Pan, J.Z.: Knowledge Graph Foundations, pp. 17–55. Springer International Publishing, Cham (2017). https://doi.org/10.1007/978-3-319-45654-6_2, https://doi.org/10.1007/978-3-319-45654-6_2
93. Wang, J., Krishnan, S., Franklin, M.J., Goldberg, K., Kraska, T., Milo, T.: A sample-and-clean framework for fast and accurate query processing on dirty data. In: Proceedings of the 2014 ACM SIGMOD international conference on Management of data. pp. 469–480 (2014)
94. Yan, Y., Meyles, S., Haghighi, A., Suciu, D.: Entity matching in the wild: A consistent and versatile framework to unify data in industrial applications. In: Proceedings of the 2020 ACM SIGMOD International Conference on Management of Data. pp. 2287–2301 (2020)
95. Zaharia, M., Chowdhury, M., Das, T., Dave, A., Ma, J., McCauly, M., Franklin, M.J., Shenker, S., Stoica, I.: Resilient distributed datasets: A fault-tolerant abstraction for in-memory cluster computing. In: 9th {USENIX} Symposium on Networked Systems Design and Implementation ({NSDI} 12). pp. 15–28 (2012)
96. Zhao, C., He, Y.: Auto-em: End-to-end fuzzy entity-matching using pre-trained deep models and transfer learning. In: The World Wide Web Conference. pp. 2413–2424 (2019)
97. Zhou, B., Svetashova, Y., Byeon, S., Pychynski, T., Mikut, R., Kharlamov, E.: Predicting Quality of Automated Welding with Machine Learning and Semantics: A Bosch Case Study. International Conference on Information and Knowledge Management, Proceedings pp. 2933–2940 (2020). https://doi.org/10.1145/3340531.3412737

Massive Data Sets – Is Data Quality Still an Issue?

Peter Filzmoser and Alexandra Mazak-Huemer

Abstract

The term "big data" has become a buzzword in the last years, and it refers to the possibility to collect and store huge amounts of information, resulting in big data bases and data repositories. This also holds for industrial applications: In a production process, for instance, it is possible to install many sensors and record data in a very high temporal resolution. The amount of information grows rapidly, but not necessarily does the insight into the production process. This is the point where machine learning or, say, statistics needs to enter, because sophisticated algorithms are now required to identify the relevant parameters which are the drivers of the quality of the product, as an example. However, is data quality still an issue? It is clear that with small amounts of data, single outliers or extreme values could affect the algorithms or statistical methods. Can "big data" overcome this problem? In this article we will focus on some specific problems in the regression context, and show that even if many parameters are measured, poor data quality can severely influence the prediction performance of the methods.

P. Filzmoser
Computational Statistics, Institute of Statistics and Mathematical Methods in Economics,
TU Wien, Vienna, Austria
e-mail: peter.filzmoser@tuwien.ac.at

A. Mazak-Huemer (✉)
Institute of Business Informatics - Software Engineering, Johannes Kepler University (JKU) Linz,
Linz, Austria
e-mail: a.mazak-huemer@rat-fte.at

1   Introduction

Industry 4.0 provides the possibility to learn more and more about industrial applications, e.g. to better understand production processes, and this in turn allows for a more refined monitoring and inspection of such processes. This is basically thanks to "digitalization", which means that data are permanently measured, with the goal to deliver a precise description of the production process. Since sensors are relatively cheap, they can be installed at many places of the production, some of them being probably even irrelevant to the final outcome. After all, the data generated in this way can be of enormous size, and the size can grow continuously over time. Of course, humans would not get a better understanding just because of having access to a lot of data; it first needs an aggregation or a summary of the data, or the relevant part of the data needs to be identified first, before humans can make the link between the digital description and the process. This, however, is exactly the difficulty: the selection of the important data part which allows to draw conclusions for the process needs to be automated. In statistics we would say that a statistical model has to provide the relevant output; in computer science one would say that a Machine Learning method needs to be used in order to learn about the machine. In any case, one needs to define a "target" appropriately before a method or a model can identify the useful information. Such a target could be the measured "quality" of the product that is produced in the process.

Independent of the size of the data, there is one permanent problem when recording data: outliers. This refers to observations that are either wrongly measured, or they are for some reason inconsistent with other observations. It is known that outliers can affect statistical estimators, and consequently they can lead to poor model prediction quality. Note that here we are not necessarily referring to outliers in one variable (univariate outliers), but outliers in the multivariate data space. Every additional sensor measurement defines a new variable in a data set, and an outlier is an observation which deviates from the bulk of the other observations with respect to the jointly measured information. It is not at all clear beforehand if such an outlier affects a statistical model, nor is it clear when we would talk about outlyingness – this would require a clearer definition. The only thing which is clear is that with more and more measurements, the risk of having errors or inconsistencies (even small ones) in the data increases. This phenomenon could be called the "curse of big data" – hopefully there is also a "blessing".

In order to be a bit more concrete, we want to illustrate our concepts with a data set from a production process. The data set is available at `https://cstat.tuwien.ac.at/filz/dat.RData`, and it has been anonymized for confidentiality reasons. It relates to a regression problem, with a continuous response, say a variable describing the quality of the production outcome, and 392 predictor variables. There are 664 observations available which are in fact different time points, and thus this is a time series data set. The task is to predict the production quality based on the predictors. It can be assumed that some or many of them are not relevant, and it is also likely that outliers are present. Overall, this is not a massive data set, but it seems to be quite complex for this task.

2   Outlier Identification

Whenever possible, it is recommended to look at the data, or at least to inspect part of the data. Figure 1 shows the response variable over time, as well as selected predictor variables. For example, variable X67 has the highest Spearman rank correlation with the response (see number on top of the plot). Already these pictures make it clear that the data set is full of outliers and artifacts, and those would very likely have an influence on regression models. A first idea would thus be to remove the outliers prior to any modeling.

Outlier detection is an important and highly developed field, and many methods are available. Basically, there are methods originating from the statistics area which assume an underlying model, and there are methods from computer science which are model-free and judge outlyingness in terms of the degree of isolation from other points. For a recent overview, see, e.g. [14].

Since we deal with time-series data, it would be natural to use outlier detection methods from this area, such as time series filtering methods, e.g. [2]. However, those methods would have to be applied to each time series separately, and the ability to identify outliers depends on several tuning parameters. Therefore, one could first consider the simplest univariate method, ignoring the time series context: for each variable, subtract the median and divide by the MAD (median absolute deviation), see [9]. According to the normal theory, values which are outside e.g. ± 3 can be considered as outliers. To be conservative, we consider here outliers if this score is outside ± 6. This is done variable-by-variable, and finally we count, how often an individual observation has been declared as outlier. Figure 2 shows the result in terms of a bar plot: each bar reports the frequency how often an observation was

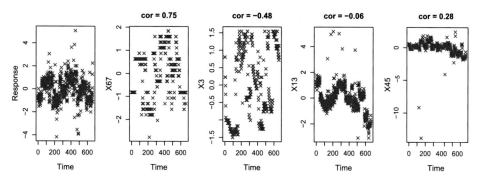

Fig. 1 Visualization of the response and some selected predictor variables; numbers on top are the Spearman rank correlations of the predictors with the response

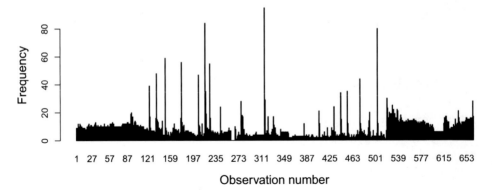

Fig. 2 Frequency of outlyingness of each observation in the different variables of the considered data set

declared as outlier. There are only 6 observations which have never been identified as outlier, but many observations are outlying in many variables, some even in extremely many.

Note that here we only computed univariate outlyingness, and multivariate outliers were not even considered. For high-dimensional data, multivariate outlier detection is quite challenging, as traditional approaches based on (robust) Mahalanobis distances would typically fail, and alternative approaches often require appropriate tuning parameters or outlier cut-off values, see, e.g. [14]. However, the question arises what to do with this outlyingness information. Obviously, it is not an option to remove observations which are outlying in any variable, since we would be left with a data set with 6 observations. Correcting the outliers is also not feasible, since this would require a statistical model. On top of that, it is not even clear if these outliers would have any influence on a regression model, since as long as they are on a linear trend with the response, they might be harmless. For this reason we proceed with regression modeling.

3   Robust Modeling

For reasons of simplicity we will ignore the time-series aspect in the data in the following, and try to model the response using linear regression. Denote the observations of the response by y_i, and of the explanatory variables by $\mathbf{x}_i = (x_{i1}, \ldots, x_{ip})^T$, for $i = 1, \ldots, n$. Then the linear regression model is given as

$$y_i = \beta_0 + \mathbf{x}_i^T \boldsymbol{\beta}_1 + \varepsilon_i \quad \text{for } i = 1, \ldots, n, \tag{1}$$

where β_0 is the intercept, $\boldsymbol{\beta}_1 = (\beta_1, \ldots, \beta_p)^T$ is the vector of regression coefficients for the predictor variables, and ε_i denotes the error terms, which are assumed to be i.i.d. random variables, distributed according to $N(0, \sigma^2)$.

Denote all unknown regression coefficients by $\boldsymbol{\beta} = (\beta_0, \beta_1, \ldots, \beta_p)^T$. Suppose now we have an estimator of $\boldsymbol{\beta}$, called $\hat{\boldsymbol{\beta}} = (\hat{\beta}_0, \hat{\beta}_1, \ldots, \hat{\beta}_p)^T$. Then we can compute the residuals as $r_i = r_i(\hat{\boldsymbol{\beta}}) = y_i - \tilde{\mathbf{x}}_i^T \hat{\boldsymbol{\beta}}$, with $\tilde{\mathbf{x}}_i = (1, \mathbf{x}_i^T)^T$, for $i = 1, \ldots, n$. The well known least-squares (LS) estimator is defined by the minimization problem

$$\hat{\boldsymbol{\beta}}_{\text{LS}} = \underset{\boldsymbol{\beta}}{\text{argmin}} \sum_{i=1}^{n} r_i(\boldsymbol{\beta})^2. \tag{2}$$

The resulting vector of squared residuals $\mathbf{r}^2(\hat{\boldsymbol{\beta}}_{\text{LS}}) = (r_1^2, \ldots, r_n^2)^T$ is often used to evaluate the fit. Here we will use a trimmed version: the order statistics of the squared residuals is $\mathbf{r}^2(\hat{\boldsymbol{\beta}}_{\text{LS}})_{1:n} \leq \ldots \leq \mathbf{r}^2(\hat{\boldsymbol{\beta}}_{\text{LS}})_{n:n}$, and the 10%-trimmed RMSE is defined as

$$\text{RMSE}_t = \sqrt{\frac{1}{t} \sum_{i=1}^{t} \mathbf{r}^2(\hat{\boldsymbol{\beta}}_{\text{LS}})_{t:n},} \tag{3}$$

where $t = [n \cdot 0.9]$ is the largest integer below $n \cdot 0.9$. This evaluation measure accounts for some robustness, since the biggest squared residuals are not accounted for.

The LS estimator is easy to compute (there is even an explicit formula), and it has several attractive statistical properties, if the model assumptions are met, e.g. [5]. It is, however, virtually impossible to check these assumption, and already a look at Fig. 1 must create doubts. Nevertheless, Fig. 3 presents some results from LS regression. Here, we first split the data randomly into training and test data, in the proportion 75% versus 25%, perform LS regression on the training data, and predict the response based on this fit for the test data. The left plot in Fig. 3 shows the fitted values from the training data versus the response, while the remaining plots show the predictions versus the response (the right-most plot is a zoom into the middle plot). The numbers on top are 10%-trimmed RMSE values for the corresponding data parts. It is surprising that we have an excellent fit with the training data, but the prediction performance of the model is very poor. The LS estimator tries to accommodate all observations of the training data, even if they are outliers, which leads to a poor predictor.

In order to downweight the effect of outliers on the regression fit, we use robust regression. The best known robust regression estimator is the M-estimator, defined as

$$\hat{\boldsymbol{\beta}}_{\text{M}} = \underset{\boldsymbol{\beta}}{\text{argmin}} \sum_{i=1}^{n} \rho\left(\frac{r_i(\boldsymbol{\beta})}{\hat{\sigma}}\right), \tag{4}$$

where $\hat{\sigma}$ is a robust scale estimator of the residuals, and $\rho(\cdot)$ is a function applied to the scaled residuals. The robustness properties of the estimator depend on the choice of this function (large absolute scaled residuals should receive a smaller contribution to the sum)

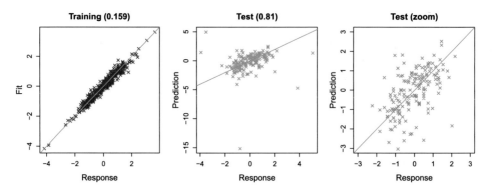

Fig. 3 Results from LS regression: fitted values versus response (*left*), prediction versus response (*middle*) and zoom (*right*); numbers on top are 10%-trimmed RMSE

and on the way how the residual scale is estimated. The so-called MM-estimator uses an iterative scheme to estimate the residuals, starting with a highly robust estimator of residual scale. For details we refer to [9]. The R package `robustbase` contains an implementation of this estimator [8].

Figure 4 shows the corresponding output of MM-regression, which can be compared to Fig. 3. The trimmed RMSE is better for the training data, but it is much worse for the test data. There seems to be a systematic bias in the prediction.

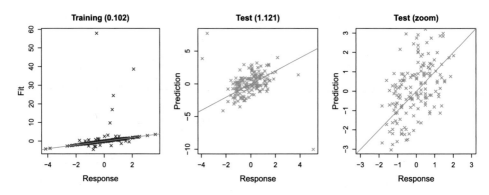

Fig. 4 Results from MM-regression: fitted values versus response (*left*), prediction versus response (*middle*) and zoom (*right*); numbers on top are 10%-trimmed RMSE

4 Variable Selection

So far, robustness did not lead to any improvement of our model. However, one should keep in mind that particularly in high-dimensional situations with many predictors, model selection is an important issue. The big difference between the trimmed RMSE for training and test data is a clear indication of overfit.

Model selection in the regression context has been widely studied, and the most simple approach is variable selection based on an appropriate criterion. Thinking in terms of robustness, it is not at all so clear, how variable selection could be carried out effectively, see [4] for some ideas.

Here we will follow an alternative path. Shrinkage estimators have become extremely popular in the last years, last but not least thanks to the development of the lasso estimator [13]. Due to the use of an L_1 penalization of the coefficient vector, the estimator has a variable selection property. A generalization of this estimator is the elastic net (enet) estimator, which combines the L_1 penalty with an L_2 penalty [15]. As an advantage, more than n variables can be selected, which is important for data sets where $n \ll p$, and whole blocks of correlated variables get a higher chance to enter the model. The estimator is defined as

$$\hat{\boldsymbol{\beta}}_{\text{enet}} = \underset{\boldsymbol{\beta}}{\operatorname{argmin}} \left\{ \sum_{i=1}^{n} \left(y_i - \tilde{\mathbf{x}}_i^T \boldsymbol{\beta} \right)^2 + \lambda \sum_{j=1}^{p} \left((1-\alpha)\frac{1}{2}\beta_j^2 + \alpha|\beta_j| \right) \right\}, \tag{5}$$

where the tuning parameter $\lambda \geq 0$ controls the strength of the penalty, and the parameter $0 \leq \alpha \leq 1$ provides a trade-off between an L_2 and an L_1 penalization.

Using the R package glmnet [3] for enet regression on the production data set gives the output shown in Fig. 5. Note that the elastic net is not robust against outlying observations, because it minimizes (a penalized) sum of squared residuals. From the fit and prediction in Fig. 5 it is hard to tell if there was any effect of outliers on the estimator. We just can see that the prediction error for the test set improved a lot when compared to the previous results, and it is in the same order of magnitude as the prediction error of the training set. This is an indication that overfit was avoided. In fact, with the optimized parameters $\hat{\lambda} = 0.15$ and $\hat{\alpha} = 0.95$, only 9 out of the 392 explanatory variables enter the model, all other variables are considered as irrelevant noise variables. This makes it possible to visually inspect the identified predictors, and a scatterplot matrix of those 9 variables is shown in Fig. 6. The numbers are Spearman rank correlations with the response. Obviously, there are many outliers in these variables. An outlier elimination would be easier in this lower-dimensional space, but it would not make any sense because with modified data the estimator would very likely end up with a different selection of predictor variables.

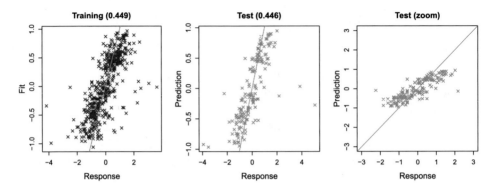

Fig. 5 Results from elastic net regression: fitted values versus response (*left*), prediction versus response (*middle*) and zoom (*right*); numbers on top are 10%-trimmed RMSE

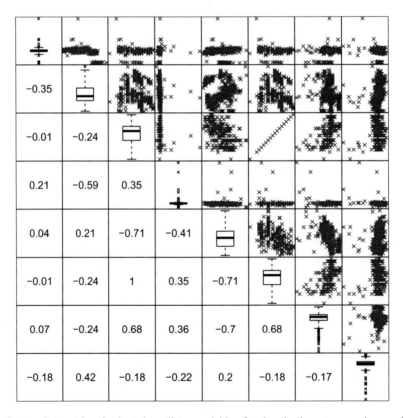

Fig. 6 Scatterplot matrix of selected predictor variables for the elastic net regression model; the numbers are the Spearman rank correlations with the response

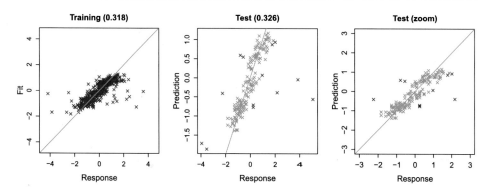

Fig. 7 Results from enetLTS regression: fitted values versus response (*left*), prediction versus response (*middle*) and zoom (*right*); numbers on top are 10%-trimmed RMSE; outliers are indicated as *red* points

A robust alternative to the elastic net regression estimator is a trimmed version, called enetLTS, proposed in [7]. The estimator is defined as

$$\hat{\boldsymbol{\beta}}_{\text{enetLTS}} = \underset{\boldsymbol{\beta}}{\text{argmin}} \left\{ \sum_{i=1}^{h} \left(\mathbf{r}^2(\boldsymbol{\beta})\right)_{i:n} + h \cdot \lambda \sum_{j=1}^{p} \left((1-\alpha)\frac{1}{2}\beta_j^2 + \alpha|\beta_j| \right) \right\}, \qquad (6)$$

with the trimming parameter $h \leq n$. This estimator is inspired by the well known LTS (least trimmed sum-of-squares) estimator, and the idea is to trim the biggest squared residuals [11]. A lasso version of this estimator has been proposed in [1]. The parameter h is usually set to $n/2$ in order to achieve high robustness, and a subsequent reweighting step improves the efficiency of the estimator.

The R implementation in the package `enetLTS` [6] searches in a cross-validation scheme for the optimal parameters, and in our application we obtain $\hat{\lambda} = 0.046$ and $\hat{\alpha} = 0.55$, which yields a model with 69 predictor variables. Figure 7 presents the results, where it can be seen that the prediction error could be improved substantially. As an additional feature, this method provides the information of outlyingness of the observations, and thus the outliers are shown by red points. This can be a very valuable information for the practitioner: these outliers are (multivariate) outliers with respect to the regression model, and not simply outliers that are abnormal in the data space. One could go back to these observations to identify the reasoning for their unusual behavior. Similarly, the model can identify outliers for new test set observations, which could indicate some irregularity in the production process.

5   Discussion and Summary

The intention of this contribution was to address the data quality problem in "big data". With increasing amounts of data, the risk of collecting erroneous or inconsistent data also increases, and this may lead to consequences for the data analysis. We have presented a data set originating from a production process in the industry, and this is for sure by far not a "big" data set. However, even this data size gives challenges to the analyst: it is no longer straightforward to "look" at the data. A visual inspection will require more sophisticated tools (see chapter on Visual Analytics), and this can be helpful to get an idea about the data quality, but probably not for outlier identification with respect to an underlying model. A promising approach in this direction is to link statistical methodology with visual analytics methods.

Data quality can be quite essential for the choice of the statistical method. Here we presented some robust alternatives to traditional regression methods, and they eventually led to a more successful regression model; more successful in terms of a smaller prediction error, and the additional advantage of the possibility to indicate outliers as observations that deviate from the regression model. There is also a price to pay: robust methods usually require more computation time. It may also depend a lot on the implementation if the method is sensitive to categorical or factor variables, or to other data specifics.

Robust methods are available for various problem settings, such as for multivariate linear regression, but also for non-linear regression, classification problems, clustering, dimension reduction, time series analysis, and many more, see [9]. As presented here for a specific data set, the benefit of such methods is not only in a robust parameter estimation, but also in robust diagnostics: outliers with respect to the underlying model are revealed, and the analyst can investigate the reason for outlyingness. The model estimation for non-robust methods, on the other hand, can be affected by the outliers themselves, and then neither the estimated parameters nor the "outlier" diagnostics is reliable.

A common practical approach when dealing with messy or noisy data is to first do a data cleaning. In case of high-dimensional data, this is done for each variable separately. There are two major drawbacks of such an approach: (a) One would miss multivariate outliers, which are outliers that are not extreme in the single coordinates, but very unusual in the multivariate space. This kind of outliers can have the same bad effect on statistical estimators as univariate outliers. (b) Especially for high-dimensional data it becomes likely that every observation contains in at least one variable an outlier. What would be the solution for practice? It makes no sense to eliminate those observations containing outliers, since at some point one is left without observations.

As a way out, especially for high-dimensional data there are some recent approaches from robust statistics. The main idea is to gain robustness not only for outliers in the observations, but also for outliers in the single cells (variables) of the observations. If only one cell of an observation is unusual, there are still $p - 1$ cells remaining which might be useful and important for modeling. As an example, [10] proposed a regression method that downweights outlying cells. The DDC (detecting deviating cells) algorithm of [12] identifies cellwise outliers in a data matrix and imputes values which are more consistent with the data. This imputed data set could be used for subsequent statistical modeling. The topic of cellwise outlyingness has become very important recently, and there are already valuable contributions available, and more of them will for sure appear in the near future.

References

1. Alfons, A., Croux, C., Gelper, S.: Sparse least trimmed squares regression for analyzing high-dimensional large data sets. The Annals of Applied Statistics **7**(1), 226–248 (2013)
2. Borowski, M., Fried, R.: Online signal extraction by robust regression in moving windows with data-adaptive width selection. Statistics and Computing **24**(4), 597–613 (2014)
3. Friedman, J., Hastie, T., Simon, N., Tibshirani, R.: glmnet: Lasso and Elastic Net Regularized Generalized Linear Models. R Foundation for Statistical Computing, Vienna, Austria (2016). http://CRAN.R-project.org/package=glmnet. R package version 2.0-5
4. Heritier, S., Cantoni, E., Copt, S., Victoria-Feser, P.M.: Robust Methods in Biostatistics. John Wiley & Sons, Chichester (2009)
5. Johnson, R., Wichern, D.: Applied Multivariate Statistical Analysis, 7th edn. Prentice Hall, Upper Saddle River, NJ (2007)
6. Kurnaz, F., Hoffmann, I., Filzmoser, P.: enetLTS: Robust and Sparse Methods for High Dimensional Linear and Logistic Regression (2018). https://CRAN.R-project.org/package=enetLTS. R package version 0.1.0
7. Kurnaz, F., Hoffmann, I., Filzmoser, P.: Robust and sparse estimation methods for high-dimensional linear and logistic regression. Chemometrics and Intelligent Laboratory Systems **172**, 211–222 (2018)
8. Maechler, M., Rousseeuw, P., Croux, C., Todorov, V., Ruckstuhl, A., Salibian-Barrera, M., Verbeke, T., Koller, M., Conceicao, E., di Palma, M.: robustbase: Basic Robust Statistics (2018). http://robustbase.r-forge.r-project.org/. R package version 0.93-3
9. Maronna, R., Martin, R., Yohai, V., Salibián-Barrera, M.: Robust Statistics: Theory and Methods (with R). John Wiley & Sons, Chichester (2019)
10. Öllerer, V., Alfons, A., Croux, C.: The shooting S-estimator for robust regression. Computational Statistics **31**(3), 829–844 (2016)
11. Rousseeuw, P.: Least median of squares regression. Journal of the American Statistical Association **79**(388), 871–880 (1984)
12. Rousseeuw, P., Vanden Bossche, W.: Detecting deviating data cells. Technometrics **60**(2), 135–145 (2018)
13. Tibshirani, R.: Regression shrinkage and selection via the lasso. Journal of the Royal Statistical Society: Series (Methodological) **58**(1), 267–288 (1996)
14. Zimek, A., Filzmoser, P.: There and back again: Outlier detection between statistical reasoning and data mining algorithms. Wiley Interdisciplinary Reviews: Data Mining and Knowledge Discovery **8**(6), e1280 (2018)
15. Zou, H., Hastie, T.: Regularization and variable selection via the elastic net. Journal of the Royal Statistical Society: Series B **67**(2), 301–320 (2005)

Modelling the Top Floor: Internal and External Data Integration and Exchange

Bernhard Wally⬤, Christian Huemer⬤ and Birgit Vogel-Heuser⬤

Abstract

Digital representations of top floor entities are inherent in higher level software suites such as enterprise resource planning (ERP) systems or manufacturing execution systems (MESs). Typical implementations utilise proprietary conceptual models that lead to a plethora of both import and export filters between different systems. In this chapter we will highlight the modelling of top floor entities by adopting international standards and discussing arising interoperability issues. With the selected standards, we outline an approach for vertical integration between the ERP and MES levels as well as horizontal integration among organisations in a value added network. We complement our structural, model-based and data-driven perspective with business process stencils that are to be customised to specific business case needs. With that, we establish a purely model-based perspective on the coupling of top floor internal and external data exchange matters.

Keywords

Model-based engineering · Enterprise resource planning · Manufacturing operations management · Business process orchestration · Interoperability · Integration · Industry 4.0

B. Wally
Business Office, Austrian Council for Research and Technology Development, Vienna, Austria
e-mail: bw@rfte.at

C. Huemer (✉)
Institute of Information Systems Engineering, TU Wien, Vienna, Austria
e-mail: huemer@big.tuwien.ac.at

B. Vogel-Heuser
Institute of Automation and Information Systems, TU München, Garching bei München, Germany
e-mail: vogel-heuser@tum.de

© The Author(s), under exclusive license to Springer-Verlag
GmbH, DE, part of Springer Nature 2023
B. Vogel-Heuser and M. Wimmer (eds.), *Digital Transformation*,
https://doi.org/10.1007/978-3-662-65004-2_12

1   Introduction

Automated production systems are no longer being considered static when it comes to their setup, function set and interrelation, but they are more and more meant to be adaptive systems of systems. This ongoing change process is also known as the fourth industrial revolution, Industry 4.0 (I4.0)—other names include smart manufacturing [25], industrial internet of things [5], and cyber-physical production systems [4]. Hardware components on all levels are becoming more and more versatile, communication systems more dynamic, faster and more powerful. The integration of heterogeneous IT and production systems plays therefore a key role in creating responsive, flexible production systems of the future [30].

Increasingly complex products, combined with significantly reduced lot-sizes as well as shorter lead times that are being demanded, culminate in requirements that are hard to fulfil. In order to meet these expectations, modern software systems, network infrastructure, and network protocols at all levels of the automation hierarchy are required: from business related software at the corporate management level, down to programmable logic controllers at the field level. For a well-designed coupling of systems that are located at different levels, it is necessary to find, define, and implement clear data conversion mechanisms—this endeavour is also known as *vertical integration*. At the same time, it is necessary to automate the inter-organisational data exchange—an aspect of *horizontal integration* [49].

For decades, the automation pyramid [24] (see left hand side of Fig. 1), used synonymously to the term "automation hierarchy" in this chapter, was in use to model different physically-oriented levels of automation, automation networks like fieldbus and cell bus and interfaces to manufacturing execution systems (MESs) and enterprise resource planning (ERP) systems. The availability of smart field devices that partly integrate programmable logic controller (PLC) functionality and the emergence of industrial Ethernet, i.e., a unified communication system that replaced the not-so-flexible field bus systems and simplified integration with higher levels of the automation hierarchy, led to migration of functions from the control level to the field level. Now it was possible to attach almost all devices to a single network and thus achieve increased flexibility. The "physical barriers" that were caused by different networks and reduced device capabilities disappeared and made place for new communication and configuration approaches [48].

In an intermediate step, an information model was developed (in the middle of Fig. 1) that enabled a rather direct coupling of the field level automation with the MES and the ERP system. In the context of Industry 4.0, the *Reference Architecture Model Industrie 4.0* (RAMI4.0, right hand side of Fig. 1) was proposed [9], including the levels of the automation pyramid, life cycle aspects as well as the layers from a business perspective. As communication interface OPC Unified Architecture (OPC UA) [21] has been proposed and is implemented for non real-time coupling of field level and MES. Recently, the asset administration shell (AAS) [38] and the digital twin [43] have been introduced trying to develop and realise the information models mentioned above.

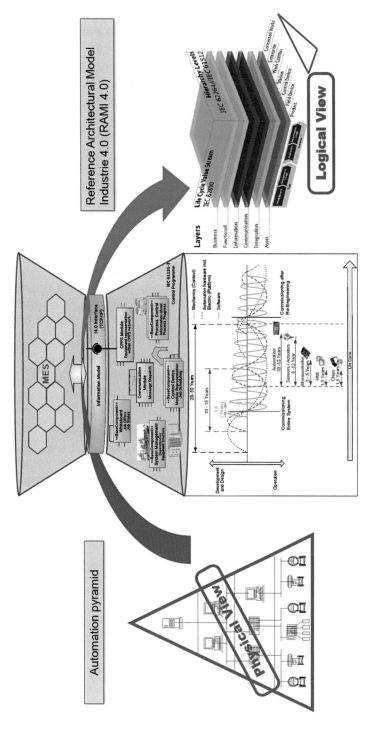

Fig. 1 Information architecture: from the automation pyramid, via the diabolo to RAMI4.0 (adapted from [48])

In this book chapter we are going to discuss the main properties of ERP (cf. Sect. 2) and MOM (cf. Sect. 3) as the *top levels* of the automation hierarchy and propose concrete approaches for both vertical (cf. Sect. 4) and horizontal (cf. Sect. 5) integration. We acknowledge the complexity of such an effort [11] and we will thus concentrate on *model-based engineering*, to provide clear and meaningful system descriptions, and *standards compliance*, so that system descriptions can be exchanged among engineering tools and business partners much more straightforward [27].

Furthermore, our focus is on the technical issues, i.e., what needs to be considered by engineers on a functional level, and not so much on strategic, societal, sustainability and resilience properties. As such, we are not discussing the European Union's Industry 5.0 approach [7], but we believe that the technical underpinning of I4.0 will be necessary to reach the goals of Industry 5.0, which we consider important, nonetheless.

Model-based engineering describes efforts to formalise systems so that they can be characterised in a structured way using a clearly defined vocabulary, specified in terms of a meta-language [6, 32]. Numerous such meta-languages have been developed and some have become accepted as formal or informal standards. E.g., in software engineering, the Unified Modeling Language (UML) [35] based on the Meta Object Facility (MOF) [36] has become the standard way of describing software systems. For general systems engineering the Systems Modeling Language (SysML) [37] can be used, and for integrated circuits the VHSIC Hardware Description Language (VHDL) [16] is the way to go. In this chapter with its focus on automated production systems, we are going to utilise (i) modelling languages that are able to describe objects (such as production equipment) and their relations as well as actions (such as production steps) and (ii) notations that allow the formulation of processes, collaborations and information exchange. We will be using these various complementary languages for shedding light on a single running example that will be used throughout this chapter—it will be presented in Sect. 2.

The organisation of this chapter is as follows: after the introduction, we'll present and discuss the main properties and corresponding international standards of the two top-most levels, namely enterprise resource planning in Sect. 2 and manufacturing operations management in Sect. 3. Thereafter, we'll explore model-based integration between these two levels in Sect. 4. In Sect. 5 our focus shifts towards horizontal integration issues that are required with respect to inter-organisational data exchange. Finally, we conclude in Sect. 6 with a brief wrap-up and an outlook for application and implementation scenarios.

2 Enterprise Resource Planning

Enterprise resource planning (ERP) is a cumulative term for activities, data, and processes that deal with business information [41]. As such, ERP systems are typically able to handle data about personnel, resources, material, products, services, and orders, as such to provide relevant information for controlling. ERP systems are developed to support business decisions in a medium to long-term planning horizon. Today, ERP systems provide the main interface to external trade partners such as suppliers and customers and as such, they are also the basis for electronic data interchange (EDI) [22] that enables a fully digitised, standardised intra-organisational information flow [40]. In this section, we will introduce a domain-agnostic business model language that can be used to model and capture the ERP level of the automation hierarchy in a well-structured way.

In the 1980s, McCarthy described a novel accounting concept "Resource-Event-Agent" (REA) that was initially meant as a radical modernisation of the rather aged double entry bookkeeping [31]. It dealt with entities that occur within a company's general ledger. It was subsequentially transformed into an expressive business model language [13, 14] that is able to handle the majority of information that is required for a typical ERP system [56, 57] and that is meanwhile available in a standardised form [23].

At its core, REA describes *resources*, which are considered economic goods of any kind, *agents*, which are people or organisations involved in business processes and *events* that interlink the former two entities with the following semantics: agent A1 and agent A2 had an exchange in ownership/possession/availability of a certain amount of resource R1 in an event E1 taking place on a specific date. In normal economic behaviour REA events come as dual events in the sense of "there is no free lunch"—giving away a specific resource in event E1 accords with an event E2 where some other resource is received in exchange. In REA this relation is reified as a *duality*, which is a key component for describing a company's business model and on an abstraction level as it is required for modelling a value chain.

In other words, a duality comprises a set of events, some of which are decreasing the value of a company, such as giving away a product that is available for sale, while others are increasing the value of the company, such as receiving money in exchange. A "sales" duality thus can be abstracted as a process that takes a product as input and outputs money. Originally, such a duality described a transfer of goods between two trade partners ("transfer duality"), but it turned out that this core rule of economic exchanges can also be applied for transformative processes, such as creating a product from raw material: the consumption of raw material and the usage of production equipment both decrease a company's value, while providing the product increases its value. Such a duality is consequently called a "transformation duality". It can be used to describe manufacturing and inventory activities such as storing and retrieving goods. In our work, we are using transformation dualities as modelling artifacts for production steps, and transfer dualities to model the interfaces to external partners, such as customers and suppliers.

It is important to mention that REA does not intrinsically provide data that produces a kind of sequence of actions. REA is not a business process model language, but a business model language. It captures the core business model without defining any kind of execution semantics. Yet, e.g., the resource stock flow in increment and decrement events and the combination of these events within dualities allows to compute resource flows between the dualities. Please, refer to Fig. 2 for a concrete example.

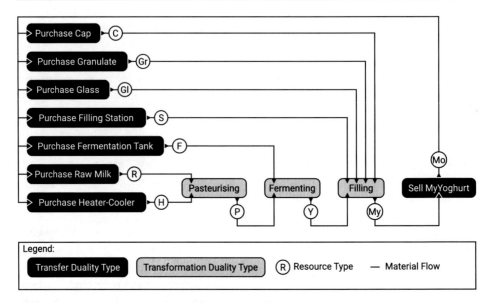

Fig. 2 Value chain view for the production of MyYoghurt [46]. Interaction with suppliers and customers is depicted with black boxes (exchange dualities), the production process itself is depicted with gray boxes (transformation dualities). The graphical syntax in use is a custom one

In the remainder of this book chapter, we will utilise the MyYoghurt production site [46] as a running example. The production process can be described as follows: raw milk is pasteurised and fermented to become yoghurt which is then filled into glasses with a customer-selected kind of granulate (such as chocolate chips). Finally, a cap is put on top of the glass. The corresponding value chain is depicted in Fig. 2, a more detailed view on the production process is given in Fig. 5 in Sec. 4.

Figure 2 depicts REA dualities as boxes with the in- and outflowing resources as their I/O ports. Since no official graphical representation for REA and its elements has been standardised, we are using a custom graphical syntax in this visualisation. It follows visualisation concepts introduced by the inventor of REA [12] and those typically used for representing resource flows in value chains or value networks [42], such as, e.g., the e^3value modelling system [15].

3   Manufacturing Operations Management

Manufacturing operations management (MOM) is responsible for the effective handling of production processes and related operations, such as maintenance, quality assurance, and intra-logistics, and represents the second-top-most level in the automation hierarchy [17]—it may be supported or driven by a manufacturing execution system (MES). MOM needs to have knowledge about various life-cycle states of the resources and processes it manages, including knowledge about the available production processes and resources, the required sequence of production processes for a specific product, the required skills of machine operators and maintenance engineers, and much more. In contrast to ERP, the planning horizon of MOM spans the short and medium term—as such its time resolution with respect to process and data granularity needs to be much narrower. Yet, with digitised ERP systems and MESs, data consistency and continuity are required in order to provide for quick reaction to potential production change requests, quick response times with respect to new manufacturing requests, and near real-time monitoring on the various levels of the automation hierarchy.

With the IEC 62264 series of standards [17], conceptual models for the digital representation of MOM data and processes have been defined. IEC 62264 is the international standard following the north american ANSI/ISA-95 (ISA-95 in short). We are using the terms IEC 62264 and ISA-95 interchangeably in this chapter. Specifically, part 4 goes well into the operations level [19], while part 2 defines what kind of information would be required for the ERP and the MOM level in order to exchange their data [18]. Its early availability— its publicly available roots go as far back as the year 2000 [2]—and convincing conceptual framework have further made it a natural ingredient of RAMI4.0. The way of its RAMI4.0 integration highlights an important aspect of the IEC 62264 standard and of MOM in general: it is focused on internal processes and functions. As such, data from external influences, e.g., just-in-time information, need to be fed into MESs from duly attached entities, such as ERP systems. This is for a reason: interaction with external parties, such as suppliers or customers, underlie certain legal regulations that might have a significant impact with respect to money-at-stake. ERP systems have, in combination with EDI, a well established and recognised track record in dealing with these external influences, and therefore such information is typically handled by the ERP as the lead system.

IEC 62264 has in the past been identified as the only international standard that takes on the task of abstracting MES structure and functionality [27], and it is being applied or supported by several MES vendors and related organisations [1]. Its versatility was successfully tested in several experimental setups and case studies, even including interfaces to automated planning solvers that enable ad-hoc production planning [33, 59–61]. Furthermore, a dedicated graphical modelling language and a model-driven engineering application have been developed to better grasp its context and expressivity [26]. The main entities that are defined in part 2 of IEC 62264 are:

Personnel: Individuals and classes of individuals that play a role in a production environment (e.g., machine operators, maintenance engineers). Personnel can be annotated with skills that might be required for certain operations.

Equipment: Machinery and classes of machinery that is required for the production of goods (e.g., conveyor, 6-axis robot). Equipment is used to model the hierarchy and layout of production sites.

Physical assets: Concrete product instances (identified by a serial number) and products (identified by a model number) that are in use at a production site (e.g., KUKA KR 500 FORTEC). Physical assets can be mapped to equipment in order to clarify which specific product is used in what location of the production site.

Material: Specific material lots or material definitions that are consumed or produced during production (e.g., powder coating in RAL 5023). IEC 62264 supports discrete, batch or continuous processes and corresponding material.

These four core entities are interwoven by the more complex *process segments* and *operations definitions*, where the former define product-agnostic production steps (what kinds of material manipulations can in principle be carried out in a plant) and the latter accumulate several process segments to define what is required to produce a certain product. Further complex entities enable the scheduling of production orders and logging of their actual performance.

In Fig. 3 we visualise the "Pasteurising" production step, as it is introduced in Fig. 2 in Sect. 2, this time using IEC 62264 vocabulary: the *process segment* Pasteurising defines *segment specifications* for *equipment class* Heater-Cooler, *personnel class* Heater-Cooler Operator and *material definition*s Raw Milk and Pasteurised Milk. The various segment specifications are depicted using annotated lines, where the annotations represent their attributes. In this example, a batch of milk to be pasteurised requires one piece of a heater-cooler in order to produce five litres of pasteurised milk from five litres of raw milk. Since the MyYoghurt production line is considered to be highly automated, no personnel is required for this process segment.

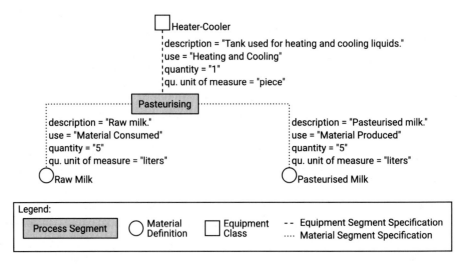

Fig. 3 The "Pasteurising" process segment, as it would be expressed using concepts of IEC 62264. This visualisation uses a custom graphical syntax since no standard notation has been defined so far

Figure 5 clarifies that the pasteurising process can itself be described by a sequence of nested production steps, such as heating the milk to 73 °C, holding this temperature for 20 s, and cooling the milk back to 45 °C. IEC 62264 part 2 provides a corresponding mechanism that allows the modelling of nested process segments. Further details that are relevant for controlling the production equipment can be refined using constructs from IEC 62264 part 4, such as *work masters*, which can be used to describe recipes or specific computerised numerical control (CNC) programmes if they are required for production operations.

4 Vertical Integration

Vertical integration between the upper levels of the automation hierarchy basically comprises integrating ERP and MOM. With this kind of integration the main effort for linking the business level with the production level is accomplished. Given such a well-defined integration, it becomes easier to couple business services with production services and to utilise data bidirectionally, when required.

4.1 Alignment of Complementary Conceptual Models

We have previously seen that parts 2 and 4 of IEC 62264 are well-suited for modelling manufacturing operations management data, while ISO 15944-4 (REA) can be used to describe the business model and provide all the required information for ERP. In this section, we are briefly describing how IEC 62264 and REA can be aligned so that data ambiguity can be mitigated and a meaningful information flow can be established.

Initial attempts for aligning core entities of REA and IEC 62264 have been undertaken in [28, 29] and re-purposed in [53] by arguing that part 4 of IEC 62264 would be well-suited for modelling the "task layer" of REA when it comes to production processes, i.e., *transformation dualities* as they are dubbed in REA. Nevertheless, we have shown that the concepts of part 2 of IEC 62264 can be well matched with REA concepts [51]. A quantitative analysis showed that about two thirds of conceptual entities can be more or less directly matched between REA and IEC 62264 [39], which tells us two things: (i) there is substantial overlap to legitimate structural mapping and (ii) there exists significant delta so that two separate domain models are indicated.

When it comes to the modelling of production systems, there exists another standard that should be discussed: AutomationML (IEC 62714) [20]. It provides a domain-agnostic modelling environment based on an XML-dialect. AutomationML is meant to be used as the exchange format between the various engineering tools that are being used throughout the engineering of a production site [10]. AutomationML propagates a separation of concerns, mainly a clear distinction between *products*, *processes*, and *resources* (PPR) [44]. This separation of concerns corresponds closely to the IEC 62264 terms *material,operations/process segment*, and *equipment* [52]. Also, a dedicated working group of the AutomationML's technical advisory council has identified useful applications of AutomationML on higher levels of the automation hierarchy [58].

If it is done well-structured, AutomationML can be used in a way that is compatible with IEC 62264 [55]. In an extensive mapping document, an AutomationML application recommendation, the modelling rules for creating IEC 62264 compliant AutomationML documents are defined [50]. Furthermore, experiments with methods from model-driven engineering indicated that it is feasible to convert data between REA, IEC 62264, and AutomationML [54]. Well-defined and unambiguous mapping and conversion rules are required so that information loss or duplication can be prevented or at least significantly reduced.

Figure 4 depicts how an alignment of elements between two conceptual models can be realised. In this example, we are aligning ERP entities from REA with MOM entities from IEC 62264. E.g., *resource types*, as they are defined in REA, correspond to four different entities in the vocabulary defined by IEC 62264: *equipment classes*, *physical asset classes*, *material classes*, and *material definitions*. REA's *transformation duality types* map to *process segments*, *stockflow types* of nested event types correspond to the various *segment specifications* that are contained therein.

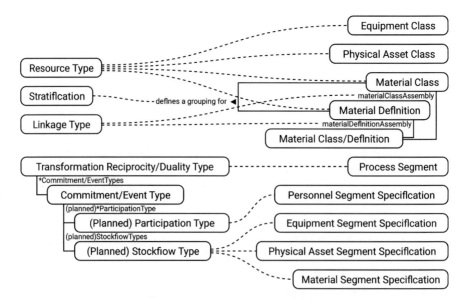

Fig. 4 Visualisation of the alignment between REA and IEC 62264 (excerpt from [51]). REA entities are displayed to the left, IEC 62264 entities to the right. Alignments are depicted as dashed lines between corresponding entities

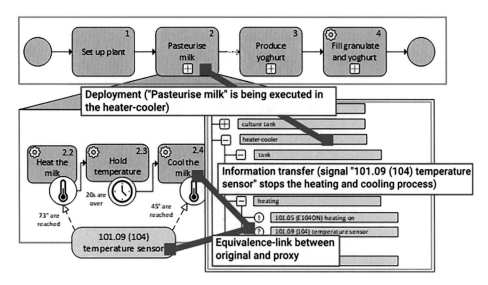

Fig. 5 Deployment of Yoghurt production process steps to parts of the MyYoghurt plant and definition of process signals to be transferred to the MES (exemplary excerpt). The notation used is MES-ML, a business process model and notation (BPMN) dialect [62]. The upper part depicts the full process, while the lower part shows the decomposition of the pasteurising process and its deployment on a specific resource

As it is typical for such alignments, not all entity mappings are unequivocal. E.g., REA's notion of a resource is much more generic than that of IEC 62264. By executing a thorough domain and conceptual model analysis, such potentially problematic constellations can be brought to light and corresponding mitigation strategies can be put in place, such as naming conventions, tagging, or adding specific discriminator attributes.

4.2   Application in the MyYoghurt Use Case

The Manufacturing Enterprise Solutions Association[1] (MESA) has standardised the core functions of MESs. Among the main challenges is the hand over of new production orders (cf. Fig. 6 in general and Fig. 5 for yoghurt production) including the required recipes and the change sequence aligned with the material flow. From the lower part of the diabolo (cf. Fig. 1 middle) process data is required to be handed over to the MES, often in aggregated form, in order to enable the calculation of key performance indicators (KPIs) (cf. Fig. 6).

The links between higher and lower levels of the automation hierarchy are bi-directional (cf. Fig. 6): (i) bottom-up from the field level to the MES and/or to the ERP system for selected field level data (potentially aggregated), especially in terms of quality assurance information,

---

[1] cf. https://mesa.org/.

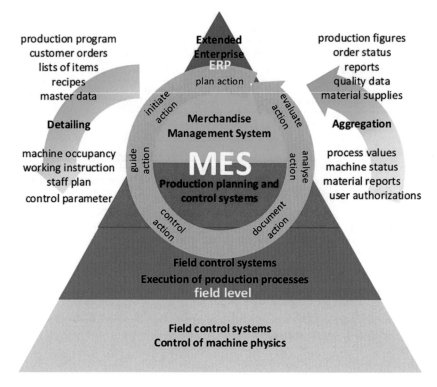

Fig. 6 Manufacturing execution systems are the link between concrete processes at the field level and ERP at the business level

machine logs and human performance, etc., (ii) top-down from the ERP system and MES to the field level for information about production changes, lot sizes as well as recipe-specific parameters (cf. Fig. 6 in general and Fig. 5 for yoghurt production). For illustration purposes, we have decomposed the "pasteurising" production step (lower part of Fig. 5) and show how the required heating and cooling is controlled using a temperature sensor. In this example, we clarify that a production process can be defined without a specific resource to be used. It is in a later refinement step that a certain process is deployed on a specific resource. Here, we would like to point out the close connection between process and resource deployment [63].

When comparing the elements of Fig. 2 and Fig. 5 it becomes clear, why structured vertical integration is a key factor especially for flexible manufacturing requirements: the resemblance between the two is striking, yet they differ. On the ERP level, information about interaction with external partners is included (and even much more prominent), on the MOM level these "transfer" activities are completely missing, but much more details that are required for the production process unveil. By meaningful vertical integration, overlapping concepts can be pointed out and data exchange between corresponding IT systems can be made more efficient and effective.

5   Horizontal Integration

In an Industry 4.0 context, horizontal integration refers to the networking between machines, services, and units on the same level of the automation pyramid. In the context of top floor modelling, the design of well coordinated processes between organisational units are of particular interest. Thereby, one may distinguish between (i) processes that are executed between multiple teams of different disciplines often within a company but also across companies to reach a common design goal and (ii) business processes between different business partners in order to realise a supply chain or network. Accordingly, this section covers two subsection, where the first one covers the first type of processes and the second one the latter.

Independent of the type of process, a common language for the specification of the processes is required. Over the last decade the *Business Process Model and Notation (BPMN)* became the first choice for modelling business processes. The latest BPMN version 2.0.2 [34] was released in 2013 and covers a wide range of concepts well-suited for modelling any kind of business process. Thus, we utilise BPMN as a base language for modelling the top floor processes and introduce useful customisations and extensions for the specific purpose in each of the two subsections.

5.1   BPMN+I to Address Multi Team Cooperation

Following the idea to model cooperation with Choreographies by Decker et al. [8], Vogel-Heuser et al. [47] adapt the interaction between two teams from different companies to an iterative cooperative engineering process. Such a process between for example the three different disciplines of a mechatronic product is modelled as double arrow with a dashed line in BPMN+I (+I for innovation). Such knowledge can also be modelled with BPMN's Conversations, where these arrows are being represented by a hexagon that is connected to affected tasks. One issue with Conversations (cf. Fig. 7) might be that they are more hierarchical and lacking the description of detailed information flows on the top level [34].

In the following a real world use case from the domain of plant operation is used to introduce the modelling of such multi-team cooperation processes. The internal teams involved include (i) the plant operators, (ii) the maintenance team with skilled electrical and mechanical workers, (iii) the repair workshop with similarly qualified staff, (iv) the purchase department, (v) the IT department, and (vi) production management. We have simplified and modelled the negotiation process between the purchase department, the maintenance department and the external device supplier in terms of a BPMN conversation (cf. Fig. 7).

Figure 7 depicts the negotiation processes involved when a piece of equipment breaks. In our case either the plant or a plant operator detects an erroneous data flow from a machine and issues a maintenance request that is taken up by the maintenance department. In case the machine cannot be easily fixed, the production line needs to be shut down and the spare part inventory checked. If no spare part is on site, the purchasing department is involved for ordering this part from a supplier. This order process might not be straightforward, e.g., if the spare part is not being produced anymore. In this case, the machine vendor needs to be contacted and an alternative compatible spare part needs to be found or engineered. In our case, the faulty sub-component C4 can be replaced by C7.

Spare parts and components of operating production units are quite often documented and managed in ERP systems. In case of a malfunction in an operating plant due to a faulty component like a valve, motor or sensor it is mandatory to identify the spare part and its availability in the local stock. As production units operate for decades often spare parts are not immediate on stock or may not be available any more. Consequently, sometimes spare parts with reduced functionality need to be used until the original part is available again. In case the spare part is available from a sub-supplier, but not on stock, it might be replaced once it is produced and available on site again.

In the process industry things are even more complex: for control valves, for instance, often sub-parts of a valve are replaced due to high product variability, which would lead to immense stock costs for every variant of a valve. In our example sub-part C7 of valve V7 can replace the faulty sub-part C4 of valve V4. Consequently, for a data update in the ERP system it is mandatory to track faulty and replaced parts on a sub-part level to document that the repaired valve V4* now includes the sub-part C7 from V7. The updated valve history for V4* would need to point to the process data history of C7 instead of C4 from the period of time of the replacement, but keep the history of V4 for all other sub-parts.

While we have exemplified in this subsection, how the information and interaction flow could be initiated from a maintenance perspective, the following subsection will focus on general process patterns with respect to horizontal integration with external partners.

5.2 Modelling Supply Chains/Networks by BPMN and UMM

In this subsection we concentrate on interactions with business partners in a supply chain or network. As outlined in Sect. 2 on ERP systems, a value chain comprises transformation dualities that lead to production processes and transfer dualities that result in interactions with business partners in the supply chain/network. The value chain of the My Yoghurt case in Fig. 2 shows the transformation dualities in grey background and the transfer dualities in black background. It is the black transfer activities that are subject to refinement by concepts introduced in this subsection. Furthermore, there is an interlink to the previous subsection on horizontal processes between different disciplines. Some activities of such a process specification may also trigger supply chain interactions. For example, consider Fig. 7 on the maintenance procedure replacing a faulty mechatronic component. The depicted process leads to the activity order new part once an appropriate alternative is found. Evidently, this triggers a supply chain interaction with a supplier to be modelled by concepts of this subsection.

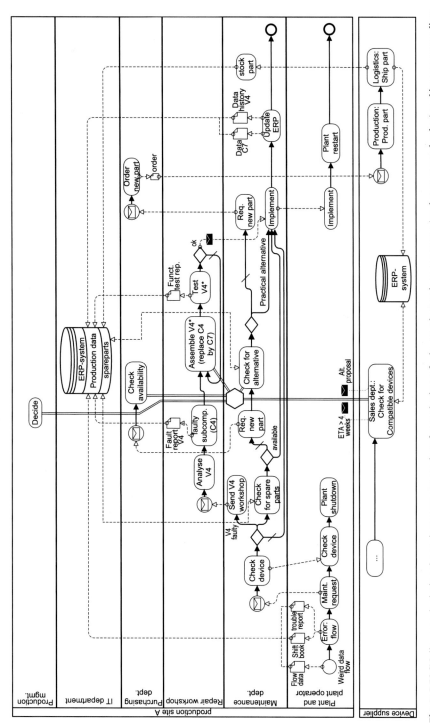

Fig. 7 BPMN diagram depicting the maintenance procedure for replacing a faulty mechatronic component (such as a control valve) and its corresponding information exchange through the ERP systems

In order to describe the inter-organisational interactions, the business scenarios between the parties involved must be identified. The *United Nation's Center for Trade Facilitation and e-Business (UN/CEFACT)* delivers international standards that facilitate the specification of such interactions between different business partners. The *UN/CEFACT Modeling Methodology* is a modelling method to specify choreographies in a business-to-business context. A choreography is a particular aspect of business processes, which relates to the way business partners coordinate their activities in a value-chain [3]. The focus is not on full orchestrations of processes operating within these partners, but rather on the collaboration that takes place between partners [64]. Accordingly, a choreography formalises the way business partners coordinate their interactions and, thus, serves as a kind of contract among the participating business partners [34].

The current UMM version 2.0 [45] released in 2011 is defined as a UML profile, that is, a set of stereotypes, tagged values, and constraints on top of the UML meta model for the special purpose of modelling choreographies. However, for business domain experts the *BPMN choreography diagrams* seem to be more intuitive than *UML activity diagrams*. Accordingly, it is advantageous to use BPMN choreographies as a notation, but extend it by concepts of UMM 2.0 that are missing in BPMN. In this subsection we describe this merger of BPMN and UMM, but limit ourselves to describing the resulting choreography and omit the requirements elicitation phase of UMM.

Business Transaction The basic building block of a UMM *business choreography* is a *business transaction*. The goal of a business transaction is synchronising the business entity states between two parties [64]. Synchronisation of states is either required uni-directionally or bi-directionally. In the former case, the initiator of the business transaction informs the other party about an already irreversible state change the other party has to accept, e.g., the notification that goods have been shipped. It follows, that responding in such a scenario is neither required nor reasonable. In the latter case, the initiating party sets a business entity to an interim state and the responding party decides about its final state - consider a request for a quote for some products that the responder answers with corresponding quotes for these products. The synchronisation takes place by exchanging business documents.

Accordingly, a *business transaction* is a basic activity in a choreography process that is executed between two parties and involves the exchange of one or two business documents. For this purpose, the BPMN meta model offers the concept of a *choreography task* which is linked to one or two *message flows*, a mandatory initiating message and an optional return message. The *choreography task* is a special kind of *choreography activity* which involves two or more – in case of a *business transaction* exactly two – *participants*, one of which is the initiating *participant*. In Fig. 8 we show an example of a UMM *business transaction* for quoting by means of a BPMN *choreography task*. The *choreography task* quoting is represented as a rounded rectangle. The two *participants* are stated at the top and bottom of the rectangle. It is irrelevant which one appears on top and which one on the bottom.

However, it is important that the initiating *participant*, in this case the buyer appears in white background and the responding *participant* in this case the seller appears in grey background. The *choreography task* quoting is linked with two *message flows*. Again a colouring mechanism is used to differentiate the message flows. The initiating message request for quote appears as a white envelope whereas the responding message quote appears as grey envelope.

In addition to the basic exchange pattern illustrated above, UMM defines certain security and timing aspects for *business transactions* that are usually part of a B2B partnership agreement, but are not part of the standard BPMN notation. Consequently, we extend appropriate BMPN elements by tagged values, i.e. key value pairs to describe these aspects. The corresponding tagged values are also shown in Fig. 8. Both the initiating message and the optional return message come with three security flags. *IsConfidential* mandates encryption so that unauthorised parties cannot view the information. *isTamperProof* requires a digital signature to check if the message has been tampered with. Is authenticated asks for a digital certificate as a proof of the signer's identity.

The following tagged values are specified for each participant in the context of a specific business transaction: *isAuthorizationRequired* forces the sender to sign a message and the receiver must validate it and send a notification in case of a failure to do so. *isNonRepudiationRequired* indicates that the participants must not be able to repudiate the document exchange. Similarly, *isNonRepudiationReceiptRequired* indicates that the participants must not be able to repudiate any sent acknowledgements of receipts of business documents. *timeToAcknowledgeReceipt* specifies the time period within which the recipient of a business document acknowledges its receipt. *timeToAcknowledgeProcessing* is the time frame within which the recipient acknowledges that the business document has passed checks

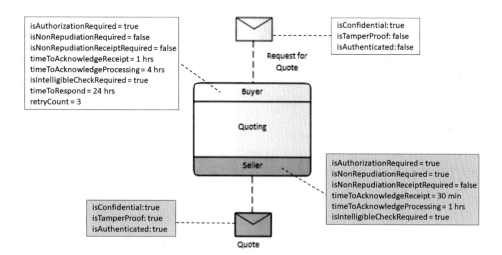

Fig. 8 Business Transaction: Quoting

against some business rules and is handed over to the application for processing. Note, the time span of the two last tagged values always ends when the sender of the business document receives the corresponding acknowledgement. *isIntelligibleCheckRequired* forces the recipient to check that a business document is not garbled (unreadable, unintelligible) before sending the acknowledgment of receipt. In addition, two more tagged values are specified for the initiating participant. *timeToRespond* specifies the maximum time within which the initiating participant expects to receive the returning business document. *retryCount* corresponds to the number of times the initiating participant must re-initiate the business transaction in case of a time-out-exception of a missing acknowledgement or response.

Given the description of the tagged values it becomes evident that not any message that is exchanged between the participants in a *business transaction* is modelled as a BPMN message flow. We distinguish *business documents* and *business signals* for acknowledging receipt or processing. *Business documents* are modelled as BPMN message flows, whereas *business signals* are only specified in the form of tagged values. Note, a null value for an acknowledgement indicates that the corresponding business signal is not needed. The *business documents* are modelled by another UN/CEFACT standard called *Core Components Technical Specification (CCTS)* which provides a modelling approach to define global business document standards as well as a mechanism to adapt these information blocks to specific requirements in a given business partnership.

Business Choreography Having discussed all the details of *business transactions*, we are able to look at a *UMM business choreography* which basically specifies a process flow among a set of *business transactions*. It is straight forward to specify a UMM *business choreography* by means of a *BPMN choreography* that is built by *choreography tasks* representing *business transactions*. Thereby, one may profit from the more expressive BPMN concepts to model process flows and the easier to understand visual representation compared to UML activity diagrams.

Each *transfer activity* (black background) in Fig. 2, should be refined by a *business choreography*, i.e. for selling the yoghurt, but also for purchasing glass, caps, granulate, filling stations, fermentation tanks, raw milk, and heater-coolers. Figure 9 shows a *business choreography* that the company uses for purchasing any product. The first *business transaction* of this *business choreography* is `quoting`, which was described in detail above. However, there is one thing more to it. We assume that if a buyer is not satisfied with a quote (s)he may change some parameters and requests another quote. In other words, the `quoting` *business transaction* may be executed multiple times. Whereas in UMM this is indicated by a tagged value, we profit from BPMN markers that may be assigned to any *choreography task* to allow multiple instances. The symbol of a non-closed cycle with an arrowhead indicates that the `quoting` process may be executed multiple times in a loop. If finally the `buyer` is satisfied with the `quote` the choreography continues with `order product` or terminates otherwise. The `order product` may lead to an existing contract for product

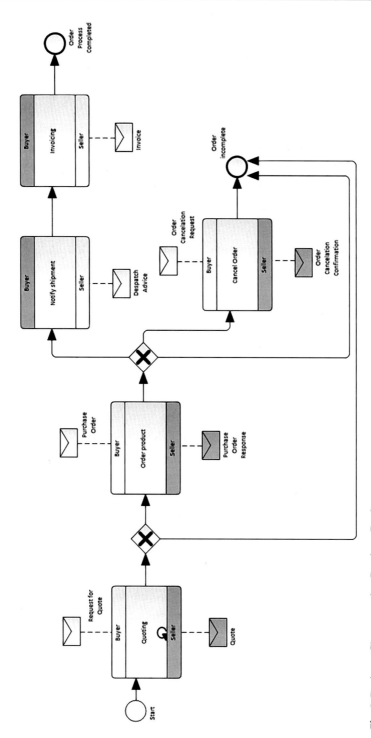

Fig. 9 Business Choreography: Purchase Product

delivery in case of positive `purchase order response` and the choreography continues. If it is a negative one the choreography terminates. In our example we assume that the buyer may cancel an order until it is shipped. Accordingly, there are two alternative paths that may be followed, either the `buyer` initiates `cancel order` which terminates the process or the `seller` starts `notify shipment` followed by `invoicing` as last *business transaction.*

Binding of Internal Processes to the Choreography Once a *business choreography* is finally negotiated, the order of sending and receiving business documents is clear for each business partner. However each partner must bind its internal process – covering also private activities that are not visible to the outside – to the negotiated choreography. Accordingly, the internal process is built by private activities and activities that send or receive business documents. The private activities usually also cover tasks to create documents that are sent and process data of documents received. It is essential that the order of sending and receiving documents is in accordance with the negotiated choreography. BPMN provides the so called *business collaboration diagram* to model the interlinking of two (or more) internal processes each of which is modelled in its own pool of activities. However, from a practical point of view of a single company it is neither desired nor possible to define the internal processes from another business partner. Thus, we recommend a different type of diagram. In this diagram the private process of the business partner under consideration is modelled in a single pool. All sending activities (marked by a little black envelope in the upper right) and all receiving activities (marked by a white little envelope in the upper right) are linked with the message flows of the *business choreography*. This is illustrated in Fig. 10 that defines the process of purchasing glass (to fill the yoghurt). Note, that the colours of the envelope of the sending/receiving activity and the linked message flow do not necessarily map because these colours present orthogonal concepts.

The internal process depicted in Fig. 10 is as follows: The process starts with the task `preparing a request for quote for glass`. The next task is `requesting a quote for glass`, which is a sending activity labelled with a little black envelope. It sends the business document `request for quote` and, thereby, corresponds to the buyer's initiating task of the choreography task `quoting`. Accordingly, we align the *requesting a quote for glass* task with the `request for quote` message flow of *quoting* in the choreography. The next task is `receive the quote for glass` which is a receiving activity with a little white envelope. Since it receives the `quote`, it is linked with the corresponding message flow of `quoting` in the choreography. Once the quote is received, it is necessary to `check the quote for glass`. If the quote is not acceptable due to major issues the process ends. In case of minor issues the process continues with `revise request for quote for glass` and afterwards enters a loop by performing *request the quote for glass* again. This loop in the internal process is consistent with the loop indicator (non closed circle with arrowhead) of the `quoting` task in the

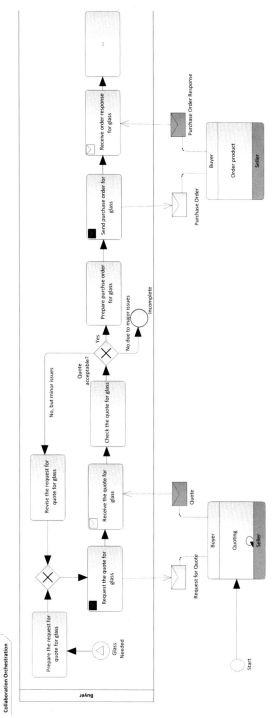

Fig. 10 Internal Process: Purchase Glass

choreography. If the quote is considered acceptable in the above mentioned check of the quote, the process continues with `prepare purchase order for glass`. It is followed by the sending activity `send purchase order for glass` and the receiving activity `receive order response for glass`. These two activities correspond to the buyer's tasks in `order product` of the choreography. Thus, these activities are linked with the two message flows `purchase order` and `purchase order response`, respectively. Due to space limits, we omit to depict the rest of the internal process. Nevertheless, we think that the interplay of an internal process and an underlying choreography becomes evident by this example and that it is of importance that the execution order of the internal process must not be in conflict to the one of the choreography.

6 Conclusion

In this chapter we have concentrated on the two top layers of the automation hierarchy, namely enterprise resource planning (ERP) and the manufacturing operations management (MOM). Following the idea of model-based systems, we used modelling languages to specify the processes to be executed on these two layers. In particular, we focus on the coupling of entities on these layers and the data exchanges between them in order to allow seamless integration and unambiguous information flow.

We suggest to use the Resource-Even-Agent (REA) ontology to describe all value activities on the ERP level. Among those the transformation activities are realised by internal production processes. These production processes may be best described by means of IEC 62264 (ISA-95) which is also used to describe exchanges between ERP and MOM. We elaborated on the mapping between REA and IEC 62264. Furthermore, we discussed the mapping of IEC 62264 and AutomationML, which is used to model production systems, in order to create IEC 62264 compliant AutomationML documents. The mapping of these different languages (REA, IEC 62264, AutomationML) targets vertical integration because it targets components at different levels of the automation hierarchy.

In addition, we have also discussed horizontal integration by targeting the business processes and the information flow between organisational units on the top level. For these purposes we have customised existing concepts of the Business Process Model and Notation (BPMN). On the one hand side we have briefly discussed the BPMN+I customisation which serves the specific purpose of collaborating units that work together towards a common output and which particularly focuses on the synchronisation of activities by means of explicit conversation dependencies. On the other hand side we introduce a BPMN notation for the UN/CEFACT Modeling Methodology (UMM) in order to model inter-organisational business processes in supply chain/network context and the corresponding business document exchanges between the business partners.

The presented model-based approaches may be used to unambiguously specify processes on the top levels of the automation hierarchy that exchange data within a company and also across companies. These process specifications assist organisational units to realise their interactions by providing precise guidelines on who has to interact with whom in which situation by sharing which data with each other. In reality, the information is not exchanged between the organisational units and the people working therein, but between applications supporting the processes of these organisational units. Thus, the goal must be an integration of these applications resulting in a seamless information exchange. Following the idea of model-driven engineering (MDE), the models should not only serve as blueprints for implementing information exchanges between these applications, but should be transformed to software artefacts realising the information exchanges. In contrast to model driven development (MDD) where code is derived from the models, the resulting artefacts may rather be interface specifications and (machine-readable) workflow specifications. Accordingly, one may derive artefacts such as representational state transfer (REST) APIs, Web services description language (WSDL) interfaces, OPC Unified Architecture information models and/or Business Process Execution Language (BPEL) processes.

References

1. Adams, M., Bangemann, T., Behnsch, J., Bienek, C., Blank, E., Hofmann, J., Küppers, U., Montino, R., Seibl, F., Theobald, C., Weinmann, M., Winkler, T., Wirt, A., Wollschlaeger, M.: Industrie 4.0: MES – prerequisite for digital operation and production management – tasks and future requirements. Position paper, German Electrical and Electronic Manufacturers' Association (ZVEI) (2017)
2. American National Standards Institute: Enterprise–control system integration part 1: Models and terminology (2000)
3. Barros, A.: Process choreography modelling. In: vom Brocke, J., Rosemann, M. (eds.) Handbook on Business Process Management 1, Introduction, Methods, and Information Systems, 2nd Ed, pp. 279–300. International Handbooks on Information Systems, Springer (2015). https://doi.org/10.1007/978-3-642-45100-3_12
4. Biffl, S., Gerhard, D., Lüder, A.: Introduction to the multi-disciplinary engineering for cyber-physical production systems. In: Multi-Disciplinary Engineering for Cyber-Physical Production Systems: Data Models and Software Solutions for Handling Complex Engineering Projects, pp. 1–24. Springer International Publishing (2017). https://doi.org/10.1007/978-3-319-56345-9_1
5. Boyes, H., Hallaq, B., Cunningham, J., Watson, T.: The industrial internet of things (IIoT): An analysis framework. Computers in Industry **101**, 1–12 (2018). https://doi.org/10.1016/j.compind.2018.04.015
6. Brambilla, M., Cabot, J., Wimmer, M.: Model-Driven Software Engineering in Practice. Morgan & Claypool, 2 edn. (2017)
7. Breque, M., De Nul, L., Petridis, A.: Industry 5.0 – towards a sustainable, human-centric and resilient European industry. R&I Paper Series – Policy Brief KI-BD-20-021-EN-N, European Commission, Directorate-General for Research and Innovation (2021). https://doi.org/10.2777/308407
8. Decker, G., Kopp, O., Leymann, F., Pfitzner, K., Weske, M.: Modeling service choreographies using BPMN and BPEL4Chor. In: Bellahsène, Z., Léonard, M. (eds.) Proceedings of the Inter-

national Conference on Advanced Information Systems Engineering (CAiSE). Lecture Notes in Computer Science, vol. 5074, pp. 79–93 (2008). https://doi.org/10.1007/978-3-540-69534-9_6

9. Deutsches Institut für Normung: Reference Architecture Model Industrie 4.0 (RAMI4.0) (2016), DIN SPEC 91345:2016-04
10. Drath, R. (ed.): AutomationML – A Practical Guide. De Gruyter Oldenbourg (2021). https://doi.org/10.1515/9783110746235
11. Ehrendorfer, M., Fassmann, J.A., Mangler, J., Rinderle-Ma, S.: Conformance checking and classification of manufacturing log data. In: 21st IEEE International Conference on Business Informatics (CBI). vol. 01, pp. 569–577 (2019). https://doi.org/10.1109/CBI.2019.00072
12. Geerts, G.L., McCarthy, W.E.: Modeling business enterprises as value-added process hierarchies with Resource-Event-Agent object templates. In: Sutherland, J., Casanave, C., Miller, J., Patel, P., Hollowell, G. (eds.) Proceedings of the OOPSLA'95 Workshop. pp. 94–113. Springer (1997). https://doi.org/10.1007/978-1-4471-0947-1_10
13. Geerts, G.L., McCarthy, W.E.: The ontological foundation of REA enterprise information systems. In: Proceedings of the Annual Meeting of the American Accounting Association. pp. 127–150 (2000)
14. Geerts, G.L., McCarthy, W.E.: An ontological analysis of the economic primitives of the extended-REA enterprise information architecture. International Journal of Accounting Information Systems 3(1), 1–16 (2002). https://doi.org/10.1016/S1467-0895(01)00020-3
15. Gordijn, J., Wieringa, R.: E^3value User Guide – Designing Your Ecosystem in a Digital World. The Value Engineers B.V., 1 edn. (2021)
16. Institute of Electrical and Electronics Engineers: IEEE standard for VHDL language reference manual (2019), IEEE 1076-2019
17. International Electrotechnical Commission: Enterprise–control system integration – part 1: Models and terminology (2013), IEC 62264-1:2013
18. International Electrotechnical Commission: Enterprise–control system integration – part 2: Objects and attributes for enterprise–control system integration (2013), IEC 62264-2:2013
19. International Electrotechnical Commission: Enterprise–control system integration – part 4: Object model attributes for manufacturing operations management integration (2015), IEC 62264-4:2015
20. International Electrotechnical Commission: Engineering data exchange format for use in industrial automation systems engineering – Automation Markup Language – part 1: Architecture and general requirements (2018), IEC 62714-1:2018
21. International Electrotechnical Commission: OPC Unified Architecture - part 1: Overview and concepts. Tech. Rep. 3.0, OPC Foundation (2020), IEC TR 62541-1:2020
22. International Organization for Standardization: Electronic data interchange for administration, commerce and transport (EDIFACT) - application level syntax rules (syntax version number: 4, syntax release number: 2) - part 10: Syntax service directories (2014), ISO 9735-10:2014
23. International Organization for Standardization, International Electrotechnical Commission: Business transaction scenarios – accounting and economic ontology (2007), ISO/IEC 15944-4:2007(E)
24. Ismail, A., Truong, H., Kastner, W.: Manufacturing process data analysis pipelines: a requirements analysis and survey. Journal of Big Data 6(1) (2019). https://doi.org/10.1186/s40537-018-0162-3
25. Kang, H.S., Lee, J.Y., Choi, S., Kim, H., Park, J.H., Son, J.Y., Kim, B.H., Noh, S.D.: Smart manufacturing: Past research, present findings, and future directions. International Journal of Precision Engineering and Manufacturing-Green Technology 3(1), 111–128 (2016). https://doi.org/10.1007/s40684-016-0015-5

26. Lang, L., Wally, B., Huemer, C., Šindelár, R., Mazak, A., Wimmer, M.: A graphical toolkit for IEC 62264-2. In: Proceedings of the 53rd CIRP Conference on Manufacturing Systems (CMS). pp. 532–537 (2020). https://doi.org/10.1016/j.procir.2020.03.049

27. Lu, Y., Morris, K., Frechette, S.: Current standards landscape for smart manufacturing systems. NISTIR 8107, National Institute of Standards and Technology (2016). https://doi.org/10.6028/NIST.IR.8107

28. Mazak, A., Huemer, C.: From business functions to control functions: Transforming REA to ISA-95. In: Proceedings of the 17th IEEE Conference on Business Informatics (CBI). vol. 1, pp. 33–42. IEEE (2015). https://doi.org/10.1109/CBI.2015.50

29. Mazak, A., Huemer, C.: HoVer: A modeling framework for horizontal and vertical integration. In: Proceedings of the 13th IEEE International Conference on Industrial Informatics (INDIN). pp. 1642–1647. IEEE (2015). https://doi.org/10.1109/INDIN.2015.7281980

30. Mazak, A., Wimmer, M., Huemer, C., Kappel, G., Kastner, W.: Rahmenwerk zur modellbasierten horizontalen und vertikalen Integration von Standards für Industrie 4.0. In: Vogel-Heuser, B., Bauernhansl, T., ten Hompel, M. (eds.) Handbuch Industrie 4.0. Springer NachschlageWissen, Springer (2015). https://doi.org/10.1007/978-3-662-45537-1_94-1

31. McCarthy, W.E.: The REA accounting model: A generalized framework for accounting systems in a shared data environment. The Accounting Review 57(3), 554–578 (1982)

32. Micouin, P.: Model Based Systems Engineering: Fundamentals and Methods. John Wiley & Sons (2014)

33. Novák, P., Vyskočil, J., Wally, B.: The digital twin as a core component for Industry 4.0 smart production planning. In: Proceedings of the 21st IFAC World Congress. pp. 10803–10809 (2020). https://doi.org/10.1016/j.ifacol.2020.12.2865

34. Object Management Group: Business process model and notation (BPMN) (2013), 2.0.2

35. Object Management Group: OMG unified modeling language (UML) (2017), 2.5.1

36. Object Management Group: OMG meta object facility (MOF) core specification (2019), 2.5.1

37. Object Management Group: OMG systems modeling language (SysML) (2019), 1.6

38. Plattform Industrie 4.0: Details of the asset administration shell (2020), part 1 - The exchange of information between partners in the value chain of Industrie 4.0 (Version 3.0RC01)

39. Polczer, P.F.: Towards Model-Driven Vertical Integration – using IEC 62264 and REA to facilitate real-time data exchange between manufacturing and enterprise systems. Diploma thesis, Faculty of Informatics, TU Wien (2021)

40. Premkumar, G., Ramamurthy, K., Nilakanta, S.: Implementation of electronic data interchange: An innovation diffusion perspective. Journal of Management Information Systems 11(2), 157–186 (1994). https://doi.org/10.1080/07421222.1994.11518044

41. Ragowsky, A., Somers, T.M.: Enterprise resource planning. Journal of Management Information Systems 19(1), 11–15 (2002). https://doi.org/10.1080/07421222.2002.11045718

42. Ricciotti, F.: From value chain to value network: a systematic literature review. Management Review Quarterly 70(2), 191—212 (2020). https://doi.org/10.1007/s11301-019-00164-7

43. Rosen, R., von Wichert, G., Lo, G., Bettenhausen, K.D.: About the importance of autonomy and digital twins for the future of manufacturing. In: Proceedings of the 15th IFAC Symposium on Information Control Problems in Manufacturing. vol. 48, pp. 567–572 (2015). https://doi.org/10.1016/j.ifacol.2015.06.141

44. Schleipen, M., Drath, R.: Three-view-concept for modeling process or manufacturing plants with AutomationML. In: Proceedings of the IEEE International Conference on Emerging Technologies and Factory Automation (ETFA) (2009). https://doi.org/10.1109/ETFA.2009.5347260

45. UN/CEFACT: UML Profile for UN/CEFACT's Modeling Methodology (UMM), Foundation Module, Version 2.0 (Apr 2011), https://unece.org/DAM/cefact/umm/UMM_Foundation_Module_V2.0.pdf

46. Vogel-Heuser, B.: Herausforderungen und Anforderungen aus Sicht der IT und der Automatisierungstechnik. In: Bauernhansl, T., ten Hompel, M., Vogel-Heuser, B. (eds.) Industrie 4.0 in Produktion, Automatisierung und Logistik: Anwendung Technologien Migration, pp. 37–48. Springer (2014). https://doi.org/10.1007/978-3-658-04682-8_2

47. Vogel-Heuser, B., Brodbeck, F., Kugler, K., Passoth, J.H., Maasen, S., Reif, J.: BPMN^{+I} to support decision making in innovation management for automated production systems including technological, multi team and organizational aspects. In: Proceedings of the 21st IFAC World Congress (2020)

48. Vogel-Heuser, B., Kegel, G., Bender, K., Wucherer, K.: Global information architecture for industrial automation. Automatisierungstechnische Praxis (atp) **51**, 108–115 (01 2009)

49. Wally, B.: Smart Manufacturing Systems: Model-Driven Integration of ERP and MOM. PhD Thesis, Faculty of Informatics, TU Wien (2020)

50. Wally, B.: Provisioning for MES and ERP. https://www.automationml.org/ (2021), Application Recommendation

51. Wally, B., Huemer, C., Mazak, A.: Aligning business services with production services: The case of REA and ISA-95. In: Proceedings of the 10th IEEE International Conference on Service Oriented Computing and Applications (SOCA). pp. 9–17 (2017). https://doi.org/10.1109/SOCA.2017.10

52. Wally, B., Huemer, C., Mazak, A.: Entwining plant engineering data and ERP information: Vertical integration with AutomationML and ISA-95. In: Proceedings of the 3rd IEEE International Conference on Control, Automation and Robotics (ICCAR) (2017). https://doi.org/10.1109/ICCAR.2017.7942718

53. Wally, B., Huemer, C., Mazak, A.: ISA-95 based task specification layer for REA in production environments. In: Proceedings of the 11th International Workshop on Value Modeling and Business Ontologies (VMBO) (2017)

54. Wally, B., Huemer, C., Mazak, A.: A view on model-driven vertical integration: Alignment of production facility models and business models. In: Proceedings of the 13th IEEE International Conference on Automation Science and Engineering (CASE) (2017). https://doi.org/10.1109/COASE.2017.8256235

55. Wally, B., Lüder, A.: AML-based enterprise control system integration by IEC 62264. In: Drath, R. (ed.) AutomationML – The Industrial Cookbook, pp. 451–466. De Gruyter Oldenbourg (2021). https://doi.org/10.1515/9783110745979-026

56. Wally, B., Mazak, A., Kratzwald, B., Huemer, C.: Model-driven retail information system based on REA business ontology and Retail-H. In: Proceedings of the 17th IEEE Conference on Business Informatics (CBI). vol. 1, pp. 116–124 (2015). https://doi.org/10.1109/CBI.2015.49

57. Wally, B., Mazak, A., Kratzwald, B., Huemer, C., Regatschnig, P., Mayrhofer, D.: REAlist – a tool demo. In: Proceedings of the 9th International Workshop on Value Modeling and Business Ontology (VMBO) (2015)

58. Wally, B., Schleipen, M., Schmidt, N., D'Agostino, N., Henßen, R., Hua, Y.: AutomationML auf höheren Automatisierungsebenen. In: Proceedings of AUTOMATION 2017. No. 2293 in VDI-Berichte, VDI-Verlag (2017)

59. Wally, B., Vyskočil, J., Novák, P., Huemer, C., Šindelár, R., Kadera, P., Mazak, A., Wimmer, M.: Flexible production systems: Automated generation of operations plans based on ISA-95 and PDDL. Robotics and Automation Letters **4**(4), 4062–4069 (2019). https://doi.org/10.1109/LRA.2019.2929991

60. Wally, B., Vyskočil, J., Novák, P., Huemer, C., Šindelár, R., Kadera, P., Mazak, A., Wimmer, M.: Production planning with IEC 62264 and PDDL. In: Proceedings of the 17th IEEE International Conference on Industrial Informatics (INDIN). pp. 492–499 (2019). https://doi.org/10.1109/INDIN41052.2019.8972050

61. Wally, B., Vyskočil, J., Novák, P., Huemer, C., Šindelář, R., Kadera, P., Mazak-Huemer, A., Wimmer, M.: Leveraging iterative plan refinement for reactive smart manufacturing systems. IEEE Transactions on Automation Science and Engineering **18**(1), 230–243 (2021). https://doi.org/10.1109/TASE.2020.3018402
62. Witsch, M.: MES-Modeling Language – Eine Beschreibungssprache für die interdisziplinäre Anforderungserhebung und Spezifikation von MES. In: Vogel-Heuser, B. (ed.) Erhöhte Verfügbarkeit und transparente Produktion, pp. 80–101. kassel university press (2011)
63. Witsch, M., Vogel-Heuser, B.: Towards a formal specification framework for manufacturing execution systems. IEEE Transactions on Industrial Informatics **8**(2), 311–320 (2012). https://doi.org/10.1109/TII.2012.2186585
64. Zapletal, M., Schuster, R., Liegl, P., Huemer, C., Hofreiter, B.: The UN/CEFACT modeling methodology UMM 2.0: Choreographing business document exchanges. In: vom Brocke, J., Rosemann, M. (eds.) Handbook on Business Process Management 1, Introduction, Methods, and Information Systems, 2nd Ed, pp. 625–647. International Handbooks on Information Systems, Springer (2015). https://doi.org/10.1007/978-3-642-45100-3_27

Data Analytics

Conceptualizing Analytics: An Overview of Business Intelligence and Analytics from a Conceptual-Modeling Perspective

Christoph G. Schuetz and Michael Schrefl

Abstract

Business intelligence and data analytics projects often involve low-level, ad hoc data wrangling and programming, which increases development effort and reduces usability of the resulting analytics solutions. Conceptual modeling allows to move data analytics onto a higher level of abstraction, facilitating the implementation and use of analytics solutions. In this chapter, we provide an overview of the data analytics landscape and explain, along the (big) data analysis pipeline, how conceptual modeling methods may benefit the development and use of data analytics solutions. We review existing literature and illustrate common issues as well as solutions using examples from cooperative research projects in the domains of precision dairy farming and air traffic management. We target practitioners involved in the planning and implementation of business intelligence and analytics projects as well as researchers interested in the application of conceptual modeling to business intelligence and analytics.

Keywords

Business intelligence · Business analytics · Data analytics · Conceptual modeling

C. G. Schuetz (✉) · M. Schrefl
Institute of Business Informatics – Data & Knowledge Engineering, Johannes Kepler University Linz, Linz, Austria
e-mail: christoph.schuetz@jku.at

M. Schrefl
e-mail: michael.schrefl@jku.at

© The Author(s), under exclusive license to Springer-Verlag GmbH, DE, part of Springer Nature 2023
B. Vogel-Heuser and M. Wimmer (eds.), *Digital Transformation*,
https://doi.org/10.1007/978-3-662-65004-2_13

311

1 Introduction

Effective data engineering separates the conceptual, logical, and physical concerns of the development of database applications. At the conceptual level, the development of database applications focuses on the representation of real-world entities and their relationships as well as query and manipulation operations in a *conceptual model* under consideration of user requirements. At the logical level, development focuses on the implementation of the conceptual model using a particular database technology. The physical level concerns specific data structures, algorithms, and performance improvements. Business intelligence and analytics projects, however, often overly emphasize the logical and physical level and, consequently, focus primarily on low-level, ad hoc data wrangling and programming. A more conceptual viewpoint that initially abstracts from low-level details fosters understanding of the domain and the problem to be solved with analytics, which facilitates the implementation and use of analytics solutions.

1.1 What Is Analytics?

"Analytics", "data analytics", and "business analytics" are elusive terms, variously defined in industry and academia alike. One definition from industry describes analytics as "data-based applications of quantitative analysis methods" [64, p. 37]. Another definition from industry describes the purpose of analytics as the "examination of information to uncover insights that give a business person the knowledge to make informed decisions" while "analytics tools enable people to query and analyze information" [56, p. 16]. A definition from academia, which defines data analytics as the "discovery and communication of meaningful patterns in data" [60, p. 329], also fits a common definition of *data mining*, another rather elusive term. In this regard, the purpose of data analytics is described as follows [60, p. 329]: "Organizations apply analytics to their data in order to describe, predict, and improve organizational performance". Similarly, business analytics can be seen as "the process of developing actionable decisions or recommendations for actions based on insights generated from historical data" [55, p. 48]. Others [10, pp. 2–3] stress the difference between the terms "analysis" and "analytics", the former referring to "the process of separating a whole problem into its parts so that the parts can be critically examined at the granular level", the latter referring to "the variety of methods, technologies, and associated tools for creating new knowledge/insight to solve complex problems". In this chapter, we do not aim to provide yet another definition of "analytics". Rather, we take a more pragmatic approach and employ the terms "analytics", "business analytics", and "data analytics" in a flexible way as an umbrella term for all kinds of data-driven processes, methods, systems, tools, etc. aimed at improving business performance. We broadly employ the term "analysis" to refer to applications of analytics for specific tasks.

Business analytics is a broad field comprising a multitude of methods. Literature distinguishes multiple subcategories of (business) analytics. In particular, business analytics can be divided into the following subcategories:

- **Descriptive analytics**: In its most basic form, analytics aims to describe what happened in the past based on historical data, providing reports and dashboards displaying various statistics and diagrams. Descriptive analytics is sometimes equated with business intelligence, e.g., by Fleckenstein and Fellows [15]. A broader view sees "business intelligence" as umbrella term that comprises analytics [64, p. 28] – the view we adopt in this chapter.
- **Predictive analytics**: Exploring what happened in the past is typically the first step in predicting the future. The term "predictive analytics" then refers to the application of statistical methods in order to develop models from historical data to predict the future. Data mining is sometimes included when referring to predictive analytics [55, p. 49]. Others equate data mining with descriptive analytics [54, pp.22–23]. In this regard, a pragmatic definition of data mining would be as follows: Data mining aims to find patterns in historical data (descriptive aspect) in order to make predictions about the future (predictive aspect).
- **Prescriptive analytics**: Accurate predictions of the future are the basis for making informed decisions in order to shape future events. For example, in precision dairy farming, based on the analysis of milk yield and feed compositions in the past, the optimal feed composition for the future can be inferred. In this regard, simulation is a common tool in prescriptive analytics. Using probabilistic models, various outcomes can be simulated in order to determine the best course of action.

Predictive analytics and prescriptive analytics are sometimes also referred to as advanced analytics [56, p. 16]. In addition to descriptive, predictive, and prescriptive analytics, other types of analytics are also sometimes distinguished, e.g., diagnostic analytics ("why did something happen?") [56, p. 16], which we do not elaborate further; we refer to Sherman [56] and Williams [64] for a comprehensive overview of the different analytics flavors.

In the context of BI and analytics, different roles can be distinguished, and just as with the definition of the term "analytics" itself, the names and characteristics of those roles may vary in practice (see [47]). Among the most commonly found roles are data engineer and data scientist. The data engineer's task is to handle possibly large amounts of data and manage the (big) data analytics architecture such that the data scientist can access those data and gather insights from the data. Involved in BI and analytics projects are also domain (business) experts and, especially in the context of cyber-physical systems, engineers familiar with sensor networks and production equipment.

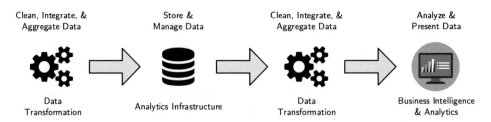

Fig. 1 BI and analytics rely on data warehouse and other types of analytics infrastructure to analyze and present data

1.2   The Bigger Picture: Business Intelligence and Analytics

We adopt a view that includes analytics in the field of business intelligence (BI). In the narrow sense, BI and analytics refer to the analysis and (visual) presentation of data. In a broader sense, "business intelligence" serves as an umbrella term that includes, besides analytics, the necessary data transformation processes as well as data warehousing and, more generally, development and management of the analytics infrastructure (Fig. 1). We employ the term "business intelligence" in both senses, depending on the situation.

In BI, data transformation comprises the cleaning, integration, and aggregation of the source data as the preparatory steps for applying analytics algorithms. Although arguably of lower prestige than analytics, data preparation, i.e., getting the data into a format suitable for the analysis, makes up a significant portion of a data scientist's workload. Different types of analytics infrastructure – e.g., data warehouse, data lake – require different types of data transformation tasks at different stages. For example, prior to loading the data into a data warehouse for permanent storage, various data sources must be integrated, erroneous and missing data eliminated, and the detailed data aggregated to the required level of detail. The necessary data transformation steps for filling the data warehouse are largely automated, the schema of the data available for the analysis is fixed. Consequently, the analyst can work with clean and consistent data in a format suitable for multidimensional analysis, reducing the number of necessary data transformation steps that the analyst must perform before conducting the actual analysis. A data lake [46], on the other hand, stores potentially valuable raw data close to the original format, thus deferring the burden of cleaning, integrating, and aggregating the available data to the point when the data analysis is actually performed – with the advantage of increased flexibility.

1.3   The (Big) Data Analysis Pipeline

In order to realize the full potential of digitization in modern manufacturing systems, the vast amounts of data originating from a multitude of sensors and databases must be put to good use by appropriate means of data analysis. The analysis of (big) data consists of multiple

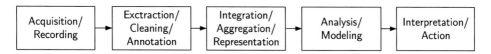

Fig. 2 The (big) data analysis pipeline (adapted from [1, p. 3])

steps (Fig. 2) – each step presenting its own challenges – starting with the recording and storage of the data, and continuing with extraction and cleaning, integration and aggregation of the generated data before actually performing the analysis and interpreting the results that can be used to take appropriate action.

On a conceptual level, big data analytics generally does not differ too much from "traditional" BI and analytics. The specific implementation technologies and algorithms required for big data analytics may indeed differ considerably from those employed in traditional BI, and some issues may be more pronounced and more difficult to solve when handling big data in comparison to the processing of the relatively small data sets in the past (cf. [9]). In this regard, existing conceptual-modeling approaches must of course evolve and new approaches be developed in order to be able to tackle newly arising challenges – but this has always been the case. In the end, however, the general tasks for big data analysis are the same as those that have always been required for performing data analysis, and the key point is that conceptual modeling still remains relevant in the age of big data analytics. An argument can even be made that BI has always been about the analysis of what constituted "big" data at the time (see [64, p. 27]). Hence, rather than referring to the workflow in Fig. 2 exclusively as the *big* data analysis pipeline, we simply consider that workflow to represent the data analysis pipeline in general.

Getting data analysis right at the conceptual level is often not only the first step to effective (big) data analysis but may also pave the way towards efficient implementations at the physical level. In the following, we present the various steps of the data analysis pipeline, describe the key challenges, and provide an overview of approaches supporting the development of solutions for the challenges associated with each step. The organization of the remainder of this chapter thus loosely follows the data analysis pipeline. Section 2 discusses data acquisition and recording. Section 3 discusses extraction, cleaning, integration, and aggregation. Section 4 discusses the actual analysis. Section 5 discusses interpretation and action based on the analysis results.

2   Acquisition and Recording

The data for the analysis originate from various sources, often at high volume and high velocity. Modern process engineering, manufacturing, and farming operations, for example, employ a multitude of sensors which produce a vast amount of data. Those data must be recorded and ultimately analyzed; the complexity of those tasks increases with the volume of the produced data. For example, in precision dairy farming, movement sensors track animal positions and activity. Assuming two position readings per second for each animal, sensors produce 172,800 records of movement data per day for each animal. Thus, on a large farm with 1,000 animals, sensors produce 63,072,000,000 movement records per year. Now suppose the goal is to detect signs of animal illness by analyzing movement data across different farms from multiple countries, possibly in relation to the data produced by other types of sensors, e.g., sensors for recording microclimate, feed intake, milk yield, and milk composition, or in relation to binary data such as images and videos obtained by on-site cameras, then storage and analysis of the data become a challenge.

Typically, the majority of the produced data are not required for meaningful analysis and the recorded data can thus be safely reduced prior to the analysis. For example, rather than storing and analyzing millions of individual location points for each animal in dairy production, a higher level of abstraction is more useful for certain use cases, as we discovered in the agriProKnow project [53]. The large number of sensor readings may be important for noise reduction and error correction but for detecting signs of animal illness in the activity patterns the number of records can be safely reduced by way of abstraction. Hence, knowing for each animal the walking distance and duration as well as the time spent lying and standing, respectively, per functional area, e.g., feeding area, resting area, or milking parlor, within each hour of day is sufficient to determine abnormal activity patterns.

The edge computing paradigm [48, 57] promotes the processing of data at the "edges" of an information system, i.e., close to where the data are actually generated, as a strategy to handling the high volume and velocity of the generated data. Accordingly, in sensor networks, the sensors themselves should take over, to some extent, various steps of data preprocessing and analysis. In a similar vein, data stream processing technology allows to preprocess and analyze the constantly generated streams of data in real time.

Stream processing and edge computing may be the technical solutions for attaining a reduction of the vast amount of generated data to a more manageable size, but the meaningful implementation of those solutions requires a conceptual understanding of the domain. Consider again the example of movement data in precision dairy farming: For each animal, walking distance and duration as well as lying and standing duration per functional area within each hour of the day are more interesting than thousands of individual position readings. The data can be reduced accordingly when recording the data. Doing so, however, requires a shared conceptualization between data scientists and domain experts – in this case veterinarians – of the different activity types and functional areas.

The *lambda architecture* [32] for big data processing describes an architecture that supports real-time data analysis through a *speed layer* for processing streams of data, in addition to supporting less time-critical processing of previously collected batches of historical data through a *batch layer*. Figure 3 illustrates an adaptation of the lambda architecture for the analysis of sensor data, which employs both a *data lake* and a *data warehouse*. A data lake is a store of heterogeneous raw data in their original format, from structured (relational, comma-separated values) and semi-structured (XML, JSON) to unstructured and binary

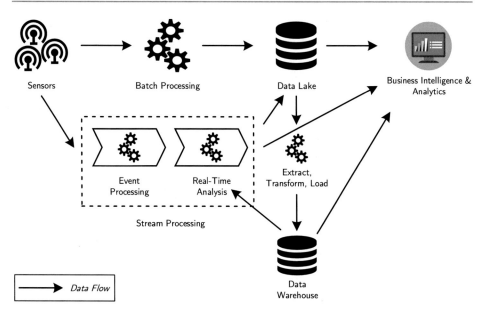

Fig. 3 A variation of the lambda architecture [32, p. 19] with a data warehouse and a data lake for the analysis of sensor data (adapted from [11, p. 7662])

data (text, images, videos). The rationale behind the data lake is twofold. First, a data lake requires minimal data preprocessing and, therefore, supposedly copes better with big data generated at high velocity than a data warehouse. Second, just because *all* the data cannot be analyzed today does not mean that those data do not hold potential value in the future. Keeping the raw data hence increases flexibility. A data warehouse, on the other hand, is an integrated, subject-oriented, non-volatile, and time-variant database [25] for a specific analytical purpose with clean data of "recognized high value" [46, p. 8]. Extract, transform, and load (ETL) processes funnel clean and consistent data, obtained from raw data of various sources via transformation routines, into the data warehouse (see Sect. 3). Both data lake and data warehouse have their place in modern data analytics.

In order to prevent a data lake from becoming a *data swamp*, "usually due to a lack of self-service and governance facilities" [21, pp. 11–12], a successful data lake infrastructure must also associate the recorded raw data with suitable metadata (see [1]). In this regard, a data lake may annotate data items with entities from a knowledge graph [9, p. 74] or ontology. A knowledge graph represents real-world entities and their relationships (cf. [29]). An ontology is essentially a machine-readable, deployable conceptual model, often based on logic, that can be applied in various practical situations. The boundaries between knowledge graph and ontology are fluent. Typically, a knowledge graph focuses more on entities (instance data) whereas an ontology describes classes of entities (schema information), although ontologies may also comprise instance data (see [23, 24] for more information). Data items in a data lake

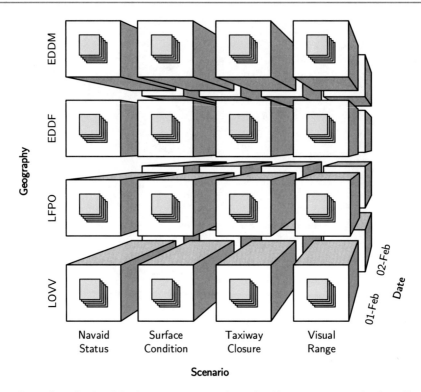

Fig. 4 Illustration of a data lake that structures raw data – in this case, messages in air traffic management – into multidimensional contexts for future analysis

can be collected into different multidimensional contexts, where each data item is associated with an entity representing, for example, geographic and temporal scope of applicability, the type of content described by the data item, or the provenance, i.e., the origin of the data item, e.g., which sensor recorded it. For example, in air traffic management (ATM), the messages exchanged between air traffic controllers and pilots may be organized by flight information region (geography), date, and air traffic scenario (Fig. 4), but also by importance for individual flights; the result is an ATM information cube [50]. In this example, a message may concern the status of a navigation aid in the LOVV (Austria) region on the 1st February, which means that the message will be collected into the corresponding container, associated with the appropriate metadata.

3   Extraction, Cleaning, Integration, and Aggregation

The collected data must be represented using adequate data models for the analysis. Loading the collected data into those data models requires various data preparation tasks.

3.1 Data Models

The data for the analysis must be stored in an adequate data analytics infrastructure in a suitable format. Regarding such infrastructure, different options exist with different requirements regarding the employed data models. A data warehouse, for example, integrates clean and consistent data extracted from various sources at a granularity suitable for a specific analytical purpose; we refer to Vaisman and Zimànyi [60] for a comprehensive introduction to data warehousing. Compared to a data lake, a data warehouse spares the analyst cumbersome data preparation work, although depending on the type of analysis, some data preparation may still be necessary. The advantage of storing the data in a form ready for analysis, however, comes at the cost of reduced flexibility. Furthermore, building a useful data warehouse requires considerable effort, although agile development methodologies may alleviate the perceived rigidity of the development process [30]. Customizable reference models also facilitate the design and implementation of a data warehouse [5, 51].

The predominant modeling paradigm in data warehousing is the *multidimensional model*, which requires extensive data cleaning and transformation prior to loading the data into the data warehouse. Figure 5 represents a multidimensional model for sensor data using Dimensional Fact Model (DFM) notation [19]. The central entity in a multidimensional model is the *fact class* – in this example: Measurement – which represents the real-world events of interest (facts) that should be analyzed. The measures quantify the events of interest represented by the fact class; in case of Measurement, the measures are value and accuracy. A fact class has multiple *dimensions*, which characterize the represented events and are hierarchically organized. The Measurement fact class, for example, has the dimensions agent, measurementType, transformation (optional), location (optional), receptionTime, and sensingTime. Each dimension consists of *levels* – represented by circles in the DFM – and *non-dimensional attributes*, which are attached to a level. The levels serve to aggregate events, the non-dimensional attributes provide additional information and serve for selecting relevant facts. The edges connecting levels represent *roll-up* relationships, which are many-to-one relationships. For example, a logical device has a logical device type but one logical device type may comprise multiple logical devices. Indyco Builder,[1] a modeling tool using the DFM, allows for grouping levels and non-dimensional attributes into groups, which can be used to structure a heterogeneous model. For example, an agent that is a person has position and age, an agent that is a process has a process type, and an agent that is a device has a nominal accuracy, a physical device, a logical device, a logical device type as well as an optional location.

Figure 6 shows example data corresponding to the multidimensional model in Fig. 5; note that some attributes are omitted for brevity. The table at the top shows facts in accordance with the Measurement fact class. For example, for the 2nd October 2018 at 2:00 pm, the data warehouse records a measurement of Type 3 reported by the agent with ID 1 with an accuracy of 0.1 and a value of 22.2, which corresponds to the average of the last ten readings.

---

[1] https://www.indyco.com/.

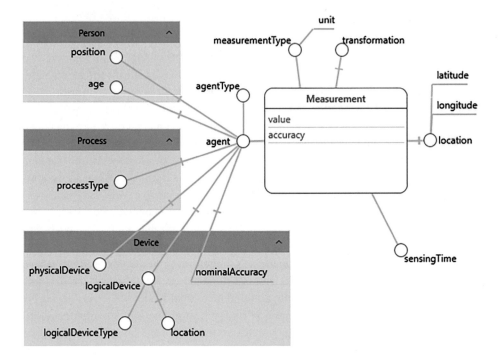

Fig. 5 Multidimensional model in DFM notation for sensor data (adapted from [11, p. 7664])

Sensing Time	Meas. Type ID	Agent ID	Trans.	Acc.	Value	Lat.	Long.
2018/10/02 14:00	3	1	AVG10	0.1	22.2	-	-
2018/10/02 14:10	3	1	AVG10	0.1	22.4	-	-
2018/10/02 14:05	2	2	AVG5	0.1	61.3	38.33	14.32
2018/10/02 14:15	2	2	AVG5	0.2	60.9	38.33	14.32
2018/10/03 10:20	2	3	-	-	62.0	-	-

Meas. Type ID	Meas. Type	Unit
1	Milk yield	kg
2	Rumination activity	Chews per cud
3	Temperature	°C

Agent ID	Agent	Agent Type	Phys. Dev.	Log. Dev.	Loc.	Dev. Type
1	THE01	Device	THE01-232	Therm. Area #1	Area #1	Thermometer
2	EAR23	Device	EAR23-143	Ear. Animal#23	-	Earmark
3	VET01	Person	-	-	-	-

Fig. 6 Example data for the multidimensional model in Fig. 5

According to the table in the middle, a measurement of Type 3 is a temperature reading, recording values are centigrade. The agent with ID 1 has the designation THE01 and is a device. The actual physical device is THE01-232, which assumes the role of Thermometer for Area #1, which is a logical device of type thermometer that is put into Area #1. Similarly, the agent with ID 2 is a device, the actual physical device is EAR23-143 which assumes the role of earmark for Animal #23. No location is associated for an earmark, the position instead being recorded under the location dimension with latitude and longitude. The agent with ID 3 is an example of a person and could be a veterinarian who manually records the chews per cud for tracking an animal's rumination activity.

The multidimensional model in Fig. 5 is a generic model accommodating various kinds of sensor data. The generic model allows the data warehouse to cope with heterogeneous data, which increases flexibility but requires the analyst to perform various data preparation tasks prior to the analysis, e.g., eliminating the NULL values caused by optional attributes. Furthermore, the generic model leaves out specific information that might be relevant in a certain domain: For example, in precision dairy farming, an animal's day of lactation within a lactation cycle would be relevant, which a domain-specific model may include.

A data warehouse often employs more specific models than the one in Fig. 5, tailored to the actual subjects of the analysis. Rather than sensor readings in general, the focus of a multidimensional model may be, for example, animal movement activity. The corresponding multidimensional model would abstract from the individual sensor readings and concentrate on the movement of animals, with specific measures such as lying and walking duration rather than a generic *sensed value* as well as specific dimensions, e.g., the farm site, animal, or functional area. Hence, the multidimensional model in Fig. 7 emphasizes animal movement activity. The measures are lying, standing, and walking duration as well as the walking distance within a functional area per hour of day for an animal on a particular farm site. Figure 8 shows example data corresponding to the multidimensional model in Fig. 7. For example, at the Pottenstein farm site, on the 2nd October 2018, in the 15th hour of the day, the animal with National ID AT0123 spent 10 min lying, 20 min standing, and 10 min walking a distance of 255 m in a feeding area. By referring to the functional area, e.g., feeding area or milking area, rather than precise positions or farm-specific location designations, e.g., Area #1, the movement activity of animals becomes comparable across farm sites. A data warehouse for precision dairy farming would also employ multidimensional models for milk yield, which would include data from external databases along with the sensor data [53].

3.2 Data Preparation

Prior to performing data analysis, the data must be extracted from the various sources, cleaned and integrated, and aggregated to a granularity suitable for the analysis task at hand. In the context of data warehousing, the data preparation steps are referred to as *extract, transform, and load* (ETL) processes, which put the data into the data warehouse. In the

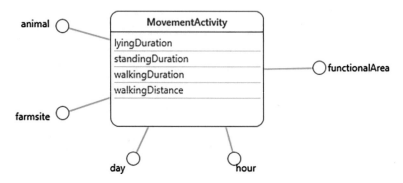

Fig. 7 A simplified multidimensional model (without dimension hierarchies) in DFM notation for the analysis of animal movement in the agriProKnow project

Farmsite	Day	Hour	Animal	FunctionalArea	LyingDur.	StandingDur.	WalkingDur.	WalkingDis.
Pottenstein	2018/10/02	15	AT0123	Feeding	10	20	10	255
Pottenstein	2018/10/02	15	AT0123	Milking	0	11	9	102
Lozorno	2018/10/02	16	SL0234	Feeding	14	2	12	127
Lozorno	2018/10/02	16	SL0234	Milking	2	16	14	145

Fig. 8 Example data for the multidimensional model in Fig. 7

following, we specifically look at ETL processes even though the general principles equally apply to data preparation for data analysis without the use of a data warehouse. For example, in case of a data lake, data preparation steps similar to the ETL processes in data warehousing must be performed on a case-by-case basis. The presented modeling approaches may also be employed in a data lake environment for performing the necessary data preparation steps for running various analytics algorithms.

Traditionally, the implementation of data preparation in general and ETL processes in particular involves extensive low-level programming, e.g., in a statistical programming language such as R or in a programming language for stored procedures such as PL/SQL. Process modeling approaches, with tool support for automatic code generation, facilitate implementation while also improving the documentation and flexibility of ETL processes. Besides proprietary modeling languages employed by commercial tools, e.g., Pentaho Data Integration (Kettle) [60, pp. 319ff.], the Business Process Model and Notation (BPMN) [13, 14] or UML activity diagrams [34] may serve for ETL process modeling. In a similar vein, modeling languages may assist with the representation of the preparatory steps in data mining [41] as well as complex data processing pipelines that include steps for the actual analysis of the data [49]. Commercial tools for data analytics, e.g., KNIME [6] and RapidMiner [27], employ graphical notations for the specification of complex data processing pipelines.

The BPMN diagram in Fig. 9 represents an ETL process for aggregating raw movement data of animals for a data warehouse, following the modeling approach proposed by El Akkaoui et al. [13, 14]. In this example, movement sensor readings are stored in a comma-

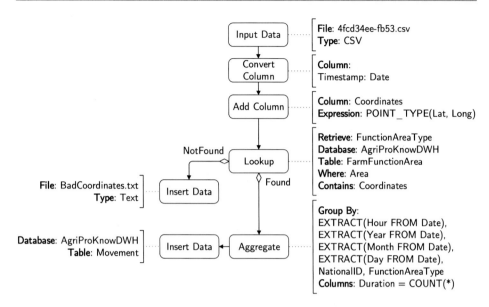

Fig. 9 A BPMN model of the necessary data transformation steps as preparation for the analysis of animal movement in precision dairy farming

separated values (CSV) file. The first step is to read a source CSV file. The timestamp of the movement records in this CSV file is subsequently converted into a Date column before adding a Coordinates column of a database-specific POINT_TYPE derived from latitude and longitude values in the CSV file. The coordinates are then used to look up the functional area from a table. In case no corresponding functional area could be found for a coordinate, a warning message is written into a log file. Individual position readings are then aggregated by hour of day for each date before the resulting tuples are inserted into a Movement table, which can be used for data analysis.

A pattern-based approach to ETL process modeling fosters code reusability and promotes best-practice solutions to commonly occurring issues in ETL, e.g., change data capture, slowly changing dimensions, and surrogate key pipelining [39]. Using common ETL patterns, the BPMN language can be extended for ETL process modeling [40]. The thus modeled ETL processes also promise better integration with an organization's business processes (see [14, 40]). ETL process modeling also faciliates a comprehensive analysis of existing ETL processes based on (process) mining for ETL patterns [59], following the Workflow Patterns Initiative [45]. ETL pattern mining may identify recurring patterns in existing ETL processes to subsequently redesign the ETL processes. Furthermore, quality metrics can be applied on ETL process models at higher level of abstraction. Finally, ETL process models may be reduced to high-level summaries in order to foster a general understanding of the models as well as to render ETL process models comparable.

Conceptual models can also be the basis for physical optimization of ETL processes. Massive distribution and parallelization of data processing is key to efficient big data analysis. In this regard, a library of common ETL tasks facilitates the parallelization of ETL processes for big data [3] – conceptual modeling becomes the fundamental to optimization. ETL processes are hence described in terms of core functionalities. Processing of the core functionalities can then be parallelized. An example of an ETL core functionality is the *look-up* of a value in a look-up table, e.g., the functional area for a coordinate.

In some cases, it may be impractical to extract all relevant data from the source systems due to volume, velocity, or volatility of the data. Approaches for on-demand ETL dynamically execute adapted ETL processes for a specific query execution [4]. Similarly, another approach would be to superimpose a multidimensional schema with mapping rules over non-owned sources that are not in a format suitable for data analysis. For example, linked open data and public knowledge graphs, e.g., Wikidata [61], are valuable sources for the analysis but do not follow the multidimensional modeling paradigm. In addition, the collaborative, volatile nature of those knowledge graphs makes the construction of a static data warehouse impractical. Superimposed multidimensional models allow for the rewriting of analytical queries into a query expression for a target query language and data model [22].

4 Analysis and Modeling

The collected data must be analyzed using the appropriate data analytics methods, the selection of which depends on the analytical task. Conceptual models may assist with the design and execution of data analytics applications.

4.1 Data Analytics

A common type of data analysis over multidimensional modeling is *online analyitcal processing* (OLAP), which is mainly descriptive in nature, answering the question: *what happened?* In OLAP, analysts leverage the hierarchical nature of multidimensional data, performing *roll-up* and *drill-down* operations in order to obtain measures for (aggregate) events at different levels of granularity. For example, the roll-up operation illustrated in Fig. 10 operates on a three-dimensional *sales* cube where the measure is the profit per product, city, and quarter. A roll-up operation aggregates the cells of the cube per category, country, and quarter, thereby obtaining a cube at a coarser granularity than the original, with the individual profits summed up. The inverse of the roll-up operation is the *drill-down* operation, e.g., a drill-down from the view of the sales per category, country, and quarter to the view of the sales per product, city, and quarter. The selection of partitions of the cubes based on various criteria is broadly referred to as *slice-and-dice*. Various other OLAP operations have been

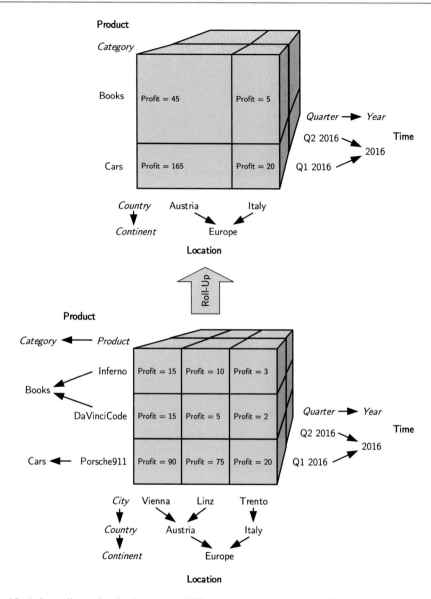

Fig. 10 A three-dimensional sales cube at different levels of granularity. The dimension hierarchies of a cube serve to obtain higher-level views on the data by performing a roll-up operation.

proposed in the past; we refer the interested reader to Vaisman and Zimányi [60] for more information on OLAP.

Advanced analytics employs statistical methods and machine-learning algorithms in order to (aim to) predict the future. In this regard, multidimensional data is only one source for

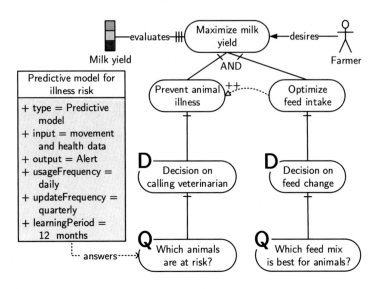

Fig. 11 Example business view for selecting a predictive model following the approach and using the notation described by Nalchigar and Yu [35]

predictive and prescriptive analytics. The choice of machine-learning algorithm depends, first, on the available data sets. Furthermore, the application of analytics must consider the necessary preparatory steps. Requirements engineering for analytics, therefore, must assume a *data preparation view*, which defines what data sets and preparatory steps are required to perform the chosen analytics [35, p. 365f.]. The data preparation view is akin to the process modeling approaches discussed in Sect. 3. In the context of requirements engineering for analytics, the explicit modeling of data preparation promotes understanding of the available data and improves the quality of the analytics solution [35, p. 365].

The application of analytics methods in general and machine-learning algorithms in particular must be aligned with an organization's strategy, necessitating a *business view* and an *analytics design view* in requirements engineering for analytics [35]. Starting from the business goals, the business view formulates a set of data analytics goals, expressed as decisions that must be taken in order to achieve the business goals. In order to take the decisions, certain questions must be answered. The data analytics view, in turn, evaluates potential methods regarding their ability to achieve the data analytics goals by comparing the various methods' strengths and weaknesses, measured by various indicators.

Figure 11 shows the business view for developing a predictive model for illness risk of an animal in precision dairy farming. The farmer pursues the goal of maximizing the milk yield. The indicator for goal achievement is the milk yield. Maximizing the milk yield depends on the subgoals of preventing animal illness and optimizing the feed intake. Optimizing the feed intake contributes considerably to preventing animal illness. Optimizing the feed intake involves decisions on changing the feed composition, which requires to ask about the

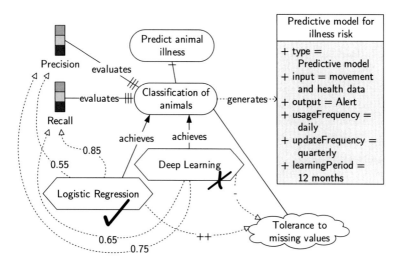

Fig. 12 Example data analytics design view for selecting a predictive model following the approach and using the notation described by Nalchigar and Yu [35]

best feed mix for animals. Preventing animal illness involves decisions on whether to call a veterinarian for examining an animal at risk, which requires to ask which animals are at risk of developing an illness, e.g., ketosis. Predictive models allow to answer the questions required to make the decisions. For example, a predictive model over movement and health data, updated quarterly using data from the last 12 months, may be used daily in order to make predictions and raise alerts about animals at risk.

Figure 12 shows a data analytics design view that corresponds to the business view in Fig. 12. The primary goal is to predict animal illness, which depends on the classification of animals. Precision and recall serve to evaluate the quality of a classifier. The precision of a classifier is the ratio of animals that are correctly classified as ill out of all animals classified as ill. The recall, on the other hand, is the ratio of animals that are correctly classified as ill out of all animals that are actually ill. Different machine-learning methods may serve for classification, e.g., logistic regression and deep learning. Different methods have specific values for precision and recall, possibly estimated over a set of test data. Depending on those values in combination with the respective importance of precision and recall – indicated by a color associated with the measure: green for high, yellow for medium, or red for low importance – as well as fulfillment of the soft goal of tolerance towards missing values, an analyst may choose logistic regression or deep learning in the example of animal illness prediction.

The rise of social media has put the focus of analysts on user-generated content on the web as a source for BI and analytics. The terms "social business intelligence" [18] and "social media analytics" [65] are typically used to refer to the application of BI and analytics on the analysis of user-generated content in social media. In this regard, social

BI is often viewed as the integration of business data with data obtained from social media whereas social media analytics may refer to the application of data analytics methods on social media data, e.g., tweets and connections. On the one hand, real-world events are often first reported on microblogging services such as Twitter. For example, an incident of racial bias in a Starbucks store in 2018 was first brought to the public's attention by a video posted on Twitter, which went "viral", spawning a multitude of online posts as a reaction to the original tweet [42]. Detection of such events can be automated [63], assisting organizations in monitoring social media in order to quickly react to incidents in the real world, thereby limiting the fallout of such an incident. On the other hand, social media analytics often involve sentiment analysis, i.e., the extraction of "subjective information" from natural language (see [43] for an overview), over comments posted on social media platforms. Marketers, for example, would be interested in questions such as: "What is the average sentiment of the public towards a company's product or a group of products?" Data obtained via sentiment analysis can be organized in data warehouses for further analysis along with business data, e.g., data relating to sales [16]. For example, a DFM model may represent cubes for the analysis of sentiment along with business data [17].

4.2   Pattern-Based Approach to Analytics

Starting from the data preparation, business, and data analytics views of an analytics problem, a *solution pattern* represents a proven approach to a common analytics problem [37]. Comprehensive design catalogs for business questions, algorithms, and data preparation tasks also assist with the design of analytics solutions [36]. By using solution patterns and design catalogs, data scientists avoid "reinventing the wheel" when faced with an analytics problem.

A pattern-based approach also benefits the formulation of individual, concrete analytical queries. In this regard, OLAP patterns [28, 52] represent interesting, frequently occurring types of analytical queries. For example, across various domains, set-to-complement and set-to-set comparison are recurring patterns of data analysis (Fig. 13). An example of a set-to-complement comparison is the computation of the ratio between the average milk yield for young animals at a farm site and the average milk yield of all other animals (except young animals) at the same farm site. An example of set-to-set comparison is the ratio between the average milk yield of animals in the first pen and the average milk yield of animals in the third pen of a farm site. In the agriProKnow project, the identification of OLAP patterns and the development of corresponding query facilities led to increased agility, which was necessary to meet the domain experts' requirements under a tight schedule (see [53]). The identification of patterns paves the way towards the documentation of best-practice solutions for satisfying generic types of information needs.

A shared conceptualization of the domain of analysis is the basis for successful communication and correct usage of analysis results. Domain ontologies, e.g., SNOMED CT [12]

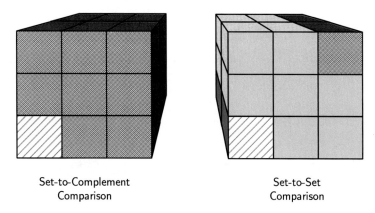

Set-to-Complement   Set-to-Set

Comparison   Comparison

Fig. 13 Generic comparative patterns for multidimensional data analysis

in the health domain or AGROVOC [8] in precision dairy farming, may serve as the basis for the definition of dimensions in a multidimensional model [2]. Multidimensional models can be enriched with explicit definitions of business terms, i.e., calculated measures and filter predicates [51]. For example, a multidimensional model in precision dairy farming defines a Milking fact class, with measures Milk Yield and Fat Content as well as dimensions Farm, Calving, Lactation, Time, and Animal (Fig. 14). Explicit definitions of filter predicates – e.g., Low Daily Milk Yield, Young Cattle, and Under Heat Stress – and calculated measures – e.g., Average Milk Yield and Average Milk Yield Ratio – may complement the multidimensional model. Expressions in a target language may serve for the definition of the semantics of those business terms [51], which are omitted in Fig. 14.

Patterns for multidimensional data analysis may be identified at various levels of abstraction, from domain-independent to domain-specific and organization-specific patterns. Figure 15 illustrates the step-wise instantiation of a domain-independent pattern using a shared conceptualization of business terms (Fig. 14), yielding first a domain-specific pattern and then a concrete query, which can be executed on a target system if translated into a query language and taking into account the logical data model. In this example, the patterns are informally described using natural language. A more formal representation of the pattern semantics in a target language allows for automatic query generation [22, 52]; the target language can be SQL, MDX, SPARQL, and theoretically any other query language or even a statistical programming language such as R.

An *analysis graph* represents interesting courses of (multidimensional) data analysis [51]. An analysis graph may serve for the proactive modeling of interesting courses of analysis [38]. Figure 16 shows a simple analysis graph for the analysis of movement data. Each box represents an *analysis situation*, which corresponds to a multidimensional query over a cube – in this example: Animal Movement. The edges of an analysis graph represent *navigation steps*, which change the parameters of the analysis situation, e.g., by performing a *drill-down* from the farm to the barn level in the Location dimension. Analysis graphs can

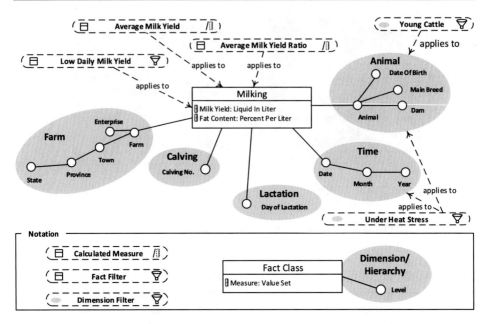

Fig. 14 A multidimensional model in DFM notation enriched with definitions of business terms

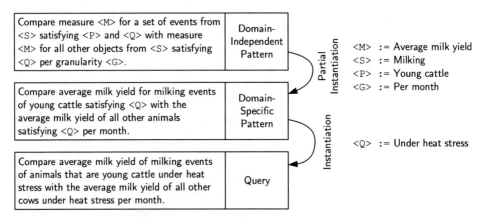

Fig. 15 Instantiation of domain-independent and domain-specific patterns for multidimensional data analysis

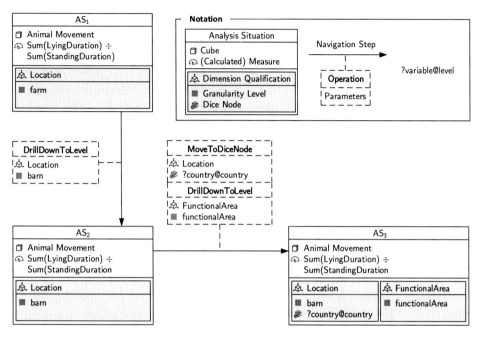

Fig. 16 An analysis graph plotting an interesting course of analysis for animal movement data in precision dairy farming

be extended with comparative analysis situations and even more generic OLAP patterns to allow for more complex queries. Other work aims to describe analytical sessions in terms of a multidimensional algebra [44] and recommend interesting queries to the user based on previous analytical sessions [31].

5   Interpretation and Action

Visualizations are a key tool in the interpretation of data. The field of *visual analytics* aims to create tools and techniques that leverage interactive visualizations to enable analysts to garner insights from large data sets [26, p. 157]. In this regard, a key decision concerns the selection of the appropriate visualization for the analytical task at hand. Adopting a conceptual, model-driven perspective may ensure that the visualization aligns with the analysis goal, desired degree of interactivity, user (casual or power user), and dimensionality/cardinality/type of the data [20]. In a sense, that model-driven perspective on visualization bears similarities to the pattern-based approaches on multidimensional data analysis [52] and machine learning [35].

VizDSL [33] is a modeling language for the definition of visual analytics workflows. VizDSL models are akin to analysis graphs [51] but with an emphasis on visualization, with both languages owing to the Interaction Flow Modeling Language (IFML) [7]. In fact,

VizDSL is based on and extends the IFML. The models in VizDSL specify visualization layouts for data sources as well as navigation flows, which makes the generated visualizations interactive.

Active data warehouses [58] aim to automate analysis tasks and decision making. For example, OLAP patterns and analysis graphs may trigger and guide, respectively, actions in day-to-day business. In the agriProKnow data warehouse [53], analysis rules based on OLAP patterns trigger specific actions, e.g., the notification of a veterinarian in case of extraordinary readings regarding an animal. Analysis graphs, on the other hand, may serve for contingency planning, i.e., mapping out a course of analysis that can be followed in case of extraordinary events, e.g., supply shortage of essential resources in manufacturing [38]. An analysis graph would then model the necessary analysis steps for finding a substitute material or an alternative supplier in case of a supply shortage of essential resources. Using the analysis graph, analysis tasks could even be partially automated.

A conceptual understanding of the application domain is the basis for the correct interpretation of analysis results and, consequently, effective decision making. For example, while social media analytics allows a company to keep track of public opinion, the company must also have a clear plan about how to respond to the public in case of a crisis, depending on the nature of the crisis and the public's reaction. In this regard, economic models may serve to choose a suitable strategy in social media crisis communication, with social media analytics providing estimates for the model's parameters [62].

6   Conclusion

BI and analytics projects require stakeholders with different backgrounds to collaborate in various ways in order to develop a common understanding of the problem as well as to decide on the appropriate data model, architecture, algorithms, and user interfaces. Hence, domain experts, business people, data scientists, and engineers must communicate with each other in a common language. Conceptual models serve as the *lingua franca* in BI and analytics projects for experts with different backgrounds. Yet, BI and analytics projects often primarily focus on low-level, ad hoc data wrangling and programming, which increases development effort and reduces usability of the developed BI and analytics solutions. Furthermore, data analytics and machine learning algorithms are often misapplied, adversely affecting the validity of analysis results in practice. A conceptual perspective on data-driven decision making ensures that analysts employ algorithms correctly, using the appropriate systems to process the available data.

Acknowledgements We thank Ilko Kovacic for permission to adapt his figures on enriched multidimensional models and OLAP patterns. We thank Median Hilal for feedback on the graphical presentation of analysis graphs.

References

1. Agrawal, D., Bernstein, P., Bertino, E., Davidson, S., Dayal, U., Franklin, M., others: Challenges and opportunities with big data – a community white paper developed by leading researchers across the United States. Tech. rep., Computing Community Consortium (2012), https://cra.org/ccc/resources/ccc-led-whitepapers/, accessed: 23 June 2020
2. Anderlik, S., Neumayr, B., Schrefl, M.: Using domain ontologies as semantic dimensions in data warehouses. In: Atzeni, P., Cheung, D.W., Ram, S. (eds.) ER 2012. LNCS, vol. 7532, pp. 88–101. Springer (2012). https://doi.org/10.1007/978-3-642-34002-4_7
3. Bala, M., Boussaid, O., Alimazighi, Z.: A fine-grained distribution approach for ETL processes in big data environments. Data & Knowledge Engineering **111**, 114–136 (2017)
4. Baldacci, L., Golfarelli, M., Graziani, S., Rizzi, S.: QETL: An approach to on-demand ETL from non-owned data sources. Data & Knowledge Engineering **112**, 17–37 (2017). https://doi.org/10.1016/j.datak.2017.09.002
5. Becker, J., Delfmann, P., Knackstedt, R.: Adaptive reference modeling: Integrating configurative and generic adaptation techniques for information models. In: Reference modeling, pp. 27–58. Springer (2007)
6. Berthold, M.R., Cebron, N., Dill, F., Gabriel, T.R., Kötter, T., Meinl, T., Ohl, P., Thiel, K., Wiswedel, B.: KNIME - the konstanz information miner: version 2.0 and beyond. SIGKDD Explorations **11**(1), 26–31 (2009). https://doi.org/10.1145/1656274.1656280
7. Brambilla, M., Fraternali, P.: Interaction flow modeling language: Model-driven UI engineering of web and mobile apps with IFML. Morgan Kaufmann (2014)
8. Caracciolo, C., Stellato, A., Morshed, A., Johannsen, G., Rajbhandari, S., Jaques, Y., Keizer, J.: The AGROVOC linked dataset. Semantic Web **4**(3), 341–348 (2013)
9. Ceravolo, P., Azzini, A., Angelini, M., Catarci, T., Cudré-Mauroux, P., Damiani, E., Mazak, A., van Keulen, M., Jarrar, M., Santucci, G., Sattler, K., Scannapieco, M., Wimmer, M., Wrembel, R., Zaraket, F.A.: Big data semantics. Journal on Data Semantics **7**(2), 65–85 (2018). https://doi.org/10.1007/s13740-018-0086-2
10. Delen, D., Ram, S.: Research challenges and opportunities in business analytics. Journal of Business Analytics **1**(1), 2–12 (2018). https://doi.org/10.1080/2573234X.2018.1507324
11. Dobson, S., Golfarelli, M., Graziani, S., Rizzi, S.: A reference architecture and model for sensor data warehousing. IEEE Sensors Journal **18**(18), 7659–7670 (2018). https://doi.org/10.1109/JSEN.2018.2861327
12. Donnelly, K.: SNOMED-CT: The advanced terminology and coding system for eHealth. Studies in Health Technology and Informatics **121**, 279 (2006)
13. El Akkaoui, Z., Mazón, J., Vaisman, A.A., Zimányi, E.: BPMN-based conceptual modeling of ETL processes. In: Cuzzocrea, A., Dayal, U. (eds.) DaWaK 2012. LNCS, vol. 7448, pp. 1–14. Springer (2012)
14. El Akkaoui, Z., Zimányi, E.: Defining ETL worfklows using BPMN and BPEL. In: Proceedings of the ACM 12th International Workshop on Data Warehousing and OLAP. pp. 41–48 (2009)
15. Fleckenstein, M., Fellows, L.: Data Analytics, pp. 133–142. Springer (2018). https://doi.org/10.1007/978-3-319-68993-7_13

16. Francia, M., Gallinucci, E., Golfarelli, M.: Social BI to understand the debate on vaccines on the web and social media: unraveling the anti-, free, and pro-vax communities in italy. Social Network Analysis and Mining **9**(1), 46:1–46:16 (2019). https://doi.org/10.1007/s13278-019-0590-x

17. Francia, M., Gallinucci, E., Golfarelli, M., Rizzi, S.: Social business intelligence in action. In: Nurcan, S., Soffer, P., Bajec, M., Eder, J. (eds.) CAiSE 2016. LNCS, vol. 9694, pp. 33–48. Springer (2016). https://doi.org/10.1007/978-3-319-39696-5_3

18. Golfarelli, M.: Design issues in social business intelligence projects. In: Zimányi, E., Abelló, A. (eds.) eBISS 2015. LNBIP, vol. 253, pp. 62–86. Springer (2016)

19. Golfarelli, M., Maio, D., Rizzi, S.: The dimensional fact model: a conceptual model for data warehouses. International Journal of Cooperative Information Systems **7**(2-3), 215–247 (1998)

20. Golfarelli, M., Rizzi, S.: A model-driven approach to automate data visualization in big data analytics. Information Visualization **19**(1) (2020). https://doi.org/10.1177/1473871619858933

21. Gorelik, A.: The enterprise big data lake: Delivering the promise of big data and data science. O'Reilly (2019)

22. Hilal, M., Schuetz, C.G., Schrefl, M.: Using superimposed multidimensional schemas and OLAP patterns for RDF data analysis. Open Computer Science **8**(1), 18–37 (2018). https://doi.org/10.1515/comp-2018-0003

23. Hitzler, P., Krötzsch, M., Rudolph, S.: Foundations of Semantic Web Technologies. Chapman and Hall/CRC Press (2010), http://www.semantic-web-book.org/

24. Hitzler, P., Krötzsch, M., Rudolph, S., Sure, Y.: Semantic Web: Grundlagen. Springer (2007)

25. Inmon, W.H.: Building the data warehouse. Wiley, fourth edn. (2005)

26. Keim, D.A., Andrienko, G.L., Fekete, J., Görg, C., Kohlhammer, J., Melançon, G.: Visual analytics: Definition, process, and challenges. In: Kerren, A., Stasko, J.T., Fekete, J., North, C. (eds.) Information Visualization - Human-Centered Issues and Perspectives, LNCS, vol. 4950, pp. 154–175. Springer (2008). https://doi.org/10.1007/978-3-540-70956-5_7

27. Kotu, V., Deshpande, B.: Data Science. Morgan Kaufmann, 2nd edn. (2019). https://doi.org/10.1016/B978-0-12-814761-0.00007-1

28. Kovacic, I., Schuetz, C.G., Schausberger, S., Sumereder, R., Schrefl, M.: Guided query composition with semantic OLAP patterns. In: Augsten, N. (ed.) Proceedings of the Workshops of the EDBT/ICDT 2018 Joint Conference. CEUR Workshop Proceedings, vol. 2083, pp. 67–74. CEUR-WS.org (2018), http://ceur-ws.org/Vol-2083/paper-11.pdf

29. Krötzsch, M., Weikum, G.: Editorial for special section on knowledge graphs. Journal of Web Semantics **37-38**, 53–54 (2016). https://doi.org/10.1016/j.websem.2016.04.002

30. Linstedt, D., Olschimke, M.: Building a scalable data warehouse with Data Vault 2.0. Morgan Kaufmann (2015)

31. Marcel, P.: OLAP query personalisation and recommendation: An introduction. In: Aufaure, M., Zimányi, E. (eds.) 2011. LNBIP, vol. 96, pp. 63–83. Springer (2011). https://doi.org/10.1007/978-3-642-27358-2_3

32. Marz, N., Warren, J.: Big Data: Principles and best practices of scalable real-time data systems. Manning Publications (2015)

33. Morgan, R., Grossmann, G., Schrefl, M., Stumptner, M., Payne, T.: VizDSL: A visual DSL for interactive information visualization. In: Krogstie, J., Reijers, H.A. (eds.) CAiSE 2018. LNCS, vol. 10816, pp. 440–455. Springer (2018). https://doi.org/10.1007/978-3-319-91563-0_27

34. Muñoz, L., Mazón, J., Pardillo, J., Trujillo, J.: Modelling ETL processes of data warehouses with UML activity diagrams. In: Meersman, R., Tari, Z., Herrero, P. (eds.) OTM 2008 Workshops. LNCS, vol. 5333, pp. 44–53. Springer (2008). https://doi.org/10.1007/978-3-540-88875-8_21

35. Nalchigar, S., Yu, E.: Business-driven data analytics: A conceptual modeling framework. Data & Knowledge Engineering **117**, 359–372 (2018). https://doi.org/10.1016/j.datak.2018.04.006

36. Nalchigar, S., Yu, E.: Designing business analytics solutions. Business & Information Systems Engineering **62**(1), 61–75 (2020). https://doi.org/10.1007/s12599-018-0555-z
37. Nalchigar, S., Yu, E.S.K., Obeidi, Y., Carbajales, S., Green, J., Chan, A.: Solution patterns for machine learning. In: Giorgini, P., Weber, B. (eds.) CAiSE 2019. LNCS, vol. 11483, pp. 627–642. Springer (2019). https://doi.org/10.1007/978-3-030-21290-2_39
38. Neuböck, T., Schrefl, M.: Modelling knowledge about data analysis processes in manufacturing. IFAC-PapersOnLine **48**(3), 277–282 (2015). https://doi.org/10.1016/j.ifacol.2015.06.094, 15th IFAC Symposium onInformation Control Problems inManufacturing
39. Oliveira, B., Belo, O.: BPMN patterns for ETL conceptual modelling and validation. In: ISMIS 2012, LNCS, vol. 7661, pp. 445–454. Springer (2012)
40. Oliveira, B., Santos, V., Belo, O.: Pattern-based ETL conceptual modelling. In: MEDI 2013, LNCS, vol. 8216, pp. 237–248. Springer (2013)
41. Ordonez, C., Maabout, S., Matusevich, D.S., Cabrera, W.: Extending er models to capture database transformations to build data sets for data mining. Data & Knowledge Engineering **89**, 38–54 (2014)
42. Peiritsch, A.R.: Starbucks' racial-bias crisis: Toward a rhetoric of renewal. Journal of Media Ethics **34**(4), 215–227 (2019). https://doi.org/10.1080/23736992.2019.1673757
43. Pozzi, F.A., Fersini, E., Messina, E., Liu, B. (eds.): Sentiment analysis in social networks. Morgan Kaufmann (2017). https://doi.org/10.1016/C2015-0-01864-0
44. Romero, O., Marcel, P., Abelló, A., Peralta, V., Bellatreche, L.: Describing analytical sessions using a multidimensional algebra. In: Cuzzocrea, A., Dayal, U. (eds.) DaWaK 2011. LNCS, vol. 6862, pp. 224–239. Springer (2011). https://doi.org/10.1007/978-3-642-23544-3_17
45. Russell, N., Van Der Aalst, W.M.P., Ter Hofstede, A.H.M.: Workflow patterns: the definitive guide. MIT Press (2016)
46. Russom, P.: Data lakes: Purposes, practices, patterns, and platforms (2017), https://tdwi.org/research/2017/03/best-practices-report-data-lakes, accessed: 05 August 2019
47. Saltz, J.S., Grady, N.W.: The ambiguity of data science team roles and the need for a data science workforce framework. In: Nie, J., Obradovic, Z., Suzumura, T., Ghosh, R., Nambiar, R., Wang, C., Zang, H., Baeza-Yates, R., Hu, X., Kepner, J., Cuzzocrea, A., Tang, J., Toyoda, M. (eds.) 2017 IEEE International Conference on Big Data. pp. 2355–2361 (2017). https://doi.org/10.1109/BigData.2017.8258190
48. Satyanarayanan, M.: The emergence of edge computing. Computer **50**(1), 30–39 (2017). https://doi.org/10.1109/MC.2017.9
49. Schäffer, M.: Modeling genome data processing pipelines. In: Plattner, H., Schapranow, M.P. (eds.) High-Performance In-Memory Genome Data Analysis: How In-Memory Database Technology Accelerates Personalized Medicine, pp. 31–53. Springer (2014). https://doi.org/10.1007/978-3-319-03035-7_2
50. Schuetz, C.G., Neumayr, B., Schrefl, M., Gringinger, E., Wilson, S.: Semantics-based summarisation of atm information: Managing information overload in pilot briefings using semantic data containers. The Aeronautical Journal (2019). https://doi.org/10.1017/aer.2019.74
51. Schuetz, C.G., Neumayr, B., Schrefl, M., Neuböck, T.: Reference modeling for data analysis: The BIRD approach. International Journal of Cooperative Information Systems **25**(2), 1–46 (2016). https://doi.org/10.1142/S0218843016500064
52. Schuetz, C.G., Schausberger, S., Kovacic, I., Schrefl, M.: Semantic OLAP patterns: Elements of reusable business analytics. In: Panetto, H., Debruyne, C., Gaaloul, W., Papazoglou, M.P., Paschke, A., Ardagna, C.A., Meersman, R. (eds.) OTM 2017. LNCS, vol. 10574, pp. 318–336. Springer (2017). https://doi.org/10.1007/978-3-319-69459-7_22

53. Schuetz, C.G., Schausberger, S., Schrefl, M.: Building an active semantic data warehouse for precision dairy farming. Journal of Organizational Computing and Electronic Commerce **28**(2), 122–141 (2018). https://doi.org/10.1080/10919392.2018.1444344
54. Seiter, M.: Business Analytics. Vahlen, 2nd edn. (2019)
55. Sharda, R., Delen, D., Turban, E.: Business intelligence, analytics, and data science: a managerial perspective. Pearson, 4th global edn. (2018)
56. Sherman, R.: Business Intelligence Guidebook. Morgan Kaufmann (2015). https://doi.org/10.1016/C2012-0-06937-2
57. Shi, W., Dustdar, S.: The promise of edge computing. Computer **49**(5), 78–81 (2016). https://doi.org/10.1109/MC.2016.145
58. Thalhammer, T., Schrefl, M., Mohania, M.K.: Active data warehouses: complementing OLAP with analysis rules. Data & Knowledge Engineering **39**(3), 241–269 (2001). https://doi.org/10.1016/S0169-023X(01)00042-8
59. Theodoroua, V., Abelló, A., Thieleb, M., Lehner, W.: Frequent patterns in ETL workflows: An empirical approach. Data & Knowledge Engineering **112**, 1–16 (2017)
60. Vaisman, A., Zimányi, E.: Data Warehouse Systems – Design and Implementation. Springer (2014)
61. Vrandečić, D., Krötzsch, M.: Wikidata: A free collaborative knowledgebase. Communications of the ACM **57**(10), 78–85 (2014). https://doi.org/10.1145/2629489
62. Wang, L., Schuetz, C.G., Cai, D.: Choosing response strategies in social media crisis communication: an evolutionary game theory perspective. Information & Management (2020). https://doi.org/10.1016/j.im.2020.103371, in press
63. Weiler, A., Grossniklaus, M., Scholl, M.H.: An evaluation of the run-time and task-based performance of event detection techniques for twitter. Information Systems **62**, 207–219 (2016). https://doi.org/10.1016/j.is.2016.01.003
64. Williams, S.: Business Intelligence Strategy and Big Data Analytics. Morgan Kaufmann (2016). https://doi.org/10.1016/C2015-0-01169-8
65. Zeng, D., Chen, H., Lusch, R., Li, S.: Social media analytics and intelligence. IEEE Intelligent Systems **25**(6), 13–16 (2010)

Discovering Actionable Knowledge for Industry 4.0: From Data Mining to Predictive and Prescriptive Analytics

Christoph G. Schuetz⬝, Matt Selway⬝, Stefan Thalmann⬝ and Michael Schrefl⬝

Abstract

Making sense of the vast amounts of data generated by modern production operations—and thus realizing the full potential of digitization—requires adequate means of data analysis. In this regard, data mining represents the employment of statistical methods to look for patterns in data. Predictive analytics then puts the thus gathered knowledge to good use by making predictions about future events, e.g., equipment failure in process industries and manufacturing or animal illness in farming operations. Finally, prescriptive analytics derives from the predicted events suggestions for action, e.g., optimized production plans or ideal animal feed composition. In this chapter, we provide an overview of common techniques for data mining as well as predictive and prescriptive analytics, with a specific focus on applications in production. In particular, we focus on association and correlation, classification, cluster analysis and outlier detection. We illustrate selected methods of data analysis using examples inspired from real-world settings in process industries, manufacturing, and precision farming.

C. G. Schuetz (✉) · M. Schrefl
Institute of Business Informatics – Data & Knowledge Engineering, Johannes Kepler University Linz, Linz, Austria
e-mail: christoph.schuetz@jku.at

M. Schrefl
e-mail: michael.schrefl@jku.at

M. Selway
Industrial AI Research Centre, University of South Australia, Adelaide, Australia
e-mail: matt.selway@unisa.edu.au

S. Thalmann
Business Analytics and Data Science Center, University of Graz, Graz, Austria
e-mail: stefan.thalmann@uni-graz.at

© The Author(s), under exclusive license to Springer-Verlag GmbH, DE, part of Springer Nature 2023
B. Vogel-Heuser and M. Wimmer (eds.), *Digital Transformation*,
https://doi.org/10.1007/978-3-662-65004-2_14

337

Keywords

Data mining • Data analytics • Predictive maintenance • Predictive quality control

1   Introduction

Modern production systems are increasingly data-driven, relying on the analysis of large amounts of data gathered from all kinds of sensors, databases, etc. Those data are collected and integrated for the purposes of "mining" for patterns in the data that allow to predict the future which, in turn, allows to shape the future by taking appropriate action. In general, the more data there are available, the more accurate become the predictions and the better decision-makers are able to act according to the circumstances, provided the appropriate methods of data analysis are applied correctly and can be run in a reasonable amount of time using the available computing resources.

Regarding terminology on data analysis, the terms "data mining", "predictive analytics", and "prescriptive analytics" are employed inconsistently throughout industry and academia. In this chapter, we use the term "data mining" to refer to the process of identifying patterns in historical data through the use of statistical methods. We use the term "predictive analytics" to refer to the employment of data mining for the prediction of future events, and we use the term "prescriptive analytics" to refer to the employment of data mining and predictive analytics to derive recommendations for action.

Before running data analysis algorithms, the data of interest for the analysis must be prepared accordingly, meaning that the data must be integrated from different data sources, cleaned by eliminating erroneous or incomplete records, and transforming the data into the required format. Data scientists typically spend a considerable amount of time with data preparation. Common tasks of data analysis are then analysis of associations and correlations in the data, classification, and cluster analysis. Analysis of associations in the data consists of the identification of patterns of events that frequently occur together. Regression analysis aims to find correlations in the data. Classification is the problem of collecting observations into various predefined categories. Cluster analysis, in turn, aims at finding different categories of observations in the data.

In this chapter, we provide a general overview and briefly discuss different aspects of data analysis using examples from precision dairy farming and manufacturing industries. In addition, we present two use cases from industry, the first describing predictive maintenance in the process industries, the second describing the application of predictive and prescriptive analytics in the context of quality control in manufacturing.

The remainder of this chapter is organized as follows. Section 2 discusses issues in data collection and preparation. Section 3 gives an introduction to common methods for data analysis. Section 4 presents a case of condition-based predictive maintenance in process industries. Section 5 presents a case of predictive quality control. Section 6 points the interested reader to further material on data mining and analytics. Section 7 concludes the paper.

2   Data Collection and Preparation

Before running data analysis algorithms, the data of interest for the analysis must be prepared accordingly. In general, data preparation consists of integrating different data sources, cleaning the data by eliminating erroneous or incomplete records, and transforming the data into the format required by a particular data analysis algorithm (Fig. 1). The specific data preparation process depends on the data sources and the employed type of data analytics infrastructure, e.g., data warehouse or data lake.

Preparing the data for the actual analysis is an important part of a data scientist's job description. Data scientists typically spend considerably more time on preparing the data than running sophisticated data analysis algorithms [29], which is sometimes dubbed the "80/20 dilemma" [38], for data preparation takes up roughly 80% of a data scientist's time. While often overlooked and arguably of lower prestige than the actual analysis task, data preparation nevertheless constitutes essential "knowledge work" [11, p. 4], being a prerequisite to gaining meaningful insights from data.

2.1   Data Cleaning, Integration, and Transformation

Collected raw data may have various quality issues, e.g., missing or inaccurate values, which distort analysis results and must be eliminated prior to data analysis. Sensor data

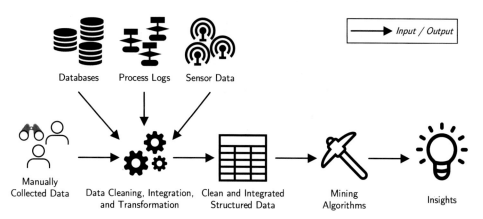

Fig. 1 Turning source data into insights requires data preparation to obtain data in a format suitable for the employed mining algorithms

in particular are subject to "noise", i.e., erroneous measurements. Statistical methods are typically required to fill gaps as well as to identify and correct wrong values in the collected data. For example, in precision dairy farming, accelerometers track animal activity over time but transmission errors cause gaps in the collected raw data. Stochastic processes may serve to fill the gaps in a series of accelerometer measurements over time [37, p. 299] while preserving the characteristics of the captured data and thus avoiding distortion of analysis results.

The data for the analysis are typically scattered over multiple sources that need to be integrated before running the actual data analysis algorithms. A single data source often does not comprise all the data required for running a certain algorithm. Databases, sensor data, process logs, but also manually collected data may all be of interest for the analysis. The various data sources typically have different data formats and follow different conventions when it comes to capturing the data, making data integration a non-trivial, time-consuming task. In precision dairy farming, for example, data of interest for the analysis of dairy farming data come from various types of sensors but also dairy herd management systems and external databases maintained by dairy herd improvement organizations [40, pp. 130–131].

Of particular importance in data integration is the consideration of data semantics and data provenance. Different data sources may capture measurements of the same phenomenon using different units of measurement, e.g., °F or °C for body temperature of dairy cattle. Furthermore, the same phenomenon may be captured using different methods, yielding results that are not easily comparable even though the same units of measurements are used. For example, the body temperature of dairy cattle can be obtained through measurement of skin temperature on various body parts using infrared cameras, or through the measurement of rectal and vaginal temperatures using thermometers [18]. When analyzing animal body temperatures obtained from different data sources, differences in measurement methods must be considered, otherwise the drawn conclusions will be wrong.

Every data analysis algorithm requires the data in a specific format, which typically differs from the data source's original format. Thus, in order to obtain data in the format required by a specific algorithm, the original data must be transformed accordingly. A common type of transformation is *data reduction*, which reduces the size of the data in order to render large data sets more manageable for computing-intensive algorithms. For example, when tracking animal movement in precision dairy farming, sensors capture the precise location of an animal every second, resulting in a large quantity of data. The high frequency of sensor readings is important to compensate for transmission errors and noise. For prediction and planning, however, working with aggregated data often suffices, or even represents a requirement for meaningful analysis. In precision dairy farming, for example, detection of extraordinary animal behavior, e.g., an animal lying down in an area where the animal is normally expected to walk around, may yield interesting results into an animal's health status. Rather than analyzing precise animal positions, however, the analyst may look at total dwelling times per functional area, e.g., feeding area or milking area, and hour of day, as well as the duration of activities, e.g., walking or lying, per functional area and hour of day.

Furthermore, having animal positions in terms of functional areas rather than coordinates renders data from different farm sites comparable.

Figure 2 illustrates some of the preparation tasks for data analysis in precision dairy farming [40]. The dairy herd management system keeps track of calving events, which determine animals' day of lactation: A calving event marks the first day of lactation of a dairy cow. Data transformation derives from the calving dates the day of lactation of each animal on dates within the time frame of interest for the analysis. For example, an animal that calved on 23 March 2020 will be in its 84th day of lactation on 12 June 2020. Movement sensors capture the position of individual animals at a high frequency. Data transformation aggregates individual position readings by animal and functional area in order to obtain the duration in minutes of an animal's dwelling in a certain functional area per hour of the day on a certain date. For example, in the 12th hour of 12 June 2020, the animal with identification AT1234 spent 14 min in a feeding area and 46 min in a resting area. The body condition score (BCS) can be determined manually by veterinarians or determined (semi-)automatically, e.g, by processing images from thermal cameras [16]. An animal's BCS will not be subject to abrupt changes, leading to a lower frequency of readings, which means that measurements of BCS values are available for certain days only and values are missing for the remaining days. In this case, data cleaning consists, first, of completing missing values by using an older BCS measurement for days with missing values. For example, the BCS measured on 11 June 2020 fills the gap for 12 June 2020, with a continuation column indicating "freshness" of the value for plausibility checking. Furthermore, the BCS can be measured manually or with assistance from thermal cameras, the former leading to subjective measurements [16]. In data analysis, comparing values obtained through different means may lead to wrong conclusions and, thus, only BCS values obtained through thermal imaging are used. Finally, the data from the various sources are integrated into a single table which can be used by data analysis algorithms.

In summary, obtaining the appropriate format of the data and thus the right representation of the problem required for a particular data mining and machine learning algorithm is a time-consuming task of paramount importance. *Representation learning* is a type of machine learning that aims to reduce the effort for manual *feature engineering*. Deep learning is a prime example of representation learning (see [15]). While deep learning might never entirely replace other types of machine learning, deep learning may play an important role in data preparation for predictive analytics [49], e.g., by improving quality of sensor readings or interpolating missing sensor readings. Furthermore, deep learning can be an important part of a larger data analysis pipeline that includes various types of machine learning, e.g., deep learning could be employed for processing thermal images in order to automatically obtain the BCS of dairy cattle while other techniques could be used to conduct further analyses with the obtained scores.

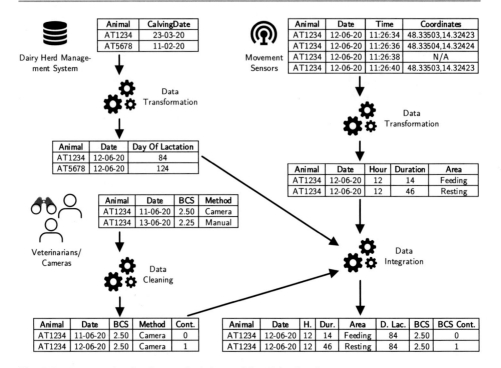

Fig. 2 Data preparation for data analysis in precision dairy farming

2.2   Data Analytics Infrastructure

Regarding the infrastructure for data collection and storage, different options exist that are mutually *non*-exclusive. Different storage solutions vary with respect to the degree to which data preparation is taken over by the system itself or left to the analyst. The choice of data analytics infrastructure depends on various considerations, e.g., the size of the data to be analyzed and the necessity for getting analysis results in real time.

A *data warehouse* stores integrated, clean and consistent data in the appropriate format for data analysis. Extract, transform, and load (ETL) processes take over data preparation, extracting the data from the sources, cleaning and transforming the data in the process, and loading the thus preprocessed data into the data warehouse. We refer to Vaisman and Zimányi [51] for a comprehensive introduction to data warehousing.

A *data lake* stores raw data in the original format for future analysis. The rationale behind the data lake is twofold. On the one hand, the sheer volume and velocity of the arising data, e.g., in sensor networks, may prohibit running in time the ETL processes necessary for populating a data warehouse. Data lake environments typically follow an approach referred to as extract, load, and transform (ELT) when it comes to getting the data into the data lake, the transformation taking place after loading, usually on demand at analysis time. Compared to traditional data warehousing, a data lake shifts the burden of data preparation largely to the analyst and the application running the analysis, respectively. On the other hand, the future value of data may not be known to an organization at present. A data lake preserves potentially valuable data which the organization may tap into in the future as new analysis methods and additional computational resources become available.

In practice, data warehouse and data lake may complement each other; the traditional data warehouse paradigm has been adapted to cope with high volume and velocity of big data. A proposed architecture for sensor data warehousing [8], for example, employs stream processing to prepare sensor data for long-term storage in a data warehouse; a data lake complements the data warehouse by storing the raw sensor data. In this setting, data stream processing enables real-time analysis using predictive models developed based on historical data stored in data warehouse and data lake. A data analytics infrastructure may likewise employ *edge computing* [41], shifting data processing to the periphery of the network, e.g., by having sensor equipment perform data preparation tasks.

3   Data Analysis

Broadly speaking, data mining aims to identify patterns in historical data which can be employed to predict future events (predictive analytics) and take corrective action in order to improve the status quo (prescriptive analytics). More specifically, historical data serve as "training" data to train (or learn) models, which are then applied to new data in order to make predictions (Fig. 3) and possibly change the course of action by reacting to the predictions accordingly. Learning a model from historical data is also referred to as *machine learning*; the distinction between data mining and machine learning, predictive and prescriptive analytics is not always clear-cut in literature and product descriptions.

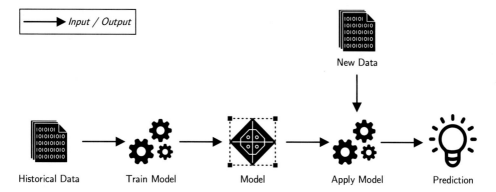

Fig. 3 Historical data serve to train models for making predictions over new data

3.1   Association and Correlation

Intuitively, association refers to events frequently occurring together whereas correlation refers to a (typically linear) trend between the occurrence of events (see [4] for more information). Both association and correlation, however, must not be confused with causation. Just because two events frequently occur together does not imply that one event causes the other.

Mining for association rules [2] and mining for sequential patterns [1] are examples for the analysis of association in data. An association rule indicates that the occurrence of one item/event frequently coincides with the occurrence of another item, e.g., if temperature and humidity in the barn are both high then an animal shows low activity. A sequential pattern is similar to an association rule but also considers the sequence of the items/events occurring together. Mining for sequential patterns has been employed for predictive maintenance [35].

Support and *confidence* are measures of how interesting an association rule is. Support is the frequency of the co-occurrence of the items/events in an association rule; low support suggests an exceptional case. For example, if the co-occurrence of high temperature and humidity on the one hand with low animal activity on the other hand is an infrequent, exceptional observation, the support of an association rule linking low animal activity to high temperature and humidity will be low. Confidence, in turn, is the certainty of the association rule, i.e., how often the association rule is true. For example, if the occurrence of high temperature and high humidity only coincides with low animal activity in a third of cases with high temperature, the association rule is quite uncertain.

In order to judge how interesting an association rule is, the correlation between events must also be taken into account besides confidence and support. In this regard, a common measure is the *lift* of an association rule. Despite high confidence and support, an association rule may actually indicate a negative correlation. For example, a confidence of 67% for an association rule linking extremely high animal activity with a normal body condition score (BCS) suggests high certainty of the link between those two variables. Yet, if 75% of all animals have a normal BCS anyway, then high animal activity negatively correlates with normal BCS.

Regression analysis aims to find correlations in the data, which can be leveraged for making predictions. In the context of machine learning, "regression" typically refers to the prediction of a continuous or categorical value from one or more independent variables. Different types of regression analysis exist, the most basic and widely known being *linear regression*. Intuitively, in a two-dimensional space, linear regression aims to find a line that best fits a given set of data points (Fig. 4). Multiple linear regression extends simple linear regression with only one predictor to allow for more than one independent variable (or predictor). Furthermore, nonlinear regression aims to find an arbitrary, nonlinear function that best fits a given set of data points. It is worth noting that correlation does not imply causality: Just because two events frequently occur together does not mean that one causes the other.

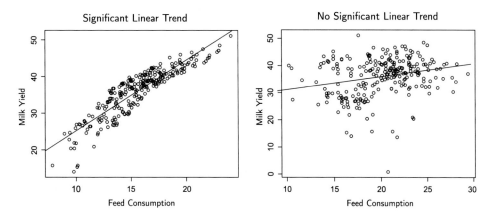

Fig. 4 Linear regression applied to two data sets: The data set on the left-hand side shows a significant linear correlation between feed consumption and milk yield whereas the data set on the right-hand side shows no apparent linear correlation and linear regression is therefore misapplied

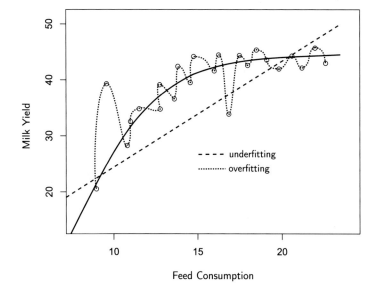

Fig. 5 Overfitting and underfitting

A general problem in data mining and machine learning is to avoid *overfitting* and *underfitting* of models to the training data, which both reduce the predictive power of models. Consider the data points in Fig. 5. The dotted line represents an overfitted model that works perfectly on the training data but will struggle with correctly predicting the milk yield from new observations of feed consumption by animals. The dashed line, on the other hand, represents a linear prediction model that does not accurately represent the distribution of the training data, and will likely fail to make correct predictions. A logarithmic function is likely the best fit for these observations. Likewise, the diagram on the right-hand side in Fig. 4 can be considered an underfitted model of the data points.

In order to detect overfitting in learned models, part of the available historical data are usually kept back as test data while the other part of the data serve as training data for learning a model. The predictive power of a learned model is then evaluated against the test data by looking at the proportion of values in the test data that the previously learned model would have predicted correctly. For example, a model that predicts milk yield dependent on feed consumption can be learned from a randomly chosen 80% of the observations in the available historical data. The remaining 20% of the data constitute the test data, which serve to evaluate how well the model predicts the actual milk yield for the known measurements of feed consumption that were not used for training the model. The coefficient of determination, known as R^2 ("R squared"), often serves to evaluate regression models, representing the proportion of the variance of the dependent variable explained by the predictors. In addition to training and test data, a validation data set is often used to evaluate different predictive models learned over the same training data; the test data then serve to evaluate the finally selected model.

The process of splitting the data into training and test data with subsequent learning and evaluation of a model can be repeated multiple times. In each iteration step, the split between training and test data varies. The final model is then trained over the entire data set. The learned model is expected to generalize well to new data if the models trained on the different subsets performed well in making predictions over the corresponding sets of test data. In this regard, *random subsampling* refers to the repeated learning of models using varying, randomly chosen samples from the available historic data for training the model, with the remainder of the data serving as test data. Similarly, *k-fold cross-validation* refers to repeated learning of models from varying sets of training data. To this end, the data set is first split into k subsets. In subsequent learning iterations, each of the k subsets alternately serves as the test data while the other subsets constitute the training data.

3.2   Classification

A *classifier* aims to predict the category that an observation belongs to based on a number of independent variables. The independent variables are also referred to as *predictors* because they can be used to predict the outcome of a dependent variable, which in case of classification is categorical, i.e., a variable, the possible values of which are categories with no sensible ordering, e.g., healthy or sick. Examples of classification problems are prediction of equipment malfunctioning or development of a disease in livestock based on sensor readings. In these cases, classifiers would aim to predict whether a piece of equipment will fall into the *malfunctioning* category based on measured machine characteristics or whether an animal will fall into the *sick* category based on the animal's behavior and measured health parameters.

A widely employed method for classification is *logistic* regression. Logistic regression aims to explain variance of a categorical dependent variable. Logistic regression with a dependent variable that can take on two possible values is referred to as *binomial* or *binary* logistic regression—arguably the most prominent type of logistic regression. Logistic regression with a dependent variable that can assume more than two possible values is referred to as *multinomial* logistic regression. The independent variables, which are supposed to explain the variance of the dependent variable, can be categorical or continuous. For example, in predictive maintenance, logistic regression may serve to predict whether a machine will *fail* or *not fail* based on the composition of oil samples drawn from the machine [34].

In precision dairy farming, logistic regression may be used to predict whether an animal is at risk of developing an illness. Prior to the diagnosis of ketosis—an illness that results from an energy deficit –dairy cattle show reduced activity and produce less milk [43], among other signs of illness. A binary logistic regression model can be developed predicting health status of animals based on measurements obtained through the use of modern sensor data tracking activity, rumination, and milk yield [42]. Figure 6 illustrates the potential use of (binary) logistic regression for prediction of an animal's health status based on walking distance and milk yield. The scatter plot shows (simulated) past measurements for walking distance and milk yield of individual animals. The color of the dots indicates the animal's determined health status a couple of days after the respective measurements were made. The line represents the *decision boundary*, slope and intercept of which have been obtained through the use of logistic regression. Intuitively, the decision boundary splits the scatter plot into two parts: One consisting mostly of observations from healthy animals, the other from sick animals. New data points that lie underneath the line in Fig. 6 will then be classified as animals that are at risk of developing ketosis.

Decision tree learning [39] aims to identify rules for the prediction of the value of a dependent variable based on a number of predictors. The outcome of performing decision tree learning on a data set is a *decision tree*. When the dependent variable is categorical the decision tree is also referred to as *classification tree*, which describes a multi-step procedure where each step corresponds to a check of a predictor's value to classify an object into a category. Consider the data set from Fig. 6, which represents observations of animals' health status, milk yield, and walking distance. From that data set, the classification tree shown in Fig. 7 can be learned, which explains variation of the dependent variable health status by the independent variables milk yield and walking distance. The decision tree subsequently may serve to predict whether an animal is at risk of developing ketosis, which boils down to the following rules.

```
1. IF Walking Distance >= 3270 THEN Healthy
2. IF Walking Distance < 3270 AND Milk Yield >= 20 THEN Healthy
3. IF Walking Distance < 3270 AND Milk Yield < 20 THEN Sick
```

Risk of Ketosis

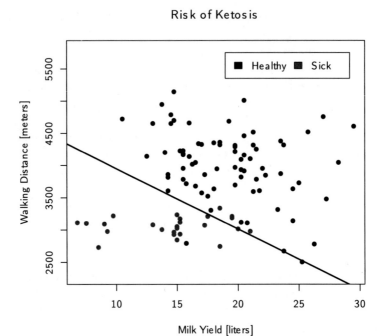

Fig. 6 Decision boundary for classification of animals into healthy and sick dependent on walking distance and milk yield using binary logistic regression

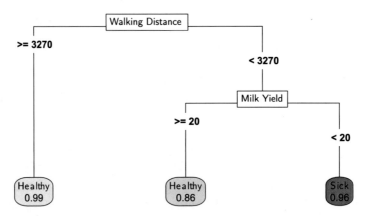

Fig. 7 Decision tree for classification of animals into healthy and sick dependent on walking distance and milk yield: 99% of animals that walked a distance greater or equal to 3270 m are healthy. Of animals that walked a distance less than 3270 m, 86% of animals with a milk yield greater or equal to 20 l are healthy while 96% of animals with a milk yield smaller than 20 l are sick

In order to construct a decision tree for classification, learning algorithms repeatedly split a data set based on one of the independent variables. Each split should most accurately partition the data set into categories from the set of possible values of the dependent variable. Different measures for deciding on the best split can be employed, e.g., Gini coefficient and information gain.

Decision tree learning is prone to overfitting, often producing complex, deep decision trees with many splits that work well on the training data but fail to classify new data correctly. In order to reduce overfitting, decision trees are often subject to *pruning*, i.e., "cutting off" branches in the tree that are apparently the result of noise in the data. Furthermore, instead of using an individual decision tree, classification may be based on a *random forest* [6], which consists of multiple decision trees, the classification results of which are combined in order to reach a final decision on an object's classification. Random forests aim to reduce the problem of overfitting. The random forest method is an *ensemble method* for classification. *Ensemble* methods, in general, aim to produce more accurate predictions through a combination of multiple learned models.

Other popular methods for classification are naïve Bayes classifiers and support vector machines. A naïve Bayes classifier is a simple form of Bayesian network (see [27] for more information), assuming independence between the predictors. A support vector machine maps observations into a multidimensional space; the method is similar to logistic regression. Support vector machines can also be employed for regression analysis in order to predict the outcome of a continuous, non-categorical dependent variable.

Another versatile machine learning technique with applications in classification, but also regression analysis, is the *artificial neural network* (ANN). Deep learning (see [15] for more information) goes beyond traditional, single-layer ANNs by introducing multiple hidden layers of neurons. Deep learning has proven effective in areas where other techniques traditionally struggle, e.g., image processing, speech recognition, and natural language processing. A general disadvantage of deep learning, however, is its "black box" nature: The predictions made by deep learning are generally not well explainable. Efforts towards *explainable AI* aim to alleviate the problem of the lack of transparency regarding the decisions, e.g., by incorporating reasoning (see [9] for an overview).

Choosing the best classifier from a range of options requires adequate evaluation methods; we refer to Han et al. [17, p. 364 ff.] and Witten et al. [52, p. 161 ff.] for a more comprehensive overview of evaluation techniques. Splitting the available historic data into training data and test data is important to properly evaluate the predictive power of classifiers. Random subsampling and k-fold cross-validation, among others, can be employed. Typical measures for evaluating classifiers include accuracy, precision, and recall. Consider a classifier that classifies observations into one of two categories, e.g., *healthy* and *not healthy* (sick). In Fig. 6, the scatter plot shows 75 observations belonging to healthy animals, 25 observations to sick animals. True positive (true negative) observations are those observations that have been correctly classified as healthy (not healthy) by the trained model. Conversely, false positive (false negative) observations are those observations that have been incorrectly classified as healthy (not healthy). The logistic regression classifier in Fig. 6 on the plotted data set would have detected 73 true positives and 20 true negatives as well as 5 false positives and 2 false negatives. Accuracy (or recognition rate) then refers to the proportion of correctly classified observations (true positive and true negative) among all observations. Conversely, error rate is the proportion of wrongly classified observations. Recall (or sensitivity) refers to the proportion of true positive classifications among all observations that are actually positive, whether classified positive or not (i.e., true positive and false negative). Precision refers to the proportion of true positive classifications among all observations that are classified as positive (i.e., true positive and false positive). Specificity refers to the proportion of true negative classifications among all observations that are actually negative, whether classified negative or not (i.e., true negative and false positive) [17, p. 365]. The logistic regression classifier in Fig. 6 has an accuracy of 93%, an error rate of 7%, a recall rate of $73 \div 75 = 97.3\%$, a precision of $73 \div (73 + 5) = 93.6\%$, and a specificity of $20 \div (20 + 5) = 80\%$.

Global measures of classifier performance may not suffice to correctly judge the reliability of individual predictions made by a classifier. For example, if an observation is unlike any observation in the training data, the classifier's prediction for that observation will not be reliable. Furthermore, common fallacies of statistical reasoning may lead to misinterpretation of the results. For example, the base rate fallacy may lead decision-makers to overestimate the probability of an outcome due to low base rate: Using a classifier for predicting scrap items with 90% recall, with a specificity of 95% and, conversely, a false positive rate of 5%, and assuming a scrap rate of 1%, i.e., only 1% of the overall produced items is actually scrap, then the probability, according to Bayes' theorem, of a predicted quality problem indicating an actual scrap item is just $(90\% \cdot 1\%) \div [(90\% \cdot 1\%) + 5\% \cdot (1 - 1\%)] = 15.4\%$.

3.3   Clustering and Outlier Detection

Clustering aims to find groups of objects or observations with similar characteristics. On the one hand, the identification of clusters in the data, e.g., groups of animals with similar feed consumption, is interesting in its own right in order to get an overview of the data. On the other hand, clustering may serve as a preparatory step for other algorithms, notably outlier or anomaly detection.

One of the most basic clustering methods is k-means clustering, which completely partitions a set of objects into a specified number k of clusters in an *unsupervised* manner, i.e., the training data are *not* manually collected into the "correct" clusters, which would then be employed for learning common characteristics of members of a cluster. The main challenge in applying data mining algorithms often lies in tweaking the parameters, and k-means clustering is no exception. In the case of k-means clustering, the challenge is to find the optimal number k of clusters. For example, the number k could be chosen such that the similarity of objects within each cluster is highest. Figure 8 shows the result of performing k-means clustering on the daily feed compositions offered to dairy cattle with the parameter $k = 4$

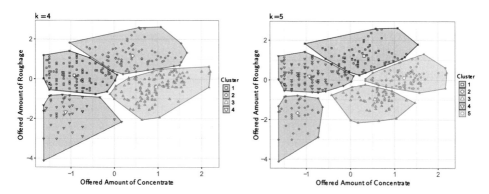

Fig. 8 Using k-means clustering to collect feed compositions into k clusters based on the contained amount of roughage and concentrate. The optimal number k of clusters can be determined using different methods

and $k = 5$, respectively. Each data point represents a composition of roughage and granulate offered to an individual animal on a particular day. Those compositions are clustered into four and five clusters, respectively, depending on the value of the parameter k.

Clustering-based techniques for outlier detection identify unusual observations by examining an observation's distance to previously identified clusters: observations that do not belong to any cluster (or a small cluster of atypical objects) are considered outliers or anomalies. Consider, for example, the industrial production of a certain type of item, e.g., a propeller [44]. Each produced individual item deviates from the ideal in some aspects. Clustering may help to find out "normal" deviations. Every item not to fall into a cluster could be problematic in terms of quality control. Statistical analysis and analysis of the "neighborhood" of an observation can also be used to determine if the observation is an outlier. Likewise, deep learning may serve for outlier detection [32].

4 Use Case: Condition-Based Predictive Maintenance

Asset maintenance is a major cost in asset intensive industries and has a large impact on business performance [30]. Condition-Based Maintenance (CBM), more recently termed Condition-Based Predictive Maintenance (CBPdM), aims to improve asset maintenance by using analytics to predict when is the best time to initiate maintenance based on equipment condition. Compared to traditional failure-driven, corrective maintenance (in which a response is made only after a failure has occurred), and time-based preventative maintenance (in which maintenance is performed periodically based on usage or *mean time between failure*), CBPdM anticipates failure events allowing timely, proactive maintenance to be performed [3]. The result is a reduction in equipment downtime and the number of urgent maintenance work-order requests, while improving asset performance, reliability, budget planning, and, ultimately, costs—both maintenance costs (through efficiency gains) and operational costs (by minimising the impact to production, such as downtime due to scheduled maintenance or unexpected failures) [23, 28].

According to [3], such an approach should be feasible as 99% of equipment failures are preceded by indications of failure. Given the current drive towards I4.0 and, when properly managed, the greater access to data that it provides, CBPdM is even more feasible now than in the past. Indeed, predictive maintenance was expected to reach an annual revenue of US$24.7 billion in 2019 [36] due to improved analytics capabilities and the data connectivity provided by I4.0 and IIoT (Industrial Internet of Things). Such is the link between CBPdM and I4.0, it has come to be known by the term Maintenance 4.0 [24].

While standards for CBM has been around since early 2000s and some aspects are well studied and even standardised (e.g., vibration monitoring of pumps [5] and various equipment categories [19]), increases in computing capabilities and the connectivity of I4.0 provide new ways of combining data and performing novel analytics that provide additional business value. For example, equipment typically incorporates multiple sensors of the same or different types. These different data sources can be combined to provide more accurate results and to differentiate sensor faults vs. equipment faults. Moreover, I4.0 provides the platform for analytics to span across multiple equipment and entire (sub-)systems, especially when supported through Digital Twins.

A digital twin provides the required context to the operational and condition history of the equipment, its usage, and performed maintenance activities [21]. This includes a broad range of contextual information, such as functional breakdown structures, equipment specifications and configuration, sensor configuration, process related information, maintenance management, etc. Such context helps to achieve better accuracy for predictions in CBPdM as analytics can be performed with a broader focus to, for example, take into account parameter configuration changes that may impact the analysis of sensor data. Moreover, simulations can be performed on the digital twins allowing real-world measurements to be tracked relative to the ideal model values. In this context CBPdM, can lead to more optimal decision-making, improved intelligent industrial operations, and creation of business value.

A broadly adopted standard for CBM is OSA-CBM (Open System Architecture for Condition-Based Maintenance) [50], an implementation specification for ISO 13374 (Condition Monitoring and Diagnostics of Machines) [20]. Together, the two standards provide a reference point for implementing CBM (equally CBPdM) systems, describing CBM in terms of six functional blocks that perform analytics at different levels, illustrated in Fig. 9. The six functional blocks cover the processing required from data collection and fault detection, through the generation of specific maintenance actions. Higher-level, strategic recommendations to operations and maintenance may also be generated, such as optimization of production planning, maintenance planning and scheduling (dynamically scheduled inspections, etc.) and spare parts management [22]. This leads to *prescriptive maintenance*, the most advanced of maintenance strategies, in which further guidance to maintenance activities is provided [26]. Prescriptive maintenance requires advanced analytics and reasoning

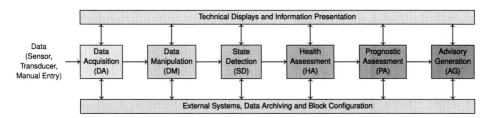

Fig. 9 CBM Data Processing Function Blocks. (Adapted from ISO 13374-1 [20])

capabilities to not only identify an appropriate pre-planned maintenance activity to be raised in the maintenance management system, but to possibly adapt or modify the maintenance activity based on contextual factors.

Briefly, the six functional blocks are as follows [20]:

DA A Data Acquisition block is closest to the data source and outputs collected data, be it manually captured or sensor-based, in conjunction with context information, such as the time, sensor configuration and calibration, data quality, etc.

DM A Data Manipulation block performs processing of the data from DA blocks, and possibly other DM blocks, such as conversions (e.g., time domain to frequency domain), signal processing, aggregation and averaging, feature extraction, and computation of virtual sensor readings (e.g., differential pressure based on inlet and outlet pressure readings).

SD State Detection identifies abnormalities (in sensor, equipment, or system) and rates them in accordance with baseline profiles and operational tolerances, possibly generating alerts and alarms based on the identified state or abnormality zone, e.g., 'normal', 'alert', or 'alarm'.

HA Health Assessment blocks diagnose faults and failures (and their likelihood), and generate assessments of the current health grade of the equipment or process; possibly including explanations and evidence.

PA Prognostic Assessment attempts to predict future health states and failures based on current health, historical information, and planned or projected usage loads; explanations and evidence may also be generated.

AG Advisory Generation blocks produce actionable items to operations and maintenance such as maintenance work requests (e.g., for inspections, repairs, or replacement of equipment) and notifications to operations of imminent or high-likelihood of failure at current workloads.

The data flow between the functional blocks is not a strict pipeline as illustrated in Fig. 9, rather, each block and its associated analytics may make use of information generated by any preceding block (including other blocks of the same type). Additionally, information from other systems (operational, maintenance, and business) may be incorporated into the analytics at each level. When paired with appropriate IT/OT architecture [25], CBPdM and Condition-based Operations (cf. [31]) is enabled in various constellations of systems to improve decision making for operations, maintenance, and the wider business. Moreover, such an architecture supports the distribution of analytics around the organisation allowing both event-driven, online analytics and slower batch analysis. Leveraging IIoT and the edge computing paradigm, for example, timely, online analytics may be performed at the 'shop-floor' from local data (sensor measurements, manual inspections, etc.), while the original data and generated outputs are published to data warehouses or data lakes through which longer-term analytics are performed. Results from higher-levels can then be pushed back down to analytics at the lower levels, creating a feedback loop. For example, a classifier used at the 'shop floor' may have its parameters tuned or be completely replaced to improve accuracy as a result of analysis performed at the higher-level—determined by subsequently scheduled inspections confirming or refuting the predictions produced at the shop-floor, which is not accessible by the on-line algorithm at the time of classification.

The different functional blocks perform analytics based on any or all of the techniques introduced in Sect. 3. Data preparation is typically performed a priori as a data scientist/analyst is needed to make sense of the available data. They will typically begin by exploring the (historical) data, identify correlations and associations, and identify appropriate analytical techniques to achieve the desired result [14]. The determined preparation steps may then be embedded into automated transformations as DM blocks, while the determined analytics and their configuration are implemented in appropriate functional blocks.

While the DA block does not perform analytics itself, the collection of data impacts the analytics performed by the subsequent blocks. For example, data collected manually via inspections will have time gaps compared to continuous monitoring from sensors. As discussed in Sect. 2, such discrepancies need to be taken into account when preparing and transforming the data for analytics purposes. DM blocks may perform such automated preparation tasks, which may be performed at the sensor (in edge computing, for example) or elsewhere depending on the architecture.

SD blocks may perform classification, clustering, and outlier detection for fault or anomaly detection. For example, logistic regression techniques may be used to determine the 'normal', 'alert', and 'alarm' states of measurements (if not imposed by operating parameters) or basic 'fault' states of the equipment. Outlier detection may be used to identify anomalous sensor readings or sensor faults. Outliers may be caused by device malfunction or sensor fault and it is important to determine which, as it may impact subsequent analytics.

HA blocks may also use classification approaches to diagnose faults and determine the health grade ('good', 'poor', etc.) of equipment. Decision tree approaches, for example, can be useful for generating explanations of fault diagnoses; in the vibration analysis of a pump, a certain path through the tree might identify bearing failure while another may identify cavitation at the impeller. This information can be presented to a user, allowing them to verify the diagnosis if necessary. The classification of health grade may be based on other states and expert knowledge of types of state or failure. For example, equipment having (mildly) dirty oil might still be rated as 'satisfactory' health while a deteriorating bearing may be considered 'poor' health.

PA blocks extend the techniques to predicting future states, particularly estimating Remaining Useful Life (RUL). Techniques such as regression and artificial neural networks (among others) may be used for this purpose, such as in [53]. Since equipment often exhibits a progression of failures until complete or catastrophic failure is reached, it is possible to predict future failure modes given appropriate precursor data. In addition, increases in the frequency of anomaly occurrences can indicate imminent failure. Therefore, ensemble approaches may be useful in combining different sources of evidence to identify the most likely future state. As with HA blocks, decision tree approaches may also be useful for generating explanations of the predictions.

AG blocks provide the highest level of functionality in CBM systems, using analytics to produce actionable items. The techniques applied here include those discussed previously (or combinations thereof) as well as others that are not discussed in this overview. For example, previous fault diagnoses and predicted failure modes may lead to the scheduling of a maintenance work request aiming to rectify an error before total failure occurs, which may require additional analytics or reasoning capabilities (e.g. planning, configuration, case-based reasoning) to adapt or synthesize maintenance tasks. Similarly, recommended actions may be posted to operations to suggest reducing the load on a piece of equipment to extend its life until the maintenance work can be fulfilled.

AG blocks may also incorporate data from other business systems to integrate external factors into its decision making, such as safety, environmental, budgetary, operational, etc., information. At this level, AG use all this information to optimize elements of operations and maintenance: for example, capability forecasts of whether a production order can be fulfilled and optimization of maintenance strategies. In the latter category, analytics can be used to determine, for example, that the historical 'run-to-failure' maintenance strategy might be the most cost effective for specific types of equipment or specific functional locations within a plant or facility.

CBPdM incorporates analytics at all levels of operations and maintenance activities by using and combining the various analytics approaches as appropriate for a given task. Analytics for CBM is dependent on the category of equipment, availability of sensors and data, the type of output that is desired, etc. There is no one size fits all approach as it requires data scientists, in conjunction with engineering experts, to explore the possibilities and apply appropriate techniques to the problem at hand. Alongside digital twins, data analysis is integral to achieving the goal of CBPdM to reduce asset maintenance costs and improve overall business performance in the era of I4.0.

5   Use Case: Predictive Quality Control

Most of the use cases in industrial settings focus on making predictions about the machines. Less attention has been brought to making predictions about the produced items. Yet, a prediction model can be directly integrated into process control that serves to adjust the production process and subsequent handling steps depending on the prediction of the quality of the produced product. This use case occurs frequently in industry and should receive more attention from academia.

Context The industrial application context of the presented use case is a high-precision metal manufacturing production line. A high-precision metal part is produced by turning and milling machines. The produced items must fulfil very high quality standards assessed in a certification procedure as they are used in safety-critical areas. Hence, intensive and challenging quality control needs to be performed for every produced item. The quality control of the produced items is conducted in an external lab specialized and certified for high-precision quality analysis. Thereby, around 40% of the produced items fail in the quality assessment as the precision criteria are very high. Overall, the transportation costs and the quality assessment costs are four times higher than the actual material and production costs. Further, prediction of low product quality during the production process could lead to an immediate sorting out of the affected piece and skipping the remaining production steps. Thus, knowing in advance if a produced item would fail could save money, resources, and time as less products need to be reproduced.

Solution Approach The core of the solution is to bring more flexibility into the management of production processes and to continuously analyze production data [48]. In contrast to other types of analysis, the aim is to continuously analyze production data for individual products not only for improvements in the design and planning phase, but also during runtime [47]. This is an important difference, as the object of analysis in this case is not the machine in the first place, it is the produced item and its quality characteristics. Research showed that complementing machine data with product or even market data is a promising way to deal with missing data [12]. To continuously collect the data from the production process and also to feed the results from the data analysis back to the process control, the CENTURIO process engine can be employed, which can adapt the process models not only during design, but also during the runtime of manufacturing processes [33]. CENTURIO manages a central data repository which is connected to the process instances and which allows customized views and aggregations for process analysis.

Based on the 37 data channels of the production machines and 13 quality measures from the quality lab, a forecasting model was created, predicting the quality of the produced items. For this purpose, the process engine requests a quality prediction at the end of the production process from this model. Based on the quality criteria and together with the domain experts from engineering, two classes were defined: 1) "Good" quality and 2) "scrap" products. It should be noted that scrap products are directly sorted out, and that all other produced items are checked in the quality lab.

The problem at hand corresponds to multivariate time series classification. In order to build a predictive model, features were extracted using the *TsFresh* feature extraction method. Different classification approaches using selected classifiers (neural networks, support vector machines, Gaussian process, random forest, naïve Bayes etc.) were tried. For the training data set (N = 600), random forest performed best.

In an evaluation with 179 items, the initial model showed an accuracy of 81%. The accuracy could be improved having larger training data sets. Further, according to the domain experts some specific sensors could provide more meaningful data about the production processes and help in increasing the production quality. An experiment was conducted involving machine operators. In the course of the experiment, machine operators should predict product quality based on observations and visual inspection of the product. In this experiment, humans had a 67.86% accuracy in classifying the same high-precision metal products compared with the outcome of the high precision lab [10]. Thus, the predictive model outperformed the production workers in terms of prediction accuracy clearly.

To produce 1000 items of reasonable quality and assuming a scrap rate of 40%, 1667 items have to be produced and checked in the quality lab. Assuming € 4 production costs and € 16 quality inspection costs, total costs are expected to be € 33 340, taking the perspective of the developed predictive model and a accuracy of 80%. Based on the assumption, that 20% of the "good" items are classified as scrap and sorted out, 2084 items (€ 8 336 production costs) have to be produced. 1250 items need to be checked in the quality lab (€ 20 000 quality inspection costs). Overall, the costs are reduced by around 15% to € 28 336 for a predictive model with 80% accuracy, with a model having 90% accuracy the reduction would be around 24% to € 25 184 for the assumed 1000 items. Please find an overview in Table 1.

Discussion and Outlook The combination of high quality standards and demand for high flexibility require increasingly high adaptivity of the production process. Hence, data-driven

Table 1 Cost overview

	No Predictive Model	Model with 80% Accuracy	Model with 90% Accuracy
Produced items	1667	2084	1852
Production costs	€ 6 668	€ 8 336	€ 7 408
Quality control items	1667	1250	1111
Quality control costs	€ 26 672	€ 20 000	€ 17 776
Quality items	1000	1000	1000
Sum of costs	€ 33 340	€ 28 336	€ 25 184

analysis of process characteristics and automated implementation is desired during run-time [33]. Using the presented approach for automatic product quality prediction, the predictability has been increased and scrap items can be sorted out more effectively. This saves time by eliminating the costly and time-consuming checks for these products. An initial model clearly outperformed human experts in predicting the product quality and that this is a non-intuitive task. Still, the classification rate of 80% shows potential for further improvement. New sensors and more test data could help to increase the accuracy further and, as an analysis of the costs showed, such an increase in accuracy offers also potential for substantial financial gains. The presented approach is suitable for production settings in which quality control and assessment procedures are more expensive than the production itself. This is the case for many safety-critical items and high-quality products. To implement the presented approach, it is necessary to collect suitable production data and to analyze them in time. Furthermore, based on the findings of the quality prediction, an improvement of the production process itself would be also desirable. To this end, a suitable visual analytics dashboard seems promising [45]. Apart from installing new sensors and tuning the prediction model further, future work also aims to consider component dependencies and introduce a multi-component model [13]. Furthermore, the prediction in different states of production is an open issue. Knowing after the first production step already that the product does not achieve the required quality anymore can lead to an early sorting out of the product. This approach could lead to further cost savings and a more efficient production.

6   Further Reading

Han et al. [17] provide a comprehensive introduction to the field of data mining. Witten et al. [52] describe basic and more advanced techniques for data mining and machine learning, with references to implementations in the WEKA tool suite[1]. Statistical methods are essential for data mining and machine learning. Daly et al. [7] provide an excellent, highly readable introduction to the most important elements of statistics. Tabachnik and Fidell [46] are the main reference when conducting advanced, multivariate statistical analysis. Goodfellow et al. [15] give a comprehensive introduction to deep learning.

7   Conclusions and Recommendations for Practice

Data analysis is key to realizing the full potential of digitization. The process of data analysis, however, is challenging. First, the data from various sources must be cleaned, integrated, and transformed into a form that is suitable for the analysis, which typically is a time-consuming and non-trivial process in itself. The choice of data analysis method or algorithm then is challenging due to the multitude of available methods or algorithms for different

---

[1] https://www.cs.waikato.ac.nz/ml/weka/

analysis purposes, each method or algorithm presenting different properties, strengths and weaknesses. Furthermore, it is important to understand that data science problems are no abstract and theoretical problems. Rather, it is crucial to understand the underlying business problem and to design the data-driven solution in such a way that it provides meaningful decision support to users in concrete situations. For practitioners, we have the following recommendations:

- The choice of appropriate data analysis algorithm depends on various factors, including the nature of the business problem and the type and volume of the data. If decisions must be justifiable then the predictions must be explainable. Likewise, when it comes to accuracy, it does not always have to be deep learning: A more traditional machine learning approach may perform better than deep learning in some cases (see Sect. 5, for example).
- Mere syntactic data integration is not sufficient. Rather, data integration must also consider the data semantics, i.e., the meaning behind the data. For example, the same measurement may be taken using different methods or making different assumptions. Comparing measurements that differ in any of these ways may lead to incorrect conclusions.
- Data cleaning must be conducted carefully. Overzealous data cleaning may discard potentially relevant but seemingly implausible observations.
- Evaluating how interesting correlations found in the data really are requires a multitude of evaluation metrics to be considered. Relying on a single measure may lead to wrong conclusions regarding the significance of a correlation. Moreover, it is important to keep in mind that correlation is not causation.
- Global measures for evaluation of a predictive model are insufficient for assessing the reliability of individual predictions. Likewise, common fallacies of statistical reasoning may influence the interpretation of analysis results.

References

1. Agrawal, R., Srikant, R.: Mining sequential patterns. In: Proceedings of the 11th International Conference on Data Engineering. pp. 3–14 (1995). https://doi.org/10.1109/ICDE.1995.380415
2. Agrawal, R., Imieliundefinedski, T., Swami, A.: Mining association rules between sets of items in large databases. In: Proceedings of the 1993 ACM SIGMOD International Conference on Management of Data. pp. 207–216 (1993). https://doi.org/10.1145/170035.170072
3. Ahmad, R., Kamaruddin, S.: An overview of time-based and condition-based maintenance in industrial application. Computers & Industrial Engineering **63**(1), 135–149 (2012)
4. Altman, N., Krzywinski, M.: Association, correlation and causation. Nature Methods **12**, 899–900 (2015). https://doi.org/10.1038/nmeth.3587
5. Birajdar, R.S., Patil, R.S., Khanzode, K.: Vibration and noise in centrifugal pumps - sources and diagnosis methods. In: Proc. 3rd International Conference on Integrity, Reliability and Failure (2009)

6. Breiman, L.: Random forests. Machine learning **45**(1), 5–32 (2001). https://doi.org/10.1023/A: 1010933404324
7. Daly, F., Hand, D.J., Jones, M., Lunn, A., McConway, K.: Elements of statistics. Addison-Wesley Publishing Company (1995)
8. Dobson, S., Golfarelli, M., Graziani, S., Rizzi, S.: A reference architecture and model for sensor data warehousing. IEEE Sensors Journal **18**(18), 7659–7670 (2018). https://doi.org/10.1109/ JSEN.2018.2861327
9. Doran, D., Schulz, S., Besold, T.R.: What does explainable AI really mean? A new conceptualization of perspectives. CoRR **abs/1710.00794** (2017), http://arxiv.org/abs/1710.00794
10. Ehrendorfer, M., Fassmann, J.A., Mangler, J., Rinderle-Ma, S.: Conformance checking and classification of manufacturing log data. In: 2019 IEEE 21st Conference on Business Informatics (CBI). vol. 1, pp. 569–577. IEEE (2019)
11. Fletcher, G.P., Groth, P.T., Sequeda, J.: Knowledge scientists: Unlocking the data-driven organization. ArXiv (2020), http://arxiv.org/abs/2004.07917
12. Gashi, M., Ofner, P., Ennsbrunner, H., Thalmann, S.: Dealing with missing usage data in defect prediction: A case study of a welding supplier. Computers in industry **132**, 103505 (2021)
13. Gashi, M., Thalmann, S.: Taking complexity into account: A structured literature review on multi-component systems in the context of predictive maintenance. In: European, Mediterranean, and Middle Eastern Conference on Information Systems. pp. 31–44. Springer (2019)
14. Gatica, C.P., Koester, M., Gaukstern, T., Berlin, E., Meyer, M.: An industrial analytics approach to predictive maintenance for machinery applications. In: 2016 IEEE 21st International Conference on Emerging Technologies and Factory Automation (ETFA). pp. 1–4 (2016)
15. Goodfellow, I., Bengio, Y., Courville, A.: Deep Learning. MIT Press (2016), http://www. deeplearningbook.org
16. Halachmi, I., Klopčič, M., Polak, P., Roberts, D., Bewley, J.: Automatic assessment of dairy cattle body condition score using thermal imaging. Computers and Electronics in Agriculture **99**, 35–40 (2013). https://doi.org/10.1016/j.compag.2013.08.012
17. Han, J., Kamber, M., Pei, J.: Data Mining: Concepts and Techniques, 3rd edition. Morgan Kaufmann (2011), http://hanj.cs.illinois.edu/bk3/
18. Hoffmann, G., Schmidt, M., Ammon, C., Rose-Meierhöfer, S., Burfeind, O., Heuwieser, W., Berg, W.: Monitoring the body temperature of cows and calves using video recordings from an infrared thermography camera. Veterinary Research Communications **37**(2), 91–99 (2013)
19. International Organization for Standardization: Condition monitoring and diagnostics of machines—vibration condition monitoring—part 1: General procedures. International Standard, ISO 13373-1:2002, ISO (2002)
20. International Organization for Standardization: Condition monitoring and diagnostics of machines—data processing, communication and presentation—part 1: General guidelines. International Standard, ISO 13374-1:2003, ISO (2003)
21. International Organization for Standardization: Automation systems and integration—oil and gas interoperability—part 1: Overview and fundamental principles. Technical Specification, ISO/TS 18101-1:2019, ISO (2019)
22. de Jonge, B., Scarf, P.A.: A review on maintenance optimization. European Journal of Operational Research **285**(3), 805–824 (2020). https://doi.org/10.1016/j.ejor.2019.09.047
23. de Jonge, B., Teunter, R., Tinga, T.: The influence of practical factors on the benefits of condition-based maintenance over time-based maintenance. Reliability engineering & system safety **158**, 21–30 (2017)
24. Kans, M., Galar, D.: The impact of maintenance 4.0 and big data analytics within strategic asset management. In: 6th International Conference on Maintenance Performance Measurement and

Management, 28 November 2016, Luleå, Sweden. pp. 96–103. Luleå University of Technology (2017)

25. Kaur, K., Selway, M., Grossmann, G., Stumptner, M., Johnston, A.T.: Towards an open-standards based framework for achieving condition-based predictive maintenance. In: Proceedings of the 8th International Conference on the Internet of Things (IoT 2018). pp. 16:1–16:8 (2018). https://doi.org/10.1145/3277593.3277608

26. Khoshafian, S., Rostetter, C.: Digital prescriptive maintenance. Internet of Things, Process of Everything, BPM Everywhere (2015)

27. Koller, D., Friedman, N.: Probabilistic graphical models: principles and techniques. MIT press (2009)

28. Lee, C., Cao, Y., Ng, K.H.: Big data analytics for predictive maintenance strategies. In: Supply Chain Management in the Big Data Era, pp. 50–74. IGI Global (2017)

29. Lohr, S.: For big-data scientists, 'janitor work' is key hurdle to insights (2014), https://www.nytimes.com/2014/08/18/technology/for-big-data-scientists-hurdle-to-insights-is-janitor-work.html, accessed: 5 May 2021

30. Narayan, V.: Business performance and maintenance: How are safety, quality, reliability, productivity and maintenance related? Journal of Quality in Maintenance Engineering **18**(2), 183–195 (2012)

31. OpenO&M For Manufacturing Joint Working Group: Condition based operations for manufacturing. Whitepaper, MIMOSA, OPC Foundation, ISA (2004), http://www.openoandm.org/files/whitepapers/2004-10-06_Condition_Based_Operations_for_Manufacturing.pdf

32. Pang, G., Shen, C., Cao, L., Hengel, A.V.D.: Deep learning for anomaly detection: A review. ACM Computing Surveys **54**(2) (2021). https://doi.org/10.1145/3439950

33. Pauker, F., Mangler, J., Rinderle-Ma, S., Pollak, C.: centurio.work—modular secure manufacturing orchestration. In: Proceedings of the Dissertation Award, Demonstration, and Industrial Track at BPM 2018 co-located with 16th International Conference on Business Process Management (BPM 2018). CEUR Workshop Proceedings, vol. 2196, pp. 164–171. CEUR-WS.org (2018), http://ceur-ws.org/Vol-2196/BPM_2018_paper_33.pdf

34. Phillips, J., Cripps, E., Lau, J.W., Hodkiewicz, M.: Classifying machinery condition using oil samples and binary logistic regression. Mechanical Systems and Signal Processing **60–61**, 316–325 (2015). https://doi.org/10.1016/j.ymssp.2014.12.020

35. Rabatel, J., Bringay, S., Poncelet, P.: Anomaly detection in monitoring sensor data for preventive maintenance. Expert Systems with Applications **38**(6), 7003–7015 (2011). https://doi.org/10.1016/j.eswa.2010.12.014

36. Research, A.: Maintenance analytics to generate $24.7 billion in 2019, driven by predictive maintenance and internet of things (March 2014), https://www.abiresearch.com/press/maintenance-analytics-to-generate-247-billion-in-2/

37. Roland, L., Lidauer, L., Sattlecker, G., Kickinger, F., Auer, W., Sturm, V., Efrosinin, D., Drillich, M., Iwersen, M.: Monitoring drinking behavior in bucket-fed dairy calves using an ear-attached tri-axial accelerometer: A pilot study. Computers and Electronics in Agriculture **145**, 298–301 (2018). https://doi.org/10.1016/j.compag.2018.01.008

38. Ruiz, A.: The 80/20 data science dilemma (2017), https://www.infoworld.com/article/3228245/the-80-20-data-science-dilemma.html, accessed: 5 May 2021

39. Safavian, S.R., Landgrebe, D.: A survey of decision tree classifier methodology. IEEE Transactions on Systems, Man, and Cybernetics **21**(3), 660–674 (1991). https://doi.org/10.1109/21.97458

40. Schuetz, C.G., Schausberger, S., Schrefl, M.: Building an active semantic data warehouse for precision dairy farming. Journal of Organizational Computing and Electronic Commerce **28**(2), 122–141 (2018). https://doi.org/10.1080/10919392.2018.1444344

41. Shi, W., Dustdar, S.: The promise of edge computing. Computer **49**(5), 78–81 (2016). https://doi.org/10.1109/MC.2016.145

42. Steensels, M., Maltz, E., Bahr, C., Berckmans, D., Antler, A., Halachmi, I.: Towards practical application of sensors for monitoring animal health; design and validation of a model to detect ketosis. Journal of Dairy Research **84**(2), 139–145 (2017). https://doi.org/10.1017/S0022029917000188

43. Steensels, M., Maltz, E., Bahr, C., Berckmans, D., Antler, A., Halachmi, I.: Towards practical application of sensors for monitoring animal health: the effect of post-calving health problems on rumination duration, activity and milk yield. Journal of Dairy Research **84**(2), 132–138 (2017). https://doi.org/10.1017/S0022029917000176

44. Stojanovic, L., Dinic, M., Stojanovic, N., Stojadinovic, A.: Big-data-driven anomaly detection in industry (4.0): an approach and a case study. In: Joshi, J., Karypis, G., Liu, L., Hu, X., Ak, R., Xia, Y., Xu, W., Sato, A., Rachuri, S., Ungar, L.H., Yu, P.S., Govindaraju, R., Suzumura, T. (eds.) 2016 IEEE International Conference on Big Data (2016). https://doi.org/10.1109/BigData.2016.7840777

45. Suschnigg, J., Ziessler, F., Brillinger, M., Vukovic, M., Mangler, J., Schreck, T., Thalmann, S.: Industrial production process improvement by a process engine visual analytics dashboard. In: Proceedings of the 53rd Hawaii International Conference on System Sciences. pp. 1320–1329 (2020)

46. Tabachnick, B.G., Fidell, L.S.: Using multivariate statistics. Pearson, 6 edn. (2014)

47. Thalmann, S., Gursch, H., Suschnigg, J., Gashi, M., Ennsbrunner, H., Fuchs, A.K., Schreck, T., Mutlu, B., Mangler, J., Kappl, G., et al.: Cognitive decision support for industrial product life cycles: A position paper. In: COGNITIVE 2019: The Eleventh International Conference on Advanced Cognitive Technologies and Applications. pp. 3–9. IARIA (2019)

48. Thalmann, S., Mangler, J., Schreck, T., Huemer, C., Streit, M., Pauker, F., Weichhart, G., Schulte, S., Kittl, C., Pollak, C., et al.: Data analytics for industrial process improvement a vision paper. In: 2018 IEEE 20th Conference on Business Informatics (CBI). vol. 2, pp. 92–96. IEEE (2018)

49. Thirumuruganathan, S., Tang, N., Ouzzani, M., Doan, A.: Data curation with deep learning. In: Proceedings of the 23rd International Conference on Extending Database Technology (EDBT 2020). pp. 277–286. OpenProceedings.org (2020). https://doi.org/10.5441/002/edbt.2020.25

50. Thurston, M., Lebold, M.: Standards developments for condition-based maintenance systems. Tech. rep., Pennsylvania State Univ University Park Applied Research Lab (2001)

51. Vaisman, A., Zimányi, E.: Data Warehouse Systems – Design and Implementation. Springer, Berlin Heidelberg (2014)

52. Witten, I.H., Frank, E., Hall, M.A., Pal, C.J.: Data mining: Practical Machine Learning Tools and Techniques. Morgan Kaufmann, 4 edn. (2017)

53. Yan, H., Wan, J., Zhang, C., Tang, S., Hua, Q., Wang, Z.: Industrial big data analytics for prediction of remaining useful life based on deep learning. IEEE Access **6**, 17190–17197 (2018)

Process Mining—Discovery, Conformance, and Enhancement of Manufacturing Processes

Stefanie Rinderle-Ma, Florian Stertz, Juergen Mangler and Florian Pauker

Abstract

Process-orientation has gained significant momentum in manufacturing as enabler for the integration of machines, sensors, systems, and human workers across all levels of the automation pyramid. With process orientation comes the opportunity to collect manufacturing data in a contextualized and integrated way in the form of process event logs (no data silos) and with that data, in turn, the opportunity to exploit the full range of process mining techniques. Process mining techniques serve three tasks, i.e., (i) the discovery of process models based on process event logs, (ii) checking the conformance between a process model and process event logs, and (iii) enhancing process models. Recent studies show that particularly, (ii) and (iii) have become increasingly important. Conformance checking during run-time can help to detect deviations and errors in manufacturing processes and related data (e.g., sensor data) when they actually happen. This facilitates an instant reaction to these deviations and errors, e.g., by adapting the processes accordingly (process enhancement), and can be taken as input for predicting deviations and errors

S. Rinderle-Ma (✉) · J. Mangler
Department of Informatics, TU München, Munich, Germany
e-mail: stefanie.rinderle-ma@tum.de

J. Mangler
e-mail: juergen.mangler@tum.de

F. Stertz
Faculty of Computer Science, Universität Wien, Vienna, Austria
e-mail: florian.stertz@univie.ac.at

F. Pauker
EVVA Sicherheitstechnologie GmbH, Vienna, Austria
e-mail: f.pauker@evva.com

© The Author(s), under exclusive license to Springer-Verlag
GmbH, DE, part of Springer Nature 2023
B. Vogel-Heuser and M. Wimmer (eds.), *Digital Transformation*,
https://doi.org/10.1007/978-3-662-65004-2_15

for future process executions. This chapter discusses process mining in the context of manufacturing processes along the phases of an analysis project, i.e., preparation and analysis of manufacturing data during design and run-time and the visualization and interpretation of process mining results. In particular, this chapter features recommendations on how to employ which process mining technique for different analysis goals in manufacturing.

Keywords

Manufacturing Processes • Process Discovery • Conformance Checking • Process Enhancement

1   Introduction

"Recent trends in automation and knowledge of the underlying processes / interactions are key to digital transformation" [14]. In manufacturing, process technology has already proven itself as a driver for digital transformation. *Manufacturing processes*—aka manufacturing orchestrations—integrate the manufacturing tasks conducted by human actors, sensors, machines, and information systems in a process-oriented way [19]. This integration, in turn, enables the contextualized collection of process-related data and hence facilitates getting full transparency on what is going on using *process mining* [24].

Figure 1a1) depicts an example manufacturing process realizing the production of a piece for a gas turbine (i.e. lowerhousing) (cmp. [9]). The manufacturing process is modeled using Business Process Modeling and Notation (BPMN).[1]

The BPMN model starts off with the sub process Turn1 represented by a complex task, followed by another turning and two milling sub processes. Afterwards a quality control (QC) task takes place at the shopfloor. If the quality is not OK the process is completed, otherwise a QC task at the customer side takes place. This decision is reflected by an alternative branching.

The BPMN model reflects the control flow of the manufacturing process, i.e., it abstracts from aspects such as data flow, resources, and time. The BPMN model can then be imported into a process execution engine such as the Cloud Process Execution Engine (CPEE)[2] where it is transformed into an executable model by specifying, for example, endpoints and data (the CPEE model is shown in Fig. 1a3). For the given manufacturing process, for example, each manufacturing process instance reflects one work piece to be produced.

Information on the execution of the manufacturing process instances is collected during run-time by the process engine and stored in *process event logs* (cf. Fig. 1a4). Note that also other information systems such as ERP systems collect (process) event logs and process event logs reflect *event-based data*. The latter means that for each manufacturing process

---

[1] bpmn.org
[2] cpee.org

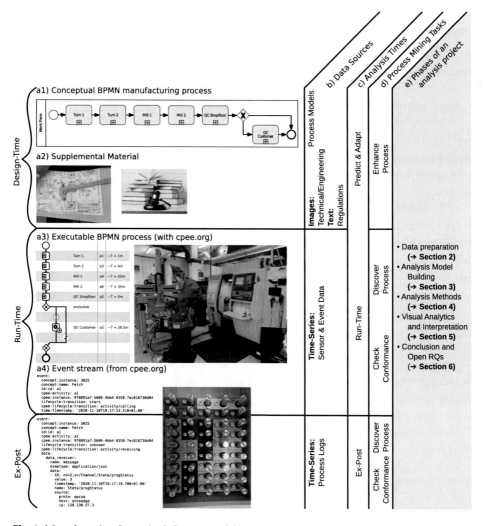

Fig. 1 Manufacturing Scenario & Structure of Chapter

task (e.g., QC Shopfloor in Fig. 1a1 and a3) at least one event reflecting its execution is stored, together with a time stamp, and information on the process instance the process task was executed for. This is the minimum information required for conducting process mining. Several extensions might be stored and analyzed such as events on start/completion of tasks, data and resources. In general, the more and better[3] data is available, the better the analysis results might turn out. The standard format for process event log is eXtensible

---

[3] "Better" here refers to the quality of the collected data. For a discussion on quality levels of process event logs see Sect. 2.

Event Stream (XES) [1]. In addition to process event logs, the following data sources might provide insights to the manufacturing process (cf. Fig. 1b): (i) sensor and machine data that is collected during the process execution, stored as *time series*; (ii) engineering drawings, stored as images, and (iii) regulations such as ISO standards, stored as text. A deeper insight into the different data sources and how to prepare them will be provided in Sect. 2.

Process event logs are the basic input for the application of *process mining*. Process mining is particularly promising for manufacturing processes as *"transparency is a prerequisite for digital transformation"* and *"process mining allows full transparency based on event logs"* as stated in [24]. Consequently, process mining seems highly promising for gaining a deeper understanding of manufacturing processes and for promoting the digitalization in the manufacturing domain. The three tasks of process mining are *process discovery, conformance checking*, and *process enhancement* [2, 6] (cf. Fig. 1d). In short, process discovery refers to detecting process models from process event logs, conformance checking to the assessment of conformance/deviations between process models and event logs, and process enhancement to the adaptation and improvement of process models based on process mining results.

Typically, process mining is conducted *ex post*, i.e., after process instances have been completed and the associated process event logs have been collected (cf. Fig. 1c). Recently, process mining during the *run-time* of process instances has gained momentum as it enables to react to analysis results, e.g., deviations, more quickly [35]. Note that for run-time process mining, literature also refers to *process event streams* instead of process event logs in order emphasize the run-time/online character of the analysis and hence to *streaming process mining* [5]. There is a fluent transition from run-time/streaming process mining to *predictions*. The latter has been addressed by the area of *predictive process monitoring* [36, 37]. These approaches focus on predicting (i) the next activity to be executed and (ii) the remaining execution time of processes. In manufacturing, *predictive maintenance*, see for example [22], has been a hot topic since several years.

Due to the integration of processes, sensors, machines, human workers, and information systems, manufacturing processes can collect data from all these sources in an *integrated and contextualized way*, i.e., process event logs/streams as event-based data, sensor and machine data as time series data, technical/engineering drawings as images, and regulations as text as depicted in Fig. 1a2 and b). In the light of this abundance of data sources and the opportunities that come with them, in the following we will examine the question of how to analyze manufacturing processes through process mining along the phases of a data analysis project as depicted in Fig. 1e): data preparation in Sect. 2, analysis model building in Sect. 3, analysis techniques in Sect. 4, as well as visualization and interpretation in Sect. 5. Finally, Sect. 6 provides a short summary of the state of the art and discusses future questions.

2   Data Preparation

The input data for process mining are *process event logs* [2]. They consist of a set of *process traces* that reflect sequences of events related to the execution of process activities. More precisely, for each process instance that was created, instantiated, and executed, a process trace reflects the event sequence produced by executing the activities of this process instance.

This section addresses the preparation of process events logs as input for process mining in the manufacturing domain. In Sect. 2.1, we comment on data quality. Section 2.2 explains which data sources are available and how they can be exploited for process mining.

2.1   Data Quality in Manufacturing

As stated for business intelligence projects in general [10] and process mining projects in particular [7], the collection and preparation of data is often the most complex and tedious task. This also holds true for the manufacturing domain where—without explicit process orientation and data collection—the data accrues over several systems and system layers along the automation pyramid as depicted in Fig. 2 (middle). The right side of Fig. 2 assigns the systems at the different layers of the automation pyramid to the quality classes of the L^* data quality model as established in the Process Mining Manifesto [7]. The L^* model features five quality levels from * (lowest) to ***** (highest) based on criteria such as event orientation, trustworthiness, and systematic collection.

Enterprise resource planning (ERP) data (***) is event-oriented and can be assumed as trustworthy. ERP data is collected in a systematic way, but provides only a specific view on the data, e.g., on the production orders. The data of lower layers of the automation pyramid is assessed as ** data meaning that event data is collected *"as a by-product of some information system"* [7]. As a consequence the information system might be bypassed resulting in missing and/or incorrect data. The lower the layer, the more unclear is the quality level (at most **). Moreover, the data collection mostly happens in a separated manner, i.e., the data is not collected across the layer in a contextualized ways. This leads to data silos. If, by contrast, the data is collected in a process-oriented way as shown on the left side of Fig. 2, the data is at least of quality level ****, i.e., event-oriented, collected in a systematic way, and trustworthy. If the data is in addition semantically annotated, it can be classified as of highest quality (*****).

Figure 3a depicts a process event log provided as eXtensible Event Stream (XES) (xes-standard.org). The `log` contains a `trace` for process instance `423`. It is crucial that traces have unique ids such that the information on the activity executions can be distinguished for different instances, e.g., activity `Turn1` was executed for instance `423` and not for instance `424`. Further on, the trace carries information on the process engine the process instance was executed with, i.e., the CPEE [2], together with a UUID. The trace contains two

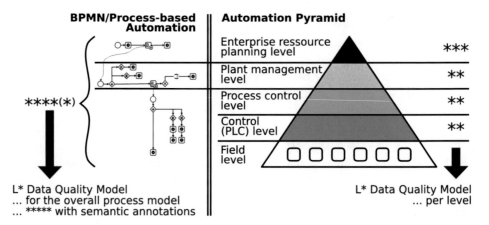

Fig. 2 Data collection along automation pyramid and in a process-oriented manner with data quality level according to L* model [7]

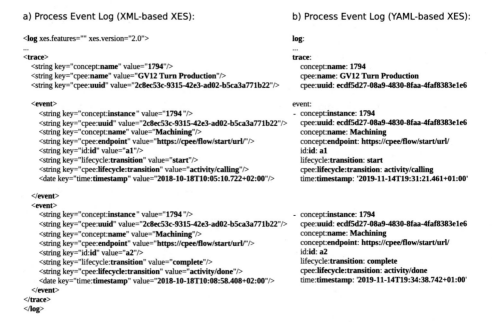

Fig. 3 Manufacturing process event logs in **a** XES-XML and **b** XES-YAML formats

events that refer to the execution of process activity with label Machining. Note that for more complex scenarios, traces may contain several thousands or more events.

The two events referring to the execution of Machining distinguish between the start and the completion of Machining, i.e., refer to two activity life cycle states [34]. Often, process event logs only store one event per activity execution, mostly for their completion. However, if life cycle states can be distinguished in the logs, more precise analysis results can be achieved, for example, on the duration of activities. In addition to `start` and `completion` of process activities life cycle states such as `interrupted` exist. This can be also seen in the more fine-granule CPEE life cycle states such as `activity/calling`. Finally, the events are equipped with a `timestamp` which is vital in order to reason about the activity orders, e.g., in a discovered process model.

However, as said before, more information can be stored, leading to more options during the analysis. The start event of activity `Machining` in Fig. 3a, for example, holds additional information on its `end point`, i.e., the service that is called for executing the activity. Further information typically stored in process event logs are `resource/originator`, i.e., the actor or machine that has performed the task.

There are several ways to represent a serialization of the XES format. XML is usually used, since it is human readable, provides a schema and can be easily processed. The example in Fig. 3a features an XML representation of an process event log. Figure 3b presents another approach to serialize the XES format in YAML. YAML is as human readable as XML [27] and offers advantages over XML for directly logging events in the XES format since the computational effort is lower because new events, instead of parsing an XML tree for the correct position, can be easily appended to a file, an operation which in many operating systems is optimized.

2.2   Data Sources and Process Mining

Table 1 summarizes existing process mining approaches along the analysis time they are applied at, i.e., Ex post, Run-Time, and Predict&Adapt, (cf. Fig. 1c), as well as along the available data sources, i.e., process event logs containing different amount of information. For the latter, we start from the minimum necessary information to be able to apply process mining, and step-by-step add information (indicated by a + in Table 1).

The different analysis methods and techniques shown in Table 1 will be explained in Sect. 4. In this section, we focus on the data preparation. The rows above the one highlighted in light blue in Table 1 refer to data sources that can be entirely captured within a process event logs. On top of the minimum necessary information, in the log we distinguish between start and end of an event or other life cycle transitions. With start and end events it becomes possible to reason about the duration of activities (e.g., [31]). Moreover, process event logs can also contain information on resources, see the organizational extension in XES.[4] This enables to analyze organizational structures connected with the process, for example, underlying work or social networks [28] as well as authorization rules (who is allowed to perform which process activity) [18]. If the process event log also stores values of process data such as temperature that are associated with process activities, these values can be exploited for finding decision rules at alternative branchings (decision discovery [16]). One

---

[4] xes.standard.org

Table 1 Data input and existing approaches

Analysis phase / Trace contains	Ex post	Runtime	Predict&Adapt
Minimum information: trace id, event name, 1 life cycle state, time stamp	Control flow discovery, conformance checking		Next activity, remaining time
+ start/end event	Activity durations, temporal profiles [31]		
+ resource/originator	Organizational mining [28], mining of actor assignments [18]		
+ (internal) process data	Decision discovery [16]	Data drift detection [33], multi perspective conformance checking [20]	
+ (external) sensor and machining data	Decision discovery [8]	Explaining and predicting concept drift [35]	
+ multimedia data (text, image, video)	Decision discovery [16]		

example would be: if the temperature is below 30 °C the machine works in normal mode, otherwise a cooling agent has to be used. We refer to this process data and the associated values as *internal data*. During run-time, process data can also be exploited for detecting drifts in the process data [33] and for multi-perspective conformance checking [20].

It is crucial to be able to establish links between process activities and the sensor/machining data. This leads to the question how to deal with sensor and machining data. This data can be used as internal data, i.e., be necessary for executing the process, but—because this is continuous or time series data—it cannot be directly stored in the process event log. This leads to the question on how to store such "internal/external" data, so it can be used for process mining. Sensors are creating data points continuously even without an active process, i.e., like the temperature in the previous example. There are at least three options available to use this data in process mining (cf. [8, 35]). The first option aims at adapting the process mining techniques to take a log containing time series data points as an additional input, so that with the temporal information of the process event log, the related data points to an event can be extracted. The other option would be to adapt the process model, to add a task, which fetches data points from a sensor. As a third option, to achieve results during run-time, this fetching task can also be inserted in a parallel loop to the process, so that the data points are inserted into the log while the process is being executed.

Manufacturing processes can be also subject to multimedia data sources (see last row in Table 1), including texts such as regulatory documents, standards, and norms, (ii) image data such as technical or engineering drawings as input or for quality control, and video data such as instructions. So far, existing approaches have provided means for exploiting data values of primitive data types, i.e., numbers and strings for determining decision rules in an ex post manner [16]. Support for more complex data types or other multimedia data is currently missing (see future work in Sect. 6).

As discussed in the Process Mining Manifesto [7], in practice, process data is often not available in an event-based format. This also holds true for the manufacturing domain where the data is often scattered over several information systems (e.g., ERP) and the machines. Process-orientation offers the opportunity to integrate these "data silos" resulting in an contextualized collection of process event logs [19].

3 Analysis Model Building

According to [10] *"[f]inding answers for analytical goals is based on models"*. Hence, modeling is an essential task in a manufacturing analysis project. Further on, [10] states that model building can be understood *"as a mapping of some part of the domain semantics of the business process into the model structure. This happens in such way that the available data enable formal analysis of questions about the process"*. Here, the domain is manufacturing, the processes are the manufacturing processes, and the available data consists of process event logs plus potentially additional data as outlined in Table 1.

For process mining, the central analysis model is a *graph* which represents the control flow of the discovered process models. The graph is typically described using existing process meta models such as BPMN in Fig. 1a1 and a3) as well as Petri Nets and Heuristic Nets in Fig. 4a and b. Note that other analysis models for discovered manufacturing processes are conceivable such as patterns [10] for, e.g., mining declarative process models (for an overview on declarative process mining approaches see [17]). Graph-based structures can be also used as analysis models beyond the control flow of processes, for example, social networks as model for organizational structures underlying a manufacturing process (cf. Fig. 4c). This underpins that depending on the data available in the process event log (cf. Table 1) and the analysis question different analysis models might become viable.

In particular for the manufacturing domain, the analysis of process event logs in combination with additional (external) data such as sensor or machining data is vital [35]. The question is which analysis model can be used as basis for the combined analysis. Sensor and machining data is available in the form of time series data. As discussed in [10], time series data is produced by collecting the states of one or several variables over time and can be analyzed based on statistical models (cf. Fig. 4d).

4 Analysis Methods

The analysis tasks of process mining are process discovery, conformance checking, and process enhancement [2]. These analysis tasks together with existing techniques are discussed in the following in the context of manufacturing processes.

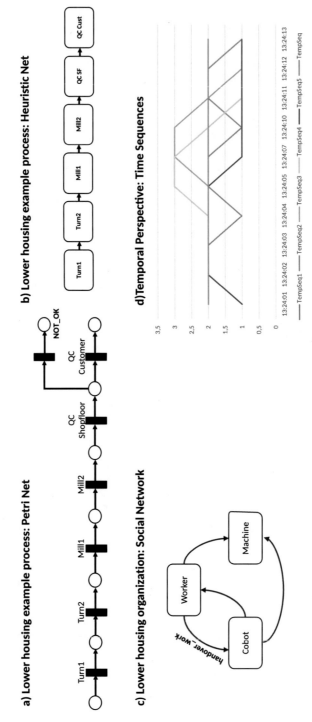

Fig. 4 Analysis models for process mining (selection): **a** Petri Nets and **b** Heuristic Nets and **c** Social Network and **d** Temporal Perspective

Discovery of manufacturing processes: Basically, the control flow of manufacturing processes can be discovered using existing algorithms such as the α miner [21], the Heuristic miner [38], or the Inductive miner [15].

Consider the `Spawn GV12 Production` process execution log[5] which consists of 81 traces. An excerpt is shown in Fig. 3b. Figure 5 depicts the Heuristic net that is discovered when applying the Heuristic miner (using PM4PY 2.2.4[6]) based on this log. The Heuristic miner considers order relations between pairs of activities a and b that can be observed in the log and calculates their relative frequency, i.e., how often does b directly follow a minus the how often the opposite happens (i.e., a directly follows b) divided by the sum of b follows a and a follows b. This leads to a number between 0 and 1. The higher the number, the more likely it is that b actually follows a. This formula is exemplary calculated for the relation between `GV12 Turn9 (start)` and `GV12 Turn9 (complete)`. The result suggests the actual order `GV12 Turn9 (start)` $\Rightarrow$ `GV12 Turn9 (complete)`.

Figure 5 shows interesting results. There are splits in the discovered Heuristic net, i.e., the order of `Manually Measure9 (start)` and `Signal Machining End9 (complete)` is often times mixed up. The same phenomenon can be witnessed with `Manually Measure9 (complete)` and `Measure with MicroVu9 (start)`. This is a clear indication to dig deeper into these cases, i.e., traces, and emphasizes the capabilities of process discovery as screening tool (this observation has been already made in the context of medical treatment processes [3]).

Figure 6 shows the Petri net resulting from applying the Inductive miner infrequent (using PM4PY) on the `Spawn GV12 Production` process event log. The discovered Petri net confirms that with this log, `Manually Measure9 (start)` and `Signal Machining End9 (complete)` appear to be executed in parallel. Based on this visual inspection of both discovered models, we can go back to the traces and dig deeper into the reason for the observation on `Measure with MicroVu`. Using, for example, the log filtering capabilities of PM4PY, we can see that only nanoseconds are separating these events. Therefore a faulty behavior in the log system is likely the origin of these errors.

Conformance of manufacturing processes: Conformance checking [6] takes as input a process model and a process event log and calculates the conformance between the log and the model or—vice versa—the difference between them. More precisely, the goal is to measure to which degree the behavior expressed by the process model (i.e., all producable traces on the model) and the behavior stored in the log (i.e., all traces in the log) match. If we understand the process model as the expression of the intended behavior and the log as collection of the actual behavior, conformance assesses how much reality is reflected by the intended/prescribed behavior in the model. Note that there might be real-world behavior that is neither reflected by the process model nor (already) stored in the log.

In manufacturing, conformance checking can be used for monitoring the process behavior over time [9], i.e., monitoring whether and how actual process execution conforms or deviates

---

[5] http://gruppe.wst.univie.ac.at/data/timesequence.zip

[6] https://pm4py.fit.fraunhofer.de/

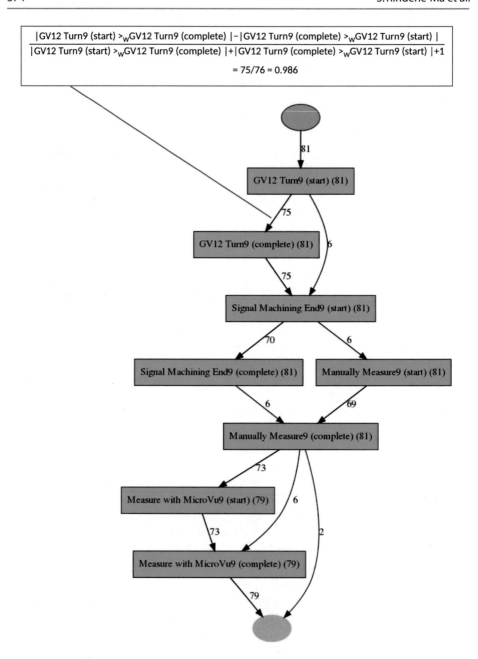

Fig. 5 Spawn GV12 Production log using Heuristic Miner yielding Heuristic net

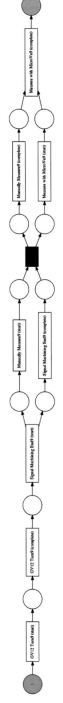

Fig. 6 Spawn GV12 Production log using Inductive Miner yielding a Petri net with infrequent transistions removed

from the originally specified model. If the manufacturing processes are already implemented and executed through a process engine, deviations from the process model, i.e., at the control flow level, can only occur due to adaptations of either single process instances (ad hoc adaptations) or due to an evolution of the entire process model [23]. Deviations in process behavior that might happen for various reasons are also referred to as *concept drift* [4, 32]. Being able to analyze, explain, and predict concept drifts is of utmost interest in general and specifically in the manufacturing domain [35].

One "component" of quantifying conformance between a process event log and a process model is *fitness* [26]. Fitness measures how much of the behavior stored in the log is reflected by the process model. This value can be calculated by virtually replaying the behavior expressed by the traces of the process event log on the given process model. For this the process model is represented as a Petri net and the replay of the traces is carried out by replaying tokens on the Petri net. Then it is counted how often the replay requires extra tokens for continuing the execution or creates missing ones. Figure 7 shows the Spawn GV12 Production example. As input we take the original process execution log containing 81 traces (for an excerpt see Fig. 3b). As discussed in the previous paragraph on process discovery, this log contains traces for which some tasks appear in the wrong order due to an inaccuracy of the logging system. The created Petri Net (cf. Fig. 6) only features frequent transitions and therefore does not guarantee perfect fitness. Hence the log does feature traces, which are not fitting the model. A small excerpt showing the results is shown in Fig. 7b. As can be seen, batch5_55 yields a fitness less than 1 and therefore is not fit.

Another measure to capture conformance is *precision* which states how much of the behavior expressed by the process model is also reflected in the log, i.e., the model should not allow additional event sequences which are not reflected in the process event log [2]. The difference between the process event log and the process model can be also measured in terms of adaptation operations that would be necessary to transform the log to the model or vice versa. As can be seen from Fig. 7c the Petri net (also called flower model) can produce any trace on the 4 contained transitions reflecting manufacturing tasks.[7] A selection of example traces is also depicted in Fig. 7c. Obviously, the fitness of, for example, the original log Spawn GV12 Production (cf. Fig. 3b) would be 1. However, the log does not reflect all traces that can be generated in the flower model. This fact can be reflected by measuring precision.

Aside other limitations as discussed in [6], fitness and precision only measure "one side of the coin", i.e., either how much behavior of the log is also reflected by the model or vice versa. In addition, in particular for the manufacturing domain, it would be more interesting to present conformance results in a more expressive way than providing a value between 0 and 1. Take, for example, a fitness value of 0.85. It means that the behavior of the log is reflected by the model in a "pretty high" manner, but it does neither allow any conclusions where deviations between model and log occur nor any assessment how severe these deviations are.

---

[7] As a simplification we only included the tasks without differentiation into start/complete tasks.

a) Original Petri Net

b) Conformance Checking Table created using PM4PY

case_id	fitness	is_fit
batch5_1	1.0	True
batch5_10	1.0	True
batch5_55	0.875	False
...	...	...

c) Flower model (Petri net) and possible traces (selection)

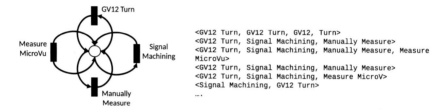

Fig. 7 Spawn GV12 Production: conformance checking on process model without infrequent transistions (transformed to Petri net model using Inductive miner, PM4PY) and original Spawn GV12 Production process event logs (using Conformance Checking, PM4PY)

Alignmnents [6] promise to overcome these limitations. At first, alignments can be used to capture deviations between log and model. More precisely, an alignment states which adaptation steps/operations are necessary to "correct" a deviation between an execution trace from the log and a possible execution sequence on the model. Taking all alignments together reflects a sequence of adaptations to transform the log and the model such that they perfectly match/conform. Moreover, the alignments can be equipped with costs.

As an example take the following execution trace t from a manufacturing log and the execution sequence s of a manufacturing process model:

```
t=<Turn1, Turn2, Mill1, Mill2, QC> and
s=<Turn1, Turn2, Mill1, Mill2, MicroVu>
```

An alignment between t and s is defined as a matrix that captures t and s and an operator $>>$. This operator indicates an adaptation that is necessary to correct a deviation. The result for t and s is shown in Table 2. We can see 4 *synchronous moves* for events in t and s referring to the same activity, e.g., Turn1. The operator $>>$ is present twice in order to balance the presence of task QC in t and task MicroVu in s.

Typically, an alignment is not unique, i.e., it is possible to transform an execution trace and an execution sequence such that they match using different alignments. Obviously, it is desirable to find a minimum alignment, i.e., an alignment with minimum number of moves.

Table 2 Alignment between process execution trace and execution sequence on manufacturing process model (example)

t	Turn1	Turn2	Mill1	Mill2	QC	»
s	Turn1	Turn2	Mill1	Mill2	»	MicroVu

This is also sufficient as moves can be assigned costs in order to assess an alignment in a quantitative way. In other words, the costs express how "expensive" it is to align the execution trace and sequence. Here, the costs for the minimum alignments are meaningful, i.e., the minimum costs for not being conformant.

A straightforward way is to assign a cost of 1 to each move. The cost of the alignment shown in Table 2 then turns out as 2. More sophisticated cost functions are conceivable, considering, for example, the influence of moves on data elements and vice versa. In [29], we presented an advanced cost function based which costs for moves indicating missing events can be reduced if the data elements still possess valid values. Such a situation might hint to a logging error where a task was actually executed and has hence manipulated the values of associated data elements, but the corresponding event has not been logged. The real-world production case in [29] underpins that control flow deviations, e.g., swapped executions of tasks, can be tolerated in case the values of relevant parameters, i.e., the final positioning of the produced parts on a tray, are in an acceptable range.

As indicated in Table 1, conformance checking can take place at either design time, i.e., ex post based on process event logs, or at run-time, i.e., online based on event streams. Experience in the manufacturing domain shows that online conformance checking is promising in order to support the continuous monitoring of the production, i.e., the manufacturing process instances. Moreover, it can be observed that control flow deviations happen due to ad-hoc adaptations of single process instances (e.g., in order to cope with exceptional situations) or due to evolution of the manufacturing process model, e.g., if errors or problem with the process occur frequently. On top of control flow deviations, shifts in sensor and machining data [35] as well as temporal deviations [31] yield valuable insights into possible deviations during run-time

Enhancement of manufacturing processes refers to (constantly) adapt and optimize the manufacturing process models and their instances. Process enhancement has gained the highest momentum recently according to the Gartner study in [14] when compared to the other process mining tasks discovery and conformance checking. However, relatively little attention has been spent on process enhancement in literature. In prior work [35], we combined results from online conformance checking in sensor data with explaining and predicting *concept drifts* (cf. Table 1) where the predicted concept drifts are realized by process evolutions. In detail, the measurements of the part (reflected by sensor data) showed significant drifts which could be explained by the occurrence of chips as by-product of the machining task. This led to the identification of several process enhancements, for example, the insertion of a dedicated `Remove Chips` task in the manufacturing process. A recent study [30] in the manufacturing domain found that conformance checking combined with the analysis of sensor data (*data-aware conformance checking*) supports the detection and explanation of problems with the quality of the workpiece whereas conformance checking focusing on time deviations (*time-aware conformance checking*) points to problems with work organization.

In general, deviations can be detected in a relatively easy and systematic manner. The main challenges are to explain/understand the deviations (i.e., to find and communicate the root cause) and to derive suitable enhancement actions. For the latter, the body of work from process change and evolution (cf. [23] for an overview) can be utilized. Approaches that aim at explaining change and supporting users in defining change operations comprise the augmentation of change operations with reasons [25] as well as the mining of change processes [11] and change tress [12] from change logs.

5 Visual Analytics and Interpretation

Visual analytics is used throughout all phases of the manufacturing analysis project (cf. Fig. 1e) including explorative analysis of the (raw) data, understanding and discussing analysis results, and presenting the final results to the domain experts and other stakeholders. According to [13], basically, *"in the Engineering domain, Visual Analytics can contribute to speed-up development time for products, materials, tools and production methods [...] One key goal [...] in the engineering domain will be the analysis of the complexity of the production systems in correlation with the achieved output, for an efficient and effective improvement of the production environments.".* This general assessment of visual analytics in the engineering domain can be directly transferred to visual process analytics in manufacturing.

According to [10], one can look at the data from different perspectives, i.e., form the customer, organization, and production perspective. These perspectives are connected with different views on the data (i.e., cross-sectional, state, and event view) and subsequently different analysis techniques (i.e., data mining, processes mining, social network analysis, etc.) based on different analysis models such as graphs. Table 3 displays the different perspectives together with the visualization options.

For the customer perspective, different types of data can be distinguished, together with different visualization options. All of these visualization options might be relevant for the manufacturing domain, particularly for visualizing parameter values. Sensor data, for example, can be visualized as time series plot, i.e., displaying the sensor values over time.

For the production perspective that corresponds to possible results of process mining in terms of process models, basically, we distinguish between design time and run-time visualization. Where design time requires the visualization of process model as graphs following some process meta model (e.g., BPMN), during run-time, the information about

Table 3 Overview on visualization options along the different analysis perspectives

Perspective	Visualizations	
Customer	Qualitative Data	Bar Charts, Pie Charts, Mosaic Plots, ...
	Quantitative Data	Contour Plots, Boxplots, ...
	Relationships	Scatter Plots, Correlation, ...
	Temporal Data	Time series Plots, Survival Plots, ...
Organization	Graph visualizations	
Production	Design Time	Visualizations of Petri Nets, BPMN
	Runtime	Equipping process models with runtime information, displaying log information additionally to process model

the execution progress/state of the different instances is depicted, as well. Here, basically, there are two options: process models can be equipped by run-time information such as colored tokens on Petri Nets or the process model is displayed alongside with process event logs.

6   Conclusion and Open Research Questions

The following recommendations for conducting a process mining project in the manufacturing domain conclude this chapter:

- Use process discovery as screening tool
- Use conformance checking for detecting deviations
- Use data-aware conformance checking (i.e., combined with the analysis of sensor data) for
  - detecting and explaining workpiece quality problems
  - explaining and predicting concept drift
  - predictive maintenance
- Use time-aware conformance checking for detecting temporal deviations and problems with work organization
- Explain and discuss detected deviations with domain experts in order to define remedy actions ($\Rightarrow$ process enhancement)

Nonetheless, several research and application-oriented questions are still open and seem promising for future work, not only for manufacturing, but for any process. Table 1 depicts research directions in light pink color. These research directions mainly refer to the analysis phase of Predict&Adapt for input log data beyond control flow, i.e., including information on resources and internal/external data and to the exploitation of external data. First approaches on exploiting sensor data in addition to log data have been presented (e.g., [35]). The observation here is that this also results in a combination of process and data mining/machine learning techniques (e.g., conformance checking in combination with dynamic time warping). The exploitation of data sources such as text or images in combination with process mining has not been explored yet, although these data sources play an important role in manufacturing (i.e., ISO standards and engineering drawings) and other domains (standard operating procedures and images in medicine).

References

1. IEEE standard for extensible event stream (XES) for achieving interoperability in event logs and event streams (Nov 2016)
2. van der Aalst, W.M.P.: Process Mining - Data Science in Action, Second Edition. Springer-Verlag Berlin Heidelberg (2016)
3. Binder, M., Dorda, W., Duftschmid, G., Dunkl, R., Fröschl, K.A., Gall, W., Grossmann, W., Harmankaya, K., Hronsky, M., Rinderle-Ma, S., Rinner, C., Weber, S.: On analyzing process compliance in skin cancer treatment: An experience report from the evidence-based medical compliance cluster (EBMC2). In: Advanced Information Systems Engineering. pp. 398–413 (2012)
4. Bose, R.J.C., Van Der Aalst, W.M., Zliobaite, I., Pechenizkiy, M.: Dealing with concept drifts in process mining. IEEE Trans. Neural Netw. Learning Syst. 25(1), 154–171 (2014)
5. Burattin, A.: Streaming process discovery and conformance checking. In: Encyclopedia of Big Data Technologies (2019)
6. Carmona, J., van Dongen, B.F., Solti, A., Weidlich, M.: Conformance Checking - Relating Processes and Models. Springer (2018), https://doi.org/10.1007/978-3-319-99414-7
7. van Der Aalst, W., et al.: Process mining manifesto. In: Business Process Management. pp. 169–194. Springer (2011)
8. Dunkl, R., Rinderle-Ma, S., Grossmann, W., Fröschl, K.A.: A method for analyzing time series data in process mining: application and extension of decision point analysis. In: CAiSE Forum. pp. 68–84 (2014)
9. Ehrendorfer, M., Fassmann, J., Mangler, J., Rinderle-Ma, S.: Conformance checking and classification of manufacturing log data. In: Business Informatics. pp. 569–577 (2019)
10. Grossmann, W., Rinderle-Ma, S.: Fundamentals of Business intelligence. Springer-Verlag Berlin Heidelberg (2015)
11. Günther, C.W., Rinderle-Ma, S., Reichert, M., van der Aalst, W.M.P., Recker, J.: Using process mining to learn from process changes in evolutionary systems. Int. J. Bus. Process. Integr. Manag. 3(1), 61–78 (2008)
12. Kaes, G., Rinderle-Ma, S.: Mining and querying process change information based on change trees. In: Service-Oriented Computing. pp. 269–284 (2015)

13. Keim, D.A., Andrienko, G.L., Fekete, J., Görg, C., Kohlhammer, J., Melançon, G.: Visual analytics: Definition, process, and challenges. In: Information Visualization - Human-Centered Issues and Perspectives, pp. 154–175 (2008)
14. Kerremans, M., Searle, S., Srivastava, T., Iijima, K.: Market guide for process mining (2020), www.gartner.com
15. Leemans, S.J.J., Fahland, D., van der Aalst, W.M.P.: Process and deviation exploration with inductive visual miner. In: BPM Demo. p. 46 (2014)
16. de Leoni, M., Mannhardt, F.: Decision discovery in business processes. In: Encyclopedia of Big Data Technologies (2019)
17. Ly, L.T., Maggi, F.M., Montali, M., Rinderle-Ma, S., van der Aalst, W.M.P.: Compliance monitoring in business processes: Functionalities, application, and tool-support. Inf. Syst. 54, 209–234 (2015)
18. Ly, L.T., Rinderle, S., Dadam, P., Reichert, M.: Mining staff assignment rules from event-based data. In: Business Process Management Workshops. pp. 177–190 (2005)
19. Mangler, J., Pauker, F., Rinderle-Ma, S., Ehrendorfer, M.: centurio.work - industry 4.0 integration assessment and evolution. In: BPM Industry Forum. pp. 106–117 (2019)
20. Mannhardt, F., de Leoni, M., Reijers, H.A., van der Aalst, W.M.P.: Balanced multi-perspective checking of process conformance. Computing 98(4), 407–437 (2016)
21. de Medeiros, A.K.A., van der Aalst, W.M.P., Weijters, A.J.M.M.: Workflow mining: Current status and future directions. In: On The Move to Meaningful Internet Systems. pp. 389–406 (2003)
22. Mobley, R.: An Introduction to Predictive Maintenance. Elsevier (2002)
23. Reichert, M., Weber, B.: Enabling Flexibility in Process-Aware Information Systems - Challenges, Methods, Technologies. Springer (2012)
24. Reinkemeyer, L.: Process Mining in Action – Principles, Use Cases and Outlook. Springer International Publishing (2020)
25. Rinderle, S., Weber, B., Reichert, M., Wild, W.: Integrating process learning and process evolution - A semantics based approach. In: Business Process Managementgs. pp. 252–267 (2005)
26. Rozinat, A., Van der Aalst, W.M.: Conformance checking of processes based on monitoring real behavior. Inf. Syst. 33(1), 64–95 (2008)
27. Shadiya, P., Haleem, P.A.: Energy efficient data formatting scheme: A review and analysis on xml alternatives. Energy 1(1) (2012)
28. Song, M., van der Aalst, W.M.P.: Towards comprehensive support for organizational mining. Decis. Support Syst. 46(1), 300–317 (2008)
29. Stertz, F., Mangler, J., Rinderle-Ma, S.: Data-driven improvement of online conformance checking. In: Enterprise Distributed Object Computing. pp. 187–196 (2020)
30. Stertz, F., Mangler, J., Rinderle-Ma, S.: The role of time and data: Process mining in the manufacturing domain. Business Information Systems Engineering (2020), (submitted to)
31. Stertz, F., Mangler, J., Rinderle-Ma, S.: Temporal conformance checking at runtime based on time-infused process models. CoRR abs/2008.07262 (2020)
32. Stertz, F., Rinderle-Ma, S.: Process histories – detecting and representing concept drifts based on event streams. In: Cooperative Information Systems. pp. 318–335 (2018)
33. Stertz, F., Rinderle-Ma, S.: Detecting and identifying data drifts in process event streams based on process histories. In: CAiSE Forum. pp. 240–252 (2019)
34. Stertz, F., Rinderle-Ma, S., Hildebrandt, T., Mangler, J.: Testing processes with service invocation: Advanced logging in CPEE. In: Service-Oriented Computing. pp. 189–193 (2016)
35. Stertz, F., Rinderle-Ma, S., Mangler, J.: Analyzing process concept drifts based on sensor event streams during runtime. In: Business Process Management. pp. 202–219 (2020)

36. Teinemaa, I., Dumas, M., Rosa, M.L., Maggi, F.M.: Outcome-oriented predictive process monitoring: Review and benchmark. ACM Trans. Knowl. Discov. Data 13(2), 17:1–17:57 (2019)
37. Verenich, I., Dumas, M., Rosa, M.L., Maggi, F.M., Teinemaa, I.: Survey and cross-benchmark comparison of remaining time prediction methods in business process monitoring. ACM Trans. Intell. Syst. Technol. 10(4), 34:1–34:34 (2019)
38. Weijters, A., van Der Aalst, W.M., De Medeiros, A.A.: Process mining with the heuristics miner-algorithm. Technische Universiteit Eindhoven, Tech. Rep. WP 166, 1–34 (2006)

Symbolic Artificial Intelligence Methods for Prescriptive Analytics

Gerhard Friedrich⊙, Martin Gebser⊙ and Erich C. Teppan⊙

Abstract

Prescriptive analytics in supply chain management and manufacturing addresses the question of *"what"* should happen *"when"*, where good recommendations require the solving of decision and optimization problems in all stages of the product life cycle at all decision levels. Artificial intelligence (AI) provides general methods and tools for the automated solving of such problems.

We start our contribution with a discussion of the relation between AI and analytics techniques. As many decision and optimization problems are computationally complex, we present the challenges and approaches for solving such hard problems by AI methods and tools. As a running example for the introduction of general problem-solving frameworks, we employ production planning and scheduling.

G. Friedrich (✉) · E. C. Teppan
Institute for Artificial Intelligence and Cybersecurity, Universität Klagenfurt, Klagenfurt, Austria
e-mail: gerhard.friedrich@aau.at

E. C. Teppan
e-mail: erich.teppan@aau.at

M. Gebser
Institute for Artificial Intelligence and Cybersecurity/Institute of Software Technology, Universität Klagenfurt/TU Graz, Klagenfurt, Austria
e-mail: martin.gebser@aau.at

© The Author(s), under exclusive license to Springer-Verlag GmbH, DE, part of Springer Nature 2023
B. Vogel-Heuser and M. Wimmer (eds.), *Digital Transformation*,
https://doi.org/10.1007/978-3-662-65004-2_16

385

First, we present the fundamental modeling and problem-solving concepts of constraint programming (CP), which has a long and successful history in solving practical planning and scheduling tasks. Second, we describe highly expressive methods for problem representation and solving based on answer set programming (ASP), which is a variant of logic programming. Finally, as the application of exact algorithms can be prohibitive for very large problem instances, we discuss some methods from the area of local search aiming at near-optimal solutions. Besides the introduction of basic principles, we point out available tools and practical showcases.

Keywords

Prescriptive Analytics • Artificial Intelligence • Problem-Solving

1   Introduction

New technologies such as the internet of things and cyber-physical systems are considered as the main drivers of a new type of industrialization called *Industrie 4.0* [56]. The massive usage of sensors in factories, transportation systems, machines and products as well as their interconnection allow a much more accurate real-time description of the state of manufacturing systems and the whole supply chain, including a detailed, comprehensive, high-quality description of their history. Based on this data, decision-making can be taken to a new level of quality. Combined with new opportunities in manufacturing technologies (such as additive manufacturing) and robotics, the next era of industrialization will be realized. This era will result in dynamic business and engineering processes, individualized competitive products and services, as well as higher resource productivity and efficiency. The significant improvements of perceiving the environment through sensors and acting upon the environment through actuators create huge opportunities for optimized decision-making based on analytics and artificial intelligence (AI). These technologies are considered vital for Industrie 4.0 [35]. To profit from the methods of these fields for implementing software solutions, we have to understand their relation.

Analytics and artificial intelligence Analytics and AI are close relatives. In analytics we distinguish descriptive, predictive, and prescriptive analytics [22, 81].

Descriptive analytics describes the state of the system under consideration at any given moment. For this task, various data and sensor information sources are consolidated, which may require data cleaning and statistical models to reconstruct missing and uncertain data. E.g., descriptive analytics is employed to provide the state of a factory.

Predictive analytics deals with the projection of future system states. This projection is based on models of the system and has to provide the computation of effects of possible actions which change the system state. E.g., predictive analytics should predict the effects of replacing a machine on the throughput or predicting the production quality of a machine at a future time point.

Prescriptive analytics generates recommendations for decisions depending on the actual state of affairs, which is provided by descriptive analytics, and on forecasting future states, which are supplied by predictive analytics. The goal of prescriptive analytics in Industrie 4.0 is to improve decision-making in supply chain management and manufacturing [82], addressing the question of *"what"* should happen *"when"* [61]. Prescriptive analytics provides recommendations for decisions in all product life cycle stages, ranging from engineering to manufacturing and finally to return and disposal.

Consequently, descriptive and predictive analytics can be seen as essential preparation steps for prescriptive analytics. E.g., prescriptive analytics provides decisions for the reconfiguration of a factory or the scheduling of operations.

There is a close relation of analytics techniques to the architecture of intelligent agents in AI, particularly to *model-based, utility-based agents* [79]. We adopt for AI the standard definition of the classical textbook on AI [79], which is the design and implementation of rational agents: *AI concentrates on general principles of rational agents and on components for constructing them.* Utility-based agents take sensor information as input and compute the likely states of a system employing models. In AI terms, this is a diagnosis task [59], for which reasoning under uncertainty or model-based diagnosis is applied [14]. Consequently, descriptive analytics is closely related to diagnosis in AI.

Based on the description of the state of a system model, utility-based agents employ planning methods to generate actions such that the objective of a system is met or optimized. For this task, models of the effects of actions and models of the system behavior are exploited to predict future states. Consequently, this model-based prediction is closely related to predictive analytics, and the planning task resembles prescriptive analytics [80].

Besides model- and utility-based agents, AI also deals with *learning agents*. Conceptually, we can distinguish two application areas of learning. (1) Agents can learn about the environment and its causalities to improve the system model. E.g., an agent may explore a workshop to learn which machines are available, how these machines operate, and when maintenance is needed [20]. (2) Agents can learn which actions are most appropriate for a particular system state to optimize or meet an objective. E.g., an agent may learn to assign operations of jobs to machines [72]. This knowledge (also termed as heuristics) can be exploited in the search for solutions. Search in this context means that an agent does not only consider a single action which should be executed in a certain system state, but the agent reasons about alternative actions that are applicable. E.g., an agent considers all applicable operations of all jobs and alternative assignments of these jobs to machines. Learning of the system model is an essential task for descriptive and predictive analytics, whereas learning to select the "right" actions is of particular importance for prescriptive analytics to deal with computationally hard decision or optimization problems.

As noted above, analytics comprises three crucial areas for optimized decision-making. While descriptive and predictive analytics are presented in accompanying chapters of this book, we focus on prescriptive analytics and symbolic AI methods built on high-level system models, which distinguish them from data-driven subsymbolic approaches (e.g., recent deep learning methods).

Prescriptive analytics and Industrie 4.0 The purpose of decision-making in Industrie 4.0 is, in many cases, the optimization of an objective, including the engineering of products and their production. For example, in the engineering and configuration process [31] of products, the objective function is usually to fulfill all the customer requests and to minimize the production costs. Various factors could complicate this engineering step. E.g., products might be manufactured at different plants with different availability and various types of subparts and materials. Consequently, engineering and configuration of products could be a highly complicated optimization problem connected to manufacturing and sourcing.

Manufacturing itself is one of the most studied areas for optimization. There are numerous variants of production processes spanning from continuous production to job, batch, or serial production. However, although the modes of production are very diverse, in all these areas, optimization is the key for efficient production processes, including the layout and the engineering of the plant and its machines. E.g., depending on the jobs, the modes of machines and the layout of the plant are subject to change to optimize the objectives. A further classical area of optimization is delivery, including vehicle routing problems.

To sum up, there is a massive demand for decision-making and optimization in all areas of production and product planning, including engineering, sourcing, making, delivering, as well as returning and disposal. Consequently, the efficient solving of decision and optimization problems is one of the most important goals for the industry. AI and analytics techniques provide a rich set of methods and tools to support these decision-making and optimization tasks.

Challenges and solutions of hard problems For solving decision and optimization problems, two major challenges have to be mastered. *First,* many decision and optimization problems turn out to be NP-hard [34]. NP-hardness means that, given the current state of the art, the worst-case runtime behavior of a problem-solving algorithm will grow exponentially with the size of the problem instance. E.g., for any known problem-solving algorithm, it might be the case that this algorithm is inapplicable to a particular scheduling problem and its instances. Formulating it positively, it might be the case that an optimization algorithm works for a particular optimization problem and its instances, but we cannot predict this in general. Consequently, there is a race for efficient algorithms and methods to solve as many NP-hard decision and optimization problems as possible in practice. The most notable fields which are participating in this race are operations research and AI. However, the borders are not strict, and parts of computer science that deal with algorithms, databases, and computational complexity must also be considered.

To cope with NP-hardness, different approaches can be pursued. (1) Some research groups focus on the development of algorithms that are as efficient as possible for all problems formulated in a domain-independent specification language, allowing for the representation of NP-hard problems [18, 47]. (2) Some research teams and vendors focus on specific domains such as vehicle routing problems, scheduling, or timetabling [66]. In this case, problems are still NP-hard, but the algorithms exploit certain properties of the application domain. E.g., in job-shop scheduling, jobs may not overlap on a machine, and scheduling algorithms take advantage of this constraint. (3) Some groups try to design specialized algorithms which are tailored to a particular NP-hard problem and its instances, i.e., a unique vehicle routing problem for a particular supermarket chain [53].

Typically, solving NP-hard problems is based on search algorithms [79]. Search algorithms make decisions that might be wrong and have to be revised. Unfortunately, there is no way to efficiently compute correct decisions that will lead to an optimum. Consequently, we are interested in designing efficient heuristics such that these decisions are correct most of the time. Depending on the scope of the algorithm, such heuristics may be designed for a specific application, a domain, or for all NP-hard problems that can be formulated in a particular domain-independent specification language.

The exploitation of heuristics is common in all three mentioned approaches. However, the three approaches differ in the utilization of various search algorithms. For approaches (1) and (2), tools exist where the search algorithm is fixed but can be adapted by parameters and heuristics. Naturally, approach (3) provides the most freedom for implementing a specific decision or optimization algorithm. In the latter case, various problem-solving methods were developed, such as informed or local search strategies [79].

The second major challenge which has to be mastered is the efficient specification of the problem and its heuristics since the decision and optimization problems differ substantially in their constraints, parameters, variables, and objective functions. Note that this also includes the adaptation of the specification because constraints and objective functions are subject to continuous changes. In fact, in the life cycle of optimization software, the effort for adaptation might be significantly higher than the effort for an initial implementation [28].

Approaches (1) and (2) provide languages that allow the declarative specification of the problem to be solved, i.e., the user provides a correct and complete specification of the properties of solutions. Ideally, no specification of heuristics and programming of algorithms are needed. However, depending on the particular problem and its instances, it might be necessary to formulate heuristics that guide the search process. In approach (3), the user has to program the search process, which in most cases results in additional implementation and adaptation effort compared to approaches (1) and (2) [33]. Moreover, the implementation of such search algorithms needs considerable knowledge about state-of-the-art methods.

Figure 1 depicts the architecture of a problem-solving component which is typical for approaches (1) and (2). The problem solver takes as inputs the specification of the problem, a specification of a particular problem instance, and optionally some heuristics. The outputs are solutions for the problem instance. E.g., the problem specification can be the description of a factory and the objective is the minimization of the makespan, so that the outputs are schedules where the makespan for completing given jobs is minimal. A specific problem instance is a set of jobs which have to be manufactured. Optionally we might specify as a heuristic that the problem solver should prefer job assignments to machines during the search process where the total remaining work of the job is maximal.

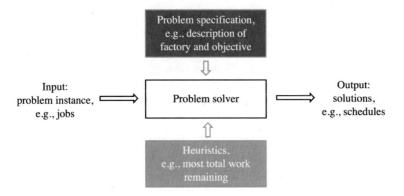

Fig. 1 Architecture of a problem-solving component

In principle, approach (3) could also follow this architecture to avoid the adaptation of program code if the problem specification changes. However, in practice, the software engineer has to decide which part of the problem specification is expressed declaratively by data and which part of the specification is implicitly contained in the program code. Consequently, all possible adaptations of the problem specification have to be anticipated.

Regarding approach (1), two broadly applied general problem-solving frameworks in AI are constraint programming (CP) [21] and logic programming (LP) [7]. CP computes solutions for constraint satisfaction problems (CSPs). Such problems are defined by variables, their value domains, and constraints that specify the allowed assignments of values to variables. Moreover, objective functions can be defined to assign numerical values to allowed variable/value assignments. Problem solvers for CSPs output allowed value assignments to variables or report unsatisfiability. In case of optimization problems, such solutions also optimize the objective function. Note that operations research problems can be encoded as CSPs. Consequently, CP systems like CP-Optimizer [60] or OR-Tools [70] integrate algorithms from AI and operations research.

LP exploits a logical description of the problem based on first-order predicate logic. The problem and its instances are encoded by rules and facts, the latter corresponding to tuples of relational database tables. Consequently, LP is closely related to relational and deductive databases. Within LP, one of the currently most popular representation and reasoning frameworks is answer set programming (ASP) [29, 49]. ASP allows to reason about the absence of knowledge and provides disjunctive conclusions, aggregation, as well as the formulation of objective functions.

In this article, we focus on the introduction of the domain-independent CP and ASP approaches. As it is the nature of domain-independent problem-solving, such methods can be applied to various decision and optimization problems in Industrie 4.0. However, we will resort to a well-known problem for the introduction of these two problem-solving frameworks, employing the scheduling of jobs in a flexible workshop as a running example.

Section 2 describes this conceptually simple flexible job-shop scheduling problem. Section 3 shows how the problem can be encoded and solved by CP, and Sect. 4 presents the ASP approach. Moreover, for applications where domain-independent approaches fail, so-called local search algorithms may still be applied successfully [52]. In Sect. 5, we present the basic concepts of local search algorithms, and we conclude with a summary in Sect. 6.

2   Running Example: Flexible Job-Shop Scheduling

As a running example, we introduce the flexible job-shop scheduling problem (FJSSP) [10]. The FJSSP is an optimization problem found in many industrial production lines. The solution of a FJSSP specifies *which* job should be processed by *which* machine *when*, and therefore FJSSP is a classical task for prescriptive analytics. FJSSP can be defined as follows:

- Given is a set $M = \{machine_1, \ldots, machine_m\}$ of machines and a set $J = \{job_1, \ldots, job_j\}$ of jobs.
- Each job $j \in J$ consists of a sequence of operations $O_j = \{j_1, \ldots, j_{l_j}\}$ whereby j_{l_j} is the last operation of job j.
  Practically, jobs can be interpreted as products, and operations can be interpreted as their production steps. For a job j and its operation j_i, the operation j_{i+1} is called successor, and the operation j_{i-1} is called predecessor.
- For each operation o, there is a set of machines $M_o \subseteq M$ representing the machines that are able (or allowed) to process operation o.
- For each operation o and machine $m \in M_o$, there is a predefined processing time $pTime_{o,m} \in \mathbb{N}$.
- A (consistent and complete) schedule consists of a machine dedication $machine_o \in M_o$ and a starting time $start_o \in \mathbb{N}$ for each operation o such that:
  - An operation's successor starts only after the operation is finished,
    i.e., for a job j and its operations j_i and j_{i+1}:
    * $start_{j_{i+1}} \geq start_{j_i} + pTime_{j_i,machine_{j_i}}$
  - Operations processed by the same machine are non-overlapping,
    i.e., concerning two operations $o1 \neq o2$ with $machine_{o1} = machine_{o2}$:
    * $start_{o1} \leq start_{o2} \rightarrow start_{o1} + pTime_{o1,machine_{o1}} \leq start_{o2}$
- Some objective function is to be minimized. One of the most classic optimization criteria in this context is the completion time C_{max}, i.e., the period needed for processing all operations:
  - $max\{start_o + pTime_{o,machine_o} : o \in O_j, j \in J\} \rightarrow min.$

Table 1 Example: job operations and processing times for all machines

OpId	JobId	Pos	Type	M-1	M-2	M-3	M-4
1	1	1	A	11	10		
2	1	2	B		6	7	
3	1	3	C				6
4	2	1	B		8	9	
5	2	2	C				8
6	2	3	A	6	5		
7	3	1	C				6
8	3	2	A	11	10		
9	3	3	B		5	6	
10	4	1	A	6	5		
11	4	2	C				3
12	4	3	B		13	14	

As a (simple) example scenario, take the following: There is a small production line with four machines, each of which supports one or two operation types.[1] Machine M-1 supports type A, M-2 supports types A and B, M-3 supports type B, and M-4 supports type C. Hence, M-2 brings in flexibility that it can process all operations that M-1 or M-3 can process. Furthermore, let us imagine that M-2 is the more recent machine to process operations slightly faster than M-1 or M-3. Jobs that can be executed in this production line consist of a sequence of operations of types A, B, and C.

Table 1 provides the input data for a small FJSSP problem instance comprising four jobs, each of which consists of three operations. E.g., the first operation (i.e., its position *Pos* in the job is 1) of job 1 (*JobId* = 1) is of type A and consequently can be processed by M-1 or M-2, whereby M-2 is a little bit faster (10 vs. 11 time units). Figure 2 depicts an optimal solution with a completion time $C_{max} = 23$, indicating the execution of operations on the machines M-1 to M-4, with their supported operation types listed in parentheses, in separate rows. Operations are marked with their *OpId* identifier and their job *JobId* followed by the position *Pos* of an operation in its job, where different colors also distinguish the jobs.

3 Constraint Programming

Constraint programming (CP) [78] is a declarative approach that builds on solving constraint satisfaction problems (CSPs). Hence, a problem at hand must be first represented as a CSP

---

[1] The concept of operation types is not needed to define the FJSSP. However, this concept is often given in real environments representing machine abilities.

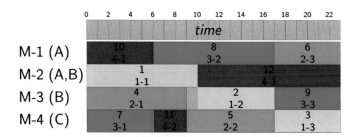

Fig. 2 Running example: optimal solution with $C_{max} = 23$

that a CP solver then solves. A CSP is defined as a triple $\langle V, D, C \rangle$, whereby V is a set of variables, D is a set of value domains so that for each $v \in V$ there is exactly one $d \in D$, and C is a set of constraints imposed on the variables.

One of the most classic variants of CSPs uses variables with finite and discrete domains. Consequently, the domain sizes are limited in the form of lower and upper bounds, and the domain values are discrete, typically integers, in which case variables are called *integer variables*. Depending on the solver used, different constraints can be employed to restrict value combinations for variables incorporated by the constraint. For example, *primitive constraints* express that some variable's value must be equal/unequal/smaller/greater than some other variable's value or constant value (e.g., $v1 \neq v2$, $v1 > 6$). *Arithmetic constraints* express arithmetic operations like $v1 + v2 = v3$. *Global constraints* restrict value combinations for a set of variables and typically implement algorithms for solving special sub-problems which, without such a global constraint, have to be represented with primitive and arithmetic constraints. Hence, global constraints significantly increase the expressive power. Take as an example the *alldifferent* global constraint, assuring that no two variables out of some predefined set of variables have the same value. For instance, having three variables $v1$, $v2$, and $v3$ with equal domains ranging from 1 to 3, *alldifferent*($\{v1, v2, v3\}$) states that exactly one variable takes the value 1, 2, and 3, respectively. With primitive constraints, three constraints would be needed to express the same, i.e., $v1 \neq v2$, $v2 \neq v3$, $v1 \neq v3$. Even when 1,000 variables should get different values, still a single *alldifferent* constraint suffices, whereas, without this global constraint, roughly half a million primitive constraints are needed.

Similarly to global constraints that lift expressive power for particular sub-problems, higher-order types of variables, e.g., *set variables*, have been introduced. Whereas a finite-domain integer variable takes values from a finite set of integers, a finite-domain set variable takes values from the power set of a finite set of integers. The expressive power of set variables is exponentially higher as they can be assigned to any of the 2^n subsets of values for a domain of size n.

Another type of higher-order variable is an *interval variable*. Interval variables are designed to meet the special requirements of scheduling problems. Interval variables are well-suited for representing time intervals, as interval variables consist of a start time from an integer domain, a duration, and an end time. An interval variable automatically enforces an arithmetic constraint, i.e., $start + duration = end$. By employing interval variables, a job operation can be represented by a single (interval) variable.

3.1 Flexible Job-Shop Scheduling: CP Formulation

In the following, we show how the flexible job-shop problem discussed in Sect. 2 can be expressed as a CSP. To this end, we use the high-level constraint language OPL from IBM.[2] To be read as an OPL problem encoding, the input data of Table 1 is represented as sets of tuples. Listing 1 displays the input file.

```
Params = <4,4>;                                                          1
                                                                         2
Ops = {<1,1,1>, <2,1,2>, <3,1,3>, <4,2,1>, <5,2,2>, <6,2,3>, <7,3,1>,    3
       <8,3,2>, <9,3,3>, <10,4,1>, <11,4,2>, <12,4,3>};                  4
                                                                         5
Modes = {<1,1,11>, <1,2,10>, <2,2,6>, <2,3,7>, <3,4,6>, <4,2,8>, <4,3,9>, 6
         <5,4,8>, <6,1,6>, <6,2,5>, <7,4,6>, <8,1,11>, <8,2,10>, <9,2,5>, 7
         <9,3,6>, <10,1,6>, <10,2,5>, <11,4,3>, <12,2,13>, <12,3,14>};   8
```

Listing 1 Sample OPL input file for the example input data in Table 1

In Line 1, some meta-information about the input is given, in particular the number of jobs and the number of machines in the problem instance. The set *Ops* (Lines 3–4) contains three-tuples of the form $\langle OpId, JobId, Pos \rangle$, i.e., the first three columns in Table 1. The set *Modes* (Lines 6–8) contains three-tuples of the form $\langle OpId, MachineId, ProcessingTime \rangle$ and, hence, represents the last four columns of Table 1. Note that *Modes* only contains tuples for possible machine assignments. Thus, the empty fields in Table 1 are not represented.

Listing 2 shows an encoding of the flexible job-shop problem in OPL. Line 1 states that the CP library is used.[3] Lines 3–18 define three tuple data types[4] (*paramsT*, *Operation*, *Mode*) to be filled with the *Params* tuple, an *Ops* tuple or a *Modes* tuple from the input file.

---

[2] https://www.ibm.com/products/ilog-cplex-optimization-studio

[3] In OPL, it is also possible to write (mixed) integer linear programs, quadratic or continuous programs.

[4] This is similar to structs in the C language.

Lines 20–22 do the actual read-in of the input data. In particular, the *Params* tuple from the input is read into the variable of type *paramsT*, the *Ops* tuples are read into a variable that takes a set of *Operation* tuples and analog for the *Modes* tuples.

In Lines 24 and 25 two integer ranges are defined for reuse with array data structures. In Line 27 an integer array is defined and initialized that saves the last job positions for all jobs, corresponding to a pointer to the last operation in a job. Up to here, the encoding is concerned with the read-in of the input data and some preparations primarily for convenience reasons.

The rest of the encoding, Lines 29–41, contains the actual declarative problem representation. In Line 29, an array of interval variables is defined so that for each *Operation* in *Ops* there is an interval variable. Note that the domains are not set explicitly such that the lower bound is 0 and the upper bound is set automatically by the solver.

In Line 30, an array of interval variables for each mode is defined. Those variables also take the processing times as input (i.e., *size md.pt*). Hence, there is an interval variable for each possible operation-to-machine combination. Recall that it is not clear beforehand on which machine a particular operation will be scheduled. For this reason, the variables are set as *optional*, which means that they can be active or not, depending on whether an operation is scheduled on a specific machine or not.

```
using CP;                                                                       1
                                                                                2
tuple paramsT {                                                                 3
     int nbJobs;                                                                4
     int nbMchs;                                                                5
};                                                                              6
                                                                                7
tuple Operation {                                                               8
  int opId;                                                                     9
  int jobId;                                                                   10
  int pos;                                                                     11
};                                                                            12
                                                                             13
tuple Mode {                                                                  14
  int opId;                                                                   15
  int mch;                                                                    16
  int pt;                                                                     17
};                                                                            18
                                                                             19
paramsT      Params = ...;                                                    20
{Operation} Ops    = ...;                                                     21
{Mode}       Modes  = ...;                                                    22
                                                                             23
range Jobs = 1..Params.nbJobs;                                                24
range Mchs = 1..Params.nbMchs;                                                25
                                                                             26
int jlast[j in Jobs] = max(o in Ops: o.jobId==j) o.pos;                       27
                                                                             28
dvar interval ops  [Ops];                                                     29
dvar interval modes[md in Modes] optional size md.pt;                         30
dvar sequence mchs [m in Mchs] in all(md in Modes: md.mch == m) modes[md]; 31
                                                                             32
minimize max(j in Jobs, o in Ops: o.pos==jlast[j]) endOf(ops[o]);             33
subject to {                                                                  34
   forall (o in Ops)                                                          35
```

```
      alternative(ops[o], all(md in Modes: md.opId==o.opId) modes[md]);       36
  forall (m in Mchs)                                                          37
    noOverlap(mchs[m]);                                                       38
  forall (j in Jobs, o1 in Ops, o2 in Ops: o1.jobId==j && o2.jobId==j &&      39
    o2.pos==1+o1.pos) endBeforeStart(ops[o1],ops[o2]);                        40
}                                                                             41
```

Listing 2 OPL encoding of flexible job-shop scheduling for input data as in Listing 1

The interval variables for operations (Line 29) and the (optional) interval variables for the modes are linked together in Lines 35–36. The *alternative* statements express that among all possible alternatives for an operation variable, i.e., all corresponding mode variables, exactly a single one can be active. The operation variable's start, end and processing time values are set equal to its active alternative. The question about which alternative to set active is part of the problem and to be determined by the solver.

In Line 31, an array of helping *sequence* variables is defined, each of which stores a list of modes for a machine. Those variables are used in Lines 37–38 for establishing *noOverlap* constraints, assuring that no two operations can be processed in parallel on the same machine.

Finally, Lines 39–40 define constraints that guarantee that each successor operation can start only after its preceding operation is finished.

3.2   Tools and Application Fields

CP has a long tradition in operations research, and consequently, there is an armada of tools on the market, and there are also many fields of application where CP has been successfully deployed.

Tools A good overview of CP solvers is given by the MiniZinc challenge.[5] This challenge is an annual event where different CP solvers that support the MiniZinc constraint language – MiniZinc is very similar to IBM's OPL constraint language – are compared based on benchmark tests. For several years Google's solver OR-Tools [70] has dominated the scene and has won the MiniZinc challenges by a large margin in almost every benchmark category.

On the other hand, some solvers do not support the MiniZinc constraint language and therefore do not participate in the MiniZinc challenge. Here in particular IBM's CP-Optimizer [60] has to be mentioned.[6]

Recent research comparing the performance of OR-Tools and CP-Optimizer (both mounted by a Java interface) indicates that CP-Optimizer performs even better than OR-Tools, at least for scheduling-like problems [18]. It has to be noted that OR-Tools is open source, whereas CP-Optimizer is proprietary.

---

[5] https://www.minizinc.org/challenge.html
[6] https://www.ibm.com/analytics/cplex-cp-optimizer

Application fields Prescriptive analytics is closely related to *planning* in the field of AI and there is a long history of CP approaches for planning. Many planning problems constitute application fields for prescriptive analytics where CP approaches are successful [80].

The whole area of production planning, including variants of open-, flow-, and (flexible) job-shop scheduling problems [10], is a very successful application area for CP (see, e.g., [17, 18]), which is becoming more and more important in the context of Industrie 4.0 and automation.

Another famous CP application area is transport and logistics, including many (vehicle) route optimization problems [73], or timetabling problems [83]. Planning tasks in robotics domains [11] constitute a sub-area important for production planning, transport, and logistics.

A further application field for CP with a direct connection to prescriptive analytics is the area of knowledge-based configuration concerned with finding suitable machine/device/process/product setups to meet or optimize certain predefined criteria [16]. A relevant application field similar to knowledge-based configuration is knowledge-based recommendation and, to put it in more general terms, expert and decision support systems [30, 32].

4   Answer Set Programming

Answer set programming (ASP) [49, 63] is a declarative branch of logic programming (LP) [7] in which a computational problem is represented by a set of logical rules such that specific truth assignments (of Boolean variables), called answer sets, satisfying the rules represent problem solutions. The typical ASP workflow is displayed in Fig. 3, where the automated reasoning process involves the two stages of grounding and solving. Grounding first instantiates a general problem encoding, comprising rules incorporating universally quantified placeholder variables, relative to facts specifying a particular problem instance. The resulting propositional problem representation in terms of Boolean variables, also called atoms, is then (internally) passed on to a solving component that searches for and optimizes answer sets. In the currently most widely used ASP systems, Clingo [39] and I-DLV [12], grounding is implemented by employing deductive database techniques [2]. Their subsequently applied

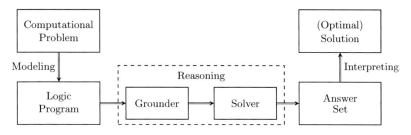

Fig. 3 ASP workflow in which automated reasoning consists of grounding and solving

back-end solvers, Clasp [42] and Wasp [3], extend conflict-driven clause learning (CDCL) [86], the state-of-the-art complete algorithm for solving Boolean satisfiability (SAT) and its optimization version MaxSAT [9], from classical propositional logic to ASP.

Similar to relational databases, a *fact* of the form $p(c_1, \ldots, c_n)$., where the predicate p constitutes the name of some n-ary relation, and the arguments $c_1, \ldots, c_n$ are constants, expresses that the relation denoted by p contains the tuple $\langle c_1, \ldots, c_n \rangle$ as an element. The latter condition is equivalently characterized by a truth assignment mapping the atom $p(c_1, \ldots, c_n)$ to true, as customarily indicated by including $p(c_1, \ldots, c_n)$ as a member in the set of true atoms. That is, a set of true atoms represents a truth assignment. At the syntactic level, predicates and constants are identifiers starting with a lowercase letter. Constants can additionally be integers. Facts and rules (see below) are completed by a period. For example, the fact $age(mary, 42)$. states that the atom $age(mary, 42)$ must be true, which is the case for the truth assignment $\{age(mary, 42)\}$ and its supersets of atoms.

A *rule* is an implication $h :- b_1, \ldots, b_m$., expressing that its head h must be satisfied if all of the body literals $b_1, \ldots, b_m$ hold. Each body literal b_i is either an atom a or its negation $not\ a$, which holds for any truth assignment including or excluding a, respectively. The head h can be an atom a or a choice of the form $l\ \{a_1 : b_{1_1}, \ldots, b_{m_1}; \ldots; a_k : b_{1_k}, \ldots, b_{m_k}\}\ u$. In the second case, $a_1, \ldots, a_k$ are atoms, each b_{j_i} is again a body literal, and l and u are optional integers specifying a lower or an upper cardinality bound, respectively. If l is omitted, it defaults to 0, whereas the default value for an omitted upper cardinality bound u is ∞. We do not write the symbol : after a_i if $m_i = 0$ (i.e., the choice contains no body literal b_{j_i}), and the head h of a rule $h :- b_1, \ldots, b_m$. can be left blank if it is the choice $1\ \{\}$, in which case the rule is also called an *integrity constraint*. As for body literals, a truth assignment satisfies a head atom a if it includes a. A choice $l\ \{a_1 : b_{1_1}, \ldots, b_{m_1}; \ldots; a_k : b_{1_k}, \ldots, b_{m_k}\}\ u$ is satisfied if we have $l \leq |A| \leq u$ for the set A of atoms a_i included in the truth assignment. Moreover, a rule is satisfied by a truth assignment if it satisfies the head h in case all of the body literals $b_1, \ldots, b_m$ hold, thus resembling the classical semantics of implications with a conjunction of literals in the antecedent. Note that a truth assignment must falsify some body literal b_i to satisfy an integrity constraint, as its head $1\ \{\}$ is always unsatisfied given that $1 > |\emptyset|$.

An ASP *program* is a set consisting of facts and rules, and it is satisfied by any truth assignment satisfying all of its facts and rules. In line with the areas of deductive databases and logic programming, the semantics of ASP programs additionally builds on the complete knowledge assumption [75], accepting an atom to be true only if it appears as a fact or can be derived by a rule whose body literals hold. Roughly speaking, an answer set of an ASP program is a truth assignment satisfying the program, where for each true atom a, the program must contain the fact a ., some rule $a :- b_1, \ldots, b_m$. for which $b_1, \ldots, b_m$ hold, or a choice rule $l\ \{a_1 : b_{1_1}, \ldots, b_{m_1}; \ldots; a : b_{1_i}, \ldots, b_{m_i}; \ldots; a_k : b_{1_k}, \ldots, b_{m_k}\}\ u :- b_1, \ldots, b_m$. such that $b_{1_i}, \ldots, b_{m_i}$ hold in addition to $b_1, \ldots, b_m$. This informal account merely gives the basic intuition of answer sets, and we refer the interested reader to [63] for a precise characterization of the semantics of ASP programs, which further addresses and denies circular derivations like of the atoms `says(a)` and `says(b)` by the rules `says(a) :- says(b).` and `says(b) :- says(a).`

Considering the program consisting of the rules `x :- y.` and `{y} :- not z.`, abbreviating the atoms `x()`, `y()`, and `z()` associated with 0-ary relations by x, y, and z, we get the empty answer set $\emptyset$ as well as $\{x, y\}$. No truth assignment including z, e.g., $\{z\}$, is an answer set because the program does not contain any rule with z in the head so that the atom z cannot be derived. A similar issue applies to the truth assignment $\{x\}$, as the body of the only rule by which x could be derived, `x :- y.`, does not hold. The same rule `x :- y.` is unsatisfied by $\{y\}$, and thus the truth assignment $\{y\}$ is not an answer set either. This yields that the truth assignments $\emptyset$ and $\{x, y\}$, which satisfy all program rules and whose atoms can be derived, are all answer sets of the program.

For *optimizing* answer sets, an ASP program can be joined with a `#minimize{`$c_{1_1}, \ldots, c_{l_1}$ `:` $b_{1_1}, \ldots, b_{m_1}; \ldots; c_{1_k}, \ldots, c_{l_k}$ `:` $b_{1_k}, \ldots, b_{m_k}$`}.` statement such that each b_{j_i} is a body literal and the constants c_{j_i} supply tuples $\langle c_{1_i}, \ldots, c_{l_i} \rangle$, whose first elements $c_{1_1}, \ldots, c_{1_k}$ are integers and remainders $c_{2_i}, \ldots, c_{l_i}$ are usually obtained by substituting placeholder variables (see below) that are appended to disambiguate multiple occurrences of the same integer. An answer set then induces the set of distinct tuples for which $b_{1_i}, \ldots, b_{m_i}$ hold. Its associated weight is the sum of integers c_{1_i} over all tuples in the set. An answer set is optimal if its weight is minimal. For example, reconsidering the answer sets $\emptyset$ and $\{x, y\}$ together with `#minimize{1 : x; 2 : not y}.`, the associated weight is 2 in case of $\emptyset$ and 1 for $\{x, y\}$ so that the latter answer set is optimal.

To make the problem modeling step, indicated on the left of Fig. 3, rapid and flexible, users typically devise general *encodings* with rules and optimization statements applicable to arbitrary problem instances given by facts. The input language of ASP systems thus supports universally quantified placeholder variables in encodings that are instantiated to a propositional problem representation in the initial grounding stage of the reasoning process. Universally quantified variables can occur in the same positions as constants and are distinguished by identifiers starting with an uppercase letter, e.g., `C` instead of `c`. For a rule `h :- `$b_1, \ldots, b_m$`.`, different substitutions by constants for variables occurring in some body literal b_i lead to separate rule instances. Similar substitutions for remaining variables in a choice `l {`a_1 `:` $b_{1_1}, \ldots, b_{m_1}; \ldots; a_k$ `:` $b_{1_k}, \ldots, b_{m_k}$`}` u or a `#minimize{`$c_{1_1}, \ldots, c_{l_1}$ `:` $b_{1_1}, \ldots, b_{m_1}; \ldots; c_{1_k}, \ldots, c_{l_k}$ `:` $b_{1_k}, \ldots, b_{m_k}$`}.` statement contribute to the respective expressions gathered in curly brackets so that the contents of choices and optimization statements also adjust to problem instances. See [37] for more information on the input language of ASP systems and the safety condition presupposed for so-called intelligent grounding.

As an example, let us investigate the following ASP program:

```
mandatory(starter,soup,4).   mandatory(starter,salad,6).
mandatory(main,stew,10).     mandatory(main,fish,14).
optional(chips).             discount(fish,chips,3).
```

```
incompatible(soup,stew).

course(C)  :- mandatory(C,D,P).

1 {select(D)  : mandatory(C,D,P)} 1  :- course(C).
  {select(D)} :- optional(D).

:- incompatible(D1,D2), select(D1), select(D2).

#minimize{P,D : mandatory(C,D,P), select(D);
    -R,D1,D2 : discount(D1,D2,R), select(D1), select(D2)}.
```

The idea is that facts of the mandatory predicate list dishes from two categories together with their prices: The starters soup and salad cost 4 or 6 dollars, respectively, and the main courses stew and fish are offered for 10 or 14 dollars. A fact of the optional predicate introduces chips as available supplement. The discount fact announces a special deal: Guests ordering fish together with chips receive a price reduction of 3 dollars. The last fact of the incompatible predicate tells us that a meal should not be composed of the starter soup together with stew as the main course.

While the facts can be viewed as the problem instance, the rules written below them constitute the general encoding. Four propositional rule instances are obtained from the first of these rules:

```
course(starter)  :- mandatory(starter,soup,4).
course(starter)  :- mandatory(starter,salad,6).
course(main)     :- mandatory(main,stew,10).
course(main)     :- mandatory(main,fish,14).
```

Since their body literals are directly given by mandatory facts, the head atoms course(starter) and course(main) must be included in every answer set. The choice rules of the encoding are then instantiated relative to starters, main courses, or optional supplements, respectively:

```
1 {select(soup)  : mandatory(starter,soup,4);
   select(salad) : mandatory(starter,salad,6)} 1 :- course(starter).
1 {select(stew)  : mandatory(main,stew,10);
   select(fish)  : mandatory(main,fish,14)} 1 :- course(main).
  {select(chips)} :- optional(chips).
```

The first two rule instances express that exactly one of the atoms select(soup) and select(salad), as well as one of select(stew) and select(fish), must be included in every answer set. Since the omitted lower and upper cardinality bounds in the third rule instance default to 0 and ∞, the inclusion of select(chips) can be decided freely. The following instance of the integrity constraint on incompatible dishes further asserts that an answer set cannot contain both `select(soup)` and `select(stew)`:

```
:- incompatible(soup,stew), select(soup), select(stew).
```

The encoding part addressed so far leads to six answer sets including the atoms derived from facts, `course(starter)` and `course(main)`, as well as different combinations of atoms of the `select` predicate: `select(soup)` and `select(fish)`, `select(salad)` and `select(stew)`, or `select(salad)` and `select(fish)`, and either of the three previous pairs along with `select(chips)`. It remains to optimize answer sets given the instance of the optimization statement:

```
#minimize{4,soup      : mandatory(starter,soup,4), select(soup);
          6,salad     : mandatory(starter,salad,6), select(salad);
          10,stew     : mandatory(main,stew,10), select(stew);
          14,fish     : mandatory(main,fish,14), select(fish);
       -3,fish,chips  : discount(fish,chips,3), select(fish), select(chips)}.
```

As it turns out, the least weight of 15 is obtained for the answer set including `select(soup)`, `select(fish)`, and `select(chips)`, where `soup` contributes its price of 4 dollars, `fish` adds another 14 dollars, and the special deal for `fish` and `chips` gives a price reduction of 3 dollars. Note that, although the integers for prices or reductions, respectively, are distinct for the problem instance under consideration, it may well be that other facts specify the same price for different dishes so that tuples like ⟨4, soup⟩ are necessary for disambiguating integers contributing to the weight of an answer set. Moreover, the grounding stage includes evaluating arithmetic expressions to integers in the propositional problem representation, e.g., the integer 3 giving the discount for `fish` and `chips` is turned into −3 employing the unary − operator. Further arithmetic operators like binary − and + are available as well. Without going into detail, we remark that back-end solvers like Clasp and Wasp, shown as *Solver* in Fig. 3, are customizable and supply several complete optimization algorithms, where model- and core-guided methods [67] are the two principal approaches.

4.1   Flexible Job-Shop Scheduling: ASP Formulation

In the following, we turn to the modeling of flexible job-shop scheduling in ASP. The facts in Listing 3 constitute a textual representation of the instance data given in Table 1. In particular, each fact of the `operation` predicate corresponds to one table row with the contents of the columns *OpId*, *JobId*, and *Pos* as arguments, e.g., `operation(4,2,1)`. states that operation 4 belongs to job 2 in position 1. Facts of the `mode` predicate additionally provide *OpId* together with entries in the columns M-1, M-2, M-3, and M-4, such as `mode(4,2,8)`. and `mode(4,3,9)`. for the machines 2 and 3 that can process operation 4 in 8 or 9 time

```
operation(1,1,1).      mode(1,1,11).      mode(1,2,10).
operation(2,1,2).      mode(2,2,6).       mode(2,3,7).
operation(3,1,3).      mode(3,4,6).
operation(4,2,1).      mode(4,2,8).       mode(4,3,9).
operation(5,2,2).      mode(5,4,8).
operation(6,2,3).      mode(6,1,6).       mode(6,2,5).
operation(7,3,1).      mode(7,4,6).
operation(8,3,2).      mode(8,1,11).      mode(8,2,10).
operation(9,3,3).      mode(9,2,5).       mode(9,3,6).
operation(10,4,1).     mode(10,1,6).      mode(10,2,5).
operation(11,4,2).     mode(11,4,3).
operation(12,4,3).     mode(12,2,13).     mode(12,3,14).
time(0..50).
```

Listing 3 Fact representation of the job operations and processing times given in Table 1

units, respectively. The remaining facts of the time predicate, where time(0..50). is a shorthand for time(0).,..., time(50)., are included to restrict the starting times of operations to the (finite) range from 0 to 50. Note that no mode facts are given for machines that cannot process an operation. Moreover, a range of starting times from zero to the sum of (longest) processing times over all operations can be taken to guarantee the existence of some schedule. While such an explicit domain restriction was unnecessary for the CP formulation in Sect. 3.1, the grounding stage of ASP must lead to a finite propositional problem representation, which is here achieved by limiting the starting times.

The general problem encoding shown in Listing 4 is the central part of the ASP formulation and applicable to arbitrary instances. Its basic idea is to assign each operation to one of the machines that can process it and fix an execution order for operations (of distinct jobs) sharing the same machine. This approach then yields the starting times for operations, where the first operations of jobs that are not preceded by any other operation on the same machine start at time 0. Other operations can start once predecessor operations of the same job and operations processed before by the same machine are completed. The latest completion time among all operations, obtained by adding the time for processing to the starting time, gives the makespan of a schedule and is subject to minimization.

In more detail, the choice rule in Line 2, below the comment after % in the first line of Listing 4, expresses that each operation must be assigned to a machine processing it. For example, the operations processed by machine 2 in the optimal schedule displayed in Fig. 2 are represented by including the atoms process(1,2,10) and process(12,2,13) in a corresponding answer set. The condition that operations 1 and 12, belonging to the distinct jobs 1 and 4, are processed by the same machine 2 is indicated by deriving ordered(1,12) by the rule in Lines 5–6. This derivation means that some order between the operations 1 and 12 must be fixed by including either order(1,12) or order(12,1) in an answer set, while the order among operations of the same job would

```
% Choose one machine per operation                                        1
1 {process(O,M,P) : mode(O,M,P)} 1 :- operation(O,J,N).                   2
                                                                           3
% Choose an order between two operations processed by the same machine    4
ordered(O1,O2) :- operation(O1,J1,N1), process(O1,M,P1),                  5
                  operation(O2,J2,N2), process(O2,M,P2), J1 < J2.         6
{order(O1,O2)} :- ordered(O1,O2).                                         7
 order(O2,O1)  :- ordered(O1,O2), not order(O1,O2).                       8
                                                                           9
% Propagate lower bounds on starting and completion times of operations   10
start_lb(O,0)   :- operation(O,J,1).                                      11
start_lb(O2,E) :- end_lb(O1,E), operation(O1,J,N), operation(O2,J,N+1).   12
start_lb(O2,E) :- end_lb(O1,E), order(O1,O2).                            13
end_lb(O,S+P)  :- start_lb(O,S), process(O,M,P), time(S).                14
                                                                           15
% Restrict starting times to finite horizon given by the time predicate   16
:- start_lb(O,S), not time(S).                                            17
                                                                           18
% Progagate lower bounds on completion times to obtain integer interval   19
makespan(E) :- end_lb(O,E), operation(O,J,N), not operation(O+1,J,N+1).   20
makespan(E) :- makespan(E+1), 0 < E.                                     21
                                                                           22
% Penalize each integer in interval by weight 1 for minimizing makespan   23
#minimize{1,E : makespan(E)}.                                            24
```

Listing 4 Logical rules encoding flexible job-shop scheduling with makespan minimization

be clear from the job sequence and needs no further machine-specific consideration (skipped in view of the built-in comparison $J1 < J2$, contained as a body literal in Line 6). The exclusive selection between order(1,12) and order(12,1) is encoded by the choice rule in Line 7 (without cardinality bounds), for which order(1,12) occurs in the head of a rule instance, and order(12,1) is in turn derived by an instance of the rule in Line 8 in case order(1,12) does not hold. For the schedule in Fig. 2, the corresponding answer set includes order(1,12), but not order(12,1)

With atoms representing a machine dedication and execution orders for machines at hand, the rules in Lines 11–14 with atoms of the start_lb and end_lb predicates in the head derive lower bounds on starting and completion times. The lower bound 0 for the first operations of job sequences is immediately derived, e.g., start_lb(1,0) is included in every answer set. Based on process(1,2,10), this derivation leads to end_lb(1,10), expressing that operation 1 is not completed before time 10. As order(1,12) signals that operation 12 succeeds 1, we derive start_lb(12,10), yet we also need to take (direct) predecessor operations of the same job 4 into account. Given that the schedule in Fig. 2 is such that the direct predecessor operation 11 of 12 is completed at time 9, represented by the atom end_lb(11,9), the corresponding answer set includes start_lb(12,9), while the greatest lower bound and thus the actual starting time of operation 12 remains 10. Given the processing time 13 specified by process(12,2,13), the greatest lower bound on the completion time of operation `12` and its job 4 can then be read off from the derived atom `end_lb(12,23)`. Note that lower bounds exceeding the range of starting times are not propagated because of `time(S)` in the body of the `end_lb(O,S+P)` rule in Line 14. The integrity constraint in Line 17 asserts that answer sets must not encompass any such invalid bound, which guarantees that the operation execution orders for machines are non-circular.

The rules for `makespan(E)` in Line 20 and 21 consider (lower bounds on) the completion times of the last operations of job sequences, as these subsume the completion times of their predecessors, and derive atoms `makespan(1),...,makespan(C_max)`, representing the maximum completion time C_{max} over all operations by an integer interval. The optimization statement in Line 24 then associates each integer in the interval with the cost 1 so that the sum obtained from the tuples $\langle 1, 1 \rangle,...,\langle 1, C_{max} \rangle$ yields C_{max} as the weight of an answer set. For the answer set corresponding to the optimal schedule displayed in Fig. 2, the resulting weight is $C_{max} = 23$, considering that no operation has a later (lower bound on its) completion time than 23. For modeling other FJSSP variants, the general encoding in Listing 4 would need to be adjusted, e.g., one could augment the `makespan` predicate with the job identifier `J` as an additional argument and replace the optimization statement by `#minimize{1,J,E : makespan(J,E)}`. in order to minimize the sum of job completion times instead of the C_{max} value.

4.2   Tools and Applications Fields

While initially conceived as a declarative semantics for negation as failure in logic programs [48], ASP has emerged as a practical paradigm for modeling and (grounding and) solving combinatorial search and optimization problems [62]. We discuss the tools and applications in the following.

Tools Users typically specify the conditions on (optimal) solutions in the input language of an ASP system's grounding component, where Clingo [39] and I-DLV [12] constitute the current state-of-the-art systems. In addition, the input language of the IDP system [19] is closely related to ASP, and the pioneering Alpha system [84] interleaves grounding with the search for answer sets. The Alpha system instantiates rules on demand relative to the search state during problem-solving. That is, a rule gets instantiated only in case a partial truth assignment satisfies its body. This approach can reduce memory consumption for inputs on which upfront grounding produces many unnecessary rule instances.

The most common back-end solvers Clasp [42] and Wasp [3] extend search and optimization algorithms from (Max)SAT [67, 86] to propositional ASP programs. Alternative approaches to ASP solving work by translation to SAT [55], SAT modulo theories [36], or integer programming [64] so that corresponding solvers can be harnessed for computing answer sets. Portfolio systems [12, 65] integrate such orthogonal approaches to select a promising solver based on instance classification. Surveys and competition reports [43, 44] present these ASP grounding and solving techniques in detail. Extensions of ASP systems for particular applications further add custom constraint propagators [15, 54], domain-specific search heuristics [23, 41], and incremental reasoning methods [13, 40].

Applications fields Given its expressive input language and the availability of off-the-shelf reasoning tools, ASP has attracted interest in various application areas from sciences to industry [26, 29]. Regarding planning and scheduling, ASP has been utilized to plan setups for space shuttle maneuvers [69], for task assignment, operation and coordination planning in robotics [27], transport logistics [1], vehicle routing [45], and automatic warehousing [68], as well as rostering [4], timetabling [6], and team building [77].

For production planning tasks like open-, flow-, and (flexible) job-shop scheduling, hybrid methods integrating CP [5] or difference logic [38] reasoning into ASP systems to represent potentially large time domains succinctly by integer variables have proven to scale up best. Further application fields with tight relationships to prescriptive analytics include product configuration [46] and reconfiguration [8] as well as decision support [71] and recommender systems [24]. The application areas of ASP and CP systems largely overlap, and the respective efficiency of problem modeling and reasoning tools considerably depends on the specific task at hand.

5   Local Search

Current CP and ASP tools implement certain types of search algorithms employing the construction of tree structures that can be memory intensive. In addition, the propositional representation of an ASP program, i.e., the grounding, can get large depending on the problem type, encoding, and/or the size of a problem instance. In most CP tools, all variables and constraints are instantiated, although they might not be required within the current search state. As a consequence, cases could occur where CP or ASP cannot be applied successfully.

A large family of approaches that are known to show very low memory consumption and provide reasonable (near-optimal) solutions for various optimization problems are *local search* methods [50]. Beside CP and ASP, local search is another core problem-solving method in AI [79].

Principle The fundamental idea in local search is to start with a sub-optimal (maybe random) solution and step-by-step improve the solution by applying small (local) changes until the solution quality is acceptable or some other stopping criterion is reached. Which possible changes are taken into account and which of them are taken as the next step are actually the main factors that lead to different local search algorithms. Listing 5 displays the general structure of a hill-climbing algorithm (e.g., [85]) as pseudo code.

```
currentSolution = getRandomSolution();                        1
WHILE stopping criterion is not fulfilled DO                  2
    changes = getLocalChanges(currentSolution);               3
    neighborSolutions = {};                                   4
    FOR EACH change in changes DO                             5
        neighborSolution = apply(change,                      6
    currentSolution);
        neighborSolutions = add(neighborSolution,             7
    neighborSolutions);
    END FOR                                                   8
    best = getBest(neighborSolutions);                        9
    IF best is better than currentSolution THEN              10
        currentSolution = best;                              11
    ELSE                                                     12
        RETURN currentSolution;                              13
    END IF                                                   14
END WHILE                                                    15
RETURN currentSolution;                                      16
```

Listing 5 Structure of hill-climbing local search algorithms

First, in Line 1, some initial solution is guessed/produced and taken as the first current solution. After that, a loop (Lines 2–15) is entered. This loop is repeated until some predefined stopping criterion is reached. Examples of such criteria are:

- the solution has reached some value, e.g., a schedule with a completion time smaller than a predefined value,
- a predefined number of loop cycles have been run through, or
- for a predefined number of loop cycles, the improvement lies below a predefined threshold.

In the loop, the following is repeated:

1. A set of local changes that could be applied to the current solution are computed (Line 3). In a job-shop scheduling problem, this set typically consists of changes in the operation order on the machines. In the simplest case, all consecutive operations, i.e., one operation is scheduled after the other on the same machine, form pairs for which the order can be flipped. Such an operation order reversal represents a local change for the current solution.
2. Based on all computed local changes, for each of those changes, a slightly modified version of the current schedule is created (Lines 4–8). These schedules are called the *neighbors* of the current schedule as each of them only differs from the current solution by one local change.
3. Then, the best neighbor solution for the optimization criterion, e.g., completion time, is selected (Line 9).

4. If the best neighbor is better than the current solution, it becomes the new current solution (Lines 10–11). Otherwise, the algorithm terminates early and returns the current solution as it is a local optimum without any better neighbor.

5. If no local optimum is found, the loop terminates when the stopping criterion is fulfilled, and the current solution is returned in Line 16.

Classical hill-climbing only accepts improving steps. A common variant for problems with too many local changes accepts the first change leading to some improvement. Furthermore, to (partially) overcome local optima, hill-climbing is often combined with random restarts, i.e., the algorithm is started several times with different initial solutions.

In contrast to accepting improving steps only, algorithms like simulated annealing [58] also accept steps that deteriorate the solution quality with a certain probability. To this end, rather than selecting the best possible change, simulated annealing randomly selects a change. If this change leads to a better neighbor, the neighbor becomes the current solution. In case the neighbor is not better than the current solution, it is accepted with some probability that decreases the worse the neighbor is and the more loop iterations have already been carried out up to this point.

Genetic and evolutionary algorithms The idea from local search of traversing the search space by applying one local change after the other, thus hopping from the current solution to neighbors, can also be found in a more general form in evolutionary algorithms that are inspired by biological evolution (e.g., [51]). Instead of one current solution, there is a set of current solutions called population. Each individual in the population encodes a solution, typically a sequence of numbers. For the generation of neighbors, which in this context are often called offspring or children, two families of operations can be applied. One possibility is to combine the solution parts of two different individuals. In genetic algorithms this is called *crossover* and the solution parts are called *genes*. Several different strategies are possible for selecting which of the individuals in the current population are used for offspring production. For example, in the *roulette wheel selection* the probability for an individual to be selected as a parent of the next child is proportional to its solution quality, called *fitness*. Besides crossover, the second possibility to change a solution is based on *mutation*. This method is nothing more than changing genes randomly from time to time (steered by a probability constant). Crossover and mutation are typically used in combination. Hence, first, some new individuals (children/offspring) are produced by using crossovers on selected individuals of the current population, and then, by chance, some of the genes are mutated.

Swarm intelligence A family of approaches very similar to evolutionary algorithms are algorithms that mimic the group dynamics of swarms like birds [76], ants [25], bees [74], or particles in particle swarm optimization (PSO) [57]. Like in evolutionary algorithms, there is a population of individuals. In PSO, the swarm consists of particles, each having a position (i.e., solution) in the search space, a moving direction (velocity), and memory of its best-known position (i.e., solution). Taking the properties (like position and moving direction) and information communicated by other particles (e.g., their best positions) into account, a particle changes its position (i.e., hops to a neighbor solution). Depending on the concrete algorithm, particles can communicate with the whole swarm or only with a subset of particles whose position is nearby. The communication in PSO plays the role of crossovers in evolutionary algorithms, and likewise swarm optimization terminates when some stopping criterion is reached (e.g., the swarm's best solution is near-optimal).

This introduction of local search methods completes the overview of major AI methods for prescriptive analytics, dealing with automated problem-solving, decision making, and optimization, and leads us to our final reflection.

6   Summary

6.1   Industrie 4.0, AI, and Analytics

The goal of Industrie 4.0 is to build smart factories considering machines, robots, logistic systems, materials, and products as an integrated cyber-physical system equipped with a massive number of sensors and actuation components. Smart factories shall implement competitive manufacturing of products for individual customer requirements, dynamic processes, and optimized decision-making. Rational AI agents and their components provide architectures and methods for realizing optimized decision-making in Industrie 4.0 and smart factories. Besides AI, the field of analytics has received considerable attention in Industrie 4.0. We have shown a close connection between the components of model-based, utility-based agents and descriptive, predictive, and prescriptive analytics. Consequently, researchers and practitioners can profit from both fields when designing and implementing software solutions.

In the book's part on analytics, this chapter complements the exposition by presenting symbolic AI methods for reasoning in prescriptive analytics. Prescriptive analytics is mainly concerned with planning tasks and provides support for deciding *"what"* should happen *"when"*. Consequently, prescriptive analytics and AI are vital technologies for automating supply chain management, manufacturing, and engineering.

6.2   AI-based Problem-Solving in Industrie 4.0

As automated problem-solving is one of the core research areas of AI with many applications in manufacturing such as planning, scheduling, engineering, and configuring, numerous methods have been developed in this field. We have introduced the most widely applied AI methods and tools for knowledge representation and reasoning as well as automated problem-solving.

AI methods aim at solving two challenges. On the one hand, problem instances have to be solved in reasonable time and satisfactory quality. On the other hand, the efficient specification and adaptation of problem descriptions are critical success factors for many practical applications. CP and ASP provide means to address both challenges. We have used flexible job-shop scheduling as a driving example to detail knowledge representation in CP and ASP, as this problem is a well-known, canonical scenario dealing with *"what"* should happen *"when"* in the production. In problem domains where such general knowledge representation and reasoning frameworks are inapplicable, AI also offers a rich set of heuristic search methods. We have provided a brief introduction to local search, a major area of AI, encompassing some of the most successfully applied search strategies for dealing with hard problems. This chapter has focused on symbolic reasoning methods, which provide sound, explainable and complete reasoning services. AI methods using machine learning techniques integrate naturally for generating search heuristics and causal models in case these heuristics and models cannot be described sufficiently with acceptable costs by engineers. Our introduction to AI methods comes with an extensive set of references, which should support further reading and the selection of suitable methods for solving practical application problems.

Acknowledgements This work has been conducted in the scope of the research project *Dyna-Con (FFG-PNr.: 861263)*, funded by the Austrian Federal Ministry of Transport, Innovation and Technology (BMVIT) under the program "ICT of the Future" between 2017 and 2020, the research project *Productive4.0*, funded by EU-ECSEL under grant agreement no737459, and the KWF project no28472, funded by cms electronics GmbH, FunderMax GmbH, Hirsch Armbänder GmbH, incubed IT GmbH, Infineon Technologies Austria AG, Isovolta AG, Kostwein Holding GmbH and Privatstiftung Kärntner Sparkasse. We thank Karen Meehan for proofreading and suggestions.

References

1. Abels, D., Jordi, J., Ostrowski, M., Schaub, T., Toletti, A., Wanko, P.: Train scheduling with hybrid ASP. In: Balduccini, M., Lierler, Y., Woltran, S. (eds.) Proceedings of the Fifteenth International Conference on Logic Programming and Nonmonotonic Reasoning (LPNMR'19). Lecture Notes in Artificial Intelligence, vol. 11481, pp. 3–17. Springer-Verlag (2019)
2. Abiteboul, S., Hull, R., Vianu, V.: Foundations of Databases. Addison-Wesley (1995)
3. Alviano, M., Amendola, G., Dodaro, C., Leone, N., Maratea, M., Ricca, F.: Evaluation of disjunctive programs in WASP. In: Balduccini, M., Lierler, Y., Woltran, S. (eds.) Proceedings of the Fifteenth International Conference on Logic Programming and Nonmonotonic Reasoning (LPNMR'19). Lecture Notes in Artificial Intelligence, vol. 11481, pp. 241–255. Springer-Verlag (2019)
4. Alviano, M., Dodaro, C., Maratea, M.: Nurse (re)scheduling via answer set programming. Intelligenza Artificiale **12**(2), 109–124 (2018)
5. Balduccini, M.: Industrial-size scheduling with ASP+CP. In: Delgrande, J., Faber, W. (eds.) Proceedings of the Eleventh International Conference on Logic Programming and Nonmonotonic Reasoning (LPNMR'11). Lecture Notes in Artificial Intelligence, vol. 6645, pp. 284–296. Springer-Verlag (2011)

6. Banbara, M., Inoue, K., Kaufmann, B., Okimoto, T., Schaub, T., Soh, T., Tamura, N., Wanko, P.:
   teaspoon: Solving the curriculum-based course timetabling problems with answer set programming. Annals of Operations Research **275**(1), 3–37 (2019)
7. Baral, C., Gelfond, M.: Logic programming and knowledge representation. Journal of Logic
   Programming **19/20**, 73–148 (1994)
8. Beck, H., Bierbaumer, B., Dao-Tran, M., Eiter, T., Hellwagner, H., Schekotihin, K.: Rule-based
   stream reasoning for intelligent administration of content-centric networks. In: Michael, L.,
   Kakas, A. (eds.) Proceedings of the Fifteenth European Conference on Logics in Artificial Intelligence (JELIA'16). Lecture Notes in Artificial Intelligence, vol. 10021, pp. 522–528. Springer-
   Verlag (2016)
9. Biere, A., Heule, M., van Maaren, H., Walsh, T. (eds.): Handbook of Satisfiability, Frontiers in
   Artificial Intelligence and Applications, vol. 185. IOS Press (2009)
10. Blazewicz, J., Ecker, K., Pesch, E., Schmidt, G., Weglarz, J.: Handbook on Scheduling: From
    Theory to Applications. Springer-Verlag (2014)
11. Booth, K., Nejat, G., Beck, C.: A constraint programming approach to multi-robot task allocation
    and scheduling in retirement homes. In: Rueher, M. (ed.) Proceedings of the Twenty-second
    International Conference on Principles and Practice of Constraint Programming (CP'16). Lecture
    Notes in Computer Science, vol. 9892, pp. 539–555. Springer-Verlag (2016)
12. Calimeri, F., Dodaro, C., Fuscà, D., Perri, S., Zangari, J.: Efficiently coupling the I-DLV grounder
    with ASP solvers. Theory and Practice of Logic Programming **20**(2), 205–224 (2020)
13. Calimeri, F., Ianni, G., Pacenza, F., Perri, S., Zangari, J.: Incremental answer set programming
    with overgrounding. Theory and Practice of Logic Programming **19**(5-6), 957–973 (2019)
14. Cordier, M., Dague, P., Lévy, F., Montmain, J., Staroswiecki, M., Travé-Massuyès, L.: Conflicts
    versus analytical redundancy relations: A comparative analysis of the model based diagnosis
    approach from the artificial intelligence and automatic control perspectives. IEEE Transactions
    on Systems, Man, and Cybernetics, Part B **34**(5), 2163–2177 (2004)
15. Cuteri, B., Dodaro, C., Ricca, F., Schüller, P.: Overcoming the grounding bottleneck due to
    constraints in ASP solving: Constraints become propagators. In: Bessiere, C. (ed.) Proceedings of
    the Twenty-Ninth International Joint Conference on Artificial Intelligence (IJCAI'20). pp. 1688–
    1694. ijcai.org (2020)
16. Da Col, G., Teppan, E.: Learning constraint satisfaction heuristics for configuration problems. In:
    Zhang, L., Haag, A. (eds.) Proceedings of the Nineteenth International Configuration Workshop.
    pp. 8–11. IESEG School of Management (2017)
17. Da Col, G., Teppan, E.: Google vs IBM: A constraint solving challenge on the job-shop schedul-
    ing problem. In: Bogaerts, B., Erdem, E., Fodor, P., Formisano, A., Ianni, G., Inclezan, D., Vidal,
    G., Villanueva, A., de Vos, M., Yang, F. (eds.) Technical Communications of the Thirty-fifth Inter-
    national Conference on Logic Programming (ICLP'19). Electronic Proceedings in Theoretical
    Computer Science, vol. 306. Open Publishing Association (2019)
18. Da Col, G., Teppan, E.: Industrial size job shop scheduling tackled by present day CP solvers.
    In: Schiex, T., de Givry, S. (eds.) Proceedings of the Twenty-fifth International Conference on
    Principles and Practice of Constraint Programming (CP'19). Lecture Notes in Computer Science,
    vol. 11802, pp. 144–160. Springer-Verlag (2019)
19. De Cat, B., Bogaerts, B., Bruynooghe, M., Janssens, G., Denecker, M.: Predicate logic as a
    modeling language: The IDP system. In: Kifer, M., Liu, A. (eds.) Declarative Logic Program-
    ming: Theory, Systems, and Applications. pp. 279–323. ACM / Morgan and Claypool Publishers
    (2018)

20. De Raedt, L., Passerini, A., Teso, S.: Learning constraints from examples. In: McIlraith, S., Weinberger, K. (eds.) Proceedings of the Thirty-Second National Conference on Artificial Intelligence (AAAI'18), the Thirtieth Conference on Innovative Applications of Artificial Intelligence (IAAI'18), and the Eighth AAAI Symposium on Educational Advances in Artificial Intelligence (EAAI'18). pp. 7965–7970. AAAI Press (2018)
21. Dechter, R.: Constraint Processing. Morgan Kaufmann Publishers (2003)
22. Delen, D., Demirkan, H.: Data, information and analytics as services. Decision Support Systems **55**(1), 359–363 (2013)
23. Dodaro, C., Gasteiger, P., Leone, N., Musitsch, B., Ricca, F., Schekotihin, K.: Combining answer set programming and domain heuristics for solving hard industrial problems. Theory and Practice of Logic Programming **16**(5-6), 653–669 (2016)
24. Dodaro, C., Leone, N., Nardi, B., Ricca, F.: Allotment problem in travel industry: A solution based on ASP. In: ten Cate, B., Mileo, A. (eds.) Proceedings of the Ninth International Conference on Web Reasoning and Rule Systems (RR'15). Lecture Notes in Computer Science, vol. 9209, pp. 77–92. Springer-Verlag (2015)
25. Dorigo, M., Birattari, M., Stützle, T.: Ant colony optimization. IEEE Computational Intelligence Magazine **1**(4), 28–39 (2006)
26. Erdem, E., Gelfond, M., Leone, N.: Applications of ASP. AI Magazine **37**(3), 53–68 (2016)
27. Erdem, E., Patoglu, V.: Applications of ASP in robotics. Künstliche Intelligenz **32**(2-3), 143–149 (2018)
28. Falkner, A., Friedrich, G., Haselböck, A., Schenner, G., Schreiner, H.: Twenty-five years of successful application of constraint technologies at Siemens. AI Magazine **37**(4), 67–80 (2016)
29. Falkner, A., Friedrich, G., Schekotihin, K., Taupe, R., Teppan, E.: Industrial applications of answer set programming. Künstliche Intelligenz **32**(2-3), 165–176 (2018)
30. Felfernig, A., Burke, R.: Constraint-based recommender systems: Technologies and research issues. In: Fensel, D., Werthner, H. (eds.) Proceedings of the Tenth International Conference on Electronic Commerce (ICEC'08). ACM International Conference Proceeding Series, vol. 342, pp. 3:1–3:10. ACM (2008)
31. Felfernig, A., Friedrich, G., Jannach, D., Zanker, M.: Intelligent support for interactive configuration of mass-customized products. In: Monostori, L., Váncza, J., Ali, M. (eds.) Proceedings of the Fourteenth International Conference on Industrial and Engineering Applications of Artificial Intelligence and Expert Systems (IEA/AIE'01). Lecture Notes in Computer Science, vol. 2070, pp. 746–756. Springer-Verlag (2001)
32. Felfernig, A., Teppan, E., Friedrich, G., Isak, K.: Intelligent debugging and repair of utility constraint sets in knowledge-based recommender applications. In: Bradshaw, J., Lieberman, H., Staab, S. (eds.) Proceedings of the Thirteenth International Conference on Intelligent User Interfaces (IUI'08). pp. 217–226. ACM (2008)
33. Fleischanderl, G., Friedrich, G., Haselböck, A., Schreiner, H., Stumptner, M.: Configuring large systems using generative constraint satisfaction. IEEE Intelligent Systems and their Applications **13**(4), 59–68 (1998)
34. Garey, M., Johnson, D.: Computers and Intractability: A Guide to the Theory of NP-Completeness. W. H. Freeman and Co. (1979)
35. Gazzaneo, L., Padovano, A., Umbrello, S.: Designing smart operator 4.0 for human values: A value sensitive design approach. Procedia Manufacturing **42**, 219–226 (2020)
36. Gebser, M., Janhunen, T., Rintanen, J.: Answer set programming as SAT modulo acyclicity. In: Schaub, T., Friedrich, G., O'Sullivan, B. (eds.) Proceedings of the Twenty-first European Conference on Artificial Intelligence (ECAI'14). pp. 351–356. IOS Press (2014)
37. Gebser, M., Kaminski, R., Kaufmann, B., Lindauer, M., Ostrowski, M., Romero, J., Schaub, T., Thiele, S., Wanko, P.: Potassco User Guide. University of Potsdam (2019)

38. Gebser, M., Kaminski, R., Kaufmann, B., Ostrowski, M., Schaub, T., Wanko, P.: Theory solving made easy with clingo 5. In: Carro, M., King, A. (eds.) Technical Communications of the Thirty-second International Conference on Logic Programming (ICLP'16). Open Access Series in Informatics (OASIcs), vol. 52, pp. 2:1–2:15. Schloss Dagstuhl–Leibniz-Zentrum fuer Informatik (2016)
39. Gebser, M., Kaminski, R., Kaufmann, B., Schaub, T.: Answer Set Solving in Practice. Synthesis Lectures on Artificial Intelligence and Machine Learning, Morgan and Claypool Publishers (2012)
40. Gebser, M., Kaminski, R., Kaufmann, B., Schaub, T.: Multi-shot ASP solving with clingo. Theory and Practice of Logic Programming **19**(1), 27–82 (2019)
41. Gebser, M., Kaufmann, B., Otero, R., Romero, J., Schaub, T., Wanko, P.: Domain-specific heuristics in answer set programming. In: desJardins, M., Littman, M. (eds.) Proceedings of the Twenty-Seventh National Conference on Artificial Intelligence (AAAI'13). pp. 350–356. AAAI Press (2013)
42. Gebser, M., Kaufmann, B., Schaub, T.: Conflict-driven answer set solving: From theory to practice. Artificial Intelligence **187–188**, 52–89 (2012)
43. Gebser, M., Leone, N., Maratea, M., Perri, S., Ricca, F., Schaub, T.: Evaluation techniques and systems for answer set programming: A survey. In: Lang, J. (ed.) Proceedings of the Twenty-seventh International Joint Conference on Artificial Intelligence (IJCAI'18). pp. 5450–5456. ijcai.org (2018)
44. Gebser, M., Maratea, M., Ricca, F.: The seventh answer set programming competition: Design and results. Theory and Practice of Logic Programming **20**(2), 176–204 (2020)
45. Gebser, M., Obermeier, P., Schaub, T., Ratsch-Heitmann, M., Runge, M.: Routing driverless transport vehicles in car assembly with answer set programming. Theory and Practice of Logic Programming **18**(3-4), 520–534 (2018)
46. Gebser, M., Ryabokon, A., Schenner, G.: Combining heuristics for configuration problems using answer set programming. In: Calimeri, F., Ianni, G., Truszczyński, M. (eds.) Proceedings of the Thirteenth International Conference on Logic Programming and Nonmonotonic Reasoning (LPNMR'15). Lecture Notes in Artificial Intelligence, vol. 9345, pp. 384–397. Springer-Verlag (2015)
47. Gebser, M., Maratea, M., Ricca, F.: What's hot in the answer set programming competition. In: Schuurmans, D., Wellman, M. (eds.) Proceedings of the Thirtieth National Conference on Artificial Intelligence (AAAI'16). pp. 4327–4329. AAAI Press (2016)
48. Gelfond, M., Lifschitz, V.: The stable model semantics for logic programming. In: Kowalski, R., Bowen, K. (eds.) Proceedings of the Fifth International Conference and Symposium of Logic Programming (ICLP'88). pp. 1070–1080. MIT Press (1988)
49. Gelfond, M.: Answer sets. In: van Harmelen, F., Lifschitz, V., Porter, B. (eds.) Handbook of Knowledge Representation. Foundations of Artificial Intelligence, vol. 3, pp. 285–316. Elsevier (2008)
50. Gendreau, M., Potvin, J.: Handbook of Metaheuristics. Springer-Verlag (2010)
51. Goldberg, D.: Genetic Algorithms in Search, Optimization and Machine Learning. Addison-Wesley (1989)
52. Hoos, H., Stützle, T.: Stochastic Local Search: Foundations & Applications. Elsevier/Morgan Kaufmann Publishers (2004)
53. Hungerländer, P., Maier, K., Pöcher, J., Rendl, A., Truden, C.: Solving an on-line capacitated vehicle routing problem with structured time windows. In: Fink, A., Fügenschuh, A., Geiger, M. (eds.) Selected Papers of the Annual International Conference of the German Operations Research Society (GOR'16). pp. 127–132. Operations Research Proceedings, Springer-Verlag (2018)

54. Janhunen, T., Kaminski, R., Ostrowski, M., Schaub, T., Schellhorn, S., Wanko, P.: Clingo goes linear constraints over reals and integers. Theory and Practice of Logic Programming **17**(5-6), 872–888 (2017)
55. Janhunen, T., Niemelä, I.: Compact translations of non-disjunctive answer set programs to propositional clauses. In: Balduccini, M., Son, T. (eds.) Logic Programming, Knowledge Representation, and Nonmonotonic Reasoning: Essays Dedicated to Michael Gelfond on the Occasion of His 65th Birthday. Lecture Notes in Computer Science, vol. 6565, pp. 111–130. Springer-Verlag (2011)
56. Kagermann, H., Wahlster, W., Helbig, J.: Recommendations for implementing the strategic initiative Industrie 4.0: Final report of the Industrie 4.0 Working Group. acatech – National Academy of Science and Engineering (2013)
57. Kennedy, J., Eberhart, R.: Particle swarm optimization. In: Proceedings of the International Conference on Neural Networks (ICNN'95). pp. 1942–1948. IEEE Computer Society Press (1995)
58. Kirkpatrick, S., Gelatt, C., Vecchi, M.: Optimization by simulated annealing. Science **220**(4598), 671–680 (1983)
59. Korbicz, J., Kowalczuk, Z., Koácielny, J., Cholewa, W. (eds.): Fault Diagnosis: Models, Artificial Intelligence, Applications. Springer-Verlag (2004)
60. Laborie, P., Rogerie, J., Shaw, P., Vilím, P.: IBM ILOG CP optimizer for scheduling – 20+ years of scheduling with constraints at IBM/ILOG. Constraints **23**(2), 210–250 (2018)
61. Lepenioti, K., Bousdekis, A., Apostolou, D., Mentzas, G.: Prescriptive analytics: Literature review and research challenges. International Journal of Information Management **50**, 57–70 (2020)
62. Lifschitz, V.: Answer set programming and plan generation. Artificial Intelligence **138**(1-2), 39–54 (2002)
63. Lifschitz, V.: Answer Set Programming. Springer-Verlag (2019)
64. Liu, G., Janhunen, T., Niemelä, I.: Answer set programming via mixed integer programming. In: Brewka, G., Eiter, T., McIlraith, S. (eds.) Proceedings of the Thirteenth International Conference on Principles of Knowledge Representation and Reasoning (KR'12). pp. 32–42. AAAI Press (2012)
65. Maratea, M., Pulina, L., Ricca, F.: Multi-engine ASP solving with policy adaptation. Journal of Logic and Computation **25**(6), 1285–1306 (2015)
66. McCollum, B., Schaerf, A., Paechter, B., McMullan, P., Lewis, R., Parkes, A., Di Gaspero, L., Qu, R., Burke, E.: Setting the research agenda in automated timetabling: The second international timetabling competition. INFORMS Journal on Computing **22**(1), 120–130 (2010)
67. Morgado, A., Heras, F., Liffiton, M., Planes, J., Marques-Silva, J.: Iterative and core-guided MaxSAT solving: A survey and assessment. Constraints **18**(4), 478–534 (2013)
68. Nguyen, V., Obermeier, P., Son, T., Schaub, T., Yeoh, W.: Generalized target assignment and path finding using answer set programming. In: Sierra, C. (ed.) Proceedings of the Twenty-sixth International Joint Conference on Artificial Intelligence (IJCAI'17). pp. 1216–1223. IJCAI/AAAI Press (2017)
69. Nogueira, M., Balduccini, M., Gelfond, M., Watson, R., Barry, M.: An A-prolog decision support system for the space shuttle. In: Ramakrishnan, I. (ed.) Proceedings of the Third International Symposium on Practical Aspects of Declarative Languages (PADL'01). Lecture Notes in Computer Science, vol. 1990, pp. 169–183. Springer-Verlag (2001)
70. Perron, L., Furnon, V.: OR-Tools https://developers.google.com/optimization/

71. Pham, T., Germano, S., Mileo, A., Kümper, D., Intizar Ali, M.: Automatic configuration of smart city applications for user-centric decision support. In: Crespi, N., Manzalini, A., Secci, S. (eds.) Proceedings of the Twentieth Conference on Innovations in Clouds, Internet and Networks (ICIN'17). pp. 360–365. IEEE Computer Society Press (2017)

72. Priore, P., Gómez, A., Pino, R., Rosillo, R.: Dynamic scheduling of manufacturing systems using machine learning: An updated review. Artificial Intelligence for Engineering Design, Analysis and Manufacturing 28(1), 83-97 (2014)

73. Rabbouch, B., Saâdaoui, F., Mraihi, R.: Constraint programming based algorithm for solving large-scale vehicle routing problems. In: Pérez García, H., Sánchez González, L., Castejón Limas, M., Quintián Pardo, H., Corchado Rodríguez, E. (eds.) Proceedings of the Fourteenth International Conference on Hybrid Artificial Intelligence Systems (HAIS'19). Lecture Notes in Computer Science, vol. 11734, pp. 526–539. Springer-Verlag (2019)

74. Rahim, M., Musirin, I., Abidin, I., Othman, M., Joshi, D.: Congestion management based optimization technique using bee colony. In: Proceedings of the Fourth International Conference on Power Engineering and Optimization (PEOCO'10). pp. 184–188. IEEE Computer Society Press (2010)

75. Reiter, R.: On closed world data bases. In: Gallaire, H., Minker, J. (eds.) Logic and Databases. pp. 55–76. Plenum Press (1978)

76. Reynolds, C.: Flocks, herds, and schools: A distributed behavioral model. In: Stone, M. (ed.) Proceedings of the Fourteenth Annual Conference on Computer Graphics and Interactive Techniques (SIGGRAPH'87). pp. 25–34. ACM (1987)

77. Ricca, F., Grasso, G., Alviano, M., Manna, M., Lio, V., Iiritano, S., Leone, N.: Team-building with answer set programming in the Gioia-Tauro seaport. Theory and Practice of Logic Programming 12(3), 361–381 (2012)

78. Rossi, F., van Beek, P., Walsh, T. (eds.): Handbook of Constraint Programming. Elsevier (2006)

79. Russell, S., Norvig, P.: Artificial Intelligence: A Modern Approach. Pearson Education (2010)

80. Salido, M., Garrido, A., Barták, R.: Introduction: Special issue on constraint satisfaction techniques for planning and scheduling problems. Engineering Applications of Artificial Intelligence 21(5), 679–682 (2008)

81. Shmueli, G., Koppius, O.: Predictive analytics in information systems research. MIS Quarterly 35(3), 553–572 (2011)

82. Souza, G.: Supply chain analytics. Business Horizons 57(5), 595–605 (2014)

83. Valouxis, C., Housos, E.: Constraint programming approach for school timetabling. Computers & Operations Research 30(10), 1555–1572 (2003)

84. Weinzierl, A., Bogaerts, B., Bomanson, J., Eiter, T., Friedrich, G., Janhunen, T., Kaminski, T., Langowski, M., Leutgeb, L., Schenner, G., Taupe, R.: The Alpha solver for lazy-grounding answer-set programming. ALP Newsletter (2019)

85. Xi, B., Liu, Z., Raghavachari, M., Xia, C., Zhang, L.: A smart hill-climbing algorithm for application server configuration. In: Feldman, S., Uretsky, M., Najork, M., Wills, C. (eds.) Proceedings of the Thirteenth International Conference on World Wide Web (WWW'04). pp. 287–296. ACM (2004)

86. Zhang, L., Madigan, C., Moskewicz, M., Malik, S.: Efficient conflict driven learning in a Boolean satisfiability solver. In: Ernst, R. (ed.) Proceedings of the International Conference on Computer-Aided Design (ICCAD'01). pp. 279–285. IEEE Computer Society Press (2001)

Machine Learning for Cyber-Physical Systems

Oliver Niggemann, Bernd Zimmering, Henrik Steude,
Jan Lukas Augustin, Alexander Windmann and Samim Multaheb

Abstract

Machine Learning plays a crucial role for many innovations for Cyber-Physical Systems such as production systems. On the one hand, this is due to the availability of more and more data in ever better quality. On the other hand, the demands on the systems are also increasing: Production systems have to support more and more product variants, saving resources is increasingly in focus and international competition is forcing companies to innovate faster. Machine Learning leverages data to solve these issues. The goal is to have self-learning systems which improve over time. There are various algorithms and methods for this, for which an overview is given here. Furthermore, this article discusses special requirements of Cyber-Physical Systems for Machine Learning processes.

O. Niggemann (✉) · B. Zimmering · H. Steude · J. L. Augustin · A. Windmann · S. Multaheb
Institute of Automation Technology, Helmut Schmidt University, Hamburg, Germany
e-mail: oliver.niggemann@hsu-hh.de
URL: http://www.hsu-hh.de/imb

B. Zimmering
e-mail: bernd.zimmering@hsu-hh.de

H. Steude
e-mail: henrik.steude@hsu-hh.de

J. L. Augustin
e-mail: lukas.augustin@hsu-hh.de

A. Windmann
e-mail: alexander.windmann@hsu-hh.de

S. Multaheb
e-mail: samim.multaheb@hsu-hh.de

B. Vogel-Heuser and M. Wimmer (eds.), *Digital Transformation*,
https://doi.org/10.1007/978-3-662-65004-2_17

415

Keywords

Machine Learning • Cyber-Physical Systems

1   Introduction

Cyber-Physical Systems (CPS) are becoming a major field for Machine Learning (ML). Challenges go far beyond a mere application of existing algorithms. What is needed are specialized algorithms which meet domain requirements such as reliability, real-time capability and maintainability. This paper gives an introduction to the opportunities and challenges of ML for CPS.

For this, potential CPS application scenarios for ML are described in Sect. 6. Sect. 3 outlines the state of the art for ML and maps features of algorithms to the uses cases from Sect. 6. From this, Sect. 4 derives specific requirements of CPS to ML. The next sections describe these requirements and the corresponding solution approaches in detail. Sect. 9 summarizes the paper content.

2   Application Scenarios

Currently ML is applied to several industrial use cases:

2.1   Condition Monitoring and Predictive Maintenance

For every plant operator, it is desirable that certain components are replaced at exactly the right time, not too early out of caution, but also not so late that the risk of failure becomes significant. The method of choice is predictive maintenance. It is made possible by condition monitoring, i.e. the continuous monitoring of the system. Nowadays, condition monitoring is based on data and observations [77, 83].

The basic idea is that various component data, such as vibration, speed and energy consumption, are continuously collected and evaluated [95]. In many cases, such data are already available anyway, but remain unused. It is only necessary in exceptional cases to install new sensors in order to generate additional data.

For condition monitoring, ML is mainly used to learn a model of the normal system behavior, including thresholds to non-normal behavior. Once this threshold is crossed, a warning is given. Predictive maintenance requires that these models can be extrapolated in time, e.g. these models predict when a threshold will be crossed.

2.2   Resource Optimization

The consumption of resources in production is becoming more and more important for companies. At the same time, attention is increasingly being paid to saving wastewater and emissions. On the one hand, these goals have financial reasons, on the other hand, increasing environmental awareness and corresponding legislation are also having an effect.

There are two variants for implementing resource optimization [99, 126]: The simple one is limited to the fact that software stores and analyzes the consumption data in detail. The employees can derive change options from this and implement them. The much more complex variant: The software not only makes the consumption data available, but also independently optimizes the control of the systems.

Here a prediction model of the resource consumption is learned. In the end, these models must have a sufficient quality to be integrated into closed-loop control loops—a very demanding requirement.

2.3   Quality Assurance of Products

ML can be used to monitor the quality of a product during manufacturing and to detect irregularities at an early stage [131].

ML systems can generate and maintain a Digital Twin [94], i.e. a virtual image of the real products and intermediate products. These Digital Twins collect all information from the engineering phase and can be used to improve the learning process by providing a-priori information [116]. Digital Twins are then enriched during the operation phase by evaluating sensor data using Artificial Intelligence (AI) and ML methods, combined with information about raw materials and production processes. This Digital Twin allows the prediction of product properties that are difficult or impossible to determine in reality. Thus, virtual measurements that are difficult to implement in reality can be carried out on a Digital Twin. Although these predictions are often less certain than real measurements, they allow an early warning in the event of quality problems.

Such Digital Twins can be used in many areas in which complex end-of-line tests or laboratory tests are currently commonly used, for example after the end of production to analyze the properties of food.

2.4   Diagnosis

AI or ML can help to identify the causes of errors in a system [15, 31, 83]. If many sensors are built into a system, as it is increasingly common today, problems are detected early on by condition monitoring and anomaly detection systems and reported as an alarm. However, the connection between a symptom and the cause of the error is often difficult to determine—this is due to the increasing complexity of modern (production) systems: The systems are getting bigger and bigger, consist of many sub-modules and are characterized by an increasingly high degree of networking and automation. An error often causes subsequent errors early in the production process and only leads to symptoms and alarms much later—a cause of error can propagate through the entire system and lead to symptoms in a wide variety of places, sometimes with a significant delay. The more complex and networked the system, the longer it takes to manually identify the cause. This means that it can take a long time before repairs can be carried out. Today this is seen as an important cost driver.

An AI/ML-based diagnostic system determines the most likely causes of failure based on the symptoms and does so within a short period of time. The user then no longer only sees the symptoms in the form of alarms and warnings—rather, possible causes of errors are displayed immediately, and repair instructions are often supplied directly.

In this use case, mainly two kinds of models are used: First a model of the normal behavior is learned and used to compute symptoms, i.e. warnings. Then a model comprising system causalities is used to identify root causes [96]. The latter is only partially learned.

3   Machine Learning

The number of ML algorithms is legion—and so are the taxonomies used to describe them [10]. Here, we will introduce two dimensions of algorithms' feature to describe algorithm: First, we will use recurrency, i.e. the ability of algorithms to handle dynamic, time-variant data. Second, we will use supervision, i.e. the degree of supervision needed by the algorithm.

Normally, ML algorithms compute a model. Just like manually created models, models can be used to predict specific system features. We start with describing the dimension "recurrency".

Static Analysis For tasks such as condition-monitoring or anomaly detection, only the signal values $\mathbf{x}_t \in \mathbf{R}^n$ at some point in time t are used. This is shown in Fig. 1 on the left hand side. In other words, for the analysis a static feature vector [35] is taken into account. Thereby, the assumption is that no information is coded in the sequence of values and all necessary information is contained in the current signal values. This assumption is true for many CPS, even for systems which have a dynamic nature. For a new data point $\mathbf{x} \in \mathbf{R}^n$, its probability $p(\mathbf{x}|X)$ given some historical data X is (at least approximately) computed. If the data is improbable, an anomaly has been identified.

Dynamic Analysis A totally different situation arises when important information is coded in the sequence of signal values over time: For a time window of data $X = \{\mathbf{x}_{t-k}, \ldots, \mathbf{x}_t\}$, its probability is computed. This is shown in Fig. 1 in the center. Again, if the time window is improbable, an anomaly has been identified.

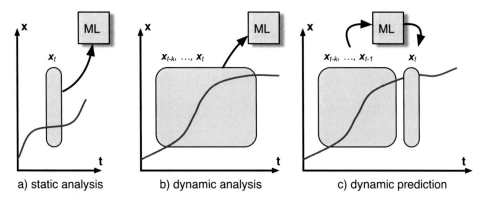

Fig. 1 Comparison of static ML (**a**), dynamic ML (**b**) and the special case of (**c**) ML by means of prediction. Signal values **x** are depicted over time

Often, especially for dynamic analyses, prediction is used to analyze the data: Let X be again a time window of data, then a prediction for the next value $\hat{\mathbf{x}}_{t+1}$ is computed. Once a real measurement for $\mathbf{x}_{t+1}$ is available, it can be compared to the prediction $\hat{\mathbf{x}}_{t+1}$—if they are significantly different, an anomalous situation has occurred. This is shown in Fig. 1 on the right hand side.

Next, the dimension supervision is described.

Supervised Machine Learning Supervised ML algorithms work on labeled data, this is shown in Fig. 2: Each data point **x** comes with a label **y**. The ML algorithm will learn a model which is able to compute suitable labels for a given input **x**. If **y** is nominal, e.g. its values are discrete classes such as "OK" or "KO", this task is called classification. If **y** is cardinal, e.g. its values are numerical values such as temperatures, this task is also called regression.

Often, these ML models are trained by using a feedback or residual signal. The algorithms start with an initial, suboptimal model and compute a prediction of $\hat{\mathbf{y}}$. The matching or fitting between the estimation $\hat{\mathbf{y}}$ and the real, wanted label **y** is assessed and provides a feedback which is used to improve the model—often the difference between $\hat{\mathbf{y}}$ and **y** is used, i.e. a residual.

Unsupervised Machine Learning Unsupervised ML uses unlabeled data (see also Fig. 3). Since again a feedback or residual signal is needed to learn a model, a generic, external criterion is used to assess the quality of a model prediction e.g. for clustering, i.e. the identification of clusters of similar data such as plant phases ("ramp-up", "normal operation", etc.). A good clustering has a high similarity between data within clusters and a low similarity between clusters.

Fig. 2 Principle idea of
supervised ML

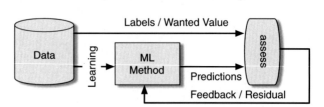

Fig. 3 Principle idea of
unsupervised ML

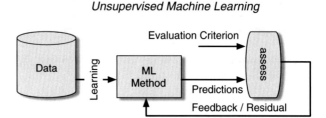

If the external criterion is given dynamically from an environment, e.g. a camera signal
for an autonomous vehicle, we speak of reinforcement learning.

We can now use these two dimensions (recurrency, supervision) to describe some common ML algorithms and describe their suitability for the use cases from Sect. 6. This is also
visualized in Fig. 4.

Static, Supervised Machine Learning (bottom left quadrant in Figure 4)

Feedforward Neural Networks: Neural networks [91] approximate complex functions by
parameterizing a network of simple, generic functions. The architecture of the network, i.e.
the so-called topology, and the chosen generic functions, i.e. the so-called neurons, decide
about the class of functions which can be approximated. The connections between neurons
comprise parameters, i.e. the so-called weights, which are used to fit the neural network to
a given data set—normally by means of optimization algorithms.

In most cases the topology of feedforward networks comprises a number of layers where
only neighboring layers are connected. The most bottom layer is fed with the input x_i where
the top most layer models the labels y_i. The network then learns the mapping from typical
inputs to corresponding labels.

Decision Trees and Random Forrests: Decision trees [105] learn a tree of decision rules
to map from a data vector x_i to labels y_i. Each decision rule splits the set of data by an
inequality on the elements in the vector x_i.

Random forests [12] extend this idea by learning a set of decision trees, decisions are
made by a majority vote.

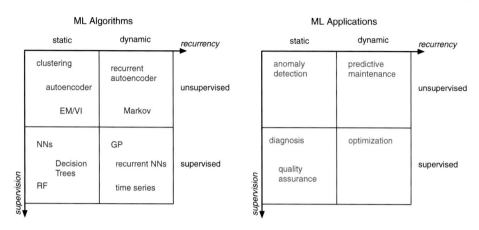

Fig. 4 Mapping of ML features to algorithms and applications

Static, Unsupervised Machine Learning (top left quadrant in Figure 4)

Clustering: Clustering groups given observations $x_i \in X$ in clusters $X_1, \ldots, X_p, p \in \mathbf{N}$ with $\bigcup_i X_i = X$. The similarity between elements within clusters is maximized while the similarity between elements in different clusters is minimized. Algorithms range from simple shaped clusters (with a given number of clusters p) such as k-means clustering [92] to approaches which are able to identify complex shapes and also identify p, e.g. DBSCAN [107]. The similarity criterion is defined externally. Clustering methods often suffer from the problem that different cluster shapes require different algorithms or algorithm configurations. *Autoencoder:* Autoencoders [40] are neural networks which remember learned data vectors x_i. The key idea is that instead of mapping from x_i to y_i, autoencoders map from x_i to x_i. Thus, autoencoders remember already observed situations and can check how similar new observations are to this memory. By checking whether a new data vector is remembered, e.g. an anomaly detection method can be implemented. A variational autoencoder (VAE) extends the idea of the classical autoencoder (AE) with concepts from probabilistic modeling and was originally introduced in [64]. *Expectation Maximization (EM) / Variational Inference (VI):* EM [91] and VI [114] are algorithms which learn probability distributions from data. For this, the structure of the probability distributions must be given. The learned distribution can be used to compute the probability for new data. This solution of course requires that the underlying distribution is already known.

Dynamic, Supervised Machine Learning (bottom right quadrant in Figure 4)

Time Series: In statistics, time series analysis [91] is a well-established field. Typical solutions such as ARMA [91] learn a given autoregressive function which expresses x_i as a linear functions of x_j, $j < i$ and of stochastic terms. The drawback is the simplicity of the used functions.

Gaussian Processes: Gaussian processes [125] model a time series as a stochastic process. The underlying probability function for a range of time steps is a multivariate normal distribution. Gaussian processes are well-suited to capture uncertainties about the learned model but often suffer from runtime challenges.

Recurrent Neural Networks, Gated Networks, Attention-based Networks: Neural networks can also be used to learn a model of time series. The simplest solution is the use of computed values of neurons at time step t as an input for the next time step $t + 1$, i.e. so-called recurrent neural networks [100]. Since such networks have problems using values from several time steps in the past, gated networks such as Long Short Term Memory (LSTM) have been used [54]. Such networks try to generate a memory of important CPS information. Gated networks must learn when to update the memory which leads to corresponding demands on the data quantity and quality. Attention-based networks [122] simplify things by only learning which past data element is relevant.

Dynamic, Unsupervised Machine Learning (top right quadrant in Figure 4)

Recurrent Autoencoder: Recurrent neural networks can also be used in an unsupervised fashion, i.e. as autoencoders [129].

Markov Models: Markov models [51] capture the time series x_i, $i = 1, \ldots, n$ as a path through a given graph. The graph comprises a set of predefined states where transitions model the probabilistic movement from one state to another state. Hidden Markov models [91] assume that the current state is not directly observable.

4   Challenges to ML for Cyber-Physical Systems

ML methods are currently a central component of many research and business activities. Despite many advances, these processes are currently mainly used in non-technical areas and are usually difficult to transfer to technical applications such as production or vehicles [120]. The reason: AI and ML methods were often developed for completely different data, such as economic data. Current ML challenges at the interface between AI / ML and engineering focus on special requirements by technical systems [95, 97].

The results of non-technical ML applications such as business data are usually interpreted, checked and used by a human [117]. The use of ML in a CPS, on the other hand, often means the application in a closed control loop: the results are interpreted by software and then automatically used for optimization. A person is usually no longer involved. This different usage scenario creates challenges [8–10, 83] which distinguish CPS from non-technical ML applications such as business data and image processing—and therefore require adapted ML solutions.

Challenge 1: Time and State: The main characteristic of all physical systems is that their behavior must be considered over time, not just at a specific point in time—therefore, for example, all ML results must also predict system behavior over time. E.g. the behavior of a chemical plant can only be understood if the history of the last hours is known or many problems of transport system arise from incorrect accelerations. The behavior over time thus includes current states, aggregating changes from the past, and information about state changes. Basic requirements for this are common, uniform time models both for the physical and software parts of a production system [74].

Challenge 2: Uncertainty and Noise: In order to use ML procedures and learned models in CPS, it is imperative to evaluate the uncertainty of the predictions of the ML systems [90]. If, for example, an ML system predicts a system failure, the degree of certainty of this forecast is decisive for the correct procedure. Uncertainties mostly arise from noise on the sensor data or from values that cannot be observed.

Challenge 3: Usage of A-Priori Knowledge: There is a lot of prior knowledge for CPS from the design phase, based on physical laws and engineering knowledge. This individual knowledge should be used to improve AI and ML practices. For ML solutions in particular, prior knowledge can alleviate the need for big amounts of data.

Challenge 4: Representations and Concepts: ML results and generated models must be explained to human operators: Why is a maintenance action necessary? Why are new parameters better than old ones? What happens if repairs are not done? For this, symbolic concepts must be learned from the (numerical) models, e.g. the concept "ramp-up phase" for some activations of a neural network. Based on these concepts, explanations and reasons are generated.

In general, it can be said that ML is in principle an interdisciplinary topic between engineering and computer science and must be approached using appropriately adapted methods. In the following, the points from above are discussed in detail.

5 Challenge 1: Time and State

In engineering or more generally speaking physical modeling, time is an essential concept, as systems change their behavior dynamically. Looking at a single point in time is insufficient, as key information about the context would be lost. A common way to encode these temporal dependencies is to introduce the latent state $\mathbf{z}(t) \in \mathbf{R}^m$, which describes the system at a given time t. The state evolves over time depending on new observations and can thus store information about the system, which can then be used for tasks like forecasting, classification or anomaly detection. Examples are automatons, where $\mathbf{z}(t) \in \{m_0, \ldots, m_i\}$ models discrete modes which are switched by events e, or ODEs, where the state $\mathbf{z}$ changes continuously over time, e.g. according to $\dot{\mathbf{z}}(t) = f(\mathbf{z}, t)$.

5.1   Approaches

A time series consists of observations over multiple time steps. In practice, there is no obvious beginning. Rather than working with all observations that are available, often a rolling window of k time steps is applied. Let $x_t \in R^n$ denote a sample at t. Then, a window consists of the data $X = \{x_{t-k}, \ldots, x_t\}$.

A general approach to describe the state of a system is the state space model [91]:

$$z_t = g\,(u_t, z_{t-1}, \epsilon_t)$$
$$x_t = h\,(z_t, u_t, \delta_t)$$

where $u_t \in R^l$ is an optional input or control signal, g is the transition model, h is the observation model and $\epsilon_t \in R^p$ and $\delta_t \in R^q$ describe noise, which is modeled as a random variable. Popular examples include ARIMA models [11, 34] and exponential smoothing [59]. For simple time series, these models work well, are interpretable and data efficient and thus widely used. Unfortunately, state space models often fail to detect complex patterns in time series and have to be tuned to every system seperately, which makes applying them labor-intensive [102].

ML methods are suited to tackle these issues. They can detect complex patterns across multiple time series and require very little engineering by hand [7, 72]. However, they lack interpretability and generally require a lot of data to work well. A model architecture that is built around that idea of learning the state of a time series is the Recurrent Neural Network (RNN). In contrast to state-space models, it does not rely on stochastic variables, but tries to model the sample distribution directly. One of the simplest forms of an RNN can be expressed as Delay Differential Equation (DDE) ([110] gives a detailed introduction). For $i \in \{t - k, \ldots, t\}$, with a randomly initialized $z_{t-k-1} \in R^m$, a DDE can be expressed as

$$z_i = W_z z_{i-1} + W_r r_{i-1} + W_x x_i + \theta_z, \tag{1}$$
$$r_i = G(z_i), \tag{2}$$

where $G(\cdot)$ is a nonlinear function (e.g. tanh) and the matrices $W_z, W_r \in R^{m \times m}$, $W_x \in R^{m \times n}$ and the bias $\theta_z \in R^m$ are trainable parameters of the network. If this RNN is applied to time series data X by shifting it from $t - k$ to t for fixed Δt between observations, the first two terms of Eq. (1) allow the network to propagate information from the past to the current time step t.

5.2   State of the Art

Gating Standard RNNs are hard to train, especially for long time series. RNNs suffer from the vanishing gradient problem [55]: The error gradient, which is needed to train neural

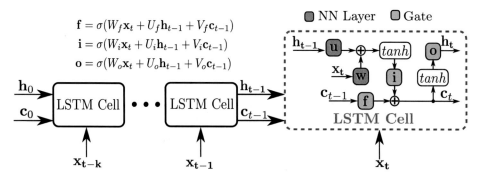

Fig. 5 Representation of an LSTM Network unfolded in time. In addition to the cell state c_t, a hidden state vector $\mathbf{h}_t$ is also formed and propagated along the time axis. The information flow along the time axis is controlled by the forget gate $\mathbf{f}$, the input gate $\mathbf{i}$ and the output gate $\mathbf{o}$. The matrices W, U and V of the respective gates are the learnable parameters of the LSTM cell that is shifted iteratively from $t - k$ to t. The LSTM is initialized with $\mathbf{h}_0$ and c_0

networks, often vanishes when flowing back along the time axis for too long. This leads to a short term behavior, as information from the past is forgotten. To overcome this problem, the Long Short-Term Memory (LSTM) [54] uses a gating mechanism as shown in Fig. 5. Its gates can decide to take new information into account or to neglect it. The gating concept is also used as a basis for further improvements of the LSTM, e.g. Gated-Recurrent Units (GRU) [25]. LSTMs are broadly used in applications like anomaly detection in CPS [47], optimizing productivity perfomance [23], smart grids [2] as well as for artificial generated sensor data [135].

Attention & Transformer Although the introduction of gates has limited the vanishing gradient problem, some core problems remain. When processing long sequences, even gated RNNs often fail to capture information from the start of the sequence properly, as all the information is crammed into one or two hidden vectors of limited size.

To mitigate this issue, the concept of attention has been introduced and modified to self-attention [4, 79]. The main idea of self-attention is to save all of the hidden states in a matrix $Z = (\mathbf{z}_{t-k}, \ldots, \mathbf{z}_t) \in \mathbf{R}^{n \times (k+1)}$ and to calculate a weighted average context vector $\tilde{\mathbf{z}} \in \mathbf{R}^n$ to work with. The original self-attention mechansim learns on what hidden states to focus on:

$$\mathbf{a} = \text{softmax}\left(\mathbf{w}^T \tanh\left(WZ\right)\right),$$

where $W \in \mathbf{R}^{l \times n}$ and $\mathbf{w} \in \mathbf{R}^l$ are learnable weights with adjustable dimension l. The attention vector determines how much a hidden state contributes to the context vector

$$\tilde{\mathbf{z}} = Z\mathbf{a}^T.$$

More recently, the hidden state vector has been divided into dedicated parts in order to dynamically identify where to pay attention with key-value attention [28] or multi-head self-attention [122].

The Transformer [122], which gets rid of the recurrent structure of RNNs altogether while solely focusing on attention, has shown spectacular results in many domains [14, 30, 32]. Transformers do not require to process samples time step by time step, as RNNs do, which accelerates training. However, this parallelized structure comes with a drawback: the attention mechanism cannot differentiate between time steps, which has to be taught to the Transformer seperately. Furthermore, the original Transformer has a quadratic runtime with respect to the input length, which makes it infeasible for long sequences. There have been various attempts to tackle these issues and to design a more efficient Transformer architecture for time series [78, 134]. While the Transformer has not seen wide adaption to CPS yet, research on the application to multivariate time series in general has been promising [80, 130].

Neural ODEs Another shortcoming of RNNs is rooted in their origin in a DDE. Standard RNNs are not suited for irregularly sampled time series, as $\mathbf{z}$ is updated once an observation $\mathbf{x}$ occurs. Defining the evolution of $\mathbf{z}_t$ continuously in form of an Ordinary Differential Equation (ODE) $\dot{\mathbf{z}}(t) = f(\mathbf{z}, t)$, where f is realized as a neural net, enables to output values for $\mathbf{z}$ whenever an observation $\mathbf{x}_t$ occurs. [20] presents an approach to efficiently learn Neural ODEs and demonstrates the advantages over RNNs on toy examples. [104] combines the idea of a continuously defined state $\mathbf{z}$ that is updated at the observation times by combining Neural ODEs and RNN, which shows impressive extrapolation capabilities. Often, physical systems show discontinuous behavior (e.g. a moving ball bouncing at the ground), therefore [18] introduces Neural Event Functions for ODEs, which are able to learn ODEs together with points in time where $\mathbf{z}(t)$ suddenly changes.

Neural ODEs have not seen wide adaption to CPS data yet, as current research focuses on architectures [33], fundamentals [86] and general properties (e.g. robustness) [3, 128].

5.3   Conclusion

Time is a fundamental concept for CPS data, yet until now there is no ideal solution that incorporates its characteristics. While RNNs and DDEs seem to be a natural fit, in practice the approach fails to produce good results. Gates, attention and the Transformer architecture try to mitigate issues of RNNs, but they introduce new challenges, as training an LSTM can be hard and Transformers are ill-suited for long sequences. Neural ODEs are another promising approach, but research is still in its early stages. Teaching models to handle time will be crucial in developing models that can work with CPS data effectively. How to do this is still largely unresolved.

6   Challenge 2: Uncertainty and Noise

Often a model is an approximation rather than a comprehensive description of every effect that takes place in the underlying system. This raises the question of how certain a prediction can be. In the literature, uncertainty is subdivided into epistemic uncertainty (model uncertainty), as well as aleatory uncertainty (data uncertainty). While epistemic uncertainty in ML models is caused by insufficiencies in model structure or training of the model, aleatory uncertainty is caused by the loss of information during the data collection of a real world system, e.g. due to noise within the path of measurement or faulty label information. Epistemic uncertainty can be reduced by improving the model or the training process, but it cannot be completely eliminated in practice. Removing aleatory uncertainty from the data is only possible to a very limited extent without further knowledge of the real world system [44, 67].

6.1   Approaches

Methods that allow an estimation of uncertainty for their predictions can be grouped into three categories: (*i*) statistical methods, (*ii*) ML approaches based on reconstruction error, and (*iii*) energy-based ML approaches [90].

(*i*) Given a set of data points X, statistical information like their distribution can be interfered. Statistical moments of e.g. first (expected value) or second order (variance) can be calculated (in case of their existence) as high level properties that describe the distribution. In case the data can be represented by a Gaussian distribution, the mean μ expresses the average value of all observed data points and the spatial distribution can be described by the standard deviation σ. By **knowing** the probability distribution, it is possible to conclude an uncertainty measure of predicted data points based on the comparison of mean and standard deviation to the training dataset.

However, capturing the probability distribution of technical systems, due to the underlying physical properties and complex interdependencies, is a difficult task. By observing the system, it is possible to estimate the likelihood, i.e. the measure how well the a-priori defined statistical model (the prior) fits the observation. This however requires extensive knowledge of the system's behavior. For tasks with simple probability distributions where an accurate prior can be found, likelihood estimation is near to the actual probability distribution, a reliable measure of confidence can be expected with uncertainty analyses (UA) and sensitivity analyses (SA) [62, 121].

However, the limitation of the described statistical methods is bound to the complexity of the data's distribution [81]. For non-trivial likelihood functions finding accurate priors is challenging: A model's complexity, i.e. the number of degrees of freedom of a model to build a function, has to match the data representation of the problem. For real-world problems with complex data distributions, this means that the epistemic uncertainty is rising as "simple" priors like Gaussian models cannot represent the data. To accomplish this, models with higher complexity, i.e. with a higher degree of freedom, e.g. neural networks, are needed.

Without knowing priors, frequentist approaches like sampling-based techniques, i.e. the selection of representative data points, can help to make statistical inference about the whole model. They are often computationally expensive, but can also be applied to NNs.

(*ii*) Reconstruction-based ML methods like autoencoders (see Sect. 3) measure the uncertainty by calculating the distance e.g. Mean-Squared-Error $MSE = \|\mathbf{x} - \hat{\mathbf{x}}\|_2$ between ground truth $\mathbf{x}$ and the reconstructed output $\hat{\mathbf{x}}$ [85]. Through learning a latent representation $\mathbf{z}$, consisting of fewer dimensions than the input sample, they are forced to learn the key features of the training data. Therefore, samples coming from a similar distribution to the learned representation will result in a smaller distance, i.e. smaller reconstruction error, whereas data far from the learned distribution would result in a higher reconstruction error.

But using reconstruction error as a measure of uncertainty is difficult as it cannot be clearly interpreted like σ of (*i*) . Therefore, the distance between the reconstruction and the grounded truth can act as an indicator for uncertainty, but not as a clear measurement of uncertainty. As it measures similarity of the training and predicted data set's distributions, it can be poorly understood as estimate for data uncertainty (e.g. noise). As the similarity of two data distributions are well suited to distinguish normal from anomalous data sets, this method is often used in anomaly detection. Here, a large reconstruction error indicates that an incoming sample is anomalous [44, 75].

(*iii*) Energy-based ML approaches use likelihood estimation instead of the reconstruction error, i.e. in addition to the distance to the mean, the variance of the data is considered for uncertainty estimation. In comparison to the statistical methods depicted in (*i*) , where a suitable statistical model has to be defined a-priori (e.g. Gaussian distribution), energy-based methods learn to fit the corresponding likelihood function to the probability of the data's occurrence. The objective is to train the model to maximize said likelihood function. Logically concluded, the more data exists to train the model, the more accurately the fitted likelihood represents the system.

6.2   State of the Art

As distributions of data sets are usually high dimensional and complex, the applicability of (*i*) statistical methods for modeling data is limited. While (*ii*) reconstruction based methods measure similarity to the training data set, (*iii*) energy based methods are considered as state of the art for estimation of uncertainty for predictions. In the following, specific techniques for the individual high-level approaches are presented.

Loss function based   While some methods incorporate statistical values by default, some methods use the idea of [98]: They try an energy-based approach to learn statistical concepts by adding σ as a second output to the architecture and by modifying the loss function. For dynamical analysis a loss function that includes a measure of uncertainty is derived in [36] using maximum likelihood approach considering three underling assumptions for an LSTM:

(1) All relevant hidden states **h** can be learned from the data.
(2) Gaussian distribution for covariance of the error $\hat{\mathbf{x}}_{t+1} - \mathbf{x}_{t+1}$ (e.g. white noise).
(3) Knowing the hidden states $\mathbf{h}_t$, the remaining noise on each sensor x_t^i is independent (e.g. white noise).

With these assumptions, the conditional probability for $\mathbf{x}_{t+1}$ is given by the multivariate Gaussian distribution in Eq. (3).

$$p_\theta(\mathbf{x}_{t+1}|\mathbf{h}_t, \mathbf{x}_t) = \prod_{i \in N}^{n} \frac{1}{\sqrt{2\pi}\sigma_{t+1}^i} \exp\left[-\frac{1}{2}\left(\frac{x_{t+1}^i - \hat{x}_{t+1}^i}{\sigma_{t+1}^i}\right)^2\right] \tag{3}$$

With regard to the maximum likelihood, Eq. (3) can be formulated as maximum-likelihood-error loss function (Eq. (4)) which can be used for training of neural nets that compute x_{t+1}^i and σ_{t+1}^i.

$$L_{t+1} = \sum_{i \in N}^{n}\left[\left(\frac{x_{t+1}^i - \hat{x}_{t+1}^i}{\sigma_{t+1}^i}\right)^2 + 2\log\sigma_{t+1}^i\right] \tag{4}$$

Equation (4) can be interpreted as an extension of the MSE loss function that further reduces the distance $\|\mathbf{x} - \hat{\mathbf{x}}\|_2$ through σ_{t+1} with respect to the log term as a penalty.

Sampling based While energy based methods explicitly learn uncertainty, frequentist approaches like sampling allow us to gather information about uncertainty. A universal approach for many applications and network types is introduced in [43]. E.g. dropout (randomly switching off neurons during the training on a NN), which is usually used to avoid overfitting, can also be used to observe uncertainty during the prediction phase. By performing predictions with dropout, the mean and standard deviation can be evaluated empirically. Another approach is training an ensemble of models [70] to estimate σ. Here several models are initialized differently and the samples of the training data that are presented during a training epoch are shuffled. During training, adversarial samples (samples that are slightly different from the original samples) are generated. Furthermore, a loss function similar to Eq. (4) is used to estimate the overall uncertainty.

Bayesian Networks Restricted Boltzmann Machines (RBM) are a type of Bayesian neural networks which is considered as energy based method. It consists of a visible layer of neurons v_i and a hidden layer of neurons h_j. These two layers are used to learn probability distributions of an unknown data distribution by taking binary states [41, 112]. Each layer uses a bias (a_i for the input layer and b_j for the hidden layer). The neurons are connected via their weights w_{ij}. The bias as well as the weights are first set randomly and then learned in accordance to the data the system is trained on. The value associated with each state of the network is referred to as the energy E of the network (Equation (5)).

$$E(\mathbf{v}, \mathbf{h}) = - \sum_{i \in visible} a_i v_i - \sum_{j \in hidden} b_j h_j - \sum_{i,j} v_i h_j w_{ij} \tag{5}$$

Equation (6) estimates the probability of a configuration $\mathbf{v}$ by the exponential energy term of the observed state divided by the sum of the exponential energy terms of all possible observations $\mathbf{v}^*$. Thus, samples going outside of the learned distribution result in a higher energy level.

$$P(\mathbf{v}) = \frac{e^{-E(\mathbf{v})}}{\sum_{\mathbf{v}^*} e^{-E(\mathbf{v}^*)}} \tag{6}$$

For a given set of parameters and data, θ and $\mathcal{D}$ respectively, the likelihood is the weighted sum of the log-probability of observed states $\mathbf{v}$.

$$\mathcal{L}(\theta, \mathcal{D}) = \frac{1}{N} \sum_{\mathbf{v}^{(i)} \in \mathcal{D}} log\ P(\mathbf{v}^i) \tag{7}$$

The loss function $\mathbf{L}$, i.e. the optimization function, being the negative log-likelihood as shown in Eq. (8), is minimized through learning, and thus the likelihood is maximized.

$$\mathbf{L}(\theta, \mathcal{D}) = -\mathcal{L}(\theta, \mathcal{D}) \tag{8}$$

6.3   Conclusion

Sources of uncertainty are of various types. While epistemic uncertainty can often be reduced by more data or a more detailed model, the occurrence of aleatory uncertainty implies an estimation of uncertainty when predictions are made with NNs. As (*i*) statistical methods are limited to non complex data distributions (*ii*) reconstruction-based methods can act as an indicator for uncertainty but do not quantify it. (*iii*) Energy-based methods such as RBMs or the modification of the loss function allow to learn a measure of uncertainty and are thus able to quantify uncertainty also for complex datasets. Furthermore, sampling based methods offer an empirical opportunity to quantify uncertainty for predictions.

7   Challenge 3: Usage of A-Priori Knowledge

The performance of a trained neural network is measured based on the expected performance on new data samples drawn from an underlying, normally unknown, distribution. While classic signal processing is typically done in up to three dimensions, the situation for high-dimensional problems dealt with in ML is substantially different. Interpolation cannot be done by techniques allowing for accurate estimation of errors. Instead, neural networks are prone to over- or underfitting and therefore limited regarding their capability to generalize to data unseen during training. A function (trained neural network) should be locally smooth with slight differences of the input resulting in similar outputs. However, if this was to be ensured solely through a sufficient amount of samples, the required amount increases exponentially as the dimensionality of the input increases. Therefore, effective priors that capture the expected regularities and complexities of the high-dimensional real-world prediction tasks need to be found and the amount and quality of training samples need to be maximized.

7.1 Approaches

Structure of the respective domain presents a source of regularity which can be utilized by making use of the corresponding symmetries, i.e. transformations leaving certain properties unchanged or invariant. Symmetries of the underlying data impose structure and are powerful priors improving learning efficiency by reducing the space of possible functions to be learned [13]. Arguably, the most illustrative examples can be found in Convolutional Neural Networks (CNNs)[73] applied to images. Convolutional filters with shared weights shifted across a grid combined with pooling layers are characteristic for the CNN network topologies exploiting translational symmetry [48]. In image classification, the image class is unchanged by shifts of the object within the image. Similarly, in time series often encountered in CPS, an anomaly is to be detected as such regardless of the point in time, so shifts are also symmetries in the problem of anomaly detection in CPS. However, whereas flipped images are often considered as equally valid samples, in the case of time series only orientation-preserving transformations may be appropriate choices. Since RNNs introduced in Sect. 5 make use of network topologies allowing to dynamically aggregate information in a way that respects the temporal progression of inputs while also allowing for online arrival of novel data-points, they are a natural choice when dealing with sequential, temporal data. One reason why shifted versions of the sequence can be treated equally is that the RNN input vectors can be seen as points on a temporal grid—a very useful prior.

Whereas in the case of images and sequences data is already recorded with inherent structure in the form of 1D or 2D grids in Euclidean space, no such structure is provided for static analysis of single time steps of multivariate CPS sensor data (see Fig. 6(a)). Typically heterogeneous sensors provide information about numerous subsystems in a non-Euclidean space. Therefore, inputs $x_t^{(a...f)}$ from sensors a through to f are typically concatenated in some arbitrary but fixed order to generate a feature vector $\mathbf{x}_t$ (see Fig. 6(b)) serving as the input for a neural network. However, domain experts such as engineers designing or maintaining such systems are aware of the underlying system structure, namely relationships and

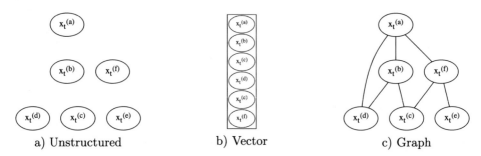

a) Unstructured   b) Vector   c) Graph

Fig. 6 Sensor inputs $x_t^{(a...f)}$ represented as **a)** an unstructured set, **b)** an input vector $\mathbf{x_t}$ and **c)** a graph G_t with verticals V and edges E adding information in the form of relations between inputs

interdependencies between information from different sensors. This source of knowledge remains to be unlocked by representing the data on a connected graph as shown in Fig. 6(c). Such graph structure can improve learning efficiency by providing additional information that limits the space of functions to be learned and enables the use of modern deep learning techniques able to operate directly on graph-structured data: Graph Neural Networks (GNNs).

Labels created by domain experts represent the most common and direct way of making use a-priori knowledge. However, with labeling being a time-consuming process resulting in quite limited amounts of training samples, supervised ML has recently been outperformed by self-supervised learning algorithms [21, 61], a subclass of unsupervised learning introduced in Sect. 3). Such techniques employ knowledge about the problem to increase the amount of training samples by obtaining labels from the unlabeled data itself. This is done by reconstructing hidden parts of the input from unhidden parts of the input or using data augmentation to learn better representations. Suitable self-supervised ML pipelines cannot be designed without a-priori knowledge allowing for appropriate choices of architecture as well as masking or augmentation techniques.

Simulations of production systems are created during the design phase to model and optimize their expected behaviour. The advantage of such models is twofold: they are interpretable by the domain expert and can be used to transfer knowledge from the expert to the learning algorithm by generating additional training samples. In contrast to real-world training samples typically covering normal system states, simulations can extend this subspace by sampling from the entire distribution of possible system states. This mitigates the issue of deep learning models often not being able to extrapolate to data unseen during training. However, real-world and simulation distributions cannot be expected to be identical without adaptation.

7.2   State of the Art

Compared to Challenges 1 and 2, usage of a-priori knowledge encompasses a set of open and heterogeneous research directions in the context of CPS. Therefore, rather than presenting specific ones in detail, an overview of highly promising methods to be explored by the community is given.

Graph Neural Networks is a collective term for deep learning approaches operating on inputs given in the form of a node feature matrix X, an adjacency matrix A and (optionally) an edge feature matrix X^e. GNNs unlock the potential of deep learning for non-Euclidian data without discarding relational information. Network layers are designed to be permutation equivariant since no canonical ordering of graph nodes is assumed [13]. Modern GNN architectures can be categorized as convolutional [65], attentional [124] or message-passing [5, 45] and are capable of operating on graphs directly processing information in the form of node features as well as edge features. Such models achieve state-of-the-art results for node, link or graph prediction tasks on protein biology [46] or detection of misinformation [89]. Notably, research on GNNs has largely been driven by the increasing availability of graph-structured data [58, 103, 108]. In the field of CPS such structure remains to be leveraged by adding it based on prior knowledge or by learning structure applying latent graph learning approaches [27, 109, 123].

Self-supervised learning has been employed to unlock the potential of the vast amounts of data available today by removing the need for human labeled data. This has led to great success in advancing the field of natural language processing, particularly when combined with transformer architectures [26, 30, 122]. These algorithms make use of knowledge about the language domain by discretizing the feature space to most common words or characters and use the inherent structure of text samples to learn about relations of characters, words or sentences by masking different parts of the input. Other approaches—some of which have already been extended to the graph domain [118]—make use of augmentations [21] or two joint slightly different architectures [50]. Successfully designing suitable self-supervised learning pipelines for CPS will require a-priori knowledge to come up with suitable masking or augmentation techniques.

Simulations model the expected behaviour of CPS even before the system is built. These simulations can be used to uniformly generate high amounts of synthetic samples covering both normal and abnormal system states [17]. Since the synthetic domain is not expected to exactly match the real-world domain of system behaviour, it is necessary to combine both by removing synthetic samples in overlap regions [16] or domain adaptation [29, 119].

7.3   Conclusion

Interdisciplinary cooperation will be a key factor for successfully incorporating a-priori knowledge. Significant parts of the store of knowledge of engineers remain to be utilized and incorporated into ML pipelines. Opportunities range from enhancing inputs by including system structure as graph-structured inputs to enable the use of GNNs, through building CPS-specific self-supervised learning techniques, to making use of often preexisting simulations. Interdisciplinary cooperation will be a key factor for successfully incorporating a-priori knowledge.

8   Challenge 4: Representations and Concepts

Recent years have shown exceptional progress in ML research, particularly deep learning [72]. This method's impact is mainly due to its successes in solving rather specific tasks, such as playing games [111], detecting diseases on medical images [38], or identifying anomalies in CPS sensor data [76]. However, little progress has been made towards general AI. Teaching machines to learn (physical) concepts from observations (e.g. sensor data) is considered a major step in this journey [69]. The emerging research field called Representation Learning (RepL) is dedicated to this objective. Clear, simple and meaningful representations of high-dimensional and complex data can enable the explainability of AI algorithms and thereby also simplify their evaluation. This is particularly relevant in the context of CPS.

8.1   Approaches

In more technical terms, the core motivation of RepL is to build models which are capable of encoding noisy real world observations of (physical) processes, into meaningful representations [6]. These representations are typically vectors of reals numbers, but might also emerge in form of other data formats such as (automate) graphs [57, 101]. Since most RepL models can be trained with unlabeled data, their most common use case is to utilize the typically lower-dimensional representations as input for downstream supervised learning tasks. Based on these representations, less complex ML models with little labeled training data can achieve satisfying performance in many cases [71]. According to [48] a good representation is one that makes it easier to solve subsequent learning tasks, as illustrated in Fig. 7. Among the frequently applied examples are the Word2Vec [88] algorithm for natural language processing and ResNet [21] for computer vision. In its most extreme form, the procedure of simplifying or enabling downstream ML-tasks with representations leads to one-shot [39] and zero-shot learning [113], where only one or even zero training examples are required respectively.

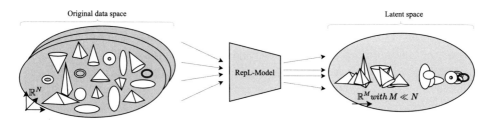

Fig. 7 Principle idea of RepL. RepL models typically encode data points from a high-dimensional space $\mathbf{R}^n$ into a lower-dimensional space $\mathbf{R}^m$. Ideally the representations also encode meaning. In this case objects of the same type are close to each other in the latent space and are clearly separated from objects of another type. Thus, the object types can be separated linearly in the latent space, while a more complex model would be needed in the original data space. This illustrates how downstream ML-tasks can be simplified using RepL models

With regards to CPS, these methods are promising for two main reasons: First, (*i*) good representation of CPS data (in most cases multivariate time series) are an important step to explainable AI. Especially among engineers who are used to working with causal system models, e.g. based on ODEs (see Sect. 5) deep learning methods are often considered as black box solutions which cannot be understood and hence also not trusted. In this context, meaningful and simple representations of the often high-dimensional data are helpful in understanding how AI algorithms work and how corresponding decisions are made. Once a RepL model is trained, downstream ML tasks can be implemented using simpler models such as linear regression with very few and interpretable parameters. This process represents the transition from sub-symbolic to symbolic AI. Engineers and humans in general think in symbols and explain causal relationships, processes and logic in symbols rather than in high-dimensional data spaces. Thus, e.g. mappings from the observation spaces (sensor data) into contexts that are understandable from an engineering point of view, such as automated graphs [57] and potentially even existing ontology models, can be very useful. Second, (*ii*) especially in regard to predictive maintenance use cases such as anomaly detection or failure predictions, the amount of labeled data is usually very limited. Thus, learning good representations from the large amounts of unlabeled data can be highly beneficial for diverse downstream analysis tasks.

However, only little research has been done on RepL for CPS. Apart from a few use cases such as computer vision methods for optical quality control, the majority of CPS related ML use cases rely on sensor data. Thus, in most cases, the training and input data are in the form of Multivariate Time Series (MTS). For this reason, in the following subsection, we will summarize the current state of research related to learning representations with a focus on MTS.

8.2   State of the Art

In line with our approach in Chap. 7, the following section provides an overview of the approaches to Challenge 4 currently discussed in the literature, rather than discussing specific solutions in detail. A very well known and comprehensive (yet slightly outdated) literature review on the field of RepL is given in [6]. According to the authors a good representation captures the underlying factors that generated the data. This definition shows that good representations are anything but unique. In order to disentangle the underlying factors, modern RepL algorithms use so-called clues or priors. In most cases, these clues are implemented in terms of the model architecture or the loss function and aim at enforcing the disentanglement of the learned representations. In a way these clues might also be interpreted as a-priori knowledge (see Sect. 7.1). A list of such clues for unsupervised RepL is provided in [6] and [48].

Time series, unlike images or other typical ML inputs, do not represent explicit features [127]. Thus, mapping MTS to meaningful representations requires particularly strong clues. Some examples of such clues that we consider to be the most important for our context are described below.

Dimensionality reduction and manifolds Learning representations, i.e. interesting and meaningful features, from MTS has a long history. A stream of research that is very closely related to RepL (and might also be considered as such) is dimensionality reduction. Well known and still frequently applied techniques for dimensionality reduction such as PCA [56] and MDS [68] have been developed decades ago. More recently, methods based on manifold learning such as Stochastic Neighbor Embedding [53], t-SNE [82], and UMAP [87] have gained popularity. A very powerful family of algorithms exploiting this clue are AEs (see Sect. 3), which can also be applied to MTS. Different model architectures utilizing RNNs and CNNs in the encoder and decoder network allow the implementation of so-called sequence to sequence models, that encode MTS into lower-dimensional representations [24, 84, 132]. In many cases the behaviour of CPS, which are observed with a large number of sensors, can be described with only a few latent variables. This concept is exploited, for example, in the artificial generation of CPS data [135].

Natural clustering The basic concept of clustering algorithms is described above in Sect. 3. The mapping of objects described in high-dimensional spaces to clusters (some cluster identifier, mostly an integer value) is a kind of representation in the sense of the definition given above. This "clue" mainly assumes, that MTS generated by similar underlying processes (factors) also have similar shapes and patterns according to some distance measure suitable for time series data such as Dynamic Time Warping (DTW) [37]. However, clustering MTS data is not a trivial problem due to the potentially high dimensionality of MTS and the challenge of defining a distance or similarity measure. A review of time series clustering methods is provided in [1]. Some examples of applications of time series clustering for CPS can be found in [42], [106], and [66].

Simple and sparse dependencies between factors The essence of this clue is to assume very simple dependencies between the underlying explanatory factors that created the data. The relationships are assumed to be so simple and general that they can be integrated into the model architecture or the loss functions. This is motivated by the fact that many physical laws can also be described in terms of simple relationships of a few quantities [6]. This "clue" is very popular because it is often used in conjunction with deep generative models, which have achieved very good results in recent years. In the context of RepL, generative models approximate the joint probability $p(\mathbf{x}, \mathbf{z})$, where $\mathbf{x} \in \mathbf{R}^n$ are the observation and $\mathbf{z} \in \mathbf{R}^m$ the latent space variables. A very basic assumption for these simple dependencies is the marginal independence of the latent space variables, such that

$$P(\mathbf{z}) = \prod_{i=1}^{m} P(z_i). \tag{9}$$

This assumption lies at the core of many famous unsupervised RepL algorithms such as Generative Adversarial Networks (GANs) [49] and Variational Autoencoders (VAEs) [64]. Many extensions or versions of VAEs and GANs have been introduced in recent years, some of which introduce other clues in addition to the marginal independence, e.g. mutual information criteria. Examples are the β-VAE [52], FactorVAE [63], β-TCVAE [19], InfoGAN [22] or the InfoVEA [133] just to name a few. Note that the dimensionality reduction and manifold assumption is also explicitly exploited in all of these algorithms.

Others assume simple causal dependencies between the latent variables and the data [115] or even between the individual univariate time series in the MTS [101].

Communicating agents A very new and still experimental approach is to train several small neural networks. These networks act as agents that perform different subtasks. The subtasks are chosen in such a way that different subsets of the underlying explanatory factors are needed to answer them. Together with a loss function that minimizes the amount of information exchanged between agents, meaningful variables can be disentangled in the latent space. These ideas are described in [93] and [60]. The main motivation is to identify physical concepts. The application of this clue in CPS use cases has not been studied yet.

8.3   Conclusion

RepL is the core area of today's deep learning and AI research. For its key challenge, the disentanglement of meaningful latent space variables, many methods have been developed. RepL is particularly relevant for CPS applications because it represents an important step towards explainable AI and because unlabeled sensor data can be used and exploited.

9   Conclusion

This paper explains the role of ML for CPS, provides an overview of the state of the art and discusses challenges that accompany the application of ML algorithms to CPS data.

To highlight the relevance of ML for CPS, Sect. 6 describes several application scenarios. ML algorithms can automatically analyze large amounts of data and can thus be used for tasks like predictive maintenance, resource optimization, the creation of Digital Twins or automated diagnosis.

An overview of ML in general is provided in Sect. 3. The ML algorithms are categorized based on two properties: how they handle dynamic, time-dependent data and how much supervision they need.

Section 4 explains why CPS require specific ML algorithms and gives an overview of the four main challenges identified in this paper: time, uncertainty, a-priori knowledge and meaningful representations.

Section 5 outlines the CPS' main characteristic: time. In order to describe a dynamically changing system, a latent state is introduced. Since traditional approaches often fail to detect complex patterns in the data, the recent surge in computational capacity has motivated an increasing interest in ML algorithms.

Uncertainty, as another important characteristic for CPS, is discussed in Sect. 6. Traditionally, uncertainty can be expressed with statistical methods. However, for high-dimensional and complex CPS data, their applicability is limited. ML algorithms can model more complex distributions and learn uncertainty directly. Furthermore, the usage of sampling approaches allows estimating uncertainty empirically.

How to use existing engineering and physical know-how to improve ML is the topic of Sect. 7. The main challenges include encoding the relationship of the CPS components, minimizing the dependence on labels set by domain experts and leveraging simulated data.

Section 8 discusses how to derive meaningful representations of high-dimensional data. Such representations are key to build explainabe ML models and to transfer knowledge from models built on different datasets.

ML has led to breakthroughs in many domains, such as computer vision, natural language processing or computational biology. The amount of data available rises rapidly, as does the computational capacity that fuels ML models. By applying more and better sensors, a CPS can generate large amounts of data as well. However, there has not been a comparable disruption for CPS yet. Incorporating successful approaches from other domains is a promising start. Yet many application problems arguably exist due to the very nature of CPS data, which are highly interwined with physical processes and have unique challenges that first have to be solved. This paper highlights these challenges and gives an outlook of possible research directions.

In the future, specialized ML algorithms are needed which work on a level of accuracy, reliability and maintainability required in the field of engineering. For this, a corresponding research field of "ML for Engineering" has to be established. The visions are ML algorithms which are fed by engineering knowledge, compute interpretable engineering models and can be deployed in closed-loop control loops and in autonomous systems.

References

1. Aghabozorgi, S., Seyed Shirkhorshidi, A., Ying Wah, T.: Time-series clustering – a decade review. Inf. Syst. **53**, 16–38 (Oct 2015)
2. Alazab, M., Khan, S., Krishnan, S.S.R., Pham, Q.V., Reddy, M.P.K., Gadekallu, T.R.: A multidirectional lstm model for predicting the stability of a smart grid. IEEE Access **8**, 85454–85463 (2020). 10.1109/access.2020.2991067
3. Anumasa, S., Srijith, P.K.: Improving robustness and uncertainty modelling in neural ordinary differential equations. In: 2021 IEEE Winter Conference on Applications of Computer Vision (WACV). pp. 4052–4060 (2021). 10.1109/WACV48630.2021.00410
4. Bahdanau, D., Cho, K., Bengio, Y.: Neural machine translation by jointly learning to align and translate, https://arxiv.org/pdf/1409.0473
5. Battaglia, P.W., Hamrick, J.B., Bapst, V., Sanchez-Gonzalez, A., Zambaldi, V., Malinowski, M., Tacchetti, A., Raposo, D., Santoro, A., Faulkner, R., Gulcehre, C., Song, F., Ballard, A., Gilmer, J., Dahl, G., Vaswani, A., Allen, K., Nash, C., Langston, V., Dyer, C., Heess, N., Wierstra, D., Kohli, P., Botvinick, M., Vinyals, O., Li, Y., Pascanu, R.: Relational inductive biases, deep learning, and graph networks. arXiv preprint arXiv:1806.01261 (Jun 2018)
6. Bengio, Y., Courville, A., Vincent, P.: Representation learning: a review and new perspectives. IEEE Trans. Pattern Anal. Mach. Intell. **35**(8), 1798–1828 (Aug 2013)
7. Bengio, Y., LeCun, Y., Hinton, G.: Deep learning for ai. Communications of the ACM **64**(7), 58–65 (2021). 10.1145/3448250
8. Beyerer, J., Kühnert, C., Niggemann, O.: Machine Learning for Cyber Physical Systems – Selected papers from the International Conference ML4CPS 2018. Springer (2019)
9. Beyerer, J., Maier, A., Niggemann, O.: Machine Learning for Cyber Physical Systems – Selected papers from the International Conference ML4CPS 2017. Springer (2020)
10. Beyerer, J., Maier, A., Niggemann, O.: Machine Learning for Cyber Physical Systems – Selected papers from the International Conference ML4CPS 2020. Springer (2021)
11. Box, G.E.P., Jenkins, G.M., Reinsel, G.C., Ljung, G.M.: Time Series Analysis: Forecasting and Control. John Wiley & Sons, Inc., Hoboken, New Jersey, USA (2015)
12. Breiman, L.: Random forests. In: Machine Learning (2001). 10.1023/A:1010933404324
13. Bronstein, M.M., Bruna, J., Cohen, T., Veličković, P.: Geometric deep learning: Grids, groups, graphs, geodesics, and gauges. arXiv preprint arXiv:2104.13478 (Apr 2021)
14. Brown, T., Mann, B., Ryder, N., Subbiah, M., Kaplan, J.D., Dhariwal, P., Neelakantan, A., Shyam, P., Sastry, G., Askell, A., Agarwal, S., Herbert-Voss, A., Krueger, G., Henighan, T., Child, R., Ramesh, A., Ziegler, D., Wu, J., Winter, C., Hesse, C., Chen, M., Sigler, E., Litwin, M., Gray, S., Chess, B., Clark, J., Berner, C., McCandlish, S., Radford, A., Sutskever, I., Amodei, D.: Language models are few-shot learners. Advances in Neural Information Processing Systems **33**, 1877–1901 (2020)
15. Bunte, A., Stein, Benno an d Niggemann, O.: Model-based diagnosis for cyber-physical production systems based on machine learning and residual-based diagnosis models. Hawaii, USA (2019)
16. Burrows, S., Frochte, J., Völske, M., Torres, A.B.M., Stein, B.: Learning overlap optimization for domain decomposition methods. In: Advances in Knowledge Discovery and Data Mining. pp. 438–449. Springer Berlin Heidelberg (2013)

17. Burrows, S., Stein, B., Frochte, J., Wiesner, D., Müller, K.: Simulation data mining for supporting bridge design. In: Proceedings of the Ninth Australasian Data Mining Conference-Volume 121. pp. 163–170 (2011)
18. Chen, R.T.Q., Amos, B., Nickel, M.: Learning neural event functions for ordinary differential equations. ICLR https://arxiv.org/pdf/2011.03902
19. Chen, R.T.Q., Li, X., Grosse, R., Duvenaud, D.: Isolating sources of disentanglement in variational autoencoders (Feb 2018)
20. Chen, R.T.Q., Rubanova, Y., Bettencourt, J., Duvenaud, D.: Neural ordinary differential equations, https://arxiv.org/pdf/1806.07366
21. Chen, T., Kornblith, S., Swersky, K., Norouzi, M., Hinton, G.: Big Self-Supervised models are strong Semi-Supervised learners (Jun 2020)
22. Chen, X., Duan, Y., Houthooft, R., Schulman, J., Sutskever, I., Abbeel, P.: InfoGAN: Interpretable representation learning by information maximizing generative adversarial nets (Jun 2016)
23. Chiu, M.C., Tsai, C.D., Li, T.L.: An integrative machine learning method to improve fault detection and productivity performance in a cyber-physical system. Journal of Computing and Information Science in Engineering **20**(2) (2020). 10.1115/1.4045663, https://asmedigitalcollection. asme.org/computingengineering/article/20/2/021009/1071865
24. Cho, K., van Merrienboer, B., Gulcehre, C., Bahdanau, D., Bougares, F., Schwenk, H., Bengio, Y.: Learning phrase representations using RNN Encoder-Decoder for statistical machine translation (Jun 2014)
25. Chung, J., Gulcehre, C., Cho, K., Bengio, Y.: Empirical evaluation of gated recurrent neural networks on sequence modeling http://arxiv.org/pdf/1412.3555v1
26. Conneau, A., Khandelwal, K., Goyal, N., Chaudhary, V., Wenzek, G., Guzmán, F., Grave, E., Ott, M., Zettlemoyer, L., Stoyanov, V.: Unsupervised cross-lingual representation learning at scale. arXiv preprint arXiv:1911.02116 (Nov 2019)
27. Cosmo, L., Kazi, A., Ahmadi, S.A., Navab, N., Bronstein, M.: Latent-Graph learning for disease prediction. In: Medical Image Computing and Computer Assisted Intervention – MICCAI 2020. pp. 643–653. Springer International Publishing (2020)
28. Daniluk, M., Rocktäschel, T., Welbl, J., Riedel, S.: Frustratingly short attention spans in neural language modeling, https://arxiv.org/pdf/1702.04521
29. Daumé III, H.: Frustratingly easy domain adaptation. arXiv preprint arXiv:0907.1815 (Jul 2009)
30. Devlin, J., Chang, M.W., Lee, K., Toutanova, K.: BERT: Pre-training of deep bidirectional transformers for language understanding. arXiv preprint arXiv:1810.04805 (Oct 2018)
31. Diedrich, A., Niggemann, O.: Model-based diagnosis of hybrid systems using satisfiability modulo theory. Hawaii, USA (2019)
32. Dosovitskiy, A., Beyer, L., Kolesnikov, A., Weissenborn, D., Zhai, X., Unterthiner, T., Dehghani, M., Minderer, M., Heigold, G., Gelly, S., Uszkoreit, J., Houlsby, N.: An image is worth 16x16 words: Transformers for image recognition at scale, https://arxiv.org/pdf/2010.11929
33. Dupont, E., Doucet, A., Teh, Y.W.: Augmented neural odes, https://arxiv.org/pdf/1904.01681
34. Durbin, J., Koopman, S.J.: Time series analysis by state space methods, Oxford statistical science series, vol. 38. Oxford Univ. Press, Oxford, 2. ed. edn. (2012). 10.1093/acprof:oso/9780199641178.001.0001
35. Eiteneuer, B., Hranisavljevic, N., Niggemann, O.: Dimensionality reduction and anomaly detection for cpps data using autoencoder. In: 20th IEEE International Conference on Industrial Technology (ICIT). IEEE, Melbourne, Australien (Feb 2019)
36. Eiteneuer, B., Niggemann, O.: Lstm for model-based anomaly detection in cyber-physical systems. In: Proceedings of the 29th International Workshop on Principles of Diagnosis. Warsaw, Poland (Aug 2018)

37. Esling, P., Agon, C.: Time-series data mining. ACM Comput. Surv. **45**(1), 1–34 (Dec 2012)
38. Esteva, A., Kuprel, B., Novoa, R.A., Ko, J., Swetter, S.M., Blau, H.M., Thrun, S.: Dermatologist-level classification of skin cancer with deep neural networks. Nature **542**(7639), 115–118 (Feb 2017)
39. Fei-Fei, L., Fergus, R., Perona, P.: One-shot learning of object categories. IEEE Trans. Pattern Anal. Mach. Intell. **28**(4), 594–611 (Apr 2006)
40. Fei-Niu, Y., Lin, Z., Jin-Ting, S., Xue, X., Gang, L.: Theories and applications of auto-encoder neural networks: A literature survey. Chinese Journal of Computers (2019)
41. Fischer, A., Igel, C.: An introduction to restricted Boltzmann machines. In: Lecture Notes in Computer Science. vol. 7441 LNCS, pp. 14–36. Springer, Berlin, Heidelberg (2012)
42. Fontes, C.H., Pereira, O.: Pattern recognition in multivariate time series – a case study applied to fault detection in a gas turbine. Eng. Appl. Artif. Intell. **49**, 10–18 (Mar 2016)
43. Gal, Y., Ghahramani, Z.: Dropout as a bayesian approximation: Representing model uncertainty in deep learning. International Conference on Machine Learning pp. 1050–1059 (2016), http://proceedings.mlr.press/v48/gal16.html
44. Gawlikowski, J., Tassi, C.R.N., Ali, M., Lee, J., Humt, M., Feng, J., Kruspe, A., Triebel, R., Jung, P., Roscher, R., Shahzad, M., Yang, W., Bamler, R., Zhu, X.X.: A survey of uncertainty in deep neural networks https://arxiv.org/pdf/2107.03342
45. Gilmer, J., Schoenholz, S.S., Riley, P.F., Vinyals, O., Dahl, G.E.: Neural message passing for quantum chemistry. In: Precup, D., Teh, Y.W. (eds.) Proceedings of the 34th International Conference on Machine Learning. Proceedings of Machine Learning Research, vol. 70, pp. 1263–1272. PMLR (2017)
46. Gligorijevic, V., Renfrew, P.D., Kosciolek, T., Leman, J.K., others: Structure-based function prediction using graph convolutional networks. bioRxiv (2020)
47. Goh, J., Adepu, S., Tan, M., Lee, Z.S.: Anomaly detection in cyber physical systems using recurrent neural networks. In: IEEE 18th International Symposium on High Assurance Systems Engineering. IEEE, Piscataway, NJ (2017). 10.1109/hase.2017.36
48. Goodfellow, I., Bengio, Y., Courville, A., Bengio, Y.: Deep learning, vol. 1. MIT press Cambridge (2016)
49. Goodfellow, I.J., Pouget-Abadie, J., Mirza, M., Xu, B., Warde-Farley, D., Ozair, S., Courville, A., Bengio, Y.: Generative adversarial networks (Jun 2014)
50. Grill, J.B., Strub, F., Altché, F., Tallec, C., Richemond, P.H., Buchatskaya, E., Doersch, C., Pires, B.A., Guo, Z.D., Azar, M.G., et al.: Bootstrap your own latent: A new approach to self-supervised learning. arXiv preprint arXiv:2006.07733 (2020)
51. Guo, G., Lu, Z., Han, Q.L.: Control with markov sensors/actuators assignment. IEEE Transactions on Automatic Control **57**(7), 1799–1804 (2012). 10.1109/TAC.2011.2176393
52. Higgins, I., Matthey, L., Pal, A., Burgess, C., Glorot, X., Botvinick, M., Mohamed, S., Lerchner, A.: beta-vae: Learning basic visual concepts with a constrained variational framework. In: ICLR (2017)
53. Hinton, G., Roweis, S.T.: Stochastic neighbor embedding. In: NIPS. vol. 15, pp. 833–840 (2002)
54. Hochreiter, S., Schmidhuber, J.: Long short-term memory. Neural computation **9**(8), 1735–1780 (1997). 10.1162/neco.1997.9.8.1735
55. Hochreiter, S., Bengio, Y., Paolo, F., Schmidhuber, J.: Gradient flow in recurrent nets: The difficulty of learning longterm dependencies. In: Kolen, J.F., Kremer, S.C. (eds.) A field guide to dynamical recurrent networks. IEEE Press and IEEE Xplore, New York and Piscataway, New Jersey (2009). 10.1109/9780470544037.ch14
56. Hotelling, H.: Analysis of a complex of statistical variables into principal components. J. Educ. Psychol. **24**(6), 417–441 (Sep 1933)

57. Hranisavljevic, N., Maier, A., Niggemann, O.: Discretization of hybrid CPPS data into timed automaton using restricted boltzmann machines. Eng. Appl. Artif. Intell. **95**, 103826 (Oct 2020)
58. Hu, W., Fey, M., Zitnik, M., Dong, Y., Ren, H., Liu, B., Catasta, M., Leskovec, J.: Open graph benchmark: Datasets for machine learning on graphs. arXiv preprint arXiv:2005.00687 (May 2020)
59. Hyndman, R.J., Koehler, A.B., Ord, J.K., Snyder, R.D.: Forecasting with Exponential Smoothing: The State Space Approach. Springer Berlin Heidelberg (2008)
60. Iten, R., Metger, T., Wilming, H., Del Rio, L., Renner, R.: Discovering physical concepts with neural networks. Phys. Rev. Lett. **124**(1), 010508 (Jan 2020)
61. Jing, L., Tian, Y.: Self-supervised visual feature learning with deep neural networks: A survey. IEEE transactions on pattern analysis and machine intelligence (2020)
62. Kennedy, M.C., O'Hagan, A.: Bayesian calibration of computer models. Journal of the Royal Statistical Society: Series B (Statistical Methodology) **63**(3), 425–464 (2001). 10.1111/1467-9868.00294
63. Kim, H., Mnih, A.: Disentangling by factorising (Feb 2018)
64. Kingma, D.P., Welling, M.: Auto-Encoding variational bayes (Dec 2013)
65. Kipf, T.N., Welling, M.: Semi-Supervised classification with graph convolutional networks. arXiv preprint arXiv:1609.02907 (Sep 2016)
66. Kiss, I., Genge, B., Haller, P., Sebestyén, G.: Data clustering-based anomaly detection in industrial control systems. In: 2014 IEEE 10th International Conference on Intelligent Computer Communication and Processing (ICCP). pp. 275–281 (Sep 2014)
67. Kiureghian, A.D., Ditlevsen, O.: Aleatory or epistemic? does it matter? Structural Safety **31**(2), 105–112 (2009). 10.1016/j.strusafe.2008.06.020
68. Kruskal, J.B.: Multidimensional scaling by optimizing goodness of fit to a nonmetric hypothesis. Psychometrika **29**(1), 1–27 (Mar 1964)
69. Lake, B.M., Ullman, T.D., Tenenbaum, J.B., Gershman, S.J.: Building machines that learn and think like people (2017)
70. Lakshminarayanan, B., Pritzel, A., Blundell, C.: Simple and scalable predictive uncertainty estimation using deep ensembles. undefined (2017), https://www.semanticscholar.org/paper/Simple-and-Scalable-Predictive-Uncertainty-using-Lakshminarayanan-Pritzel/802168a81571dde28f5ddb94d84677bc007afa7b
71. Le Paine, T., Khorrami, P., Han, W., Huang, T.S.: An analysis of unsupervised pre-training in light of recent advances (Dec 2014)
72. LeCun, Y., Bengio, Y., Hinton, G.: Deep learning. Nature **521**(7553), 436–444 (May 2015)
73. LeCun, Y., Bottou, L., Bengio, Y., Haffner, P.: Gradient-based learning applied to document recognition. Proceedings of the IEEE **86**(11), 2278–2324 (1998)
74. Lee, E.A.: Cps foundations. In: Proceedings of the 47th Design Automation Conference. DAC '10, ACM, New York, NY, USA (2010)
75. Legrand, A., Trannois, H., Cournier, A.: Use of uncertainty with autoencoder neural networks for anomaly detection. In: IEEE Second International Conference on Artificial Intelligence and Knowledge Engineering. pp. 32–35. Conference Publishing Services, IEEE Computer Society, Los Alamitos, California and Washington and Tokyo (2019). 10.1109/AIKE.2019.00014
76. Li, D., Chen, D., Jin, B., Shi, L., Goh, J., Ng, S.K.: MAD-GAN: Multivariate anomaly detection for time series data with generative adversarial networks. In: Artificial Neural Networks and Machine Learning – ICANN 2019: Text and Time Series. pp. 703–716. Springer International Publishing (2019)
77. Li, P., Niggemann, O.: Improving clustering based anomaly detection with concave hull: An application in condition monitoring of wind turbines. In: 14th IEEE International Conference on Industrial Informatics (INDIN 2016). Poltiers (France) (2016)

78. Li, S., Jin, X., Xuan, Y., Zhou, X., Chen, W., Wang, Y.X., Yan, X.: Enhancing the locality and breaking the memory bottleneck of transformer on time series forecasting. Advances in Neural Information Processing Systems **32** (2019)

79. Lin, Z., Feng, M., Santos, C.N.d., Yu, M., Xiang, B., Zhou, B., Bengio, Y.: A structured self-attentive sentence embedding, https://arxiv.org/pdf/1703.03130

80. Liu, M., Ren, S., Ma, S., Jiao, J., Chen, Y., Wang, Z., Song, W.: Gated transformer networks for multivariate time series classification, https://arxiv.org/pdf/2103.14438

81. Loucks, D.P., van Beek, E., Loucks, D.P., van Beek, E.: System Sensitivity and Uncertainty Analysis. In: Water Resource Systems Planning and Management, pp. 331–374. Springer International Publishing (2017)

82. van der Maaten, L., Hinton, G.: Visualizing data using t-SNE. J. Mach. Learn. Res. **9**(86), 2579–2605 (2008)

83. Maier, A., Schriegel, S., Niggemann, O.: Big data and machine learning for the smart factory - solutions for condition monitoring, diagnosis and optimization. Industrial Internet of Things: Cybermanufacturing Systems (2016)

84. Malhotra, P., Vishnu, T.V., Vig, L., Agarwal, P., Shroff, G.: TimeNet: Pre-trained deep recurrent neural network for time series classification (Jun 2017)

85. Mallidi, S.H., Ogawa, T., Hermansky, H.: Uncertainty estimation of DNN classifiers. In: 2015 IEEE Workshop on Automatic Speech Recognition and Understanding, ASRU 2015 - Proc. pp. 283–288. I. of Electrical and Electronics Engineers Inc. (2 2016). 10.1109/ASRU.2015.7404806

86. Massaroli, S., Poli, M., Park, J., Yamashita, A., Asama, H.: Dissecting neural odes. In: H. Larochelle, M. Ranzato, R. Hadsell, M. F. Balcan, H. Lin (eds.) Advances in Neural Information Processing Systems. vol. 33, pp. 3952–3963. Curran Associates, Inc (2020), https://proceedings.neurips.cc/paper/2020/file/293835c2cc75b585649498ee74b395f5-Paper.pdf

87. McInnes, L., Healy, J., Melville, J.: UMAP: Uniform manifold approximation and projection for dimension reduction (Feb 2018)

88. Mikolov, T., Chen, K., Corrado, G., Dean, J.: Efficient estimation of word representations in vector space (Jan 2013)

89. Monti, F., Frasca, F., Eynard, D., Mannion, D., Bronstein, M.M.: Fake news detection on social media using geometric deep learning. arXiv preprint arXiv:1902.06673 (Feb 2019)

90. Multaheb, S., Zimmering, B., Niggemann, O.: Expressing uncertainty in neural networks for production systems. at - Automatisierungstechnik **63**(3), 221–230 (2021)

91. Murphy, K.: Machine Learning: A Probabilistic Perspective. MIT Press, Cambridge, Massachusetts, USA (2012)

92. Na, S., Xumin, L., Yong, G.: Research on k-means clustering algorithm: An improved k-means clustering algorithm. In: 2010 Third International Symposium on Intelligent Information Technology and Security Informatics. pp. 63–67 (2010). 10.1109/IITSI.2010.74

93. Nautrup, H.P., Metger, T., Iten, R., Jerbi, S., Trenkwalder, L.M., Wilming, H., Briegel, H.J., Renner, R.: Operationally meaningful representations of physical systems in neural networks (Jan 2020)

94. Niggemann, O., Diedrich, A., Pfannstiel, E., Schraven, J., Kühnert, C.: A generic digitaltwin model for artificial intelligence applications. In: IEEE International Conference on Industrial Cyber-Physical Systems (ICPS) (2021)

95. Niggemann, O., Frey, C.: Data-driven anomaly detection in cyber-physical production systems. at - Automatisierungstechnik(63) **63**, 821–832 (2016)

96. Niggemann, O., Lohweg, V.: On the diagnosis of cyber-physical production systems - state-of-the-art and research agenda. Austin, Texas, USA (2015)

97. Niggemann, O., Schüller, P.: IMPROVE - Innovative Modelling Approaches for Production Systems to Raise Validatable Efficiency. Springer Vieweg (2018)

98. Nix, D.A., Weigend, A.S.: Estimating the mean and variance of the target probability distribution. In: The 1994 IEEE International Conference on Neural Networks. pp. 55–60 vol.1. IEEE Neural Networks Council, New York and Piscataway, NJ (1994). 10.1109/ICNN.1994.374138

99. Otto, J., Vogel-Heuser, B., Niggemann, O.: Automatic parameter estimation for reusable software components of modular and reconfigurable cyber physical production systems in the domain of discrete manufacturing. IEEE Transactions on Industrial Informatics (2018)

100. Pascanu, R., Mikolov, T., Bengio, Y.: On the difficulty of training recurrent neural networks. In: Dasgupta, S., McAllester, D. (eds.) Proceedings of the 30th International Conference on Machine Learning. Proceedings of Machine Learning Research, vol. 28, pp. 1310–1318. PMLR, Atlanta, Georgia, USA (17–19 Jun 2013), http://proceedings.mlr.press/v28/pascanu13.html

101. Pineau, E., Razakarivony, S., Bonald, T.: Unsupervised ageing detection of mechanical systems on a causality graph. In: ICMLA (2020)

102. Rangapuram, S.S., Seeger, M.W., Gasthaus, J., Stella, L., Wang, Y., Januschowski, T.: Deep state space models for time series forecasting. In: Bengio, S., Wallach, H., Larochelle, H., Grauman, K., Cesa-Bianchi, N., Garnett, R. (eds.) Advances in Neural Information Processing Systems. vol. 31. Curran Associates, Inc. (2018), https://proceedings.neurips.cc/paper/2018/file/5cf68969fb67aa6082363a6d4e6468e2-Paper.pdf

103. Rossi, R., Ahmed, N.: The network data repository with interactive graph analytics and visualization. AAAI **29**(1) (Mar 2015)

104. Rubanova, Y., Chen, R.T.Q., Duvenaud, D.: Latent ordinary differential equations for irregularly-sampled time series (2019), https://openreview.net/forum?id=HygCYNSlLB

105. Safavian, S., Landgrebe, D.: A survey of decision tree classifier methodology. IEEE Transactions on Systems, Man, and Cybernetics **21**(3), 660–674 (1991). 10.1109/21.97458

106. Schmidt, T., Hauer, F., Pretschner, A.: Automated anomaly detection in CPS log files. In: Computer Safety, Reliability, and Security. pp. 179–194. Springer International Publishing (2020)

107. Schubert, E., Sander, J., Ester, M., Kriegel, H.P., Xu, X.: Dbscan revisited, revisited: Why and how you should (still) use dbscan. ACM Trans. Database Syst. **42**(3) (Jul 2017). 10.1145/3068335, https://doi.org/10.1145/3068335

108. Sen, P., Namata, G., Bilgic, M., Getoor, L., Galligher, B., Eliassi-Rad, T.: Collective classification in network data. AIMag **29**(3), 93–93 (Sep 2008)

109. Shang, C., Chen, J., Bi, J.: Discrete graph structure learning for forecasting multiple time series. arXiv preprint arXiv:2101.06861 (2021)

110. Sherstinsky, A.: Fundamentals of recurrent neural network (rnn) and long short-term memory (lstm) network. Physica D: Nonlinear Phenomena **404**(8), 132306 (2020). 10.1016/j.physd.2019.132306, https://arxiv.org/pdf/1808.03314

111. Silver, D., Hubert, T., Schrittwieser, J., Antonoglou, I., Lai, M., Guez, A., Lanctot, M., Sifre, L., Kumaran, D., Graepel, T., Lillicrap, T., Simonyan, K., Hassabis, D.: Mastering chess and shogi by Self-Play with a general reinforcement learning algorithm (Dec 2017)

112. Smolensky, P.: Information Processing in Dynamical Systems: Foundations of Harmony Theory, p. 194-281. MIT Press, Cambridge, MA, USA (1986)

113. Socher, R., Ganjoo, M., Sridhar, H., Bastani, O., Manning, C.D., Ng, A.Y.: Zero-Shot learning through Cross-Modal transfer (Jan 2013)

114. Sun, X., Bischl, B.: Tutorial and survey on probabilistic graphical model and variational inference in deep reinforcement learning. In: 2019 IEEE Symposium Series on Computational Intelligence (SSCI). pp. 110–119 (2019). 10.1109/SSCI44817.2019.9003114

115. Suter, R., Miladinovic, D., Schölkopf, B., Bauer, S.: Robustly disentangled causal mechanisms: Validating deep representations for interventional robustness. In: Chaudhuri, K., Salakhutdinov,

R. (eds.) Proceedings of the 36th International Conference on Machine Learning. Proceedings of Machine Learning Research, vol. 97, pp. 6056–6065. PMLR (2019)

116. Talkhestani, B.A., Jung, T., Lindemann, B., Sahlab, N., Jazdi, N., Schloegl, W., Weyrich, M.: An architecture of an intelligent digital twin in a cyber-physical production system:. at - Automatisierungstechnik **67**(9), 762–782 (2019). https://doi.org/10.1515/auto-2019-0039

117. Tan, P.N., Steinbach, M., Karpatne, A., Kumar, V.: Introduction to Data Mining, 2nd Edition. Pearson Education, New York, NY, USA (2019)

118. Thakoor, S., Tallec, C., Azar, M.G., Munos, R., Veličković, P., Valko, M.: Bootstrapped representation learning on graphs. arXiv preprint arXiv:2102.06514 (2021)

119. Tzeng, E., Hoffman, J., Saenko, K., Darrell, T.: Adversarial discriminative domain adaptation. In: Proceedings of the IEEE conference on computer vision and pattern recognition. pp. 7167–7176. openaccess.thecvf.com (2017)

120. University, S.: Artificial Intelligence Index Report 2021. HAI Human-centered Artificial Intelligence (2021)

121. Uusitalo, L., Lehikoinen, A., Helle, I., Myrberg, K.: An overview of methods to evaluate uncertainty of deterministic models in decision support (jan 2015). 10.1016/j.envsoft.2014.09.017

122. Vaswani, A., Shazeer, N., Parmar, N., Uszkoreit, J., Jones, L., Gomez, A.N., Kaiser, Ł., Polosukhin, I.: Attention is all you need. Advances in Neural Information Processing Systems **30** (2017)

123. Veličković, P., Buesing, L., Overlan, M.C., Pascanu, R., Vinyals, O., Blundell, C.: Pointer graph networks. arXiv preprint arXiv:2006.06380 (2020)

124. Veličković, P., Cucurull, G., Casanova, A., Romero, A., Liò, P., Bengio, Y.: Graph attention networks. arXiv preprint arXiv:1710.10903 (Oct 2017)

125. Wang, J.M., Fleet, D.J., Hertzmann, A.: Gaussian process dynamical models. In: Proceedings of the 18th International Conference on Neural Information Processing Systems. p. 1441-1448. NIPS'05, MIT Press, Cambridge, MA, USA (2005)

126. Windmann, S., Niggemann, O., Stichweh, H.: Energy efficiency optimization by automatic coordination of motor speeds in conveying systems (2015)

127. Xing, Z., Pei, J., Keogh, E.: A brief survey on sequence classification. SIGKDD Explor. Newsl. **12**(1), 40–48 (Nov 2010)

128. Yan, H., Jiawei, D., Tan, V.Y.F., Feng, J.: On robustness of neural ordinary differential equations. In: 2020 International Conference on Learning Representations (2020), https://arxiv.org/pdf/1910.05513

129. Yang, Y., Sautière, G., Ryu, J.J., Cohen, T.S.: Feedback recurrent autoencoder. In: ICASSP 2020 - 2020 IEEE International Conference on Acoustics, Speech and Signal Processing (ICASSP). pp. 3347–3351 (2020). 10.1109/ICASSP40776.2020.9054074

130. Zerveas, G., Jayaraman, S., Patel, D., Bhamidipaty, A., Eickhoff, C.: A transformer-based framework for multivariate time series representation learning. In: Zhu, F., Chin Ooi, B., Miao, C. (eds.) Proceedings of the 27th ACM SIGKDD Conference on Knowledge Discovery & Data Mining. pp. 2114–2124. ACM, New York, NY, USA (08142021). 10.1145/3447548.3467401

131. Zhang, F., Pinkal, K., Wefing, P., Conradi, F., Schneider, J., Niggemann, O.: Quality control of continuous wort production through production data analysis in latent space (2019)

132. Zhang, J.X., Ling, Z.H., Liu, L.J., Jiang, Y., Dai, L.R.: Sequence-to-Sequence acoustic modeling for voice conversion. IEEE/ACM Transactions on Audio, Speech, and Language Processing **27**(3), 631–644 (Mar 2019)

133. Zhao, S., Song, J., Ermon, S.: InfoVAE: Balancing learning and inference in variational autoencoders. AAAI **33**(01), 5885–5892 (Jul 2019)

134. Zhou, H., Zhang, S., Peng, J., Zhang, S., Li, J., Xiong, H., Zhang, W.: Informer: Beyond efficient transformer for long sequence time-series forecasting, https://arxiv.org/pdf/2012.07436

135. Zimmering, B., Niggemann, O., Hasterok, C., Pfannstiel, E., Ramming, D., Pfrommer, J.: Generating artificial sensor data for the comparison of unsupervised machine learning methods. Sensors **21**(7) (2021). 10.3390/s21072397, https://www.mdpi.com/1424-8220/21/7/2397

Visual Data Science for Industrial Applications

Tobias Schreck◉, Belgin Mutlu and Marc Streit◉

Abstract

Advances in sensor and data acquisition technology and in methods of data analysis pose many research challenges but also promising application opportunities in many domains. The need to cope with and leverage large sensor data streams is particularly urgent for industrial applications due to strong business competition and innovation pressure. In maintenance, for example, sensor readings of machinery or products may allow to predict at which point in time maintenance will be required and allow to schedule service operations respectively. Another application is the discovery of the relationships between production input parameters on the quality of the output products. Analysis of respective industrial data typically cannot be done in an out-of-the-box manner but requires to incorporate background knowledge from fields such as engineering, operation research,

T. Schreck
Computer Graphics and Knowledge Visualization, TU Graz, Graz, Austria
e-mail: tobias.schreck@cgv.tugraz.at

B. Mutlu
Cognitive Decision Support, Pro2Future GmbH, Graz, Austria
e-mail: belgin.mutlu@pro2future.at

M. Streit (✉)
Visual Data Science Lab, Johannes Kepler University Linz, Linz, Austria
e-mail: marc.streit@jku.at

B. Vogel-Heuser and M. Wimmer (eds.), *Digital Transformation*,
https://doi.org/10.1007/978-3-662-65004-2_18

and business to be effective. Hence, approaches for interactive and visual data analysis can be particularly useful for analyzing complex industrial data, combining the advantages of modern automatic data analysis with domain knowledge and hypothesis generation capabilities of domain experts.

In this chapter, we introduce some of the main principles of visual data analysis. We discuss how techniques for data visualization, data analysis, and user interaction can be combined to analyze data, generate and verify hypotheses about patterns in data, and present the findings. We discuss this in the light of important requirements and applications in the analysis of industrial data and based on current research in the area. We provide examples for visual data analysis approaches, including condition monitoring, quality control, and production planning.

Keywords

Visual Data Science • Industrial Data Analysis • Production Data

1   Introduction

The Industry 4.0 is considered as the "fourth industrial revolution" that may fully automatize the production process in the manufacturing industry. In essence, it is based on large-scale digitalization of manufacturing, where machines and humans are connected as a collaborative community, generating large volumes of complex data—often referred to as *big data*. The core idea is to collect data, consolidate, and analyze it across the entire manufacturing process in real-time. The hope is that an analysis of data captured from industrial processes can lead to a better understanding of production processes, and in turn, support their improvement. Predicting the required maintenance intervals of production machinery, for example, can lower operational costs. Another example is the identification of influence factors on production quality, which may be used for quality control.

Current production facilities are typically equipped with sensors that can collect relevant process data. Together with communication and processing capabilities, sensor systems can also exchange data from machine-to-machine or machine-to-human, or perform on-the-fly data analysis at the sensor (edge computing). This data should support the machines in the task of identifying (e.g., trace parts and sub-assemblies), to adapt the production to changing requirements and individual needs, and ensuring product flexibility. Yet, raw production data alone does not provide valuable information and requires domain experts to extract valuable insights. The volume of data generated in a production process, however, can be overwhelming. First, the data has to be monitored and recorded using methods that can handle huge datasets. The next stage includes the analysis of the data (often in real-time) in order to, e.g., (i) identify undetected process correlations, (ii) forecast the production quality, and (iii) perform root cause analysis of failures or problems. A promising application for understanding production data is the use of *visual data science* tools and methods that effectively combine machine intelligence with human intelligence and interactive data exploration. Such interactive approaches can help domain experts to gain insight from manufacturing data, identify interesting patterns, and extract actionable information.

Recently, Zhou et al. [65] identified application opportunities and benefits of visual data science in industrial applications in different industrial sectors, such as automotive and energy, and key operations, such as replacement and creation. Researchers in visual data science address this application space with an increasing set of techniques that effectively exploit the power of data analytics methods and human information processing capabilities. Due to this intelligent use of human visual system and analytical methods, it becomes possible to shift the limiting factors of the analysis of management and production processes data in complex industrial scenarios. In this chapter, we first introduce the visual data science methodology and goals and discuss the infrastructure to implement the technology in concrete applications in Sect. 2. In Sect. 3, we review example visual data science solutions for selected industrial applications, such as production planning, quality control, and condition monitoring. In Sect. 4, we discuss future directions and conclude this chapter.

2 Foundations of Visual Data Science and Challenges

The main idea in visual data science (VDS), also discussed as visual analytics (VA) or visual data analysis, is to support the exploration, understanding, and explanation/communication of relevant patterns in data. It is often used for tasks such as data-driven decision-making, monitoring, and steering of analytical processes. It builds among others on concepts from data visualization, human-computer interaction, and computational data analysis approaches. The latter include approaches from statistics (e.g., from exploratory data analysis) and machine learning, including specific deep learning techniques and artificial intelligence. For this article, we subsume the latter approaches simply as computational data analysis. For a more detailed discussion of respective terminology, we refer to the overview given by Cao [12].

Visual data science approaches, typically integrate data visualization with computational data analysis techniques into interactive systems. In these systems, users explore data to solve tasks in an interactive, sometimes open-ended process. We start by surveying fundamental concepts of interactive data visualization (Sect. 2.1) and visual data science (Sect. 2.2). Then, we give an overview of the data to be analyzed, its complexities (Sect. 2.3), and challenges (Sect. 2.4)—all through the perspective of visual data analysis. In addition, we review the technical infrastructure, ranging from off-the-shelf tools to libraries, which can help in building visual data science solutions for industrial applications (Sect. 2.5).

2.1   Interactive Data Visualization

Interactive data visualization aims to find suitable visual representations of data, such that important data properties can be effectively and efficiently perceived by users [13, 36, 57]. The idea is to leverage the human visual perception capabilities to perceive large amounts of visual information and to link to cognitive processes ("Using vision to think") [49]. The integration of interaction techniques [64] allow users to dynamically explore the data by interacting with its visual representations. This way, users can, for instance, verify hypotheses about the data, dynamically select and filter data items, and change visual encodings to reveal patterns.

The capabilities of visualizations are determined by their specific *visualization designs* that are composed of visual marks and their geometric and appearance properties as their main elements. Marks are essentially geometric shapes of type point, line, area, or three-dimensional marks, as illustrated in Fig. 1a. The appearance of the marks is controlled by visual channels that encode the properties of the marks. The classification of the visual channels was proposed by Bertin [8], who originally called them visual variables. Bertin's list of visual channels comprised position, size, shape, value, color, orientation, and texture. Later, Mackinlay [34] extended the list with the channel length, angle, slope, area, volume, density, color saturation, color hue, connection, and containment. Figure 1 shows a selection of marks whose appearance is modified by means of visual channels. Users can create visualizations by mapping data dimensions to these marks and channels.

Effective visual designs should convey the data as accurately as possible to users, allowing a focus on the most important data features, while scaling with the volume of data (see also Sect. 2.4). Visual design is both a science, in measuring and comparing the effectiveness and efficiency of design, and an art, in creating designs. Over the years, researchers in cartography, statistics, and computer science have formalized perceptual guidelines which a visual designer should consider when defining effective visual representations [8, 23, 34]. Guidelines comprise criteria for visual encoding, perceptually-motivated rankings, and characteristics of the visual channels. Mackinlay, for instance, developed a formal visual encoding language to generate graphical presentations for relational information [34]. He defined expressiveness and effectiveness as the main principles for following in visualization. The expressiveness reflect how accurately a graphical language can encode the desired information. The effectiveness describes how well the graphical language exploits the output medium's capabilities and the human visual system. Based on studies, the most effective visual channels for interpreting data from visualization include position, length, orientation, area, and depth.

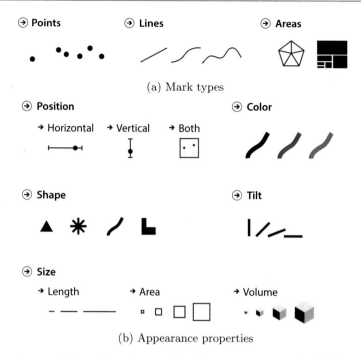

(a) Mark types

(b) Appearance properties

Fig. 1 Elementary mark types (**a**) and geometric and appearance properties (**b**). By mapping data properties to marks, visualization designs can be constructed. Figures reused from [36] courtesy of Tamara Munzner

2.2 Integrating Data Visualization with Data Science

Visualization techniques allow users to get an overview of data sets, interactively explore data, and gain insights. Building on visualization techniques, *visual analytics* or *visual data science* systems bridge between interactive data visualization on the one hand, and algorithmic approaches for data analysis (*data mining, data science*) [30, 54, 55] on the other hand. Both components can be integrated to form highly interactive, powerful data analysis systems (see Fig. 2, left part). The goal is to leverage the joint capabilities of analysts, their background knowledge, and the advantages of automatic data analysis such as pattern search, clustering, and classification. The inclusion of Machine Learning techniques into visualization aims to address several goals, including scalability to large data sets that could not be exhaustively inspected by visualization of the raw data. The result is an encompassing analytical process, going from discovering single findings in data, to forming insights and hypotheses about data, and eventually, arriving at actionable results and decision support [44] (see also Fig. 2, right part).

To date, many data science methods have been proposed—including techniques for data reduction, clustering, classification, and prediction [22]—and combined with visualization

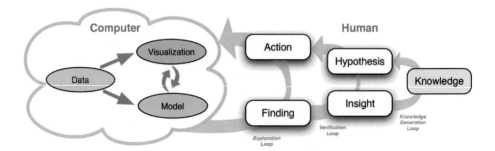

Fig. 2 The visual knowledge generation model. It suggests that analysts can obtain knowledge from data by integration of data analysis (modeling) and interactive visualization. The knowledge generation process works by obtaining findings and insights, which can lead to hypotheses and actions. Figure reused from [44] with permission

to form effective visual analysis systems [19, 33]. Recently, also *artificial intelligence* (AI)-based methods, in particular deep neural networks, gained much popularity. These methods have been successfully applied in many real-time decision-making use cases, including autonomous driving, recommender systems, and automated diagnosis. AI is attracting a growing interest in the industry too, since it is perceived as a powerful technology for (i) identifying undetected process correlations, (ii) forecasting the production quality, and (iii) performing root-cause analysis of failures or problems. AI-based approaches can show superior results in many learning domains compared to more traditional approaches.

How an AI algorithm operates is, in many cases, a black box. Even for experts, it can be hard to answer why AI algorithms make certain decisions. Usually, this is due to the input of a large number of parameters and complex data structures. Yet, this lack of transparency can be a key problem in industrial practices and may hinder AI-based methods from being actively used as part of real decision-making processes. However, this issue can be addressed by making machine knowledge *explainable* and *understandable*. Recent work focuses on two methods to provide *explainable AI* (xAI): transparency and post-hoc interpretation [61]. The former renders the AI algorithm transparent, hence showing how the algorithm functions internally. The latter provides explanations of its behavior. As a result, practitioners are able to understand, e.g., what the algorithm has predicted and why. However, the biggest challenge for xAI is to provide explanations that are interpretable by humans. Visualization offers many opportunities to help AI-based systems become understandable and explainable, as discussed by Hohman et al. [24].

2.3   Data Types and Characteristics

A key prerequisite to employ visual analytics techniques for practical data analysis is to model the input data. In order to select suitable analysis and visualization techniques, it needs to be clear which types of data are input to the analysis. In the following, we review a selection of data types that are typically found in industrial applications and for which the scientific community has developed a broad variety of visual representations and interaction techniques. For additional details and visualization examples, see the linked references to textbooks and surveys. Note that this selection of data is not intended to be complete but follows a pragmatic approach linking to applications. The distinction of visualization techniques by data type is often used in visualization research, dating back to Shneiderman's data type by task taxonomy [49].

Tabular data consists of rows and columns. In a simple flat table, each row shows an item, and each column is an attribute describing the item. An item is an entity representing a city, a person, or a shop, for example. An attribute is a specification that can be measured, observed, or logged, such as age, price, and temperature. A combination of a row (=item) and a column (=attribute) represents a cell that contains a value of that pair. However, multidimensional tables have a more complex structure: the data values are ordered in hierarchies [36]. Appropriate sorting of tables is typically required to detect patterns and relationships in the data [6].

Time-dependent data ubiquitously occurs in many applications, for example, when measuring resource consumption over a production cycle, or in sales data. In the context of visualization, the temporal aspect is particularly challenging because of the unique characteristics of time. Time has hierarchical levels of granularity ($60\,s = 1\,h$, $24\,h = 1\,day$, etc.) with irregular divisions (a month consists of 28 to 31 days), cyclic patterns (e.g., seasons of a year), and cannot be perceived by humans directly [2]. Depending on the size, dimensionality, and resolution of time series data, again many scalable visualization techniques have been proposed [2]. Examples are time-charts, pixel-oriented, and glyph-based representations.

Spatial data comprises data that is localized in a reference frame of some kind. For instance, geographic maps can encode information on land usage, transportation facilities, or addresses of customers. 2D and 3D shaped information can be used for describing parts to be produced in an assembly line. Spatial data often arises in engineering or simulation tasks to describe the physical properties of a space or product, for example, air turbulence caused by a jet engine design. Note that spatial data can include any specific data type, as long as it is localized. Air turbulence, for instance, can be described as a time-dependent flow (or vector) field. Spatially moving objects give rise to trajectories, the visual analysis of which is an important application [39],

Graphs (or networks) represent data by nodes, links connecting the nodes, and descriptive data associated with both. They are an important model to represent structured information. Examples in industrial applications include flows of input factors, product hierarchies, and marketing networks. Visual analysis techniques for graphs include representations as node-link diagrams based on various layout techniques and more abstract representations, including adjacency matrices [31, 40]. Many analysis tasks in real-world graphs depend on analyzing and exploring changes in both the topology of the graph and data attributes associated with the nodes and edges [40]. Dynamic graph visualization is required to address this aspect [4].

Textual data may stem from textual communication, reporting, feedback captured from customers, technical specifications of products and patents, etc. Text visualization techniques [27] tackle the challenging problem of identifying key information in a large text collection. Different analysis perspectives are possible, focusing on text features, text structure, or names and entities, for example.

Image, video, and audio data also regularly occur in industrial processes, e.g., during quality control for optical, X-ray, ultrasonic material inspection, or visual surveillance of production processes. Appropriate techniques, e.g., from multimedia signal processing and computer vision, can be applied to transform the input data to one or several of the above data types, for example, to represent numeric measurements or events. For an overview of video visualization methods, see the survey by Borgo et al. [9].

In practice, data analysis often requires integrating data of different types, sources, and qualities. In many cases, data transformations are needed to apply data analysis and visualization, e.g., to describe customer transactions by selected numeric features by which they can be grouped and compared. In the following, we review important challenges in the context of visual data science.

2.4   Challenges in Creating Visual Data Analysis Applications

We next discuss a set of important challenges pertaining to data properties and complexity of user tasks in visual data analysis.

The Three (or four) V's of (big) Data As discussed in Sect. 2.3, application data often comes in different forms and in large volumes, often referred to as **big data**. It is a term that is not only used by domain experts, but that has also entered the mainstream vocabulary. In fact, big data has become the subject and driving factor of enormous importance in both academia and industry. The first association to big data is often related to the data volume, although it covers a much broader set of aspects. Even though many definitions have been proposed over the years, the most established one is the **3 Vs of big data**—**volume**, **velocity**, and **variety**—formulated by the Gartner Group [32].

The first V, **data volume**, is a moving target. As technology develops, we continue to increase the volume of data we collect. Scientists and data science practitioners have for many decades been faced with the challenge of making sense of more data than they can process using the methods and technologies available. This implies that a specific visualization will most likely not be able to display the data at the highest level of detail on the given output device. Consequently, data aggregation and filtering methods are needed to deal with this aspect.

The second V, **data velocity**, refers to the speed at which the data is generated and needs to be processed. For instance, analyzing live streaming data from production machines or social media is more difficult than processing hourly measurements acquired by weather stations stored in a static file.

The third V, **data variety**, addresses data heterogeneity in terms of type, source, and format. Detecting patterns by automatically clustering a homogeneous table with thousands of rows and columns might be more straightforward than finding correlations in a small but heterogeneous table in which the columns have different semantics and attribute types. The problem becomes even more difficult when one needs to make sense of multiple interconnected datasets—possibly even mixing various data types. An example is high-dimensional production data that needs to be investigated along the production process.

In addition, **data veracity** may be considered a fourth aspect—extending the definition to become the **4 Vs of Big Data**. It pertains to the quality of the data, e.g., precision and completeness of measurements, as well as its trustworthiness and origin. It is another decisive factor, as only analysis of accurate and relevant data can lead to relevant insights and decisions.

User Tasks and Personalization Besides technical challenges in data visualization and analysis, there is also the challenge of supporting the many different possible user tasks (or goals) in data analysis. Models of the visual data analysis process identify a variety of user goals in understanding data [36], e.g., identifying trends and outliers, classification, prediction, comparison and many more. In addition, the background knowledge and expertise of users may vary. Hence, visual data analysis systems must be flexible and support different tasks and types of users. To this end, it is interesting to develop **adaptive** systems. These custom visual analytics systems build upon the idea that the choice of visual representation depends to a large extent on the visual perception and interests of the user. Visual data analysis design considering user's preferences and interest is a research topic that receives increasing attention in recent years. Current research mainly focuses on using either explicit user feedback [5, 10, 38] provided in the form of ratings (representing user's visual preferences) or tags (representing user's topic of needs), or gaze movements [48, 50, 51], to help identify visualizations that best address the users' preferences, expertise, and tasks.

User Guidance Visual data science approaches often require a significant level of visual and data analytical skills from the user. However, users may not possess such skills right away, and hence have difficulties while analyzing the data. Also, domain experts within a specific area may have little or no knowledge about visual data analysis. Therefore, companies or institutes employ data analysts with visualization and analysis skills, but little domain knowledge in manufacturing processes, which as a result might impede the decision-making process. One possible solution to tackle this issue is to **guide** user throughout the data exploration process. Ceneda et al. [14, 15] define guidance as a "computer-assisted process that aims to actively resolve a knowledge gap during an interactive visual analytics session". The key factor in guided analytics is to exactly figuring out what the users' needs and preferences are and which steps to take to address them. In this context, guidance can be provided to recommend the appropriate visualizations and analytical steps [37].

2.5   Under the Hood: Visual Data Science Infrastructure

Technology Stack   As in any sector, technology is moving fast. So any discussion of specific tools and libraries will be outdated quickly. Therefore, in this section our aim is to briefly outline the technology stack to help readers better understand the spectrum that ranges from using off-the-shelf tools, through employing declarative libraries, to using programming languages for creating tailored visual analysis tools designed for a specific, well-defined purpose. Figure 3 illustrates the technology stack. Off-the-shelf tools at the top of the stack, such as: Tableau,[1] Microsoft Power BI,[2] and Spotfire,[3] have the advantage that they are easy to use and do not require programming skills by domain experts. However, they are limited in the sense that they support a fixed set of data types, visualization techniques, and analytical capabilities—although extension and plug-in mechanisms alleviate this limitation. Behrisch et al. [7] provide a comprehensive overview of commercial visual analysis tools, evaluating the performance, available features, and usability.

An alternative to using off-the-shelf tools is to employ static and interactive plotting libraries, as part of Notebook environments such as Jupyter Notebook[4] or R Markdown,[5] for instance. Example libraries are Vega[6] and Vega-lite,[7] Chart.js,[8] and plot.ly.[9] Finally, making use of high-level programming libraries for creating specialized visual analysis tools is the most expressive option, but also the one that requires the highest effort and skills. For instance, D3.js[10] is a popular JavaScript library that allows developers to flexibly create web-based visualizations.

General Purpose vs. Tailored Tools   General purpose visualization tools are effective for answering a broad set of analysis questions. In contrast, answering (domain-)specific questions often requires tailored tools that are designed for a small set of specialized users. This

---

[1] https://www.tableau.com/

[2] https://powerbi.microsoft.com/

[3] https://spotfire.tibco.com/

[4] https://jupyter.org/

[5] https://rmarkdown.rstudio.com/

[6] https://vega.github.io/vega/

[7] https://vega.github.io/vega-lite/

[8] https://www.chartjs.org/

[9] https://plotly.com/

[10] https://d3js.org/

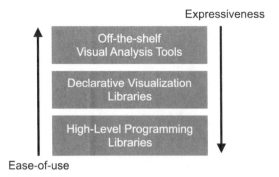

Fig. 3 The visual analysis technology stack with off-the-shelf tools at the top, declarative static and interactive visualization libraries in the middle, and high-level programming libraries at the bottom. While the ease of use and skills required increase from the bottom to the top, the expressiveness and tailoring possibilities decrease. The figure is adapted from Jeff Heer's keynote given at the OpenVis Conference in 2015

is frequently the case for ill-defined domain-specific problems that need to be investigated by means of interrelated, heterogeneous datasets, as often encountered in data-driven sciences. Visual analytics researchers can contribute to the solving of domain-specific research questions by designing and building tailored visual analysis solutions and tools. Figure 4 illustrates the relationship between the type of questions to be asked and the number of potential users. The more specific the questions are, the lower the number of users that can benefit from the tool becomes. Our advice is to rely on proved and tested general purpose tools. Nevertheless, if that is not possible and the problem to be solved is important enough, it is worth to invest time and money in creating highly specialized domain-specific solutions.

Dashboards: Multiple Coordinated Views An important requirement for the visual analysis of heterogeneous data is that the analyst is able to evaluate, compare, and interpret related data subsets shown in various visual representations and at different levels of granularity (i.e., complete datasets, groups of items, or single items). Multiple Coordinated Views (MCV) [43] is an established and powerful concept that addresses this requirement by linking multiple juxtaposed views. Nowadays, this concepts is colloquially referred to as dashboards.

The coordination of views refers to the principle that operations triggered in one view are immediately reflected in all other views. This coordination can concern data operations, such as filters and selections that result in a synchronized highlight of items (known as linking & brushing) or synchronized view operations, such as pan, rotate, and zoom. The views to be linked can show the same data subset using different visualization techniques, different data subsets encoded by the same visualization technique, or combinations thereof—also at various levels of granularity.

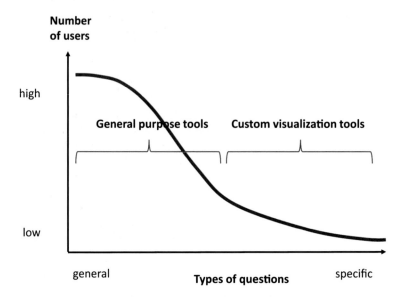

Fig. 4 Relationship between the type of analysis questions and the number of users (adapted from [45]). The more specific the question, the lower the number of potential users. General-purpose tools are designed for answering general questions, while customized visualization tools are able to address specific questions that are only relevant for a small set of highly specialized domain experts

Dashboards are an integral part of many state-of-the-art visual analysis tools. However, designing effective dashboards is not a trivial task and many different factors need to be considered for this purpose [20, 46]. In the following section, we discuss selected visual analysis solutions that are specifically tailored to the needs of real-world industrial use cases—going beyond standard off-the-shelf dashboards.

3   Selected Visual Data Science Approaches for Industrial Data Analysis

Visual data science methods are increasingly applied to solve problems in many domains and disciplines. Stakeholders from industry also show strong interest in these methods, and researchers have started to develop concepts and applications for visual analysis of industrial data. An overview of a number of techniques is provided by Zhou et al. [65]. The authors group the approaches by industrial sectors (automotive, energy, etc.), and phases of the process (production, service, etc.) and discuss a number of representative works.

In this section, we also give an exemplary overview of approaches. To this end, we chose a number of key industrial operation tasks, and for each one, we selected exemplary approaches from the literature and our own research. While we cannot claim completeness of the operations or approaches, we have striven to achieve a representative overview. Operations partly overlap and several of the approaches could be applied to more than one operation. For example, production planning operations also depend on and influence condition monitoring and quality control operations. Research in this area is active and the space of known solutions is steadily growing.

3.1   Production Planning

In production planning, the goal is to plan resource allocation to provide efficient production or service, subject to dependencies and constraints. In one application example from the metal industry, Wu et al. [58] propose to use abstract graphical elements to represent the smelting furnace and heating oven for metal ingots casting in order to support the engineers involved to achieve a better understanding of the synchronous relationship of scheduled capacity and the load between these two components. Jo et al. [29] focus on visualizing manufacturing schedules (i.e., plans to manufacture a product) used in semiconductor facilities. They use LiveGantt, a novel interactive schedule visualization tool that supports temporal filtering, product filtering, and resource filtering to help users explore large and highly concurrent production schedules from various perspectives. Although a case study demonstrated the efficacy of LiveGantt, it suffers from scalability issues when applied to large manufacturing schedules. To tackle this issue, the authors proposed more advanced visualization techniques, such as horizon graphs.

ViDX [62] is a visual analytics system that visualizes the processing time and status of work stations in automatic assembly lines, allowing production planners to explore the production data for identifying inefficiencies, locating anomalies, and defining hypotheses about their causes and effects. Their solution maps the production lines and dependencies to a dense flow chart enriched with time series and metadata on the production (see Fig. 5a). VIDX scales well to real-time for small data volumes, but not to year long data. Moreover, the visualizations are not responsive and do not easily adapt to different user devices.

A common problem in production planning is that there are typically multiple objectives (goals) given, e.g., cost, time, and quality, but trade-offs exist which prevent all goals from being optimized simultaneously. Hence, production planners need to decide on a single solution from the efficient solution space (in Pareto set). The PAVED system [16] was created following a design study in the motor construction industry. The approach supports the visual exploration of solutions, using multidimensional data visualization and appropriate interaction facilities (see Fig. 5b).

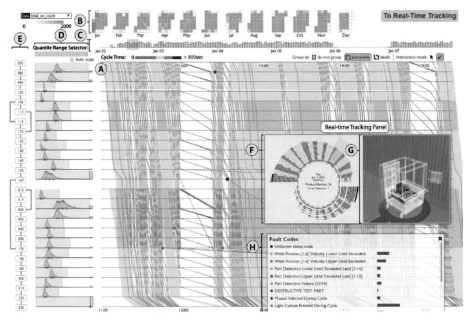

(a) Assembly sequence analysis

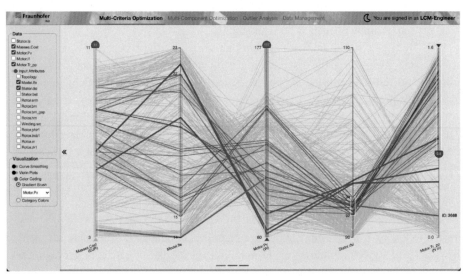

(b) Exploration of multi-criterion optimization solutions

Fig. 5 Example visual analytics approaches for selected industrial data analysis problems. (**a**) Visual analysis of assembly sequences. Figure reused from [62] with permission. (**b**) The PAVED system [16] uses interactive parallel coordinate plots for the exploration of engineering solutions in a multi-criterion optimization process. The approach resulted from a design study with domain experts. Figure courtesy of Lena Cibulski

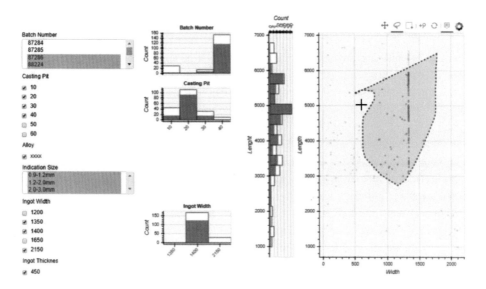

Fig. 6 Inspection of unwanted material inclusions for product quality monitoring (right) and identification of possible influencing factors (left) in the ADAM system [28]. Figure courtesy of Nikolina Jekic

3.2   Quality Control

Inspection of product and service quality is another important task that needs to be done regularly to ensure reliable output. As an example from the metal industry, the ADAM system [28] provides visual analysis of quality properties of aluminum plates produced by a casting and rolling process. The quality is measured by ultrasonic analysis of the metal plates, which record indications of inclusions in the metal by position and size. The ADAM tool represents the inclusion data in a scatter plot (see Fig. 6, right) from which densities and distributions of inclusions can be readily perceived. The interface allows to efficiently browse through large data sets and compare inclusion distributions for different production runs. ADAM also enables users to compare inclusion analysis results with production parameters, such as alloy recipe and cast control parameters. To this end, an array of data displays (see Fig. 6, left) are linked to the inclusion view, and interactive selection and highlighting allow users to search for outstanding patterns and possible correlations. However, ADAM has not been fully evaluated yet. Therefore, we cannot drive any conclusions about the efficacy and scalability of this tool.

3.3   Equipment Condition Monitoring

Equipment monitoring plays an important role in industrial applications and refers to observing a process, system, or machine with the goal to guarantee its expected functioning. Pure monitoring tasks allow operators or other personnel to inspect live streaming data coming from sensors or other data sources. Monitoring solutions frequently present the latest data in the context of historical data or provide predictions to assist users in judging or planning future operations (see section on predictive maintenance below). As an example, Wu et al. [59] present an interactive visual analytics system with a semi-supervised framework that supports equipment condition monitoring (see Fig. 7a). The idea is two-fold. Monitoring of operations is supported by visualizing the correlations of sensors, where changes can be noted in near-real-time and can inform domain experts in an exploratory way. Furthermore, this also involves a semi-supervised approach that learns about normal states of operations from user labeling of production sensor data. A classifier is trained, which can report deviations from the normal situation, again informing condition monitoring experts, or potentially triggering detection measures. However, the tool can face scalability issues when there are too many sensors. To tackle this issue, the authors propose to first group sensors into modules, then modules into super modules, and encode the statistical information of these super modules. Only one of the super modules will be visualized when selected by the user.

Anomaly detection is an important aspect of monitoring the condition of equipment or the whole system. Currently, the challenge of anomaly detection is to specify possible anomalies in advance, since its detection is situation-dependent, and previously unknown or unexpected anomalies may occur. Anomalies frequently need to be detected in real-time. Dutta et al. [17] use a comparative visualization technique to analyze the spatio-temporal evolution (variations of distributions, statistically anomalous regions in the data) of rotating stall in a jet engine simulation. They use a heat map to visualize the anomaly detection results and a 2D plot to show the evolution of anomalous regions of the jet engine stall, both allowing the engineers to verify the performance of the jet engine design prototype and improve its structure. Janetzko et al. [26] provide a visual analytics tool to report on anomalies found in multi-variate industrial energy consumption data and guide the user to the important time points. The algorithm used to detect the anomalies does not require computationally expensive calculations, which makes it possible to recognize sudden unexpected changes in power consumption. Likewise, Maier et al. [35] present a visual anomaly detection approach that guides the user in detecting anomalies found in time series of production plants. To reduce the users' workload in identifying the anomalies, the dimensionality of the dataset is reduced using principal component analysis. While this approach reduces the information space to the most important dimensions, it may also result in not all anomalies being represented in the visualization.

Sedelmair et al. [47] proposed a visual analytics tool for the automotive industry which combines visualization techniques with anomaly detection algorithm that allows engineers to explore anomalies in messages from the in-car communication network. Although effective, a field study has revealed scalability issues of the tool that hinder its use by engineers in their daily work. In our own previous work, we considered anomaly detection in multivariate times series of engine test bed cycles [52]. Anomalies are declared if the measurements in one cycle differ, more than a certain degree, from expected values based on previous cycles. Various anomaly detection methods are implemented. Users can interactively explore the detections on multiple levels of detail, including cycle glyphs and correlation matrices (see Fig. 7b, top), as well as line-chart-based details (see Fig. 7b, bottom). However, plotting up to thousand of cycles can be an issue for glyph and matrix representations. To tackle this scalability issue, advanced filtering and reordering techniques could be used. Also, in order to include the users' preferences and interests in this selection process, methods for preference elicitation could be implemented.

A common technique in condition testing are acoustic testing procedures. The IRVINE system [18] is based on acoustic analysis of test objects, e.g., engines during analysis in a testbed. The system visualizes obtained spectrograms, and allows to compare these across different tests to find deviations which may explain product errors. IRVINE features an annotation tool which allows users to record observations for future reference. In addition, the system provides cluster analysis to cope with increasingly large amounts of acoustic test data, supporting scalability. Figure 8 illustrates the interactive views provided.

Another relevant work conducted in the area on condition monitoring is proposed by Post et al. [42], who provide a series of interactive visualizations that shows the process data generated by a complex production system. In addition, the visualizations are used to highlight the bottlenecks or excess machining capacities, hence guiding the user to interesting locations and events. The guidance is particularly important, as it helps users to focus on single products or machines and readily identify the critical issues that affect them.

3.4 Predictive Maintenance

Predictive maintenance (PdM) refers to monitoring the performance and condition of equipment and machines in industrial environments during the normal production process and implementing methods to reduce malfunctions, failures, and errors. In order to detect such failures and performance issues, PdM uses condition monitoring tools that provide early warnings of fault or degradation (see Sect. 3.3). As an application example, the design limitations of complex products, such as products in aerospace, are usually found after the physical prototype is manufactured or is in use. It stands to reason that this causes delays and high costs in production. To tackle this issue, Peng et al. [41] and Abate et al. [1] propose visual analytics frameworks that are designed to help product designers and maintainability technicians to simulate and evaluate the product life cycle. As a result, the product designer can iteratively adjust the product schemes and hence reduce the development cycle time and costs. Peng et al. present a systematic approach of a visualization system that lacks an evaluation with end users. Abate et al. [1] present a virtual reality (VR) system that

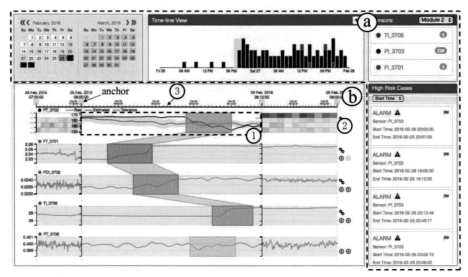

(a) Time series-based equipment condition monitoring

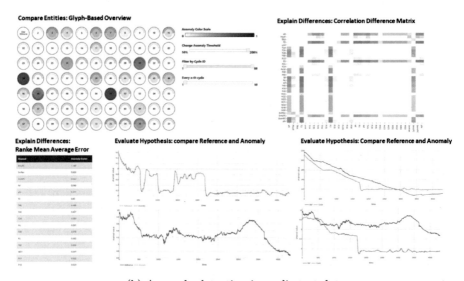

(b) Anomaly detection in cyclic test data

Fig. 7 (**a**) Equipment condition monitoring based on machine state classification. Figure reused from [59] with permission. (**b**) Anomaly detection in sensor data streams using glyphs, a correlation matrix, and line chart inspection for details [52]. Figure courtesy of Josef Suschnigg

supports interactions with the environment in which maintenance activities can be simulated. The usability and the usefulness of the system has been assessed with a user study, revealing increased user performance and satisfaction when performing a maintenance task when using the proposed system. Canizo et al. [11] provide methods to predict failures on wind turbines. The methods are executed in a cloud computing environment, to tackle the scalability issues and make predictions in real-time. Given that each wind farm has its own configuration and turbines, Canizo et al. further provide a visualization dashboard that visualizes the geographical location of the turbines together with the notification about the predictions and the status information of the turbine in real-time. Finally, Wörner et al. [60] propose a visual analytics tool that visualizes diagnostic machine data from the manufacturing domain, hence helping experts to judge whether or not specific parts or elements of the machines are behaving as expected or need to be repaired or replaced. The system has only been tested with a small data set and is therefore difficult to assess with regard to its scalability and usefulness.

3.5   Causality Analysis

Machine learning is being used increasingly for industrial scenarios to investigate the statistical associations between the production parameters for generating predictions or event forecasts. Although powerful, the method often fails to answer the critical question "What influences X?" [21]. Such questions can be answered with causality tools. Nevertheless, the casual structure of a production process is often too complex for users to follow and understand by only looking at the outcome of the algorithms. The visualization community aims to address this problem by providing tools that help users to easily perceive and understand the complex structure of causal relations. A commonly used tool to display the causal relations between parameters is a directed graph [25, 63], whereas the current research showed promising results for interactive path diagrams and parallel coordinates to be used to expose the flow of causal sequences [56] (see Fig. 8). Graph visualizations, path diagrams, and parallel coordinates are common visualization techniques for visualizing high-dimensional data. As a result, they achieve higher efficiency and scalability for visualizing causal relationships between parameters when compared to other visualization techniques.

4   Conclusions and Future Directions

We gave a compact overview of goals, fundamental concepts, and existing solutions in visual data science in the light of industrial applications. Such approaches integrate computational data analysis methods with interactive data visualization, aiming to support data understanding and pattern discovery. There are many application opportunities in industrial data analysis, pertaining to tasks such as production planning, equipment condition monitoring,

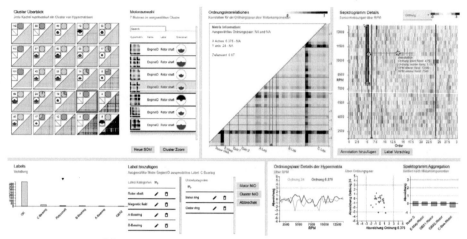

(a) Inspection and annotation of acoustic test data

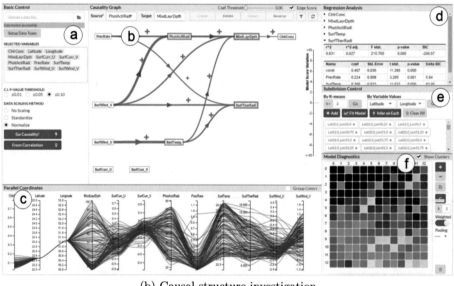

(b) Causal structure investigation

Fig. 8 (**a**) The IRVINE system [18] supports inspecting large amounts of acoustic test data, overviewing clusters of acoustic profiles of test objects. Test engineers can compare acoustic profiles, annotate observations for hints of errors, and compare measurement details. Figure courtesy of Joscha Eirich. (**b**) The causal structure investigator interface with interactive path diagrams (see mark b) for visualizing the causal relations, and parallel coordinates (see mark c) for observing data partitions and identifying causal models potentially hidden in the data. Figure reused from [56] with permission

quality control, and anomaly detection. We discussed the main data types and challenges in data acquisition. By means of application examples, we highlighted selected results from the state of the art, and closed with a selection of promising future directions.

There are many opportunities to leverage the potential of visual data science approaches for industrial data analysis problems. In order to advance in this field, many research challenges sill remain to be tackled. In the context of generic challenges in visual data science and application opportunities, we refer to a recent overview by Andrienko et al. [3].

In the following, we outline a number of challenges specific to industrial visual data analysis. First, we observe that the trend to Industry 4.0 is characterized by continuous digitalization of industrial operations. Notably, there is a continuous process to add new data sources for monitoring production and operation. Hence, when developing visual data science applications for industry, we cannot work with a stable set of data sources, but need to be able to flexibly integrate additional data types into running processes. This means that agile development environments are needed, and long-lasting research planning processes may not be applicable. We observe from our project work that, due to continuous integration of data sources and sensors, data heterogeneity is increasing in terms of data formats, sizes, resolutions, and qualities. While industrial standards already exist, data heterogeneity and complexity is expected to remain a permanent challenge due to the heterogeneity of equipment manufacturers.

Data ownership and governance problems may also occur in projects. For example, both equipment manufacturers and operators are interested in obtaining and analyzing equipment data. However, access to the data may be complicated due to competition and data privacy relationships.

In addition to data-driven approaches for understanding industrial processes, rich knowledge also exists in the form of industrial process management engines, containing domain-specific production and engineering knowledge as production programs and historic production records. In our opinion, how to link and integrate process engines and operational data analysis applications is a challenge that offers promising application potential. For more discussion on this, we refer to the work by Thalmann et al. [53].

Acknowledgements This work has partially been supported by the FFG, Contract No. 881844: "Pro^{2}Future is funded within the Austrian COMET Program Competence Centers for Excellent Technologies under the auspices of the Austrian Federal Ministry for Climate Action, Environment, Energy, Mobility, Innovation and Technology, the Austrian Federal Ministry for Digital and Economic Affairs and of the Provinces of Upper Austria and Styria. COMET is managed by the Austrian Research Promotion Agency FFG."

References

1. Abate, A., Guida, M., Leoncini, P., Nappi, M., Ricciardi, S.: Ahaptic-based approach to virtual training for aerospace industry. Journal of Visual Languages & Computing **20**, 318–325 (2009).

https://doi.org/10.1016/j.jvlc.2009.07.003
2. Aigner, W., Miksch, S., Schumann, H., Tominski, C.: Visualization of Time-Oriented Data. Human-Computer Interaction Series, Springer (2011). https://doi.org/10.1007/978-0-85729-079-3
3. Andrienko, G.L., Andrienko, N.V., Drucker, S.M., Fekete, J., Fisher, D., Idreos, S., Kraska, T., Li, G., Ma, K., Mackinlay, J.D., Oulasvirta, A., Schreck, T., Schumann, H., Stonebraker, M., Auber, D., Bikakis, N., Chrysanthis, P.K., Papastefanatos, G., Sharaf, M.A.: Big data visualization and analytics: Future research challenges and emerging applications. In: Proceedings of the Workshops of the EDBT/ICDT 2020 Joint Conference (2020), http://ceur-ws.org/Vol-2578/BigVis1.pdf
4. Beck, F., Burch, M., Diehl, S., Weiskopf, D.: A taxonomy and survey of dynamic graph visualization. Comput. Graph. Forum **36**(1), 133–159 (2017). https://doi.org/10.1111/cgf.12791
5. Behrisch, M., Korkmaz, F., Shao, L., Schreck, T.: Feedback-driven interactive exploration of large multidimensional data supported by visual classifier. In: 2014 IEEE Conference on Visual Analytics Science and Technology (VAST). pp. 43–52 (2014). https://doi.org/10.1109/VAST.2014.7042480
6. Behrisch, M., Bach, B., Riche, N.H., Schreck, T., Fekete, J.: Matrix reordering methods for table and network visualization. Computer Graphics Forum **35**(3), 693–716 (2016). https://doi.org/10.1111/cgf.12935
7. Behrisch, M., Streeb, D., Stoffel, F., Seebacher, D., Matejek, B., Weber, S.H., Mittelstädt, S., Pfister, H., Keim, D.A.: Commercial visual analytics systems-advances in the big data analytics field. IEEE Trans. Vis. Comput. Graph. **25**(10), 3011–3031 (2019). https://doi.org/10.1109/TVCG.2018.2859973
8. Bertin, J., Berg, W., Wainer, H., of Wisconsin Press, U.: Semiology of Graphics. University of Wisconsin Press (1983), https://books.google.at/books?id=luZQAAAAMAAJ
9. Borgo, R., Chen, M., Daubney, B., Grundy, E., Heidemann, G., Höferlin, B., Höferlin, M., Leitte, H., Weiskopf, D., Xie, X.: State of the art report on video-based graphics and video visualization. Comput. Graph. Forum **31**(8), 2450–2477 (2012). https://doi.org/10.1111/j.1467-8659.2012.03158.x
10. Bouali, F., Guettala, A., Venturini, G.: Vizassist: An interactive user assistant for visual data mining. Vis. Comput. **32**(11), 1447–1463 (Nov 2016). https://doi.org/10.1007/s00371-015-1132-9
11. Canizo, M., Onieva, E., Conde, A., Charramendieta, S., Trujillo, S.: Real-time predictive maintenance for wind turbines using big data frameworks. In: 2017 IEEE International Conference on Prognostics and Health Management (ICPHM). pp. 70–77 (2017)
12. Cao, L.: Data science: A comprehensive overview. ACM Comput. Surv. **50**(3) (2017). https://doi.org/10.1145/3076253
13. Card, S., Mackinlay, J., Shneiderman, B.: Readings in information visualization: using vision to think. Morgan Kaufmann Publishers Inc. (1999)
14. Ceneda, D., Gschwandtner, T., May, T., Miksch, S., Schulz, H., Streit, M., Tominski, C.: Characterizing guidance in visual analytics. IEEE Trans. Vis. Comput. Graph. **23**(1), 111–120 (2017). https://doi.org/10.1109/TVCG.2016.2598468
15. Ceneda, D., Gschwandtner, T., Miksch, S.: A review of guidance approaches in visual data analysis: A multifocal perspective. Comput. Graph. Forum **38**(3), 861–879 (2019). https://doi.org/10.1111/cgf.13730
16. Cibulski, L., Mitterhofer, H., May, T., Kohlhammer, J.: PAVED: Pareto Front Visualization for Engineering Design. Computer Graphics Forum (2020). https://doi.org/10.1111/cgf.13990
17. Dutta, S., Shen, H., Chen, J.: In situ prediction driven feature analysis in jet engine simulations. In: 2018 IEEE Pacific Visualization Symposium (PacificVis). pp. 66–75 (2018)

18. Eirich, J., Bonart, J., Jackle, D., Sedlmair, M., Schmid, U., Fischbach, K., Schreck, T., Bernard, J.: Irvine: A design study on analyzing correlation patterns of electrical engines. IEEE Transactions on Visualization & Computer Graphics (01), 1–1 (sep 2021). https://doi.org/10.1109/TVCG. 2021.3114797
19. Endert, A., Ribarsky, W., Turkay, C., Wong, B.L.W., Nabney, I.T., Blanco, I.D., Rossi, F.: The state of the art in integrating machine learning into visual analytics. CoRR **abs/1802.07954** (2018), http://arxiv.org/abs/1802.07954
20. Froese, M., Tory, M.: Lessons learned from designing visualization dashboards. IEEE Computer Graphics and Applications **36**(2), 83–89 (2016). https://doi.org/10.1109/MCG.2016.33
21. Guo, R., Cheng, L., Li, J., Hahn, P.R., Liu, H.: A survey of learning causality with data: Problems and methods. ACM Computing Surveys (CSUR) (2018)
22. Han, J., Kamber, M., Pei, J.: Data mining concepts and techniques, third edition (2012)
23. Heer, J., Bostock, M.: Crowdsourcing graphical perception: Using mechanical turk to assess visualization design. p. 203-212. CHI '10, Association for Computing Machinery, New York, NY, USA (2010). https://doi.org/10.1145/1753326.1753357
24. Hohman, F., Kahng, M., Pienta, R., Chau, D.H.: Visual analytics in deep learning: An interrogative survey for the next frontiers. CoRR **abs/1801.06889** (2018), http://arxiv.org/abs/1801. 06889
25. Holst, A., Pashami, S., Bae, J.: Incremental causal discovery and visualization. In: Proceedings of the Workshop on Interactive Data Mining. WIDM'19, Association for Computing Machinery, New York, NY, USA (2019). https://doi.org/10.1145/3304079.3310287
26. Janetzko, H., Stoffel, F., Mittelstädt, S., Keim, D.A.: Anomaly detection for visual analytics of power consumption data. Computers & Graphics **38**, 27–37 (2014). https://doi.org/10.1016/j.cag.2013.10.006, http://www.sciencedirect.com/science/article/pii/ S0097849313001477
27. Jänicke, S., Franzini, G., Cheema, M.F., Scheuermann, G.: Visual text analysis in digital humanities. Comput. Graph. Forum **36**(6), 226–250 (2017). https://doi.org/10.1111/cgf.12873
28. Jekic, N., Mutlu, B., Faschang, M., Neubert, S., Thalmann, S., Schreck, T.: Visual analysis of aluminum production data with tightly linked views. In: 21st Eurographics Conference on Visualization, EuroVis 2019 - Posters, Porto, Portugal, June 3–7, 2019. pp. 49–51 (2019). https:// doi.org/10.2312/eurp.20191143
29. Jo, J., Huh, J., Park, J., Kim, B., Seo, J.: Livegantt: Interactively visualizing a large manufacturing schedule. IEEE Transactions on Visualization and Computer Graphics **20**(12), 2329–2338 (2014)
30. Keim, D., Kohlhammer, J., Ellis, G., Mansmann, F. (eds.): Mastering the information age : solving problems with visual analytics. Goslar: Eurographics Association (2010), https://diglib.eg.org/ handle/10.2312/14803
31. von Landesberger, T., Kuijper, A., Schreck, T., Kohlhammer, J., van Wijk, J.J., Fekete, J., Fellner, D.W.: Visual analysis of large graphs: State-of-the-art and future research challenges. Computer Graphics Forum **30**(6), 1719–1749 (2011). https://doi.org/10.1111/j.1467-8659.2011.01898.x
32. Laney, D.: 3-D Data Management: Controlling Data Volume. Velocity and Variety, META Group Original Research Note (2001)
33. Lu, Y., Garcia, R., Hansen, B., Gleicher, M., Maciejewski, R.: The state-of-the-art in predictive visual analytics. Comput. Graph. Forum **36**(3), 539–562 (2017). https://doi.org/10.1111/cgf. 13210
34. Mackinlay, J.: Automating the design of graphical presentations of relational information. ACM Transactions on Graphics **5**(2), 110–141 (Apr 1986)
35. Maier, A., Tack, T., Niggemann, O.: Visual anomaly detection in production plants. In: Ferrier, J., Bernard, A., Gusikhin, O.Y., Madani, K. (eds.) ICINCO 2012 – Proceedings of the 9th Interna-

tional Conference on Informatics in Control, Automation and Robotics, Volume 1, Rome, Italy, 28–31 July, 2012. pp. 67–75. SciTePress (2012)

36. Munzner, T.: Visualization Analysis and Design. CRC Press (2014)

37. Mutlu, B., Gashi, M., Sabol, V.: Towards a task-based guidance in exploratory visual analytics. In: 54th Hawaii International Conference on System Sciences, HICSS 2021, Kauai, Hawaii, USA, January 5, 2021. pp. 1–9. ScholarSpace (2021), http://hdl.handle.net/10125/70789

38. Mutlu, B., Veas, E., Trattner, C.: Vizrec: Recommending personalized visualizations. ACM Transactions on Interactive Intelligent Systems 6(4), 31:1–31:39 (2016)

39. Nara, A.: Visual analytics of movement, by gennady andrienko, natalia andrienko, peter bak, daniel keim and stefan wrobel, berlin heidelberg, springer-verlag, 2013, xviii + 387 pp., us$129 (hardcover), ISBN 978-3-642-37582-8. Ann. GIS 21(1), 91–92 (2015). https://doi.org/10.1080/19475683.2015.992828

40. Nobre, C., Meyer, M.D., Streit, M., Lex, A.: The state of the art in visualizing multivariate networks. Comput. Graph. Forum 38(3), 807–832 (2019). https://doi.org/10.1111/cgf.13728

41. Peng, G., Hou, X., Gao, J., Cheng, D.: A visualization system for integrating maintainability design and evaluation at product design stage. The International Journal of Advanced Manufacturing Technology 61 (2011). https://doi.org/10.1007/s00170-011-3702-y

42. Post, T., Ilsen, R., Hamann, B., Hagen, H., Aurich, J.C.: User-Guided Visual Analysis of Cyber-Physical Production Systems. Journal of Computing and Information Science in Engineering 17(2) (2017)

43. Roberts, J.C.: State of the art: Coordinated multiple views in exploratory visualization. In: Fifth International Conference on Coordinated and Multiple Views in Exploratory Visualization (CMV 2007). pp. 61–71 (2007)

44. Sacha, D., Stoffel, A., Stoffel, F., Kwon, B.C., Ellis, G., Keim, D.A.: Knowledge generation model for visual analytics. IEEE Transactions on Visualization and Computer Graphics 20, 1604–1613 (2014)

45. Sakai, R.: Biological data visualization: Analysis and design (2016)

46. Sarikaya, A., Correll, M., Bartram, L., Tory, M., Fisher, D.: What do we talk about when we talk about dashboards? IEEE Transactions on Visualization and Computer Graphics 25(1), 682–692 (2019)

47. Sedlmair, M., Isenberg, P., Baur, D., Mauerer, M., Pigorsch, C., Butz, A.: Cardiogram: Visual analytics for automotive engineers. pp. 1727–1736 (2011). https://doi.org/10.1145/1978942.1979194

48. Shao, L., Silva, N., Eggeling, E., Schreck, T.: Visual exploration of large scatter plot matrices by pattern recommendation based on eye tracking. In: Proceedings of the 2017 ACM Workshop on Exploratory Search and Interactive Data Analytics. pp. 9–16. ESIDA '17, ACM, New York, NY, USA (2017). https://doi.org/10.1145/3038462.3038463

49. Shneiderman, B.: The eyes have it: a task by data type taxonomy for information visualizations. In: Proc. IEEE Symposium on Visual Languages. pp. 336–343. IEEE (1996)

50. Silva, N., Schreck, T., Veas, E., Sabol, V., Eggeling, E., Fellner, D.W.: Leveraging eye-gaze and time-series features to predict user interests and build a recommendation model for visual analysis. In: Proceedings of the 2018 ACM Symposium on Eye Tracking Research & Applications. pp. 13:1–13:9. ETRA '18, ACM, New York, NY, USA (2018). https://doi.org/10.1145/3204493.3204546

51. Steichen, B., Carenini, G., Conati, C.: User-adaptive information visualization: Using eye gaze data to infer visualization tasks and user cognitive abilities. In: Proceedings of the 2013 International Conference on Intelligent User Interfaces. p. 317–328. IUI '13, Association for Computing Machinery, New York, NY, USA (2013). https://doi.org/10.1145/2449396.2449439

52. Suschnigg, J., Mutlu, B., Koutroulis, G., Sabol, V., Thalmann, S., Schreck, T.: Visual exploration of anomalies in cyclic time series data with matrix and glyph representations. Big Data Research **26**, 100251 (2021). https://doi.org/10.1016/j.bdr.2021.100251

53. Thalmann, S., Mangler, J., Schreck, T., Huemer, C., Streit, M., Pauker, F., Weichhart, G., Schulte, S., Kittl, C., Pollak, C., Vukovic, M., Kappel, G., Gashi, M., Rinderle-Ma, S., Suschnigg, J., Jekic, N., Lindstaedt, S.: Data analytics for industrial process improvement – a vision paper. In: IEEE 20th Conference on Business Informatics (CBI). vol. 02, pp. 92–96 (2018)

54. Thomas, J.J., Cook, K.A.: Illuminating the Path: The Research and Development Agenda for Visual Analytics. National Visualization and Analytics Ctr (2005)

55. Tominski, C., Schuman, H.: Interactive Visual Data Analysis. AK Peters/CRC Press (2020), forthcoming

56. Wang, J., Mueller, K.: Visual causality analysis made practical. In: 2017 IEEE Conference on Visual Analytics Science and Technology (VAST). pp. 151–161 (2017)

57. Ward, M., Grinstein, G., Keim, D.: Interactive Data Visualization: Foundations, Techniques, and Applications. A. K. Peters, Ltd., USA (2010)

58. Wu, P.Y.F.: Visualizing capacity and load in production planning. In: Proceedings Fifth International Conference on Information Visualisation. pp. 357–360 (2001)

59. Wu, W., Zheng, Y., Chen, K., Wang, X., Cao, N.: A visual analytics approach for equipment condition monitoring in smart factories of process industry. In: 2018 IEEE Pacific Visualization Symposium (PacificVis). pp. 140–149 (2018)

60. Wörner, M., Metzger, M., T.Ertl: Dataflow-based Visual Analysis for Fault Diagnosis and Predictive Maintenance in Manufacturing. In: Pohl, M., Schumann, H. (eds.) EuroVis Workshop on Visual Analytics. The Eurographics Association (2013). https://doi.org/10.2312/PE.EuroVAST. EuroVA13.055-059

61. Xu, F., Uszkoreit, H., Du, Y., Fan, W., Zhao, D., Zhu, J.: Explainable AI: A Brief Survey on History, Research Areas, Approaches and Challenges, pp. 563–574 (2019)

62. Xu, P., Mei, H., Ren, L., Chen, W.: Vidx: Visual diagnostics of assembly line performance in smart factories. IEEE Transactions on Visualization and Computer Graphics **23**(1), 291–300 (2017)

63. Yen, C., Parameswaran, A., Fu, W.: An exploratory user study of visual causality analysis. Computer Graphics Forum **38**, 173–184 (06 2019). https://doi.org/10.1111/cgf.13680

64. Yi, J.S., ah Kang, Y., Stasko, J.: Toward a deeper understanding of the role of interaction in information visualization. IEEE Trans. Vis. Comput. Graph. **13**(6), 1224–1231 (2007)

65. Zhou, F., Lin, X., Liu, C., Zhao, Y., Xu, P., Ren, L., Xue, T., Ren, L.: A survey of visualization for smart manufacturing. Journal of Visualization **22**, 419–435 (2019)

Digital Transformation towards Industry 5.0

Self-Adaptive Digital Assistance Systems for Work 4.0

Enes Yigitbas⦿, Stefan Sauer⦿ and Gregor Engels⦿

Abstract

In the era of digital transformation, new technological foundations and possibilities for collaboration, production as well as organization open up many opportunities to work differently in the future. The digitization of workflows results in new forms of working which is denoted by the term Work 4.0. In the context of Work 4.0, digital assistance systems play an important role as they give users additional situation-specific information about a workflow or a product via displays, mobile devices such as tablets and smartphones, or data glasses.

Furthermore, such digital assistance systems can be used to provide instructions and technical support in the working process as well as for training purposes. However, existing digital assistance systems are mostly created focusing on the "design for all" paradigm neglecting the situation-specific tasks, skills, preferences, or environments of an individual human worker. To overcome this issue, we present a monitoring and adaptation framework for supporting self-adaptive digital assistance systems for Work 4.0. Our framework supports context monitoring as well as UI adaptation for augmented (AR) and virtual reality (VR)-based digital assistance systems. The benefit of our framework is shown based on exemplary case studies from different domains, e.g. context-aware maintenance application in AR or warehouse management training in VR.

E. Yigitbas · S. Sauer · G. Engels (✉)
Institute of Computer Science, Paderborn University, Paderborn, Germany
e-mail: engels@upb.de

E. Yigitbas
e-mail: enes@mail.upb.de

S. Sauer
e-mail: sauer@uni-paderborn.de

B. Vogel-Heuser and M. Wimmer (eds.), *Digital Transformation*,
https://doi.org/10.1007/978-3-662-65004-2_19

Keywords

Digital Assistance Systems • Work 4.0 • Industry 4.0 • Self-Adaptive • Situation-Aware

1 Introduction

Nowadays we are witnessing a rising trend of digital transformation which is shaping our everyday life, value creation processes, and the way we are working. Especially in the context of production processes, the increasing amount of digitization and interconnection of production systems is sometimes referred to as *Industry 4.0*. As a result of this industrial (r)evolution, the role of human work changes significantly, which is often denoted with the term *Work 4.0* [2]. This means that in the context of Work 4.0 due to the digitization of workflows each individual worker will face a variety of challenges and problems to solve, mostly related to high cognitive activities.

To overcome this problem, digital assistance systems play a crucial role to support human workers to execute their tasks in an efficient, effective, and pleasant manner. For this purpose, digital assistance systems assist human workers by providing additional situation-specific information about a workflow or a product via displays, mobile devices such as tablets and smartphones, or data glasses. Such digital assistance systems can be used to provide instructions and technical support in the working process as well as for training purposes.

In the last decades, various digital assistance systems have been proposed in different application domains such as manufacturing [14], assembly [9], or maintenance [16]. However, existing digital assistance systems are created focusing on the "design for all" paradigm neglecting the situation-specific tasks, skills, preferences, or environments of an individual human worker. In most cases, the existing digital assistance systems are system-centred in a way that they primarily focus on the industrial task they are supporting. Certainly, in this connection, the context-of-use which is crucial for the interaction of the user with the production system is not considered.

To overcome this issue, we present a monitoring and adaptation framework for supporting self-adaptive digital assistance systems (SADAS) for Work 4.0. According to Laddaga et al., a "Self-adaptive Software System evaluates its own behavior and changes behavior when the evaluation indicates that it is not accomplishing what the software is intended to do, or when better functionality or performance is possible" [18]. We make use of this definition and transfer the idea of self-adaptive software systems to self-adaptive digital assistance systems (SADAS). For this purpose, our framework supports context monitoring and UI adaptation for AR/VR-based SADAS. The benefit of our framework is shown based on example case studies from different domains, e.g. context-aware maintenance application in augmented reality or warehouse management training in virtual reality.

The remainder of this book chapter is structured as follows: In Sect. 2, we present background information on Industry&Work 4.0 as well as digital assistance systems. In Sect. 3, we discuss the main challenges in developing self-adaptive digital assistance systems. Based on these challenges, in Sect. 4, we describe and discuss related approaches. Section 5 is dedicated to presenting our monitoring and adaptation framework for SADAS. In Sect. 6, we present case studies to show the applicability of our framework. Finally, Sect. 7 concludes our work with an outlook on future work.

2   Background

In this section, we introduce basic concepts of *Industry 4.0* and *Work 4.0* as well as the main idea behind *Digital Assistance Systems*.

2.1   *Industry* and *Work 4.0*

Since the beginning of industrialization, technological advancements have led to paradigm shifts which today are named "industrial revolutions": in the field of mechanization (the so-called 1st industrial revolution), of the intensive use of electrical energy (the so-called 2nd industrial revolution), and of the widespread digitization (the so-called 3rd industrial revolution) [19]. On the basis of an advanced digitization within factories, the combination of internet technologies and future-oriented technologies in the field of "smart" objects (machines and products) the term "Industry 4.0" was established for a planned "4th industrial revolution" [19].

According to [25], the term *Industry 4.0* stands for the fourth industrial revolution which is defined as a new level of organization and control over the entire value chain of the life cycle of products. For realizing the future of productivity and growth in manufacturing industries, *Industry 4.0* includes several enabling technologies such as cyber physical systems (CPS), internet of things (IoT), cloud computing, or novel forms of human-computer interaction. Over the last few years, Industry 4.0 has emerged as a promising technology framework used for integrating and extending manufacturing processes at both intra- and inter-organizational levels. This emergence of Industry 4.0 has been fuelled by the recent development in ICT. The developments and the technological advances in Industry 4.0 provide a viable array of solutions to the growing needs of digitization in manufacturing industries [27].

On the other hand, the process of digitization and incorporation of new technologies and intelligent systems in various sectors and domains, are the core enablers for the changes which are about to come with the new way of work [1]. Nowadays, the business processes of every corporation and organization are supported by powerful IT systems which become more enhanced by the introduction of sophisticated robotic and sensor technologies, Cyber-Physical Systems, 3D printing technologies and intelligent software systems. As a consequence of the process of rapid digitization and the technology fluctuations, the requirements and demands for the working individuals in the workplace are changing.

Therefore, the term *Work 4.0* was introduced in November 2015 by the German Federal Ministry of Labour and Social Affairs (BMAS) when it launched a report entitled Re-Imagining Work: Green Paper Work 4.0 [26]. This initiative envisions new ways of work where the focus will be on the human workers, taking into account their individual abilities, characteristics, and preferences while aiming at allowing greater flexibility and ensuring work-life balance. Considering the current predictions, it becomes necessary to focus on the human as an important part in the sector of industrial production. Therefore, the need to develop digital assistance systems which are able to adapt to the personal abilities, needs, and individual characteristics of the working individuals is emerging.

2.2   Digital Assistance Systems

The term *Digital Assistance System* was introduced in [10] as the primary interface to optimally integrate humans into a production environment during task execution. Based on this definition, a DAS can be seen as a technical system for dynamically delivering digitally prepared information. Informational assistance systems record data via sensors and inputs, then process this data to provide employees the right information ("what") at the right time ("when") in the desired format ("how") [22]. The main goals of a DAS are to avoid uncertainty and mental stress for users, warn them of dangers, as well as increase of productivity, e.g., reduction of training time, search times, or operating errors [10]. DAS can be divided into stationary assistance systems, mobile assistance systems, handheld devices (such as tablet PCs), and wearables [22]. While stationary assistance systems are permanently installed at a work station (such as a mounted projection device), mobile assistance systems, in contrast, are moved to the assembly object via a mobile solution. Wearables can be classified by the body part on which they are worn (such as "smart glasses," "smart gloves," "smart watches").

In the past, DAS have often been used to create standardized instruction manuals to be used by all employees working on the assembly system (design for all) - independent of their individual features. To tackle this limitation, providing personalized and situation-specific assistance is a promising alternative to empower the workers while supporting them in performing complex physical and cognitive tasks. However, in order to provide such assistance, the DAS needs to be enriched with capabilities concerning continuous monitoring and self-adaptation which are known from the area of *Autonomic Computing*. The term Autonomic Computing (also known as AC) refers to the self-managing characteristics of distributed computing resources, adapting to unpredictable changes while hiding intrinsic complexity to operators and users [15]. The AC system concept is designed to monitor and adapt a *Managed Element*, using high-level policies. It will constantly check and optimize its status and automatically adapt itself to changing conditions by using the Monitor, Analyze, Plan, Execute-Knowledge loop. Based on the ideas of context-awareness and self-adaptation, we aim to bring classical DAS to a new level of Self-adaptive Digital Assistance Systems (SADAS).

3   Challenges

In the course of two different research projects, we have analyzed the challenges in the application and adoption of digital assistance systems in the industry setting. The first project was related to the manual assembly of an Electrical Cabinet (E-cabinet), while the second one was dealing with the process of manual assembly of a concrete product in a Smart Factory [12]. For identifying the requirements and needs of the human workers in the industrial sector with regard to the usage of DAS, we have conducted semi-structured interviews with experts from different research fields: Psychology, Sociology, Didactic, Economics, Computer Science, Electrical and Mechanical Engineering. Based on this investigation, we have identified the following main challenges:

Challenge 1: Information Presentation
For many years traditional graphical user interfaces (GUIs) have been successfully adopted for mobile platforms. e.g. through the integration of multi-touch interaction and responsive layout algorithms that adapt the visual display to different device sizes. However, in applications that rely on spatial information related to a real-world environment GUIs are not ideal, because the information displayed in the interface is removed from its real-world context and interaction is effected indirectly through the interface [24]. Especially digital assistance systems in the context of manufacturing and assembly using current sensor data and the user's current location are examples that rely heavily on such spatial information which can be reached through interaction technologies like *Augmented Reality* or *Virtual Reality*. Associated with the aspect of information presentation is the question of the computing platform how a digital assistance system works and can be accessed by the end-users. There are several target devices for DAS on the market which are developed by different companies and organizations. Target devices could be smartphones, tablets, or HMDs for Augmented Reality (e.g. Microsoft Hololens, RealWear Glasses, or Google Glass Enterprise Edition) or Virtual Reality (e.g. HTC Vive, Oculus Quest, or Valve Index). The cost of a device, its comfort in using it, and the ability of a device to help the user accomplish her task are some of the reasons that influence the type of equipment that different users and organizations use to acquire them. Each computing platform can have different properties regarding hardware and sensor, operating system, used SDKs, etc. Given the heterogeneous span of various devices for DAS, it is essential to have multi-platform support so that a digital assistance system can be deployed and used across varying computing platforms.

Challenge 2: Monitoring

The acceptance and successful application of a digital assistance system highly depends on the quality of the information that is shown to the end-users in guiding them through their task. With this regard, it is important to provide situation-aware information for the end-users so that they can accomplish their tasks in an efficient, effective, and satisfying manner. For this purpose, a digital assistance system (DAS) should enable context monitoring features to the end-users to inform them about dynamically changing characteristics of the working environment. With this regard, an important challenge is to continuously observe the context-of-use of a DAS through various sensors. The context-of-use can be described through different characteristics regarding the user (physical, emotional, preferences, etc.), platform (Hololens, Handheld, etc.), and environment (real vs. virtual environmental information). Due to the rich context dimension which is spanning over the real-world and virtual objects, it is a complex task to track and relate the relevant context information to each other. The mixture of real (position, posture, emotion, etc.) and virtual (coordinates, view angle, walk-through, etc.) context information additionally increases the aspect of context management compared to classical context-aware applications like in the web or mobile context.

Challenge 3: Adaptation

Based on the collected context information, a decision-making process is required to analyze and decide whether conditions and constraints are fulfilled to trigger specific adaptation operations on the DAS. In general, an important challenge is to cope with conflicting adaptation rules which aim at different adaptation goals. This problem is even more emphasized in the case of AR-based digital assistance systems as we need to ensure a consistent display between the real-world entities and virtual overlay information. For the decision-making step, it is also important to decide about a reasoning technique like rule-based or learning-based to provide a performant and scalable solution.

As AR/VR-based digital assistance systems consist of a complex structure and composition, an extremely high number of various adaptations is possible. The adaptations should cover text, symbols, 2D images, and videos, as well as 3D models and animations. In this regard, many adaptation combinations and modality changes increase the complexity of the adaptation process.

4   Related Work

In previous work, different approaches were introduced to address the above-mentioned challenges *Information Presentation, Monitoring,* and *Adaptation* within the scope of digital assistance systems.

In [3], the authors present a framework for assistance systems to support work processes in smart factories. They argue that, due to the large spectrum of assistance systems, it is hard to acquire an overview and to select an adequate digital assistance system based on meaningful criteria. Therefore, they suggest a set of comparison criteria in order to ease the selection of an adequate assistance system. Compared to our framework, this work is rather supporting the process of selecting a suitable digital assistance system while the above-mentioned challenges are not explicitly covered.

Similar work is presented in [5] where the authors present solution ideas for the techno-logical assistance of workers. Besides technological means for supporting human-machine interaction in the Industry 4.0 era, the authors describe how the novel role of human work-ers in the context of Industry 4.0 should be addressed. As concrete examples for digital assistance systems, they focus on web-based and mobile apps incorporating AR features for Work 4.0 scenarios. They also focus on the aspects of context monitoring and adaptive UIs for the AR-based assistance app. However, the main focus is on hand-held AR devices, while the usage of head-mounted displays in AR or VR is not covered.

A more formal approach in guiding different stakeholders to choose an adequate digital assistance system for their organization, domain, or application field is presented in [16]. This work proposes a process-based model to facilitate the selection of suitable DAS for supporting maintenance operations in manufacturing industries. Using this approach, a dig-ital assistance system is selected and linked to maintenance activities. Furthermore, they collect user feedback by employing the selected DAS to improve the quality of recommen-dations and to identify the strength and weaknesses of each DAS in association with the maintenance tasks. While this approach supports the selection of an adequate DAS in the context of Industry 4.0 it is not focusing on the aspects of monitoring and adaptation.

Besides the above-described approaches, some other works in the field of digital assis-tance systems apply the human-centered design process in order to design and develop assistance systems that fulfill the needs of the user requirements and the context-of-use. An example work in this direction is presented in [21] where the authors develop a digital assistance system for production planning and control. Similar to our work, this approach is focusing on the aspects of context monitoring and UI adaptations within DAS, however, they do not apply AR/VR interfaces to cover spatial information and interaction with a DAS.

Another type of work related to digital assistance systems is presented in [4] where the authors propose a lightweight canvas method to foster interdisciplinary discussions on DAS. While this approach is primarily focusing on interdisciplinary discussions in the early stages of requirements understanding and design of DAS, it is not addressing a development process for situation-aware digital assistance systems in AR or VR.

The most related approach to our monitoring and adaptation framework for SADAS is presented in [12]. In this work, the authors introduce a digital-twin based multi-modal UI adaptation framework for assistance systems in Industry 4.0. This approach characterizes a predecessor solution of our presented work here. While this work covers aspects of context-awareness and adaptation for DAS, the scope of targeted applications and devices remains restricted so that AR and VR technologies for example are not covered.

While the above-described approaches highly focus on the selection and development of digital assistance systems, they are not fully covering the novel aspects of context-awareness and adaptation especially in the combination of AR and VR applied for DAS. Therefore, in the following, according to the challenges introduced in Sect. 3, we analyze further approaches which focus more on the topic of context monitoring and adaptation within the scope of AR and VR applications.

Augmented Reality (AR) and Virtual Reality (VR) have been a topic of intense research in the last decades. In the past few years, massive advances in affordable consumer hardware and accessible software frameworks are now bringing AR and VR to the masses. AR enables the augmentation of real-world physical objects with virtual elements and has been already applied for different aspects such as robot programming [34], product configuration (e.g., [6, 7]), prototyping [13], planning and measurements [40] or for realizing smart interfaces (e.g., [17, 35]). In contrast to AR, VR interfaces support the interaction in an immersive computer-generated 3D world and have been used in different application domains such as training [36], prototyping [38], robotics [37], education [41], healthcare [31], or even for collaborative software modeling [30].

While context-awareness has been exploited in various types of applications including web [28], mobile (e.g., [33] or [32]), and cross-channel applications (e.g., [39] or [29]) to improve the usability of an interactive system by adapting its user interface, only a few existing works are focusing on the topic of context-awareness in AR and VR.

In [8], the concept of Pervasive Augmented Reality (PAR) is introduced. A taxonomy for PAR and context-aware AR that classifies context sources and targets is presented. The context sources are classified as human, environmental, and system factors. As apparent in the title, Grubert's work treats Augmented Reality, here with special regards to pervasive Augmented Reality.

Context-aware Mobile Augmented Reality (CAMAR) [23] is an approach on context-awareness in mobile AR focusing on user context, which is measured using the user's mobile device. It enables the user to customize the presentation of virtual content and to share this information with other users selectively, depending on the context. Furthermore, a framework called UCAM (Unified Context-aware Application Model) [11] can be used to create CAMAR-enabled applications. UCAM is a framework which besides the acquisition, process, and awareness of contextual information provides also a unified way of representation with respect to user, content, and environment.

The framework presented in [20] focuses on the context-aware adaptation of interfaces in mixed reality, with the main adaptation points being what content is displayed, where it is shown and how much information of it is displayed. It is designed to adjust the content display depending on the user's tasks and their cognitive load and archives this using a combination of rule-based decisions and combinatorial optimization. The framework uses parameters about the applications that are to be displayed as input additionally to the context-specific parameters to achieve a fitting layout optimization. The framework is mentioning mixed reality as its base, but regarding that, it shows contents in the real world and does not create a whole new virtual world, it can safely be said that AR is supported.

Apart from the above-mentioned approaches which address the development of context-aware applications in general without directly focusing on digital assistance systems in the context of Industry 4.0, there are also specific approaches that use AR in the smart factory context. One example of such an approach is presented in [24]. Here, AR is used for supporting workers in an Industry 4.0 environment where they have to accomplish assembly tasks. The work presents the initial experience with the AR-based assistance systems.

5 Monitoring and Adaptation Framework

In order to address the described challenges, we present a monitoring and adaptation framework for supporting self-adaptive digital assistance systems (SADAS). Our framework which is based on the *MAPE-K* architecture [15], is depicted in Fig. 1.

It is basically divided up into two main components, the *Autonomic Manager* and the *Managed Element*. The *Autonomic Manager* is responsible for continuously monitoring the *Managed Element* through *Sensors* and to automatically react to changing conditions by adapting the *Managed Element* through *Effectors*. For this purpose, the *Autonomic Manager* consists of a control loop that is called MAPE-K, while this acronym represents the starting letters of the main sub-components: *Monitor, Analyze, Plan, Execute,* and *Knowledge.*

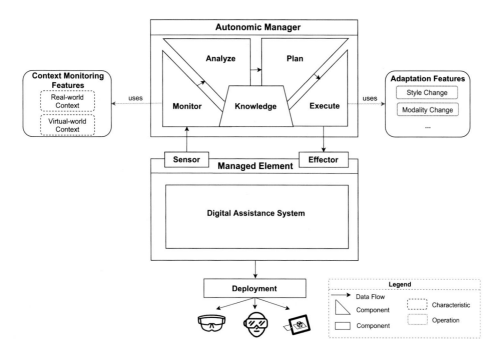

Fig. 1 Architectural overview of the monitoring and adaptation framework

The *Managed Element* represents in our case the *Digital Assistance System (DAS)* that is deployed on an execution platform that can be accessed through different devices such as VR HMDs, AR Smart Glasses, or Tablets. Furthermore, the DAS is characterized through context information that can be observed through *Context Monitoring Features*. This context information can consist either of *Real-world Context* information which is gathered by sensing existing sensors in the real physical world (in the case of AR) or *Virtual-world Context* information when context information such as gestures, pose, or virtual environment information are continuously monitored in the VR world. Besides context information that can be observed through the sensors of the DAS, there are *Adaptation Features* to characterize the adaptation operations which are executed with the means of the *Effectors* of the DAS. The *Adaptation Features* can contain various adaptation operations to adjust the DAS interface through run-time adaptations, e.g., changing modality or layout.

A refined architectural overview of our monitoring and adaptation framework for virtual and augmented reality (MAVAR) based SADAS is depicted in Fig. 2. It consists of three main components: *Context Monitoring*, *Decision Making*, and *Adaptation*.

The *Context Monitoring* component is responsible for constantly collecting information about different kinds of context to enable the framework to react to them appropriately. All the information on the context is measured by sensors; partially real sensors like the camera or the inertial sensors like gyroscopes, partially in a more figurative sense like measuring information about the app such as positioning of virtual objects or usage. The

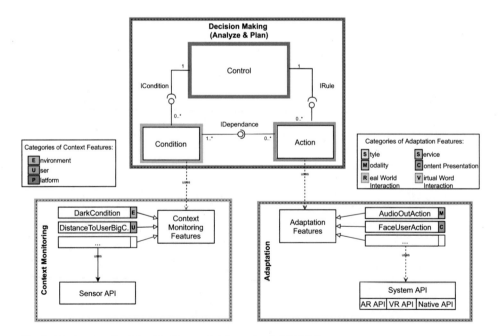

Fig. 2 *Monitoring, Decision Making*, and *Adaptation* in MAVAR

sensor data is read out through *Sensor APIs* and used by several sub-components which are each responsible for checking on one specific context feature (e.g. *DarkCondition* and *DistanceToUserBigCondition* in Fig. 2). The entirety of the *Context Monitoring Features* is used by the *Condition* and each of the context features has its own component responsible for monitoring it. The context observed by the *Condition* sub-classes is categorized into *Environment, User,* and *Platform Context.* The *Environment Context* includes everything that impacts the system from the outside, such as noise or objects in the real world (in AR) or virtual environmental information (in VR). *User Context* denotes any information available about the user, such as age or experience, but also the social context the user is currently in. Any information about the platform on which the system is running, like the availability of sensors or the compatibility with different software kits, is summarized in the *Platform Context.*

The *Decision Making* is led by the *Control* component. This component supervises the active conditions and rules from the *Context Monitoring* and *Adaptation* components. The connections between the *Control, Rule,* and *Condition* component make it possible for the components to work together closely and to incorporate and connect the *Context Monitoring* and *Adaptation* component.

The *Adaptation* component is responsible for the execution of actions in response to a captured context change. The *Adaptation Features* component consists of several sub-components (like the *AudioOutRule* or the *FaceUserRule* in Fig. 2) specified to each execute a respective adaptation. To do so, the sub-components make use of different parts of the *System API* to access the needed functions. Some of these APIs are for example for *AR API,* which is necessary to influence the AR part of the application, or the *Native API,* which is used to influence native system functions as the language. The adaptation features, which are executed by the *Rule* sub-classes, are divided into the *Style, Modality, Service, Content Presentation, Real-World,* and *Virtual-World* changes. The *Style* adaptation operation changes the look or behavior of single elements, the *Modality* operation adjusts the sensory input and output the user utilizes to interact with the application, and *Content Presentation* treats the way contents are presented to the user on the screen. Furthermore, the *Service* change operation describes changes made on the device level regarding the type of device or features it offers, the *Real-World* changes treat actions that are tied to objects from the real, physical surroundings of the device and the *Virtual-World* changes describe adjustments of virtual objects either in an AR or VR scene.

To achieve the required functionality of monitoring and adaptation of DAS, it is necessary to connect adaptations to the context changes that should trigger them. In MAVAR, this is done by adding one or more conditions to a rule, which will react to changes in the context features monitored by the conditions with the adaptation which is implemented for it. To be more specific, a rule will become active and get executed as soon as all of its conditions are fulfilled at the same time. For cleanup purposes, there is also an *unexecute* method, which will be called when one or more of a rule's conditions are not fulfilled anymore (making the rule inactive) and which is supposed to be used to reverse any effects of the rules execution that should only be active as long as its conditions are true.

The constant monitoring of the context through the conditions and the prompt execution of the adaptations through the rules is ensured by the control component. The control component acts as an observer for all of the condition and rule components. Conditions report a change to the control when they detect a change in the context feature which they monitor, therefore they report either on it newly being true or newly being false. Rules report a change to the control when they are either executed (thereby executing an adaptation) or unexecuted (reversing an adaptation). As the rules depend on their registered conditions for changing, they report to the control if either all their conditions are true and were not all true before (rule gets executed), or if all conditions were true before and at least one newly turned false (rule gets unexecuted).

To make sure the context changes are detected, the control component regularly updates itself. If the update method is called, each of the registered rules is evaluated and in turn evaluates its respective conditions to check whether a context change has happened since the last evaluation and returns the new state of the context to the rule. If a change occurred, the condition also notifies the observing control component, causing a new update process. The rules receive the result of each of their respective conditions and react accordingly (for example by being calling their execute method). If a rule is executed or unexecuted, it notifies the observing control component, so a new update will be executed in case any context features were impacted by the rule's actions.

6 Case Studies

In this section, two case studies are presented which show the benefit of our monitoring and adaptation framework for digital assistance systems. The first case study deals with an AR-based SADAS for maintenance scenarios. The second case study shows an example of a VR-based SADAS which supports warehouse management training in a virtual environment.

6.1 Example 1: AR-Based Context-Aware Assistance for Maintenance Tasks

As an example application of our monitoring and adaptation framework for DAS, a multi-platform and context-aware AR app for printer maintenance was developed. The app guides its user through the process of exchanging the ink cartridges of a printer step by step, with each step being described in a text window and illustrated by 3D arrows and other elements arranged on the printer.

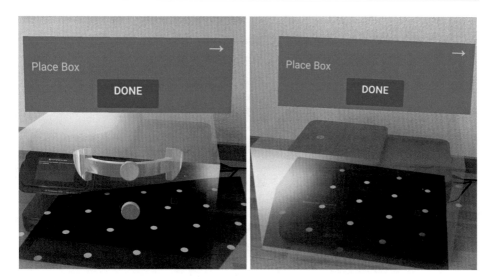

Fig. 3 Experience-level adaptation: Instructions on how to turn elements are only shown to new users

Some of the monitoring and adaptation features are illustrated with screenshots in the following figures, with the left image illustrating the state before the adaptation and the right image showing it after adapting.

In Fig. 3, the effect of an adaptation responding to the user's experience is shown. It displays example control elements to illustrate how to do different transformations on 3D objects and thereby shows the user how they can, for example, rotate objects. As the action responsible for displaying the illustrations is connected to a condition monitoring the number of app uses, it is only executed on the first five uses of the app, so the user can use the app undisturbed once they got used to the controls.

While working on the printer, the user has to access the printer from several angles. This can cause them to look at the message with the instructions from a very steep angle, which makes it hard or impossible to read. Of course, the user could move away from the printer, read the message, and then go back to the printer again, but that would be rather inconvenient and disruptive for the workflow. For this reason, the application uses an action that rotates the specified object, in this case, the message window, towards the user at all times (see Fig. 4) to make sure it is always readable.

There is also a condition that turns true whenever the camera receives very little light input, which is usually the case when the device was laid down, for example, if the user needs their hands free. As this prevents the user from seeing any objects of the AR application, a voice interface is activated (see Fig. 5). It makes sure the description of the current task is read out to the user and it enables them to interact with the application using voice commands.

Fig. 4 View-angle adaptation: The window is dynamically rotating to face the user regardless of their position

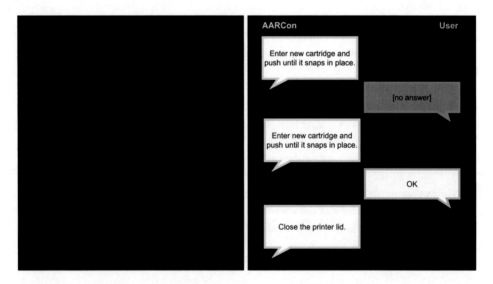

Fig. 5 Modality adaptation: The application switches to a conversational UI if the AR camera is not available

This allows the user to interact with the application even if they are currently not able to hold the device to use it in AR mode.

Figure 6 shows a change in the instruction window's level of detail. This is achieved using the action for this in combination with a condition that reacts to the user's distance to the printer, so the detail is lowered if the user is more than for example 1.2 m away from the printer. Adjusting the level of detail can help the user to focus on the currently important task. It can also be used to make the user come closer to important objects or to create a simpler and tidier AR environment by removing information that is currently unnecessary.

These features, amongst others, aim to enhance the user's experience using the printer maintenance application. As they are all executed automatically based on context information, the user does not have to do anything to get an application that is at all times customized to the current situation.

Furthermore, in Fig. 7, screenshots from the same digital assistance system application are shown for a different target platform. Instead of an Android-based AR app like shown before, we now have the same app running on the HoloLens with the same context-awareness and UI adaptation features supported for the new target platform.

In summary, the implementation of the above described AR-based digital assistance system shows how our MAVAR framework supports the monitoring and adaptation process of a DAS on different target platforms.

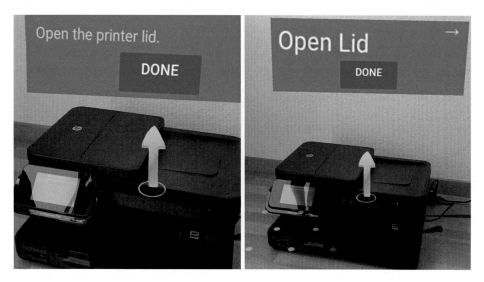

Fig. 6 Distance-based adaptation: When the user moves away from the printer, the level of detail is decreased while the text size is increased

(a) Modality Adaptation: Control of Assistance System via Speech.

(b) Distance-based Adaptation: When the user is near to the printer, status of the ink cartridges is visualized.

Fig. 7 Context-aware AR Printer Maintenance App on HoloLens

6.2   Example 2: VR-Based Context-Aware Assistance for Warehouse Management Training

As a further application scenario for our monitoring and adaptation framework for DAS, we present an example from the logistics domain. A typical task in this domain is warehouse management where employees have to pursue pick and order operations. As shown in Fig. 8(a), a digital assistance system is used for supporting the employees in their tasks. In most cases, the picking process is paper-based in a classical sense, or sometimes there is a digital assistance system in the form of an application running on a tablet. In both cases, still, logistics executives often detect stock discrepancies and misplaced wares. This is due to the different levels of expertise of employees with the warehouse management system and order picking process.

(a) Digital Assistance System for Warehouse Management running on a Tablet.

(b) VR-based Assistance System for Warehouse Management Training in a virtual Environment.

Fig. 8 Warehouse Management

In addition, the differences in performance between order pickers are great. Some pick quickly and precisely, others with errors or even damage wares. Furthermore, in a real physical setting, training of the employees is expensive and time-consuming. It is often also the case that companies do not have a spare warehouse where they can train new employees and relocate moved wares to the original position after the training. Furthermore, it is desirable to offer repeatable and comparable training, especially for new or short-term staff.

To overcome these issues, we have developed a VR-based assistance system to support the training of the picking process. As shown in Fig. 8(b), the same warehouse management application running on the tablet is provided in the virtual training environment. In addition to that, the core application logic of the warehouse management app can be further augmented through virtual elements that can be displayed in the virtual environment. This is, for example, used to realize a step by the guidance of the user of the VR training application. In Fig. 9(a), one can see how additional textual descriptions help the user to accomplish the task. Similar to the AR-based assistance system, we have also here monitoring and adaptation features to guide the learning process in the warehouse management training app in the most suitable way. In Fig. 9(b), for example, it is shown how the location of the wares which the user has to pick are highlighted by a green box. Based on such situation-aware information, the users can be guided through step-by-step context-aware information so that the effect of learning can increase.

Furthermore, our VR-based assistance system is designed in such a way that different workflows in the area of warehouse management (Single-order and Multi-order picking with different kinds of exceptions) can be supported. For this purpose, the different training workflows are specified based on a process model (in our case BPMN) which can be edited according to the needs of the VR training application.

To sum up, the illustration of the above described VR-based digital assistance system for warehouse management training shows that our monitoring and adaptation framework

(a) Step by step guidance of the user in the VR Training Application.   (b) Context-aware guidance: Location of the wares to pick are highlighted by a green box.

Fig. 9 VR Warehouse Management Training Application

is applicable for different kinds of digital assistance systems and flexible to cover various workflows in this area.

6.3   Discussion

While the above-described case studies illustrate the benefit of our monitoring and adaptation framework, there is still room for improvement of such self-adaptive digital assistance systems so that they can find their way to industrial practice. With this regard, it has to be mentioned that the implementation of our framework currently is in a prototypical state where several improvements regarding visualization of the AR/VR interfaces can be done to increase the usability and user experience (UX) of the end-users. In this context, a usability study should be conducted to assess the usability and UX of the resulting DAS based on our framework. Besides that, the efficiency and effectiveness of our monitoring and adaptation framework should be analyzed to check its applicability and benefit in further domains beyond maintenance and training.

7   Conclusion and Outlook

As a consequence of ongoing digital transformation, new technologies are emerging which change the way we are working and communicating with humans and machines. In this context, digital assistance systems play a crucial role as they provide means for supporting human-to-human and human-to-machine interactions. Furthermore, such digital assistance systems can be used to provide instructions and technical support in the working process as well as for training purposes.

In this book chapter, we argue that existing digital assistance systems are mostly created focusing on the "design for all" paradigm neglecting the situation-specific tasks, skills, preferences, or environments of an individual human worker. To overcome this issue, we first discuss the main challenges in developing self-adaptive AR/VR-based digital assistance systems. After that, we present a monitoring and adaptation framework for supporting self-adaptive AR/VR-based digital assistance systems for Work 4.0. Our framework supports context monitoring as well as UI adaptation for AR/VR-based digital assistance systems. The benefit of our framework is shown based on exemplary case studies from different domains, e.g. context-aware maintenance application in augmented reality or warehouse management training in virtual reality.

Although the presented framework makes a further step to support and improve working processes of humans in times of Industry 4.0, further research according to assistance systems has to be done to reach a better degree of acceptance. For reaching this, further improvements in the areas of hardware and display technology are required so that dynamic 3d objects and information can be visualized on smart glasses or similar wearables. Beyond that, holographic 3d displays are emerging which could lead to the future of holographic working environments. Furthermore, intelligent techniques are necessary to improve object detection at run time that is a key enabler in assisting humans in working processes. In its current state, the implemented adaptation process in our framework follows a rule-based approach. Further optimization of UI adaptations can be reached through extending the adaptation manager by machine learning algorithms. This way, log data (context information, previous adaptations, and user feedback) can be analyzed to learn the most suitable adaptations for future context-of-use situations. In general, a broader acceptance of AR/VR technologies needs to be reached so that these technologies can be used to augment human abilities and thus improve their cognitive and physical tasks.

References

1. Bonekamp, L., Sure, M.: Consequences of industry 4.0 on human labour and work organisation. Journal of Business and Media Psychology **6**(1), 33–40 (2015)
2. De Vos, M.: Work 4.0 and the future of labour law. Available at SSRN 3217834 (2018)
3. Fellmann, M., Robert, S., Büttner, S., Mucha, H., Röcker, C.: Towards a framework for assistance systems to support work processes in smart factories. In: Holzinger, A., Kieseberg, P., Tjoa, A.M., Weippl, E.R. (eds.) Machine Learning and Knowledge Extraction – First IFIP TC 5, WG 8.4, 8.9, 12.9 International Cross-Domain Conference, CD-MAKE 2017, Reggio di Calabria, Italy, August 29 – September 1, 2017, Proceedings. Lecture Notes in Computer Science, vol. 10410, pp. 59–68. Springer (2017). https://doi.org/10.1007/978-3-319-66808-6_5
4. Fischer, H., Senft, B., Rittmeier, F., Sauer, S.: A canvas method to foster interdisciplinary discussions on digital assistance systems. In: Marcus, A., Wang, W. (eds.) Design, User Experience, and Usability: Theory and Practice – 7th International Conference, DUXU 2018, Held as Part of HCI International 2018, Las Vegas, NV, USA, July 15–20, 2018, Proceedings, Part I. Lecture Notes in Computer Science, vol. 10918, pp. 711–724. Springer (2018). https://doi.org/10.1007/978-3-319-91797-9_49
5. Gorecky, D., Schmitt, M., Loskyll, M., Zühlke, D.: Human-machine-interaction in the industry 4.0 era. In: 12th IEEE International Conference on Industrial Informatics, INDIN 2014, Porto Alegre, RS, Brazil, July 27–30, 2014. pp. 289–294. IEEE (2014). https://doi.org/10.1109/INDIN.2014.6945523
6. Gottschalk, S., Yigitbas, E., Schmidt, E., Engels, G.: Model-based product configuration in augmented reality applications. In: Bernhaupt, R., Ardito, C., Sauer, S. (eds.) Human-Centered Software Engineering – 8th IFIP WG 13.2 International Working Conference, HCSE 2020, Eindhoven, The Netherlands, November 30 – December 2, 2020, Proceedings. Lecture Notes in Computer Science, vol. 12481, pp. 84–104. Springer (2020). https://doi.org/10.1007/978-3-030-64266-2_5
7. Gottschalk, S., Yigitbas, E., Schmidt, E., Engels, G.: Proconar: A tool support for model-based AR product configuration. In: Bernhaupt, R., Ardito, C., Sauer, S. (eds.) Human-Centered Software Engineering – 8th IFIP WG 13.2 International Working Conference, HCSE 2020, Eindhoven, The Netherlands, November 30 – December 2, 2020, Proceedings. Lecture Notes in Computer Science, vol. 12481, pp. 207–215. Springer (2020). https://doi.org/10.1007/978-3-030-64266-2_14
8. Grubert, J., et al.: Towards pervasive augmented reality: Context-awareness in augmented reality. IEEE Trans. Vis. Comput. Graph. **23**(6), 1706–1724 (2017)

9. Hinrichsen, S., Bendzioch, S.: How digital assistance systems improve work productivity in assembly. In: International Conference on Applied Human Factors and Ergonomics. pp. 332–342. Springer (2018)

10. Hold, P., Erol, S., Reisinger, G., Sihn, W.: Planning and evaluation of digital assistance systems. Procedia Manufacturing **9**, 143–150 (2017)

11. Hong, D., Shin, C., Oh, S., Woo, W.: A new paradigm for user interaction in ubiquitous computing environment. ISUVR 2006 pp. 41–44 (2006)

12. Josifovska, K., Yigitbas, E., Engels, G.: A digital twin-based multi-modal UI adaptation framework for assistance systems in industry 4.0. In: Kurosu, M. (ed.) Human-Computer Interaction. Design Practice in Contemporary Societies – Thematic Area, HCI 2019, Held as Part of the 21st HCI International Conference, HCII 2019, Orlando, FL, USA, July 26–31, 2019, Proceedings, Part III. Lecture Notes in Computer Science, vol. 11568, pp. 398–409. Springer (2019). https://doi.org/10.1007/978-3-030-22636-7_30

13. Jovanovikj, I., Yigitbas, E., Sauer, S., Engels, G.: Augmented and virtual reality object repository for rapid prototyping. In: Bernhaupt, R., Ardito, C., Sauer, S. (eds.) Human-Centered Software Engineering – 8th IFIP WG 13.2 International Working Conference, HCSE 2020, Eindhoven, The Netherlands, November 30 – December 2, 2020, Proceedings. Lecture Notes in Computer Science, vol. 12481, pp. 216–224. Springer (2020). https://doi.org/10.1007/978-3-030-64266-2_15

14. Keller, T., Bayer, C., Bausch, P., Metternich, J.: Benefit evaluation of digital assistance systems for assembly workstations. Procedia CIRP **81**, 441–446 (2019)

15. Kephart, J.O., Chess, D.M.: The vision of autonomic computing. Computer **36**(1), 41–50 (2003). https://doi.org/10.1109/MC.2003.1160055

16. Kovacs, K., Ansari, F., Geisert, C., Uhlmann, E., Glawar, R., Sihn, W.: A process model for enhancing digital assistance in knowledge-based maintenance. In: Beyerer, J., Kühnert, C., Niggemann, O. (eds.) Machine Learning for Cyber Physical Systems, Selected papers from the International Conference ML4CPS 2018, Karlsruhe, Germany, October 23–24, 2018. pp. 87–96. Springer (2018). https://doi.org/10.1007/978-3-662-58485-9_10

17. Krings, S., Yigitbas, E., Jovanovikj, I., Sauer, S., Engels, G.: Development framework for context-aware augmented reality applications. In: Bowen, J., Vanderdonckt, J., Winckler, M. (eds.) EICS '20: ACM SIGCHI Symposium on Engineering Interactive Computing Systems, Sophia Antipolis, France, June 23–26, 2020. pp. 9:1–9:6. ACM (2020). https://doi.org/10.1145/3393672.3398640

18. Laddaga, R., Robertson, P.: Self adaptive software: A position paper. In: SELF-STAR: International Workshop on Self-* Properties in Complex Information Systems. vol. 31, p. 19. Citeseer (2004)

19. Lasi, H., Fettke, P., Kemper, H.G., Feld, T., Hoffmann, M.: Industry 4.0. Business & information systems engineering **6**(4), 239–242 (2014)

20. Lindlbauer, D., Feit, A.M., Hilliges, O.: Context-aware online adaptation of mixed reality interfaces. In: Proceedings of the 32nd Annual ACM Symposium on User Interface Software and Technology, UIST 2019, New Orleans, LA, USA, October 20–23, 2019. pp. 147–160 (2019)

21. Nelles, J., Kuz, S., Mertens, A., Schlick, C.M.: Human-centered design of assistance systems for production planning and control: The role of the human in industry 4.0. In: IEEE International Conference on Industrial Technology, ICIT 2016, Taipei, Taiwan, March 14–17, 2016. pp. 2099–2104. IEEE (2016). https://doi.org/10.1109/ICIT.2016.7475093

22. Nikolenko, A., Sehr, P., Hinrichsen, S., Bendzioch, S.: Digital assembly assistance systems–a case study. In: International Conference on Applied Human Factors and Ergonomics. pp. 24–33. Springer (2019)

23. Oh, S., Woo, W., et al.: Camar: Context-aware mobile augmented reality in smart space. Proc. of IWUVR **9**, 48–51 (2009)

24. Paelke, V.: Augmented reality in the smart factory: Supporting workers in an industry 4.0. environment. In: Grau, A., Martínez, H. (eds.) Proceedings of the 2014 IEEE Emerging Technology and Factory Automation, ETFA 2014, Barcelona, Spain, September 16–19, 2014. pp. 1–4. IEEE (2014). https://doi.org/10.1109/ETFA.2014.7005252

25. Rüßmann, M., Lorenz, M., Gerbert, P., Waldner, M., Justus, J., Engel, P., Harnisch, M.: Industry 4.0: The future of productivity and growth in manufacturing industries. Boston Consulting Group **9**(1), 54–89 (2015)

26. Salimi, M.: Work 4.0: An enormous potential for economic growth in germany. ADAPT Bulletin **16** (2015)

27. Xu, L.D., Xu, E.L., Li, L.: Industry 4.0: state of the art and future trends. International Journal of Production Research **56**(8), 2941–2962 (2018)

28. Yigitbas, E., et al.: Self-adaptive UIs: Integrated model-driven development of UIs and their adaptations. In: Proc. of the ECMFA 2017. pp. 126–141 (2017)

29. Yigitbas, E., Anjorin, A., Jovanovikj, I., Kern, T., Sauer, S., Engels, G.: Usability evaluation of model-driven cross-device web user interfaces. In: Bogdan, C., Kuusinen, K., Lárusdóttir, M.K., Palanque, P.A., Winckler, M. (eds.) Human-Centered Software Engineering – 7th IFIP WG 13.2 International Working Conference, HCSE 2018, Sophia Antipolis, France, September 3–5, 2018, Revised Selected Papers. Lecture Notes in Computer Science, vol. 11262, pp. 231–247. Springer (2018). https://doi.org/10.1007/978-3-030-05909-5_14

30. Yigitbas, E., Gorissen, S., Weidmann, N., Engels, G.: Collaborative software modeling in virtual reality. CoRR **abs/2107.12772** (2021), https://arxiv.org/abs/2107.12772

31. Yigitbas, E., Heindörfer, J., Engels, G.: A context-aware virtual reality first aid training application. In: Alt, F., Bulling, A., Döring, T. (eds.) Proc. of Mensch und Computer 2019. pp. 885–888. GI/ACM (2019)

32. Yigitbas, E., Hottung, A., Rojas, S.M., Anjorin, A., Sauer, S., Engels, G.: Context- and data-driven satisfaction analysis of user interface adaptations based on instant user feedback. Proc. ACM Hum. Comput. Interact. **3**(EICS), 19:1–19:20 (2019). https://doi.org/10.1145/3331161

33. Yigitbas, E., Josifovska, K., Jovanovikj, I., Kalinci, F., Anjorin, A., Engels, G.: Component-based development of adaptive user interfaces. In: Proceedings of the ACM SIGCHI Symposium on Engineering Interactive Computing Systems, EICS 2019, Valencia, Spain, June 18-21, 2019. pp. 13:1–13:7 (2019)

34. Yigitbas, E., Jovanovikj, I., Engels, G.: Simplifying robot programming using augmented reality and end-user development. In: Ardito, C., Lanzilotti, R., Malizia, A., Petrie, H., Piccinno, A., Desolda, G., Inkpen, K. (eds.) Human-Computer Interaction – INTERACT 2021 – 18th IFIP TC 13 International Conference, Bari, Italy, August 30 – September 3, 2021, Proceedings, Part I. Lecture Notes in Computer Science, vol. 12932, pp. 631–651. Springer (2021). https://doi.org/10.1007/978-3-030-85623-6_36

35. Yigitbas, E., Jovanovikj, I., Sauer, S., Engels, G.: On the development of context-aware augmented reality applications. In: Abdelnour-Nocera, J.L., Parmaxi, A., Winckler, M., Loizides, F., Ardito, C., Bhutkar, G., Dannenmann, P. (eds.) Beyond Interactions – INTERACT 2019 IFIP TC 13 Workshops, Paphos, Cyprus, September 2–6, 2019, Revised Selected Papers. Lecture Notes in Computer Science, vol. 11930, pp. 107–120. Springer (2019). https://doi.org/10.1007/978-3-030-46540-7_11

36. Yigitbas, E., Jovanovikj, I., Scholand, J., Engels, G.: VR training for warehouse management. In: Teather, R.J., Joslin, C., Stuerzlinger, W., Figueroa, P., Hu, Y., Batmaz, A.U., Lee, W., Ortega, F.R. (eds.) VRST '20: 26th ACM Symposium on Virtual Reality Software and Technology. pp. 78:1–78:3. ACM (2020)

37. Yigitbas, E., Karakaya, K., Jovanovikj, I., Engels, G.: Enhancing human-in-the-loop adaptive systems through digital twins and VR interfaces. In: 16th International Symposium on Software Engineering for Adaptive and Self-Managing Systems, SEAMS@ICSE 2021, Madrid, Spain, May 18–24, 2021. pp. 30–40. IEEE (2021). https://doi.org/10.1109/SEAMS51251.2021.00015

38. Yigitbas, E., Klauke, J., Gottschalk, S., Engels, G.: VREUD - an end-user development tool to simplify the creation of interactive VR scenes. CoRR **abs/2107.00377** (2021)

39. Yigitbas, E., Sauer, S.: Engineering context-adaptive uis for task-continuous cross-channel applications. In: Human-Centered and Error-Resilient Systems Development – IFIP WG 13.2/13.5 Joint Working Conference. pp. 281–300 (2016)

40. Yigitbas, E., Sauer, S., Engels, G.: Using augmented reality for enhancing planning and measurements in the scaffolding business. In: EICS '21: ACM SIGCHI Symposium on Engineering Interactive Computing Systems, virtual, June 8–11, 2021. ACM (2021), https://doi.org/10.1145/3459926.3464747

41. Yigitbas, E., Tejedor, C.B., Engels, G.: Experiencing and programming the ENIAC in VR. In: Alt, F., Schneegass, S., Hornecker, E. (eds.) Mensch und Computer 2020. pp. 505–506. ACM (2020)

Digital Transformation—Towards Flexible Human-Centric Enterprises

Burkhard Kehrbusch and Gregor Engels⬤

Abstract

Our society is progressing from an industrial society to a knowledge society and thereby establishing constant changes with unprecedented extent and speed. This is due to the urge of mankind to improve quality of life by gaining knowledge and insights, and to the steadily increased power of information technology. For enterprises, the changing environment constantly opens new chances and existential risks, which force them to adapt to their changing contexts on time. So, to survive and succeed, enterprises must organize digital transformation as a process to steadily shape their future, and they must consider their context in a wider scope than usual. Also, entrepreneurs are facing increasing challenges. With these insights, we propose a novel human-centric view on enterprises, their digital transformation, and their position in the society. It combines technical and business levers with enterprise culture. We introduce a reference model-based approach for a continuous, holistic enterprise evolution and focus on the orchestrated solution provider (OSP) as the future enterprise model. It supports the entrepreneur and self-responsible teams to master digital transformation and to sustain the success of their enterprise in the knowledge society. In this sense, the OSP follows the vision of Industry 5.0 for a sustainable, human-centric and resilient European industry, while going far beyond with its holistic view.

B. Kehrbusch
Kehrbusch Management Consulting, Krefeld, Germany
e-mail: bk@kehrbusch-online.com

G. Engels (✉)
Computer Science Institute, Paderborn University, Paderborn, Germany
e-mail: engels@upb.de

B. Vogel-Heuser and M. Wimmer (eds.), *Digital Transformation*,
https://doi.org/10.1007/978-3-662-65004-2_20

497

Keywords

Digital Transformation · Human-centric · Knowledge Society · Enterprise Design · Continuous Evolution · Reference Model · Industry 5.0

1   Introduction

Recent studies show that almost every entrepreneur of any industry agrees on the need of digital transformation for her/his enterprise, of continuous change, and of a scope, which goes beyond technology and includes the human factor, since effective digital transformations require shifts in mindset and behavior [5]. Practiced approaches differ and it seems that no one can fully understand the rich set of methods and practices, which emerge or are ready to support these transformations. It has become best practice and provides some guidance for *digital transformations*, to differentiate the use of digital technology for optimization of existing business models (*digitization*) and for enabling new business models (*digitalization*). Nevertheless, the human factor is often reduced to expectations of customers or motivation of employees.

Instead of adding the human factor to the dominating business and technology views, we place humans in the center of our consideration, and take a broad and holistic view on enterprises and their environment.

This holistic view reveals a fundamental and sustainable change of our society, which implies new business rules. It is the progress from an industrial to a knowledge society which establishes a "new normal" of constant changes with unprecedented extent and speed [11]. This is due to two complementary drivers: On the one hand, mankind longing to improve quality of life by gaining knowledge and insights which constantly changes human needs and behaviors. On the other hand, information technology, whose increasing power turns everything into software-defined items, changeable on demand, available everywhere for everyone, at any time.

Knowledge has always been critical. But how it is treated has changed: first slowly over a long period and restricted to privileged people, now boosted by information technology, providing any kind of data, highly available for everyone. But mankind has learnt reality differs from individual perception and there is no absolute truth. Everyone must find own insights for a meaningful life. Hence, knowledge and insights are steadily questioned and individually reflected. Mankind has also learnt the limited ability to grasp reality, which leads to the acceptance of not-knowing and the insight to handle it respectfully. This results in reassessment of risk we impose on our environment and society. Thus, the new type of society is also characterized as risk society [6], society of multiple options [20] or reflexive society [7].

Consequently, the half-life of acquired knowledge decreases and uncertainty, complexity, and ambiguity increase. This means greater freedom for every individual in shaping her/his own life. Communities of people must establish values and behaviors that provide social cohesion and thereby respect that enterprise cultures are as specific as the personalities and ideas of people building the enterprise teams. Constant search for fulfilment of meaning also converges professional and private world and fosters new solutions for work-life balance [29]. The nature of work is also changing. Digital technologies are not only increasing the usability and availability of information, but also the speed of acquisition and change. This turns information into a substantial production factor. Knowledge workers replace industrial workers and ask for novel working models, which resonate with their personal work-life-ideas [13].

These changes hold true on global scale. So, do changing habits from generation to generation. Globally classified as baby boomers, X, Y, Z [26], each generation sets its own priorities, and often takes as granted, what is just gradually accepted from the previous generation (e.g. mobile computing). Enterprises must not only realize the impacts on customer behavior (e.g. sharing instead of owning) or on workplace requirements (e.g. usage of social media), but also on environment conditions.

Summing up, the change to a knowledge society requires enterprises to transform, to leverage new digital business models, and to master disruptive changes. Digital transformations must meet specific requirements:

- Being *flexible* and *adaptive* to seize new opportunities or respond to even unexpected changes in an enterprise or its context. This might raise trade-offs between flexibility and optimality with respect to short-term business goals.
- Being *human-centric* to understand motives and changing behaviors, since humans are drivers and contributors as entrepreneurs and employees within an enterprise or as customers, partners, and non-customers outside an enterprise.
- Being *service-oriented* to meet the individual needs of the customers and to increase customer loyalty. This might comprise products to enable services.
- Being *value-based* to reflect targets and behavior, and to respect the request for meaningful life. This is backed by insights into dependencies between successful business models and the achievement of value propositions. See e.g. the UNESCO sustainable development goals [49] and the relevance of Corporate Social Responsibility (CSR) [39].

As a solution to these requirements, we present in this article a novel form of the future enterprise, the *Orchestrated Solution Provider (OSP)* as a reference enterprise model and the corresponding *OSP Evolution Method (OSP-EM)*. In this sense, the OSP follows the vision of Industry 5.0 for a sustainable, human-centric and resilient European industry, while going far beyond with its holistic view [14].

An OSP is a flexible, human-centric enterprise. It acts on a solid value base towards a meaningful purpose and complies with a reference set of *Critical Success Factors (CSFs)*, which address the characteristics of the knowledge society, the chances it offers and the risks it imposes. In OSP, the *entrepreneur* has a strengthened, but changed role from commander and controller to moderator and motivator, setting guidelines and providing conditions for the *enterprise teams*. They steadily reflect own targets and behavior and act in far-reaching self-organization. Processes are executed iteratively. Structures are built from loosely coupled elements to establish the *flexibility* required.

The OSP evolution method OSP-EM is holistic, since it respects any tangible and intangible aspect, which is relevant for the evolution of an enterprise. Thus, OSP-EM also covers digital transformation and enhances this towards continuous evolution. OSP-EM follows a *reference model-based evolution approach*, which integrates all concepts used into a consistent solution, to document, analyze, and simulate impacts, to allow stepwise introduction into an enterprise, to utilize latest technological developments. It supports the enterprise team in managing complexity and enables sustainable success. Like map and compass, OSP-EM navigates the entrepreneur and the team on their digital transformation journey, and spots areas of activities.

This introduction is followed by 4 sections. Sect. 2, *Foundations and Related Work*, introduces and explains the main notions used in this article and relates them to existing approaches. Sect. 3, *Constituents of the OSP Evolution Method*, introduces the taxonomy used to build OSP models. Sect. 4, Reference *Model-based Evolution of OSP*, proposes and explains the orchestrated solution provider as the new type of enterprise and OSP-EM for its continuous evolution. Sect. 4, *Digital Transformation of Enterprises*, describes how to realize a digital transformation by introducing OSP-EM into an existing enterprise. Sect. 5, *Conclusions and Future Perspectives*, finalizes this article and gives an outlook on future work.

2   Foundations and Related Work

This section introduces the main notions regarding orchestration, an enterprise as such, enterprise design, and its digital transformation, since these notions build the foundation for the solution presented in the subsequent sections of this article. The definitions given reflect our human-centric perspective on enterprises and their context. The second subsection positions the introduced notions in the context of related work as well as within our approach.

2.1   Notions

Orchestration
The notion of orchestration refers to the abstract capability of an enterprise to flexibly compose its products and services as well as the way they are produced and provided according to changing business opportunities and threats. The capability addresses the time needed from the first trigger to the availability of the new solution, the degree of novelty over the existing solution and its complexity (number of elements involved).

Enterprise

An enterprise is an open social system [30]. It consists of people, who agree to collaborate, based on defined rules, under the mutual influence of the context, towards a defined purpose. An enterprise has an enterprise life cycle and a related lifetime. The enterprise life starts with an entrepreneur's idea and continues over its foundation until its termination. It goes through several enterprise life phases.

The *enterprise team* are the people (*employees*) the enterprise consists of, with membership clearly defined by their agreement to the rules of their collaboration, i.e. contract of employment. Depending on her/his role, the *entrepreneur* can also be a member of the enterprise team. *Purpose* refers to the general reason, why the enterprise exists, at all. It gives the enterprise team quite a stable *orientation* during the enterprise life cycle ("north star"). The *enterprise rules* of an enterprise consist of agreed *design constraints* and *design factors*. Fig. 1 gives an overview of our enterprise understanding.

This definition of an enterprise covers organizations of all kind, commercial companies in any industry or any legal construct as well as non-commercial institutions or organizations (government, non-profit, e.g.), of any size (small, medium, large) and at any phase of its existence (start-up, established, e.g.).

Enterprise Design

Each enterprise has a specific enterprise design. It characterizes the evolving collaboration of the enterprise team holistically through its culture, strategy, and structure.

There is various evidence from industry and science perspective that culture must be taken seriously and that culture, strategy, and structure are mutually dependent and should be evolved comprehensively [22, 23] (cf. Fig. 2).

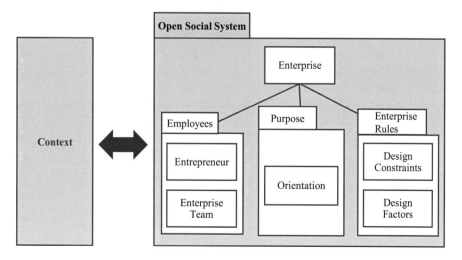

Fig. 1 Main notions of our enterprise understanding

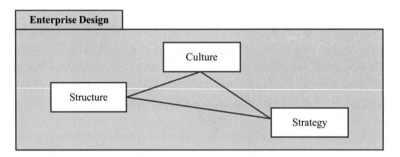

Fig. 2 Elements of an Enterprise Design

Here, the *enterprise culture* are the *intangible* rules how the enterprise team shapes their collaboration, i.e. values, norms, expectations, and the respective evolution. The *enterprise structure* are the *tangible* rules how the enterprise team shapes their collaboration, i.e. processes, responsibilities and roles, and the respective evolution. The *enterprise strategy* is the approach of how the enterprise team pursuits the enterprise purpose, i.e. goals, benefit creation model, and the comprehensive evolution of the collaboration rules and the benefit creation model. The enterprise strategy contains both tangible and intangible elements.

The enterprise culture is called *reflexive*, if the enterprise team constantly reflects their strategy ("Are we doing the right things?"), their structure ("Are we acting in the best possible way?"), and their culture ("Are we obeying what is important to us?") against changing context conditions and if they draw conclusions out of it.

The enterprise design has design *states*, which are defined by the combined states of its design factors culture, strategy, and structure.

An enterprise design state is *related* to a certain point in time (past, current, future), and to a defined subject (e.g., the enterprise life cycle or a dedicated scenario to resist certain impacts from the context of the enterprise). The design state is called

- *solid*, if opportunities and threats (*risk profile*) are under full control of the management, i.e. analyzed, evaluated and measures defined,
- *flexible*, if any actual design state can be changed within a given time into a new solid state which meets changing context conditions,
- *harmonized*, if the states of the design factors culture, strategy, and structure are synchronized with respect to their mutual dependencies.
- *unharmonized*, if the design states of at least two design factors are not synchronized (e.g. when developed isolated).
- *digitized*, if the business model of the enterprise is improved by using digital technology (e.g. increased profitability by process automation),
- *digitalized*, if the business model of the enterprise is enabled by using digital technology (e.g. digital sales channel by internet portal technology).

The *enterprise operation* is the execution of enterprise activities according to an enterprise design state given. The *maturity* of an enterprise design state expresses to which degree the design state enables enterprise operations to fulfill a defined set of enterprise constraints and rules.

An enterprise design can be described by an *enterprise model* [19]. Such an enterprise model might conform to an *enterprise reference model* where enterprise constraints are predetermined.

Enterprise Evolution

Enterprise evolution is any sequence of enterprise design states which are related to the same enterprise life cycle (cf. Fig. 3).

Enterprise transformation is any managed transition from a starting design state to a target design state. The transition may comprise multiple *transformation steps*. A *transformation roadmap* is a description of any planned sequence of transformation steps. Any transformation state can be characterized by the state of the enterprise design and its maturity.

An enterprise evolution is called

- *holistic*, if each state of the evolution is harmonized. An enterprise transformation is called holistic, if the target design state is harmonized.
- *continuous*, if a transformation process is established which at any point in time transforms a given design state into a subsequent design state.

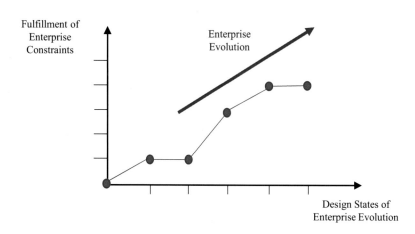

Fig. 3 Enterprise Evolution

- *enterprise model-based*, if an enterprise model is used to describe the states of the enterprise design within the evolution and to describe the respective transformation measures taken.

Digital Transformation

An enterprise design is called *digital* if it is based on a digitalized design stage. The digital maturity of an enterprise design expresses to which degree it enables the enterprise operations to fulfill its digitalized business model. An enterprise transformation roadmap is called a *digital transformation* if it increases the digital maturity of the enterprise design state it starts with. An enterprise design state is called *digital ready* if the enterprise has defined a digital transformation to establish a digital design state.

2.2   Related Work

Designing an enterprise has been studied from different viewpoints within several scientific disciplines. Informatics and Business Informatics have investigated Architectural Frameworks which range from a business over an IT application landscape towards an execution platform perspective. As enterprises are economic units, Economics has studied e.g. business strategies, models and processes, organizational structures, enterprise evolution and change processes. As enterprises employ people, psychological issues have been studied in Economics and Work Psychology. As enterprises operate in a society, sociological results like social systems theory are important.

Architectural frameworks have gained in importance as they allow to integrate into one framework business as well as information technology aspects. This enables to handle the interdependencies between business goals and constraints on one side with software solutions and underlying technology infrastructures on the other side. Well-known examples are The Open Group Architectural Framework TOGAF [35], the Integrated Architectural Framework (IAF) [50], the Generalized Enterprise Reference Architecture and Methodology (GERAM) [8], or the OASIS Reference Model for Service Oriented Architecture (SOA-RM) [33].

Some of these architectural frameworks come with a corresponding architecture development method, as e.g. the TOGAF *Architecture Development Method* (ADM) or *Quasar Enterprise* for IAF [15]. They give concrete guidelines how to develop an IT architecture that meets the needs of an enterprise. Also, the ArchiMate® Enterprise Architecture Modeling Language, a standard of The Open Group, has to be mentioned here [36]. It supports to describe the construction and operation of business processes, organizational structures, information flows, IT systems, and technical infrastructures. The resulting models help stakeholders to design, assess, and communicate the consequences of decisions and changes within and between business domains.

All these frameworks and modeling languages take a quite technical view on relations of business activities and information technology. This is also the case in recent research results on agile developments of software ecosystems, where software and business aspects can be adapted on-the-fly during system enactment [54].

Our approach on developing future enterprises is on a higher abstraction level, and thus less detailed than these model-based approaches. We take a holistic, reference model-based view and consider culture and strategy aspects equal to business and technical aspects. We deploy a three-layered approach as known from traditional database schemes, which distinguish the three levels termed conceptual, logical, and physical data base scheme [3]. This distinction has also been reused by the OMG (Object Management Group) within their Model Driven Architecture® (MDA®) approach which differentiates the Computation-Independent Model (CIM), the Platform-Independent Model (PIM), and the Platform-Specific Model (PSM). They are stepwise refined in a system development process [34].

We will reuse such a three-layered refinement approach to refine high-level enterprise models into our novel *Orchestrated Solution Provider (OSP)* reference enterprise model. This reference model can then be used to design a concrete enterprise model by deploying the corresponding OSP evolution method OSP-EM.

There is nowadays a common agreement that any kind of development and evolution process should be done in an iterative and agile way. Our approach, too, is based on basic principles of the Agile Manifesto [31] like

- close cooperation between all involved stakeholders,
- openness to any kind of changing requirements and context influences,
- self-reflexive and self-organized team structures and
- cultural values like highly motivated individuals and trusted relationships.

A management-oriented approach to enterprise development and evolution has been developed since several decades at University St. Gallen. The current St. Gallen Management Model (SGMM) [41] has a systemic and entrepreneurial orientation and differentiates management into operational, strategic and normative aspects. At the same time, it emphasizes that management and organization are in a dynamic interaction with the context and that management is a reflective design practice. While there are a number of similarities with our approach, we aim at concepts dedicated to a digital transformation, give culture, structure, and strategy equal relevance and thus do not focus on management issues.

The SGMM and our approach have in common that we follow Luhmann's system theory [30]. He claims that a system in principle distinguishes itself from its environment. So, there is always something that belongs to the system and something that does not (environment). Other systems also belong to the environment. This difference system/environment is the basis of the whole system theory by Luhmann.

There is a series of recent work on enterprise design from a scientific as well as from industrial experience point of view each focusing on certain aspects. Thus, they are less holistic in their approach as we are but influenced our work. Some examples are

- the "Design of enterprise systems" approach by Giachetti [19], who proposes a quite concrete engineering process, where a dedicated enterprise engineer guides all aspects of an enterprise development process,
- the "Enterprise Architecture as Strategy" approach by Ross et al., who shows how constructing the right enterprise architecture enhances profitability and time to market, and improves strategy execution [45] or
- the investigations by Alwadain et al., who identified the factors which influence an enterprise architecture evolution [1].

In most of these technology- or economics-driven approaches, culture and human aspect are underrepresented. Due to the discussion on new working formats in enterprises, these aspects are gaining a higher relevance. In agile and lean process approaches, e.g., all kind of stakeholders are handled as first-class entities. Another example is the increasing care about workers' welfare, which is strongly influenced by Seligman's Positive Psychology [46]. This, and recent discussions on changes in work-life-balance as well as the "new normal" of future work underline that culture and human-centricity have grown to equal importance with pure strategic or organizational issues. This is reflected by our holistic approach to enterprise design and evolution.

3 Reference Model-Based Evolution of OSP

This section introduces our three-layered enterprise evolution approach. Based on an underlying taxonomy, we will introduce concepts and constituents of a conceptual enterprise model, our reference enterprise model OSP as well as the specifics of a concrete enterprise model.

The OSP enterprise reference model enables a comprehensive and continuous evolution of all tangible and intangible elements of an enterprise and their interdependencies within the enterprise and with its context. It supports managing the complexity, which is driven by the vast number of relevant elements and their changing interdependencies. Furthermore, the evolution increases the flexibility of the enterprise to a level, where it is able to conduct fundamental and far-reaching changes with increasing speed after relevant events or insights have occurred.

The corresponding OSP evolution method OSP-EM meets these requirements by separating three concerns:

1. the evolution of a value-based targeted enterprise with people at its core,
2. the flexible evolution of an enterprise, which respects concern 1 and meets the special conditions of the knowledge society,
3. the evolution of a concrete, existing enterprise, which respects concern 2 and the special situation of a real enterprise.

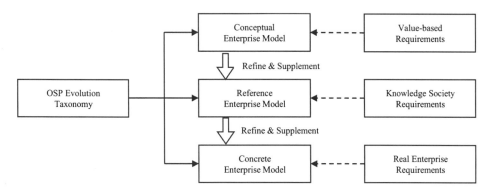

Fig. 4 Three-layered structure of enterprise evolution models

Each of these concerns is covered by a dedicated three-layered *enterprise evolution approach*: (1) the *conceptual enterprise model*, (2) *the reference enterprise model* (OSP model), and (3) the *concrete enterprise model* (cf. Fig. 4). Each of them is introduced in detail in the subsequent subsections.

All of them rely on the same taxonomy and build upon each other by well-defined refinements and supplements. Refinements and supplements are methods for enterprise evolution and artifacts for the description of relevant facts. This supports sustainability by adaptability to latest methods, technologies, and specific knowledge as needed, while the constraints given by the taxonomy ensure effective integration. The enterprise model-based approach supports to document, analyze, and simulate impacts [10]. But the restriction remains that a model-based approach only approximates the real world and focuses on aspects relevant for the evolution of a concrete enterprise [48].

Subsequently, we describe the taxonomy and each enterprise model in detail.

3.1   OSP Evolution Taxonomy

The OSP evolution taxonomy as the base for our three-layered enterprise evolution approach integrates the concept of an enterprise as an open social system (cf. Sect. 2.1) and a fundamental scheme of human interaction comprehensively into a network of impacts on humans and their behavior. It is given by a *structural view* as well as a *dependency view* on this network (cf. Fig. 5).

The interaction scheme represents human behavior causing impacts, which then influence human behavior again. *Personas* group human behavior according to roles humans take. *Impact factors* categorize impacts depending on the related roles and the kind of influence. This scheme provides a general description for all aspects of enterprise evolution within the enterprise and its interdependencies with its context. The structural view is given by an *information model*, the dependency view by a *dependency matrix*. Our

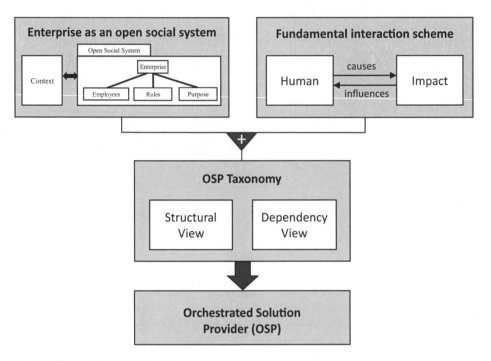

Fig. 5 Views of OSP Evolution Taxonomy

new reference enterprise model, the Orchestrated Solution Provider (OSP), is derived from this network of impacts.

Structural View of OSP taxonomy

The structural view (cf. Fig. 6) defines the environment of an enterprise as the *social system, consisting of the society* all humans belong to. Their *behavior* as the *context* is in mutual influence with the enterprise. *Personas* and *impact factors* provide a seamless integration of the enterprise and the society. *Personas* consistently describe the roles people have in the society and the enterprise. They are differentiated into those, who have an external role related to the enterprise (like customers, partners, competitors), and those which build the enterprise team and are enterprise employees. Depending on her/his role, the *entrepreneur* is external and can also be an employee of the enterprise.

Impact factors are the *behavior* of humans in the society, the *purpose* of the enterprise, and the *enterprise rules*. The *context factors* describe which opportunities and risks arise for the enterprise (trends) through the behavior of humans in the society. The *orientation factor* motivates the general direction the enterprise takes. *Constituent factors* and *Critical Success Factors* (CSFs) define the constraints for enterprise design. The *constituent factors* define the starting point for the enterprise design and the allowed corridor for their evolution. The CSFs define the conditions the enterprise design must meet

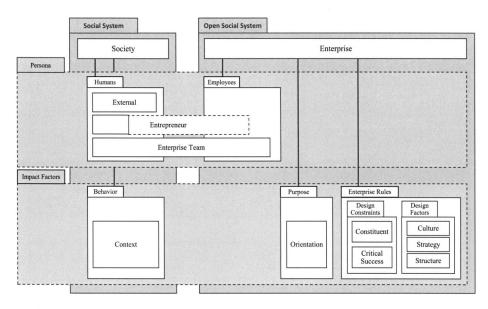

Fig. 6 Structural View of OSP taxonomy

under the given or developing context factors to enable fulfillment of the enterprise's purpose. Since culture, strategy, and structure comprehensively describe the enterprise design, determine the collaboration of the enterprise team and the evolution of the enterprise, they are called the *design factors* of the enterprise.

Dependency View of OSP Taxonomy

Dependencies are between personas and impact factors, and between impact factors.

The external humans change the context factors. In this regard, e.g. the buying behavior of customers is also abstracted to context factors. The entrepreneur sets the constituent factors and decides on the level of their details. The enterprise team reflects and develops the design factors and enacts them in its daily operations.

The context factors influence the behavior of the external humans and define requirements to the constituent factors, the design factors, and the CSFs by providing opportunities and threads. The constituent factor defines the orientation factor, and the starting point for an enterprise design as well as the allowed corridor for its evolution. The design factors guide the behavior of the enterprise team towards the orientation factor according to the CSFs. They define further evolution of the orientation factor and the CSFs. The orientation factor promises benefits to the external humans (cf. Table 1).

Table 1 Dependencies between notions of the OSP Taxonomy

	affects ➡	Personas			Impact Factors				
		External	Entre-preneur	Enterprise Team	Context	Consti-tuent	Critical Success	Design Factors	Orien-tation
Personas	External				➡				
	Entre-preneur					➡			
	Enterprise Team							➡	
Impact Factors	Context	➡				➡	➡	➡	
	Constituent							➡	➡
	Critical Success							➡	
	Design Factors			➡			➡		➡
	Orientation	➡							

3.2 Conceptual Enterprise Model

The *conceptual enterprise model* represents the first of the three model layers (cf. Fig. 4). It describes the fundamental rules according to which a value-based targeted enterprise with people at its core operates and how it is transformed, without considering special requirements from the knowledge society. In the conceptual enterprise model, e.g. CSFs and the role of values as such are introduced, and the strategic management process is introduced through its tasks and outcomes without paying any attention to flexible execution. The conceptual enterprise model consists of the building blocks personas and impact factors (context, constituent, critical success, design, orientation), as introduced in the OSP-EM taxonomy (cf. Fig. 6).

Personas

Personas structure the information which group (*segment*) of people has which influence on the course of the enterprise and how does it change over time. How do they behave? What is their expectation and perception of the enterprise? Which data provides insight and allows which degree of prediction? External people, entrepreneur, enterprise team are basic segments. Additional subsegments of external people are e. g. those with direct impact on the enterprise (like customers or suppliers), and those with indirect impact (like various social groups without any touchpoint with the enterprise).

Impact Factors

Context Factors The *context factors* handle opportunities and risks from outside the enterprise. They are evaluated for their relevance to the enterprise within the strategy management process. The enterprise defines how to capture the information and which methods to apply for its evaluation. The context factors are aligned with the segmentation of the external people. For those without direct touchpoints with the enterprise, PESTEL (political, economic, socio-cultural, technological, ecological) e.g. provides a general structure, which can be further expanded by special research methods and trend analyses [2]. For customers, insights into buying behavior and experience, product and service perception, loyalty, and willingness to pay are relevant criteria.

Orientation Factor The *orientation factor* motivates the reason why the enterprise does exist and provide guidance for the definition of enterprise goals and transformation initiatives.

Constituent Factors The *constituent factors* describe the guidelines the entrepreneur sets for the enterprise team to fulfill her/his business idea; they are considered quite stable and can only be changed in agreement with the entrepreneur. They typically comprise

- the purpose as general orientation,
- the value base on which the enterprise operates as guidelines, which behavior is right (expected) or wrong (not tolerated) and what to prioritize,
- goals the enterprise shall achieve while pursuing its purpose,
- general conditions to be met (e.g. finances, risk profile, governance model),
- the role of the entrepreneur and her/his influence on the enterprise,
- the participation in decision making and the degree of profit-sharing.

Critical Success Factors The *Critical Success Factors (CSFs)* are initially derived from the constituent and context factors at the foundation of the enterprise. They are related to the business model and updated with insights from the strategy management process. Measurable criteria are set for each success factor, to evaluate the degree an enterprise fulfills its CSFs and to balance and control evolution measures.

Design Factors The *design factors culture, strategy, and structure* are initially derived from the context, constituent, and critical success factors. They are updated with insights from the strategy management process and synchronized towards the purpose pursuit. Measures to evolve the design factors must respect their specific nature. The enterprise culture consists of intangible topics like expectations and tolerance. The enterprise structure consists of tangible topics like formalized responsibilities. The enterprise strategy comprises a mixture of both, with purpose and goals intentionally characterizing where to go, approximated by clearly defined performance indicators and initiatives. The enterprise culture is as individual as personalities in the enterprise team, and their behaviors.

All three design factors are described by appropriate models. In this sense, our approach is based on the OSP reference enterprise model as well as on prescriptive models in the classical sense of model-based development [10].

Culture Factor The *culture factor* describes intended and actual behavior of the enterprise team, and initiatives to evolve it. Leitbild documents and code of conducts formalize aspects of intended behavior through values, norm, and rules, and align understanding in the team.

From a human-centric perspective, the evolution of culture is based on the acceptance that there remains an intangible part, which can neither be fully defined nor directly be measured, or which is left intentionally open. Examples of the remaining part are informal communication channels and "hidden rules", which the enterprise team is not even aware of. Creativity is an example of a typical Leitbild value, which can be stimulated, whose outputs can be measured and sometimes be forecasted, but not be determined in advance. The evolution relies on stimulation through framework conditions, and intense communication (e.g. reflection on intended culture, feedback on behavior, positive and negative examples, or role models). Rituals and symbols support the perception of a certain style and the forming of a team identity.

Due to its intangible nature, the evolution of the enterprise culture progresses slowly and requires steady impulses and great foresight. Alignment with the evolution of strategy and structure is mandatory (e.g. which project supports which values and should be highlighted appropriately, which project might be perceived as conflicting with certain values and must be adjusted or strongly supported), but might cause trade-offs with regard to timelines. In general, leadership teams must be aware, if their actions are conclusive. In many Leitbild documents as models of the culture factor, one can find the values openness, credibility, and motivated employees. But if, e.g., open-space offices are introduced and cost saving measures let the employees sit together so close that the true motive becomes obvious, but the leadership team keeps referring to employee motivation and denying the cost saving motive, the culture is harmed and the Leitbild better had never been written.

If CSR activities (corporate social responsibility) are considered important, they should also be included into the culture design and their value contribution be linked to enterprise goals instead of handling them as a social fig leaf [39].

Strategy Factor The *strategy factor* details how the enterprise fulfills its purpose and comprises, and might be described by several types of modeling approaches:

- value creation design (incl. purpose, value proposition, value chain, income cost ratio), visualized e.g. by business model canvas [37],
- goals and objectives, expressed e.g. with a balanced score card [28],

- evaluation of impact factors, e.g. according to SWOT matrix [24],
- CSFs [32] including maturity states of enterprise design,
- visualized target picture (scenarios for CSFs or customer journeys [16]),
- the most limiting factors of further prosperity [17]
- transformation roadmap based on program portfolio management techniques [38] including priority setting and resource allocation,
- communication concept, e.g. using story-telling techniques [9].

These results and the implementation of the transformation roadmap are provided by the *strategy management process*, which is organized within the design factor structure. The strategy management process also identifies relations with culture and structure, to ensure that restrictions and needs for transformation are identified, that appropriate decisions are made, and that measures are synchronized and included into the transformation roadmap. With culture e.g., the way customers are addressed, ethical products and production, targeted enterprise image, communication strategy. With structure e.g., elements of the value chain to be organized, restrictions by risk profile.

Structure Factor The *structure factor* implements the value creation design of the strategy factor into how the enterprise team provides these values. It comprises process flows and conditions, responsibilities and communication, roles, and their owners. The used models are architectures, which are kept mutually aligned:

- business architecture,
- process map (management process, core processes, support processes),
- operating model (policies on variability vs. stability, centralization vs. decentralization, differentiating vs. non-differentiating)
- organization and governance structure,
- information architecture (data, systems, networks),
- infrastructure architecture (physical locations and capacities).

The design must not stop with formal elements. Also, decisions must be made, how informal communication and informal communities shall be handled in the enterprise and to which extent they shall be involved into the structure design. Guidelines and policies help to stimulate e.g. reflection workshops or engagement in social media groups and clarify budgets for provisioning of required infrastructure. In case of designing skill profiles, either for special roles or in a general way, attention must also be paid to soft skills like social behavior, communication style, teamwork, etc.

The management process comprises the strategy management process, covering the evolution of the enterprise, and the operation management process, covering the execution of the enterprise activities. The strategy management process is divided into the two subprocesses *strategy development* and *strategy implementation* [43]. Outcomes of the first subprocess are the current and planned status of the design factors and the synchronized transformation roadmap. The second subprocess comprises the execution of the roadmap and the accompanying organizational change and risk management.

3.3   Reference Enterprise Model

The *reference enterprise model* embodies the idea of how to evolve the Orchestrated Solution Provider (OSP) as the new type of enterprise in the knowledge society. It represents the second of the three enterprise model layers (cf. Fig. 4). It also consists of the building blocks personas and impact factors (context, constituent, critical success, design, orientation), as introduced in the OSP-EM taxonomy (cf. Fig. 6). It is built upon the idea of a value-based targeted enterprise and thus refines and supplements the respective elements of the conceptual enterprise model. The driving refinement is a reference set of CSFs, which is derived from the general impact of the knowledge society and further on translated in respective refinements of all impact factors. Since the OSP can only fulfill its CSFs by sophisticated use of digital technologies, the enterprise design of the OSP is digital. Therefore, the evolution of an enterprise design according to the OSP enterprise model is a digital transformation. This goes clearly beyond improvement of the existing business model.

Personas

Personas change their expectations and behavior as society progresses into the knowledge era. The absence of absolute truth, the increase of multiple options, and of intense self-reflection foster individuality and self-responsibility. The personas elements of the conceptual enterprise model are refined towards the reference enterprise model by profiles to study expectations and behavior. Subsequently, basic profile descriptions are proposed.

Customers as *external personas* are more and more demanding. Not least because of social networks and of ubiquitously available data, they are well-informed, and they use multiple channels to connect. Thus, winning new customers and intensifying their loyalty is an increasing challenge.

The role of the *entrepreneur* also changes. Increase of dynamics, knowledge and options turn command and control styles into bottlenecks and request collaborative styles like moderator and motivator, to unleash team-intelligence [44, 47]. This is amplified through self-organized networks inside the enterprise and in its context. The entrepreneur must be aware that even the purpose of the enterprise must not be considered as solid. Hence, he must cultivate her/his sense, if and when to change the purpose.

Enterprise teams do not function as deterministic input–output units. Respective work is replaced by machines. Instead, social, and intellectual skills become more important, and the borderline between private and professional life is diminishing. The workplace and the purpose of work become an integral part of personal life. Enterprise teams constantly self-reflect their work and consider measures to improve.

Impact Factors

Context Factors The dynamics of the *context factors* increase, and events anywhere can have an immediate impact everywhere else due to strong and far-reaching network connections. Thus, detection and reaction better occur instantly. This calls for predictive data analysis and measures to improve resilience. There are relevant trends in all PESTEL dimensions. Sustainability in ecology, growing demand for healthiness in society, accelerating innovations in technology. This leads to new products and services. Thus, one-time analysis is not sufficient. But methods are established (e.g. trend radars or big data analysis), which support steady analysis and continuously updated predictions and scenarios.

Orientation Factor In the dynamic context of the knowledge society, the importance of the enterprise's purpose increases for orientation and motivation ("north star"). Nevertheless, it must allow fundamental questioning if there is a need to do so. Otherwise, an enterprise might lose its elasticity and resistance to change grows. This is a topic hardly to be formalized but left to the sure instinct of the entrepreneur.

Constituent Factors In dynamic contexts, *constituent factors* are rather guidelines than detailed prescriptions. They give the enterprise team orientation and leave the right amount of freedom to act. The purpose is essential for the orientation and the value base provides a solid foundation for the team's behavior. The entrepreneur should be clear in expectations and limitations regarding chances and risk and her/his space for intervention.

Critical Success Factors The *Critical Success Factors (CSFs)* are individual for each enterprise according to the impact of the context factor on its concrete business model. But the fundamental changes to the knowledge society imply criteria which become relevant for most likely any enterprise. They form a starting point for evaluation and continuous evolution and characterize the *orchestrated solution provider*, who

- pursues a purpose with focus on customer benefit and social responsibility,
- offers individual solutions from integrated products and services as close to real-time provision as possible (e.g. real-time tracing of transportations),
- continuously shapes its strategy and operational behavior to open and utilize new purpose fulfilling opportunities,
- uses state-of-the-art technology and guides its behavior by intense analysis of data and by exhaustive usage of performance indicators,
- orchestrates dynamic networks of self-organized teams internally and independent business partners externally, and allocates resources as needed,
- establishes and develops working conditions which are perceived as attractive and which promote creativity and open collaboration.

Design Factors In the reference enterprise model, the evolutions of the *design factors* are organized as steady processes each, synchronized into a holistic transformation roadmap, fueled by the reflective behavior of the enterprise team, and embedded into a rapidly changing context. Numerous options, continuous change, self-organization, and increasing creativity can let the enterprise team get bogged down, not allocate limited resources to the right priorities, and finally miss a successful evolution. To mitigate this risk, we introduce the *Targeted Adaptive Evolution (TAE) principle* as a general habit and as logic of procedures, and *entrepreneurial cells* as concept for organization structures.

TAE Principle The *TAE principle* is the underlying concept for procedures to manage progress in dynamic, complex situations and requires cultural prerequisites to be applied successfully. It enhances known techniques for stepwise, iterative progress with guidelines to keep track of progress and resource control: *hypotheses-based evolution loop, life phases and maturity grades, 4 horizons leadership calendar.*

(1) **hypotheses-based evolution loop**

In a complex environment, iterative, stepwise approaches are required, which also include systematic reflection and learning (see e.g. PDCA cycle [12], the lean startup method [40]). We structure the often-used build-measure-learn evolution loop into 4 steps *Evaluate, Plan, Prepare, Execute*:

- *Evaluate* reflects the objectives, evaluates insights into the subject domain, its context and into the solution requirements, evaluates the learnings made so far and solution options, updates the task backlog and decides on the option to follow (if at all), on the solution approach to take or a hypothesis to be tested [21] and on performance measures to be implemented.
- *Plan* defines the next solution by selecting tasks from the backlog, either as a solution update or as a tested hypothesis. The scope is limited for fast execution as it is e.g., practiced with sprints in Scrum. This step is aligned with the overall program management to identify and resolve dependencies between various initiatives, and to agree on resource allocation. It ensures preparation of required change management activities; it may add dedicated reflection activities and initiatives which foster experience of cross-team successes and promote enterprise-wide solidarity. Dependent on the extent of change, the subsequent execution can require further steps to adjust or modify the solution approach. It defines the communication model and the performance indicators, to control the progress of the initiative and the result achievement.
- *Prepare* ensures an efficient execution, provides required infrastructure and resources, involves all affected people, informs about the initiative, empowers and enables the right people [4, 27].
- *Execution* implements the defined scope or validates the hypothesis. Intense communication keeps the team aligned.

(2) **life phases and maturity grades**

In a dynamic environment, any enterprise object (product, process design, machine, IT systems, business model) can fulfill its requirements only during a certain period. To actively manage a healthy state at any point in time, a life phase and a maturity grade are assigned to any enterprise object. This supports to qualify development needs, control progress, and to balance resource allocation.

(3) **4 horizons leadership calendar**

Enterprises follow the *principle of economic efficiency* and activities compete for scarce resources. Especially in dynamic contexts or when cannibalizing initiatives are driven, the enterprise team must give the right attention to each initiative and spend sufficient time on reflecting its activities. Therefore, a scheme is applied to allocate and control resources along 4 horizons: (1) daily operations, (2) optimizing daily operations and removing ballast of the past, (3) ongoing initiatives, (4) future initiatives after ongoing are finalized. This scheme is used to set up a leadership calendar which distributes the available leadership time to these 4 horizons.

The hypotheses-based evolution loop is not about process organization only. It is embedded into a cultural context which fosters agile mindset and behavior. Teams trust each other for open communication and direct feedback, which is practiced intensely. Joint striving for benefit endowing results nurtures performance orientation and collaboration "across siloes". Teams are equipped with as broad a range of competencies as necessary and have autonomy to make rapid decisions. Early results are preferred over long analyses, experimentation is encouraged, and failures are accepted as learning opportunities, and quickly corrected. Establishing such a context can be a challenge.

Entrepreneurial Cells An entrepreneurial cell (short: cell) is the approach to yield a flexible structure in the OSP enterprise model. The cell implements a single element of the value chain and provides its solution results with a maximum of self-containment, from side-effects with other cells as free as possible [51]. The cell may be composed of smaller cells and their relationships, and it may have relationships with other cells for solution reception and provision. These relationships are based on clear result and performance commitments. The enterprise is seen as the top-level cell, which has external relationships with its business partners, suppliers, and customers.

The cell has attached a life stage and a maturity grade, to support enterprise-wide allocation of resources and prioritization of initiatives. They are derived from the business models, which the implemented element of the value chain belongs to.

Self-containment of the cell is supported mainly by the three factors *mixed teams, service-oriented enterprise architecture, service-oriented performance controls*:

- The *mixed teams* are built upon all skill groups required to deliver the results committed. The teams follow the enterprise value base and organize themselves, choose methods and tools to their needs, compliant with enterprise-wide standards, ensure cross-team communication and collaboration. They participate in cross-cell skill groups for knowledge exchange and to develop their individual skills [53].

- The *service-oriented enterprise architecture* [1] provides the structural alignment of business and information architecture. Information architecture provides self-contained information services, which are ubiquitously available, up and down scalable, modifiable and exchangeable, secure and robust as required. They can be flexibly composed to orchestrate new information services and they limit dependencies between cells by avoiding an intensely mashed information architecture.
- *Service-oriented performance controls* measure value contribution against the performance level committed and according to actual service consumption, which is supported by the end-to-end cell structure.

Based on the TAE principle and the concept of entrepreneurial cells, we describe how the design factors are refined in the reference enterprise model of the Orchestrated Solution Provider (OSP) as the new type of enterprise in the knowledge society.

Culture Factor For the *culture factor* values like customer-orientation, trust, curiosity, and courage along with social responsibility are eminent. Creating enthusiasm is accepted as a major leadership task and the evolution of the respective culture is stimulated accordingly. Collaboration is strengthened "cross siloes", and performance increased by honest, intense, and direct feedback. CSR is fully integrated into the enterprise value chain [39] and intensifies the relevance of the purpose. The leadership team, starting with the entrepreneur her/himself, acts as a role model of how to behave according to the enterprise's values and change their leadership style from commander and controller to moderator and motivator [47]. They demonstrate personal engagement for trust, open communication, and continuous search for the best way to fulfill the joint purpose of the enterprise [25]. This also fosters a performance culture and supports constant change becoming a natural habit of enterprise life. The evolution of the culture factor is driven by a process which follows the TAE principle and fosters steady stimulation, and steady reflection of behavior.

Strategy Factor The steady evolution of the strategy factor by an established strategy management process, which integrates the TAE principle, is the leading mechanism for digital transformation and continuous evolution. It steadily evaluates the impact factors with a broad sense and adjusts the backlog of initiatives for strategy implementation. The risk profile is steadily managed with a wide lookahead, to keep the balance between opening new opportunities and mitigating risks. Implementations of initiatives are organized as quick sprints and respect MVP (Minimal Viable Product) criteria [40]. Program management techniques are applied for the synchronization of ongoing initiatives and for the management of the backlog. The steps *plan* and *prepare* of the TAE loop synchronize with organizational change management and respect self-reflecting and self-organizing teams. They are involved early and openly, to raise understanding and support of initiatives planned [4, 27].

Since the business model of the enterprise cannot be considered as stable, neither the business model becomes also subject to a steady management process following the TAE principle. A single business model is replaced by a portfolio of business models, which represents different life stages and multiple options to enable sustainable success [18]. Also, the purpose might be questioned, but with an incredibly careful mindset.

Structure Factor In a "Modern Firm" [42], the evolution of the *structure design* is a steady and holistic task, closely interwoven with the evolution of the culture and the strategy, steadily implementing new requirements and insights, improving flexibility, resilience, transparency and efficiency to fulfill the CSFs of the enterprise. New products, new customer segments, new production methods, new supply chains, or growth beyond existing capacity are exemplary triggers which require to reflect an existing organization and to potentially change it. The OSP model organizes the evolution of the structure design by a steady process which follows the TAE principle.

In the OSP enterprise model, the structure design is an *orchestrated network of loosely coupled entrepreneurial cells*, which maximizes adaptability of the enterprise and minimizes side-effects of changes [51]. The orchestration of this network is subject to the enterprise-wide structure design process and starts with level 1 elements of the value chain. The major task is to define orchestration guidelines and to design the layout of the next level of cells. Proven orchestration guidelines build on the operating model criteria described by [45] and require decisions, which services become an enterprise standard for efficient reuse and stability, and which services will be independent for increased variability, which services are differentiating and kept inhouse, and which services are not, thus sourced from partners. Customer-oriented criteria like the life states of business models and maturity of the provided solutions help to resolve these trade-offs. Enabling functions are linked with the primary value creating cells by staffing the mixed teams of the cells with appropriate skills, and coordinate skills development across all cells. This is a steady and dynamic task in an enterprise-wide responsibility. Enabling functions also provide requested standards and policies. HR, e.g., takes a special focus on developing soft skills and culture. Finance, e.g., provides performance data in line with the cell structure, to support solid decisions, e.g., on external service sourcing, consumption-based pricing models. There is also an enhanced role of managing the enterprise's eco system, to build enterprise-wide strategic partnerships and to establish the right conditions for dynamic collaboration models.

Informal communication, communities of practice, social workshops, participation in external social media groups, or creativity events like hackathons are supported by clear guidelines and policies [52]. Since communication and value-based behavior are essential, evolution of soft skills is critical.

3.4   The Concrete Enterprise Model

The concrete enterprise model describes the design of a real enterprise and enables its continuous evolution under the conditions of the knowledge society, according to OSP-EM and the flexible human-centric enterprise. It represents the third model layer (cf. Fig. 4).

Enterprise Model Derivation The derivation of the concrete enterprise model from the reference enterprise model is called the *implementation* of the OSP enterprise model. The implementation is done by detailing the elements of the reference enterprise model, to represent at least the current and a target design state of a real enterprise as required to support the enterprise transformation roadmap.

The enterprise model is also implemented stepwise by applying the TAE principle. This leads to a model evolution roadmap which must be aligned with the enterprise evolution roadmap to ensure that the evolving model always covers, what the next enterprise evolution step requires, and to comprise a sufficiently detailed *big picture* of the enterprise to provide guiding context information for the next enterprise evolution steps. Thus, organizing this alignment follows the pattern of a *co-evolution process* (cf. Fig. 7).

The *big picture* of the enterprise answers at least on a high level:

- Which purpose is pursued and how is it communicated?
- Which values and behaviors characterize the culture?
- Which strategy is followed, what are the most important messages?
- What are the major strategic initiatives and their priorities?
- What is critical for success?

Questions which drive the enterprise evolution and are answered through strategic initiatives which require appropriate model support for affected domains are e.g.:

- Do cultural and infrastructural conditions support New Work solutions?
- What is required to experiment with a new service idea?

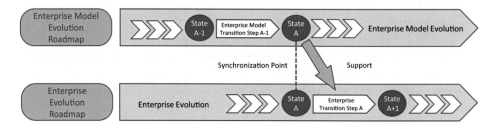

Fig. 7 Co-Evolution Process

- Which bottlenecks avoid half cutting the time to market of new services?
- What is the impact of providing omnichannel experience to customers?
- What is required to introduce predictive maintenance into production lines?
- Where can entrepreneurial cells be piloted best? Who must be involved?

Value Contribution OSP-EM enables the enterprise team to identify and define transformation initiatives for an *effective* evolution of their enterprise in the knowledge society:

- The strategy management process utilizes a network of impacts which comprehensively respects opportunities and threats from the knowledge society to translate these influences via CSFs into objectives for the enterprise evolution which enable fulfillment of the enterprise purpose.
- The design factors culture, strategy, and structure allow to build a holistic, human-centric transformation roadmap that meets the evolution objectives.
- The evolution loop of the TAE principle establishes an enterprise-wide iteration rhythm which defines the backbone for the enterprise's ability to act swiftly and integrates a mechanism for reflection and continuous optimization.
- Loosely coupled cells enable flexible orchestration.
- The CSF maturity grades foster a differentiation between digitized and digitalized design states, hence mitigate the risk to mistake digitization measures (like process automation or mobile workplaces) already as digitalization.
- Co-evolution ensures that enterprise evolution and model evolution are aligned towards the same evolution objectives and are synchronized in small, iterative transformation steps.
- The 4 horizons of the leadership calendar support the leadership team to allocate their required attention to future-oriented tasks.

OSP-EM enables the enterprise team to *efficiently* conduct transformation initiatives:

- The end-to-end traceable value contribution, from initiatives over design factors to CSF fulfillment and purpose achievement, fosters team motivation and supports setting the right priorities.
- Loosely coupled structure, performance culture and self-organized teams unleash available energy and allow a high degree of simultaneous activities.
- The progress of all initiatives, including the evolution of OSP-EM itself, is consistently managed by a comprehensive roadmap which keeps available resources under control, including management attention (4 horizons of the leadership calendar), and respects dependencies of the initiatives.
- Small iterative steps including holistic change management provide early results and allow for swift adaptation to changing requirements.

- Steady analysis of environment factors, CSFs and transformation status provide the enterprise team continuously with a realistic big picture of the actual enterprise design state and hence support managing complexity.

The *success criteria* met by the implementation process of OSP-EM are:

1) The entrepreneur and the leadership team do actively support the approach and accept OSP-EM as a strategic instrument for themselves.
2) The concrete enterprise model has an appointed owner and full transparency is given for the evolution of the enterprise model, its outcomes and the resources needed.
3) Benefits are made measurable by linking OSP-EM to other initiatives supported and by measuring leadership time spent for the "4 horizons".

4   Digital Transformation of Enterprises

OSP-EM realizes the digital transformation of a specific enterprise as a roadmap of transformation steps towards a digital enterprise design and embeds this roadmap into the framing process for continuous evolution of the enterprise. This corresponds to the execution of the strategy management process in the concrete model of the enterprise. Thus, applying the concrete enterprise model for digital transformation and implementing the concrete enterprise model must be aligned via co-evolution, which basically follows three phases (cf. Fig. 8):

Phase 1: Prepare the Ground
Phase 1 aligns the ideas and the understanding of how to apply OSP-EM to the evolution of the enterprise with the entrepreneur and the leadership team. It allows agreement on success factors and on how to proceed. Since the entrepreneur, her/his expectations, her/his ideas, her/his behavior are decisive for the success of the transformation it must be clarified in this step if he is aware of this significance and of the impact on her/his current role and if he is willing to change personal behavior as required. This is a prerequisite to gain the necessary leadership support which then allows to communicate and organize the next steps.

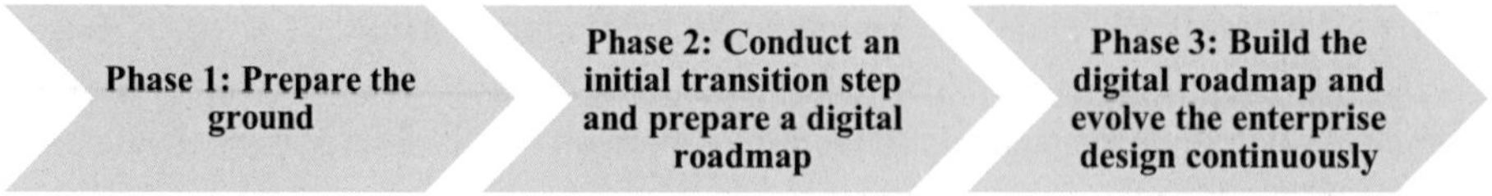

Phase 1: Prepare the ground	Phase 2: Conduct an initial transition step and prepare a digital roadmap	Phase 3: Build the digital roadmap and evolve the enterprise design continuously

Fig. 8 Phases of Digital Transformation of Enterprises

Phase 2: Conduct an Initial Transition Step and Prepare a Digital Roadmap

Phase 2 lets the leadership team experience how to apply OSP-EM to a small transition with a real topic, introduces the 4-step-approach according to the TAE principle, which will be continuously iterated. It provides insights in how to tailor OSP-EM to the specific enterprise and prepares the initial digital roadmap of phase 3.

First step, select a strategic question of interest, which is assumed to be answered within a few weeks and which is limited to a small range of enterprise domains. Introduce the concrete model of the enterprise for the current and the target design state by a "light weighted" execution of the strategy process. This establishes the initial state of the concrete enterprise model and provides a big picture of the enterprise and of the scope to be investigated down to the details the description of the solution context for the selected question requires. The big picture covers at least working hypotheses for all impact factors and design factors according to the taxonomy terms.

Second step, build the roadmap of steps to solve the strategic question selected, communicate the initiative, and establish what is minimally needed for the first step.

Third step, apply the model built to conduct the first transition step of the roadmap.

Forth step, draw conclusions, tailor and improve for next iteration steps.

Phase 3: Build the Digital Roadmap and Evolve the Enterprise Design Continuously

Phase 3 is the entry into the continuous evolution process by reiterating the steps of phase 2, enhancing and detailing the results to an initial digital roadmap in the first iteration. This roadmap shows by which transition steps the enterprise design is supposed to reach a digital state, which defines the enterprise as *digital ready*.

The first reiteration of step 1 focuses on impacts caused by the knowledge society and the definition of the respective CSFs for the targeted business model of the enterprise. Thereby, it defines at least an initial targeted digital enterprise design. It also defines the (digital) maturity grades of the CSFs, which fit into the situation of the enterprise. It sets the structure, and objectives for the construction of the roadmap of transitions steps, and initiatives to reach the various maturity grades. Subsequent iterations will evaluate new insights into any impact factors (e.g. by focused investigations conducted or by successful innovation initiatives) and update the business model, the CSFs, and the roadmap, respectively.

5   Conclusions and Future Perspectives

Development of society, enterprises and technology are interdependent. They are triggered and followed by changing human behavior. Therefore, enterprises must be evolved holistically and centered by humanity. Also, importance of values, purpose, self-realization, and self-organization is increasing. Thus, the role of the entrepreneur also changes, from a commander and controller towards a moderator and motivator of enterprise teams and eco-systems. But the role as such will remain of eminent importance, while facing a new level of complexity.

The presented OSP enterprise model and the methodical, model-based approach OSP-EM consider all these aspects and support the entrepreneur and the enterprise team to conduct a digital transformation and to further evolve the enterprise. OSP enterprise model and evolution method are based on scientific results, on deep industrial experience, and on a conviction that human centricity is key as the people make the difference. In this sense, the OSP follows the vision of Industry 5.0 for a sustainable, human-centric and resilient European industry, while going far beyond with its holistic view [14].

Investigation of enterprise and transformation patterns as well as the creation of OPS-specific modeling languages will further evolve the approach. They will increase efficiency especially in overly complex situations and will uncover further interdependencies between impact factors introduced.

Our holistic approach is based on scientific insights from different disciplines. We see a strong need in intensifying such an interdisciplinary approach where sociological system theory becomes stronger aligned with economic and technological viewpoints. This will also sharpen and extend the competence and activity profile of enterprise engineering teams and their responsibility for digital transformation.

References

1. Alwadain, A.S., Alqahtani, F.H.: A Model of the Factors Influencing Enterprise Architecture Evolution in Organizations. The International Conference on Information Technology, USA, 2015.
2. Aguilar, F.J.: General Managers in Action: Policies and Strategies, Oxford University Press, 1992.
3. ANSI/X3/SPARC Study Group on Data Base Management Systems: Interim Report. FDT, ACM SIGMOD bulletin. Volume 7, No. 2, 1975.
4. Anderson J.: The Lean Change Method: Managing Agile Transformation through Kanban, Kotter, And Lean Startup Thinking, book.leanchangemethod.com, 2014.
5. Antonizzi, J., Smuts H. (2020): The Characteristics of Digital Entrepreneurship and Digital Transformation: A Systematic Literature Review. In: Hattingh, M., Matthee, M., Smuts, H., Pappas, I., Dwivedi, Y., Mäntymäki, M. (eds): Responsible Design, Implementation and Use of Information and Communication Technology. LNCS 12066. Springer, Cham, 2020.
6. Beck, U.: Risk Society: Towards a New Modernity. London: Sage, 1992.
7. Beck, U., Giddens, A., Lash, S.: Reflexive Modernization. Politics, Tradition and Aesthetics in the Modern Social Order. Cambridge: Polity, 1994.
8. Bernus, P., Noran, O., Molina, A.: Enterprise Architecture: Twenty Years of the GERAM Framework, IFAC Proceedings Volumes, 47(3), 3300–3308, 2014.
9. Boje, D.: Storytelling Organizations, SAGE Publications Ltd, 2008.
10. Combemale, C., Kienzle, J., Mussbacher, G., Ali, H., Amyot, D., Bagherzadeh, M., Batot, E., Bencomo, N., Benni, B., Bruel, J.-M., Cabot, J., Cheng, B., Collet, Ph., Engels, G., Heinrich, R., Jézéquel, J.-M., Koziolek, A., Mosser, S., Reussner, R., Sahraoui, H., Rijul Sainiy, R.,

Sallou, J., Stinckwich, S., Syriani, E., Wimmer, M.: A Hitchhiker's Guide to Model-Driven Engineering for Data-Centric Systems, IEEE Software, 2020.

11. Clair, G. St., Levy, B.: The Knowledge Services Handbook – A Guide for the Knowledge Strategist, De Gruyter, 2020.

12. Deming, W. E.: Out of the Crisis, Center for Advanced Engineering Study, Massachusetts Institute of Technology, Cambridge, Massachusetts., 1982.

13. Drucker, P. F.: Management Challenges of the 21st Century. New York: Harper Business, 1999.

14. European Commission: Industry 5.0, https://ec.europa.eu/info/publications/industry-50_en, last accessed 2021/11/14.

15. Engels, G., Hess, A., Humm, B., Juwig, J., Lohmann, M., Richter, J.-P., Voß, M., Willkomm, J.: Quasar Enterprise: Anwendungslandschaften serviceorientiert gestalten. dpunkt-Verlag, München, 2008.

16. Følstad, A., Kvale, K.: Customer journeys: a systematic literature review, Journal of Service Theory and Practice, Vol. 28 No. 2, pp. 196–227, 2018.

17. Friedrich, K., Malik, F., Seiwert, L.: Das große 1x1 der Erfolgsstrategie. Gabal Verlag, Offenbach, 2009.

18. Gassmann, O., Frankenberger, K., Csik, M.: The Business Model Navigator: 55 Models That Will Revolutionise Your Business, Upper Saddle River, NJ: FT Press, 2014.

19. Giachetti, R.E.: Design of Enterprise Systems, Theory, Architecture, and Methods, CRC Press, Boca Raton, FL, 2010.

20. Gross, P.: Die Multioptionsgesellschaft, Frankfurt a. M., Suhrkamp, 1994.

21. Gottschalk, S., Yigitbas, E., Engels, G.: Model-based Hypothesis Engineering for Supporting Adaptation to Uncertain Customer Needs, Business Modeling and Software Design, Springer International Publishing, pp. 276–286, 2020.

22. Hamel, G., Breen, B.: The Future of Management, Harvard Business Press, 2007.

23. Hartl, E., Hess, T.: The Role of Cultural Values for Digital Transformation: Insights from a Delphi Study. In Proc. of the 23rd Americas Conference on Information Systems (AMCIS 2017), Boston, Massachusetts, USA, 2017.

24. Humphrey, A.: SWOT Analysis for Management Consulting, SRI International, 2005.

25. Hurley R.F.: The Decision to Trust: How Leaders Create High-Trust Organizations, John Wiley & Sons Inc, 2011.

26. Kasasa: Boomers, Gen X, Gen Y, and Gen Z Explained, www.kasasa.com/articles/generations/gen-x-gen-y-gen-z, last accessed 2021/02/15.

27. Kotter J.P., Cohen D.S.: The Heart of Change: Real-Life Stories of How People Change Their Organizations, Harvard Business School, Boston, MA, 2002.

28. Kaplan, R.S., Norton, D.P.: The Balanced Scorecard: Translating Strategy into Action, Boston, MA: Harvard Business Press, 1996.

29. Las Heras, M., Chinchilla, N., Grau-Grau, M. (Eds.): The New Ideal Worker – Organizations Between Work-Life Balance, Gender and Leadership, Springer, 2020.

30. Luhmann, N.: Social Systems, Stanford University Press, 1995.

31. Manifesto for Agile Software Development, 2001, https://agilemanifesto.org/, last accessed 2021/02/15.

32. Mesly, O.: Project Feasibility: Tools for Uncovering Points of Vulnerability, CRC Press, 2016.

33. Reference Model for Service Oriented Architecture 1.0 OASIS Standard, 2006, http://docs.oasis-open.org/soa-rm/v1.0/soa-rm.pdf, last accessed 2021/02/15.

34. OMG, Model-driven Architecture (MDA), https://www.omg.org/cgi-bin/doc?ormsc/14-06-01, last accessed 2021/11/14.

35. Open Group, The TOGAF® Standard, Version 9.2 Overview, https://www.opengroup.org/togaf, last accessed 2021/02/15.
36. Open Group, The ArchiMate® Enterprise Architecture Modeling Language, https://www.open-group.org/architecture-forum, last accessed 2021/02/15.
37. Osterwalder A., Pigneur, Y.: Business Model Generation, John Wiley & Sons, Hoboken, NJ, 2010.
38. Project Management Institute: The Standard for Portfolio Management, 2017.
39. Porter, M.E., Kramer, M.R.: Strategy and Society: The link between competitive advantage and Corporate Social Responsibility, Harvard Business Rev., 78–93, 2006.
40. Ries, E.: The Lean Startup: How Today's Entrepreneurs Use Continuous Innovation to Create Radically Successful Businesses, Crown Business, New York, 2011.
41. Rüegg-Stürm, J., Grand, S.: Das St. Galler Management-Modell, 2. Auflage, utb GmbH, Stuttgart, 2020.
42. Roberts, J.: The Modern Firm – Organizational Design for Performance and Growth, Oxford University Press, 2007.
43. Rothaermel, F.T.: Strategic Management, McGraw Hill, 2019.
44. Runsten, Ph.: Team Intelligence: The Foundations of Intelligent Organizations – A Literature Review, SSE Working Paper Series in Business Administration 2017:2, Stockholm School of Economics, 2017.
45. Ross, J.W., Weill, P.D., Robertson, D.C.: Enterprise Architecture as Strategy – Creating a Foundation for Business Execution, Harvard Business School Press, 2006.
46. Seligman, M.E.P.: Authentic Happiness: Using the New Positive Psychology to Realize your Potential for Lasting Fulfillment, Atria Books, 2002.
47. Seddon, J.: Beyond Command and Control, Vanguard Consulting Ltd, 2019.
48. Stachowiak, H.: Allgemeine Modelltheorie, 1973.
49. UNESCO sustainable development goals, https://en.unesco.org/sustainabledevelopmentgoals, last accessed 2021/02/15.
50. van't Wout, J., Waage, M., Hartman, H., Stahlecker, M., Hofman, A.: The Integrated Architecture Framework Explained – Why, What, How, Springer, 2010.
51. Weick, K.E.: Educational Organizations as Loosely Coupled Systems, Administrative Science Quarterly, Vol. 21, No. 1, pp. 1–19, 1976.
52. Wenger, E., McDermott, R., Snyder, W.M.: Cultivating Communities of Practice. HBS press, 2002.
53. Wiedemann, A., Wiesche, M., Krcmar, H.A.O.: Integrating Development and Operations in Cross-Functional Teams – Toward a DevOps Competency Model, Proc. of the 2019 on Computers and People Research Conference, pp. 14–19, 2019.
54. Huma, Z., Gerth, Chr., Engels, G.: On-The-Fly computing: automatic service discovery and composition in heterogeneous domains. Comp. Sci. Res. Dev. 30(3–4): 333–361, 2015.

Printed in the United States
by Baker & Taylor Publisher Services